LIVERPOOL
JOHN MOORES UNIVERSITY
AVRIL ROBARTS LRC
TITHEBARN STREET
LIVERPOOL L2 2ER
TEL. 0151 231 4022

WITHDRAWN
LIVERPOOL JMU LIBRARY
3 1111 00914 5663

Principles of Engineering Mechanics

Principles of Engineering Mechanics

Second Edition

H. R. Harrison BSc, PhD, MRAeS

Formerly, Department of Mechanical Engineering
and Aeronautics,
The City University, London

T. Nettleton MSc, MIMechE

Department of Mechanical Engineering and Aeronautics,
The City University, London

Edward Arnold
A member of the Hodder Headline Group
LONDON MELBOURNE AUCKLAND

First published in Great Britain 1978
Second edition 1994

British Library Cataloguing in Publication Data

Harrison, Harry Ronald
Principles of Engineering Mechanics. –
2Rev.ed
I. Title II. Nettleton, T.
620.1

ISBN 0–340–56831–3

Typeset in 10/11 Times by Wearset, Boldon, Tyne and Wear.
Printed and bound in Great Britain for Edward Arnold, a division of Hodder Headline Plc, 338 Euston Road, London NW1 3BH
by Butler & Tanner Limited, Frome, Somerset.

Contents

Preface

This book covers the basic principles of the Part 1, Part 2 and much of the Part 3 Engineering Mechanics syllabuses of degree courses in engineering. The emphasis of the book is on the principles of mechanics and examples are drawn from a wide range of engineering applications.

The order of presentation has been chosen to correspond with that which we have found to be the most easily assimilated by students. Thus, although in some cases we proceed from the general to the particular, the gentler approach is adopted in discussing first two-dimensional and then three-dimensional problems.

The early part of the book deals with the dynamics of particles and of rigid bodies in two-dimensional motion. Both two- and three-dimensional statics problems are discussed. Vector notation is used initially as a label, in order to develop familiarity, and later on the methods of vector algebra are introduced as they naturally arise.

Vibration of single-degree-of-freedom systems are treated in detail and developed into a study of two-degree-of-freedom undamped systems.

An introduction to automatic control systems is included extending into frequency response methods and the use of Nyquist and Bode diagrams.

Three-dimensional dynamics of a particle and of a rigid body are tackled, making full use of vector algebra and introducing matrix notation. This chapter develops Euler's equations for rigid body motion.

It is becoming common to combine the areas usually referred to as mechanics and strength of materials and to present a single integrated course in solid mechanics. To this end a chapter is presented on continuum mechanics; this includes a study of one-dimensional and plane stress and strain leading to stresses and deflection of beams and shafts. Also included in this chapter are the basic elements of fluid dynamics, the purpose of this material is to show the similarities and the differences in the methods of setting up the equations for solid and fluid continua. It is not intended that this should replace a text in fluid dynamics but to develop the basics in parallel with solid mechanics. Most students study the two fields independently, so it is hoped that seeing both Lagrangian and Eulerian co-ordinate systems in use in the same chapter will assist in the understanding of both disciplines.

There is also a discussion of axial wave propagation in rods (12.9), this is a topic not usually covered at this level and may well be omitted at a first reading. The fluid mechanics sections (12.10–16) can also be omitted if only solid mechanics is required.

The student may be uncertain as to which method is best for a particular problem and because of this may be unable to start the solution. Each chapter in this book is thus divided into two parts. The first is an exposition of the basic theory with a few explanatory examples. The second part contains worked examples, many of which are described and explained in a manner usually reserved for the tutorial. Where relevant, different methods for solving the same problem are compared and difficulties arising with certain techniques are pointed out. Each chapter ends with a series of problems for solution. These are graded in such a way as to build up the confidence of students as they proceed. Answers are given.

Numerical problems are posed using SI units, but other systems of units are covered in an appendix.

The intention of the book is to provide a firm basis in mechanics, preparing the ground for advanced study in any specialisation. The applications are wide-ranging and chosen to show as many facets of engineering mechanics as is practical in a book of this size.

We are grateful to The City University for permission to use examination questions as a basis for a large number of the problems. Thanks are also due to our fellow teachers of Engineering Mechanics who contributed many of the questions.

July 1993

H.R.H.
T.N.

1

Co-ordinate systems and position vectors

1.1 Introduction

Dynamics is a study of the motion of material bodies and of the associated forces.

The study of motion is called kinematics and involves the use of geometry and the concept of time, whereas the study of the forces associated with the motion is called kinetics and involves some abstract reasoning and the proposal of basic 'laws' or axioms. Statics is a special case where there is no motion. The combined study of dynamics and statics forms the science of mechanics.

1.2 Co-ordinate systems

Initially we shall be concerned with describing the position of a point, and later this will be related to the movement of a real object.

The position of a point is defined only in relation to some reference axes. In three-dimensional space we require three independent co-ordinates to specify the unique position of a point relative to the chosen set of axes.

One-dimensional systems

If a point is known to lie on a fixed path – such as a straight line, circle or helix – then only one number is required to locate the point with respect to some arbitrary reference point on the path. This is the system used in road maps, where place B (Fig. 1.1) is said to be 10 km (say) from A along road R. Unless A happens to be the end of road R, we must specify the direction which is to be regarded as positive. This system is often referred to as a path co-ordinate system.

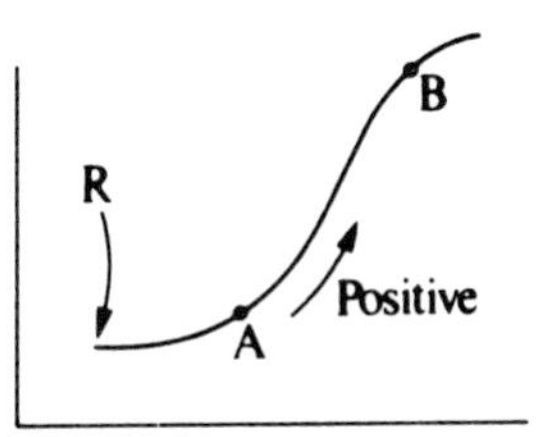

Figure 1.1

Two-dimensional systems

If a point lies on a surface – such as that of a plane, a cylinder or a sphere – then two numbers are required to specify the position of the point. For a plane surface, two systems of co-ordinates are in common use.

a) *Cartesian co-ordinates.* In this system an orthogonal grid of lines is constructed and a point is defined as being the intersection of two of these straight lines.

In Fig. 1.2, point P is positioned relative to the x- and y-axes by the intersection of the lines $x = 3$ and $y = 2$ and is denoted by P(+3, +2).

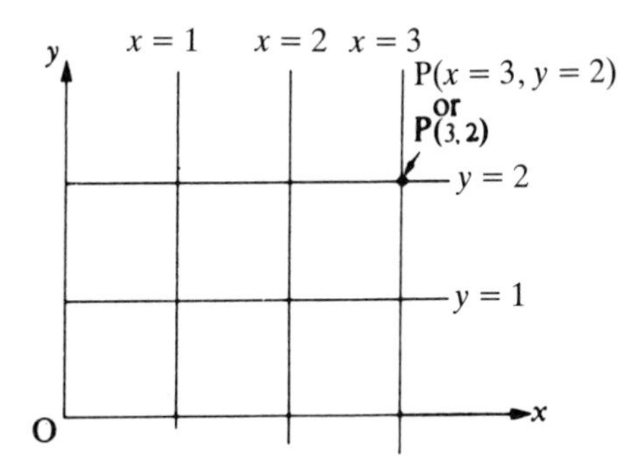

Figure 1.2

b) *Polar co-ordinates.* In this system (Fig. 1.3) the distance from the origin is given together with the angle which OP makes with the x-axis.

If the surface is that of a sphere, then lines of latitude and longitude may be used as in terrestrial navigation.

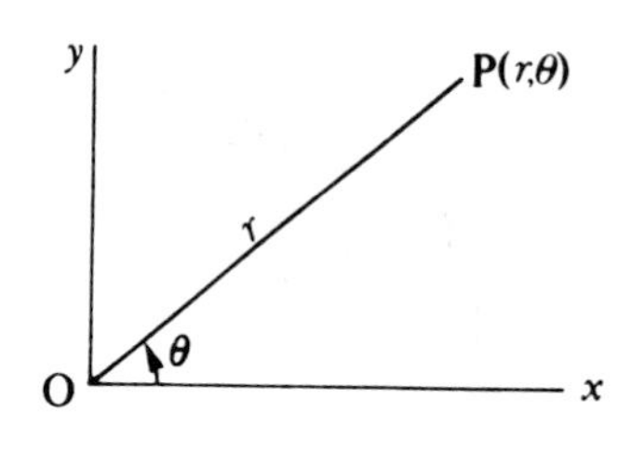

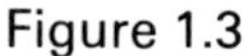
Figure 1.3

Three-dimensional systems
Three systems are in common use:

a) *Cartesian co-ordinates.* This is a simple extension of the two-dimensional case where a third axis, the z-axis, has been added. The sense is not arbitrary but is drawn according to the right-hand screw convention, as shown in Fig. 1.4. This set of axes is known as a normal right-handed triad.

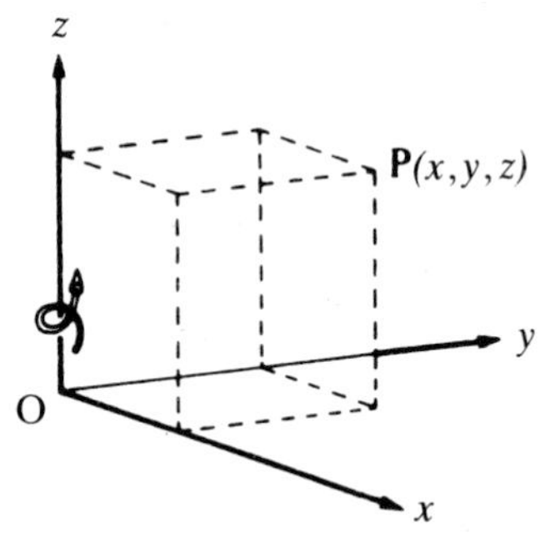

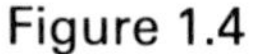
Figure 1.4

b) *Cylindrical co-ordinates.* This is an extension of the polar co-ordinate system, the convention for positive θ and z being as shown in Fig. 1.5. It is clear that if R is constant then the point will lie on the surface of a right circular cylinder.

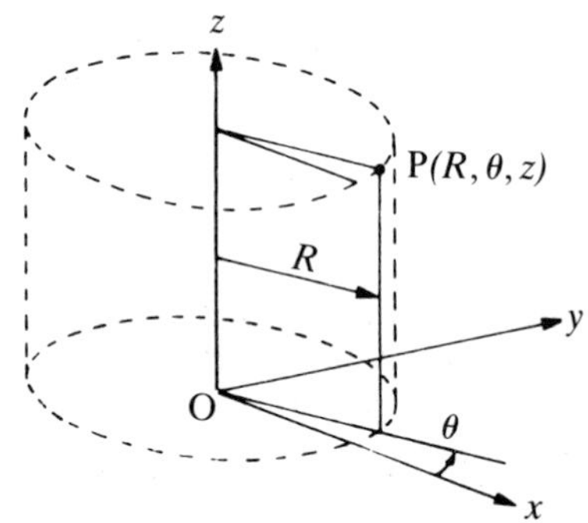

Figure 1.5

c) *Spherical co-ordinates.* In this system the position is specified by the distance of a point from the origin, and the direction is given by two angles as shown in Fig. 1.6(a) or (b).

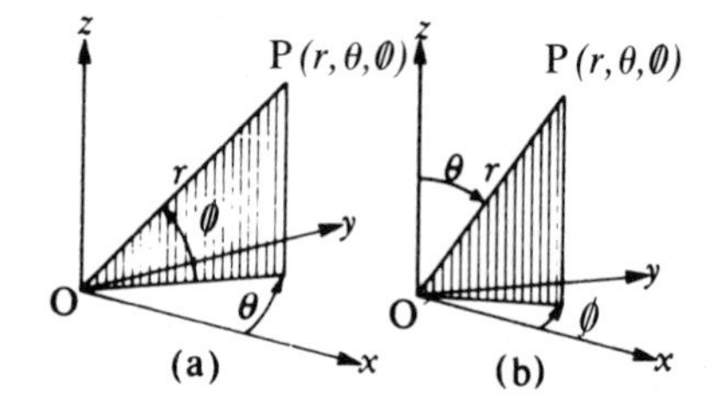

Figure 1.6

Note that, while straight-line motion is one-dimensional, one-dimensional motion is not confined to a straight line; for example, path co-ordinates are quite suitable for describing the motion of a point in space, and an angle is sufficient to define the position of a wheel rotating about a fixed axis. It is also true that spherical co-ordinates could be used in a problem involving motion in a straight line not passing through the origin O of the axes; however, this would involve an unnecessary complication.

1.3 Vector representation

The position vector
A line drawn from the origin O to the point P always completely specifies the position of P and is independent of any co-ordinate system. It follows that some other line drawn to a convenient scale can also be used to represent the position of P relative to O (written $\overrightarrow{OP}$).

In Fig. 1.7(b), *both* vectors represent the position of P relative to O, which is shown in 1.7(a), as both give the magnitude and the direction of P relative to O. These are called free vectors. Hence in mechanics a vector may be defined as a line segment which represents a physical quantity in magnitude and direction. There is, however, a restriction on this definition which is now considered.

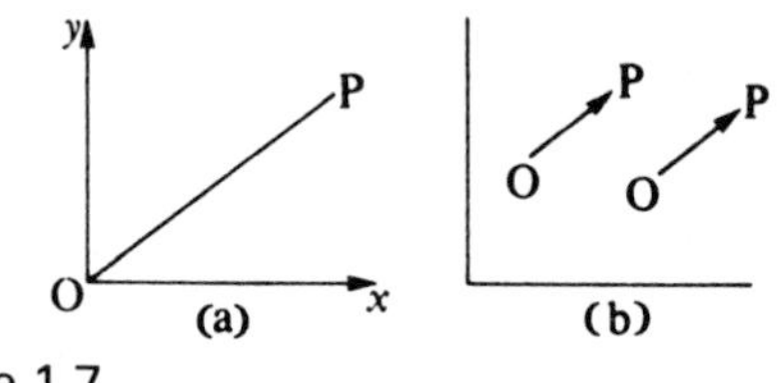

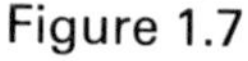
Figure 1.7

Addition of vectors
The position of P relative to O may be regarded as the position of Q relative to O plus the position of P relative to Q, as shown in Fig. 1.8(a).

The position of P could also be considered as the position of Q′ relative to O plus that of P relative to Q′. If Q′ is chosen such that OQ′PQ is a parallelogram, i.e. OQ′ = QP and OQ = Q′P, then the corresponding vector diagram will also be a parallelogram. Now, since the position vector represented by oq', Fig. 1.8(b), is identical to that represented by qp, and oq is identical to $q'p$, it follows that the sum of two vectors is independent of the order of addition.

Conversely, if a physical quantity is a vector then addition must satisfy the parallelogram law. The important physical quantity which does not obey this addition rule is finite rotation, because it can be demonstrated that the sum of two finite

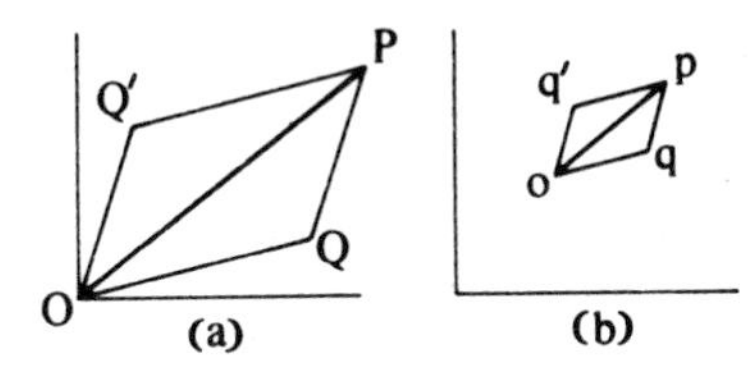

Figure 1.8

rotations depends on the order of addition (see Chapter 10).

The law of addition may be written symbolically as

$$\overrightarrow{OP} = \overrightarrow{OQ} + \overrightarrow{QP} = \overrightarrow{QP} + \overrightarrow{OQ} \tag{1.1}$$

Vector notation
As vector algebra will be used extensively later, formal vector notation will now be introduced. It is convenient to represent a vector by a single symbol and it is conventional to use bold-face type in printed work or to underline a symbol in manuscript. For position we shall use

$$\overrightarrow{OP} \equiv \boldsymbol{r}$$

The fact that addition is commutative is demonstrated in Fig. 1.9:

$$\boldsymbol{r} = \boldsymbol{r}_1 + \boldsymbol{r}_2 = \boldsymbol{r}_2 + \boldsymbol{r}_1 \tag{1.2}$$

Unit vector
It is often convenient to separate the magnitude of a vector from its direction. This is done by introducing a unit vector $\boldsymbol{e}$ which has unit

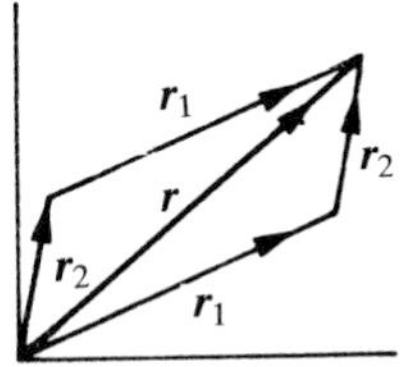

Figure 1.9

magnitude and is in the required direction. Hence $\boldsymbol{r}$ may be written

$$\boldsymbol{r} = r\boldsymbol{e} \tag{1.3}$$

where r is the magnitude (a scalar). The modulus, written as $|\boldsymbol{r}|$, is the size of the vector and is always positive. In this book, vector magnitudes may be positive or negative.

Components of a vector
Any number of vectors which add to give another vector are said to be *components* of that other vector. Usually the components are taken to be orthogonal, as shown in Fig. 1.10.

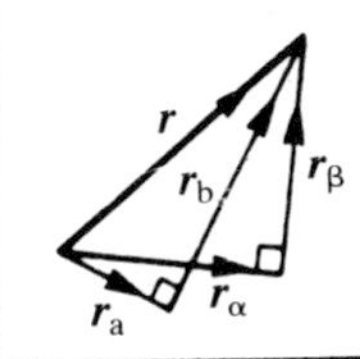

Figure 1.10

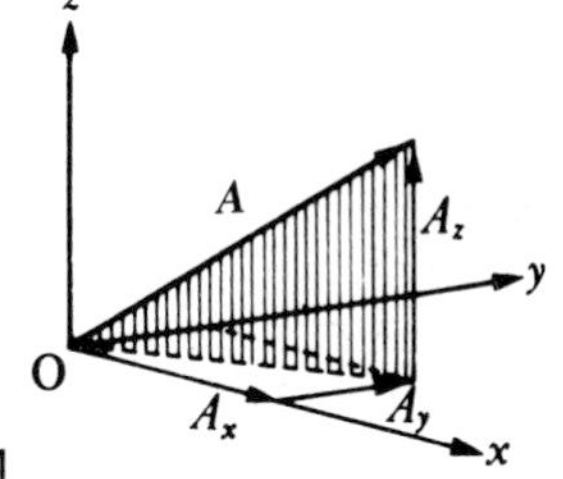

Figure 1.11

In Cartesian co-ordinates the unit vectors in the x, y and z directions are given the symbols $\boldsymbol{i}, \boldsymbol{j}$ and $\boldsymbol{k}$ respectively. Hence the components of $\boldsymbol{A}$ (Fig. 1.11) may be written

$$\boldsymbol{A} = A_x\boldsymbol{i} + A_y\boldsymbol{j} + A_z\boldsymbol{k}, \tag{1.4}$$

where A_x, A_y and A_z are said to be the components of $\boldsymbol{A}$ with respect to the x-, y-, z-axes.

It follows that, if $\boldsymbol{B} = B_x\boldsymbol{i} + B_y\boldsymbol{j} + B_z\boldsymbol{k}$, then

$$\boldsymbol{A} + \boldsymbol{B} = (A_x + B_x)\boldsymbol{i} + (A_y + B_y)\boldsymbol{j} + (A_z + B_z)\boldsymbol{k} \tag{1.5}$$

It is also easily shown that

$$(\boldsymbol{A}+\boldsymbol{B})+\boldsymbol{C}=\boldsymbol{A}+(\boldsymbol{B}+\boldsymbol{C})$$

and also that

$$a\boldsymbol{A}=aA_x\boldsymbol{i}+aA_y\boldsymbol{j}+aA_z\boldsymbol{k} \tag{1.6}$$

where a is a scalar.

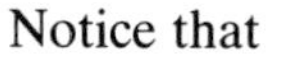

Notice that

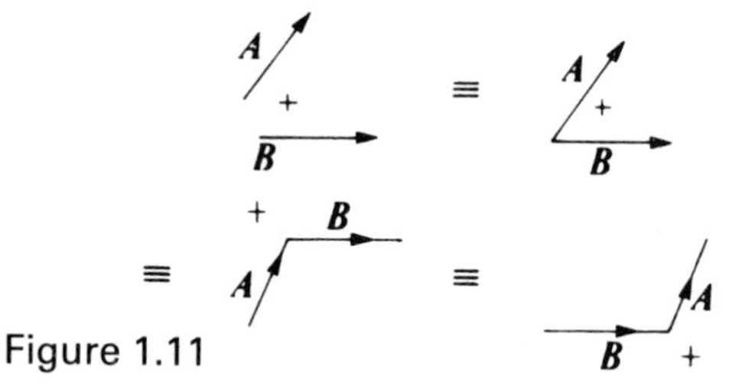

Figure 1.11

because $\boldsymbol{A}$ and $\boldsymbol{B}$ are free vectors.

Scalar product of two vectors

The scalar product of two vectors $\boldsymbol{A}$ and $\boldsymbol{B}$ (sometimes referred to as the *dot product*) is formally defined as $|\boldsymbol{A}||\boldsymbol{B}|\cos\theta$, Fig. 1.12, where θ is the smallest angle between the two vectors. The scalar product is denoted by a dot placed between the two vector symbols:

$$\boldsymbol{A}\cdot\boldsymbol{B}=|\boldsymbol{A}||\boldsymbol{B}|\cos\theta \tag{1.7}$$

It follows from this definition that $\boldsymbol{A}\cdot\boldsymbol{B}=\boldsymbol{B}\cdot\boldsymbol{A}$.

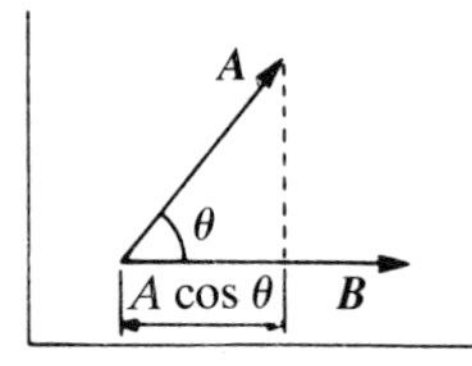

Figure 1.12

From Fig. 1.12 it is seen that $|\boldsymbol{A}|\cos\theta$ is the component of $\boldsymbol{A}$ in the direction of $\boldsymbol{B}$; similarly $|\boldsymbol{B}|\cos\theta$ is the component of $\boldsymbol{B}$ in the direction of $\boldsymbol{A}$. This definition will later be seen to be useful in the description of work and power. If $\boldsymbol{B}$ is a unit vector $\boldsymbol{e}$, then

$$\boldsymbol{A}\cdot\boldsymbol{e}=|\boldsymbol{A}|\cos\theta \tag{1.8}$$

that is the scalar component of $\boldsymbol{A}$ in the direction of $\boldsymbol{e}$.

It is seen that

$$\boldsymbol{i}\cdot\boldsymbol{i}=\boldsymbol{j}\cdot\boldsymbol{j}=\boldsymbol{k}\cdot\boldsymbol{k}=1$$

and $\boldsymbol{i}\cdot\boldsymbol{j}=\boldsymbol{i}\cdot\boldsymbol{k}=\boldsymbol{j}\cdot\boldsymbol{k}=0$

Direction cosines

Consider the vector $\boldsymbol{A}=A_x\boldsymbol{i}+A_y\boldsymbol{j}+A_z\boldsymbol{k}$. The modulus of $\boldsymbol{A}$ is found by the simple application of Pythagoras's theorem to give

$$|\boldsymbol{A}|=\sqrt{(A_x^2+A_y^2+A_z^2)} \tag{1.9}$$

The direction cosine, l, is defined as the cosine of the angle between the vector and the positive x-axis, i.e. from Fig. 1.13.

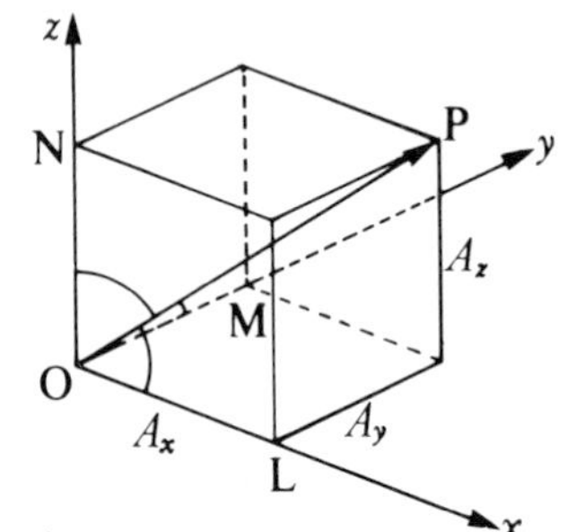

Figure 1.13

$$l=\cos(\angle\mathrm{POL})=A_x/|\boldsymbol{A}| \tag{1.10a}$$

similarly
$$m=\cos(\angle\mathrm{POM})=A_y/|\boldsymbol{A}| \tag{1.10b}$$

$$n=\cos(\angle\mathrm{PON})=A_z/|\boldsymbol{A}| \tag{1.10c}$$

From equations 1.3 and 1.10,

$$\boldsymbol{e}=\frac{\boldsymbol{A}}{|\boldsymbol{A}|}=\frac{A_x}{|\boldsymbol{A}|}\boldsymbol{i}+\frac{A_y}{|\boldsymbol{A}|}\boldsymbol{j}+\frac{A_z}{|\boldsymbol{A}|}\boldsymbol{k}$$
$$=l\boldsymbol{i}+m\boldsymbol{j}+n\boldsymbol{k}$$

that is the direction cosines are the components of the unit vector; hence

$$l^2+m^2+n^2=1 \tag{1.11}$$

Discussion examples

Example 1.1

See Fig. 1.14. A surveying instrument at C can measure distance and angle.

Relative to the fixed x-, y-, z-axes at C, point A is at an elevation of 9.2° above the horizontal (xy) plane. The body of the instrument has to be rotated about the vertical axis through 41° from the x direction in order to be aligned with A. The distance from C to A is 5005 m. Corresponding values for point B are 1.3°, 73.4° and 7037 m.

Determine (a) the locations of points A and B in Cartesian co-ordinates relative to the axes at C, (b) the distance from A to B, and (c) the distance from A to B projected on to the horizontal plane.

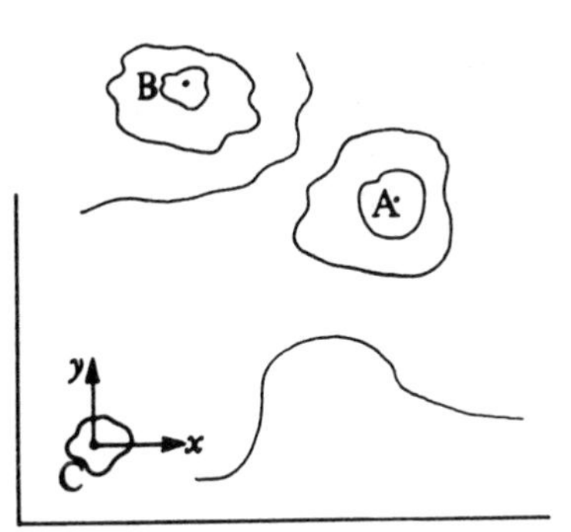

Figure 1.14

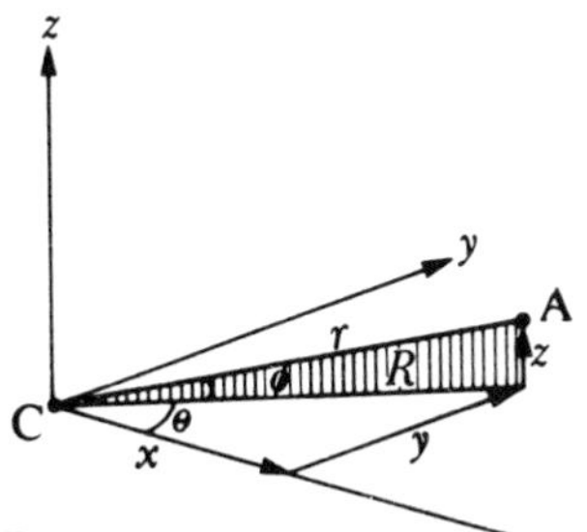

Figure 1.15

Solution See Fig. 1.15. For point A, $r = 5005$ m, $\theta = 41°$, $\phi = 9.2°$.

$$\begin{aligned} z &= r\sin\phi = 5005\sin 9.2° = 800.2\text{ m} \\ R &= r\cos\phi = 5005\cos 9.2° = 4941.0\text{ m} \\ x &= R\cos\theta = 4941\cos 41° = 3729.0\text{ m} \\ y &= R\sin\theta = 4941\sin 41° = 3241.0\text{ m} \end{aligned}$$

so A is located at point (3729, 3241, 800.2) m.

For point B, $r = 7037$ m, $\theta = 73.4°$, $\phi = 1.3°$; hence B is located at point (2010, 6742, 159.7) m.

Adding the vectors $\overrightarrow{CA}$ and $\overrightarrow{AB}$, we have

$$\overrightarrow{CA} + \overrightarrow{AB} = \overrightarrow{CB}$$

or

$$\begin{aligned} \overrightarrow{AB} &= \overrightarrow{CB} - \overrightarrow{CA} \\ &= (2010\boldsymbol{i} + 6742\boldsymbol{j} + 159.7\boldsymbol{k}) \\ &\quad - (3729\boldsymbol{i} + 3241\boldsymbol{j} + 800.2\boldsymbol{k}) \\ &= (-1719\boldsymbol{i} + 3501\boldsymbol{j} - 640.5\boldsymbol{k})\text{ m} \end{aligned}$$

The distance from A to B is given by

$$\begin{aligned} |\overrightarrow{AB}| &= \surd[(-1791)^2 + (3501)^2 + (-640.5)^2] \\ &= 3952\text{ m} \end{aligned}$$

and the component of $\overrightarrow{AB}$ in the xy-plane is

$$\surd[(-1719)^2 + (3501)^2] = 3900\text{ m}$$

Example 1.2

Point A is located at (0, 3, 2) m and point B at (3, 4, 5) m. If the location vector from A to C is (−2, 0, 4) m, find the position of point C and the position vector from B to C.

Solution A simple application of the laws of vector addition is all that is required for the solution of this problem. Referring to Fig. 1.16,

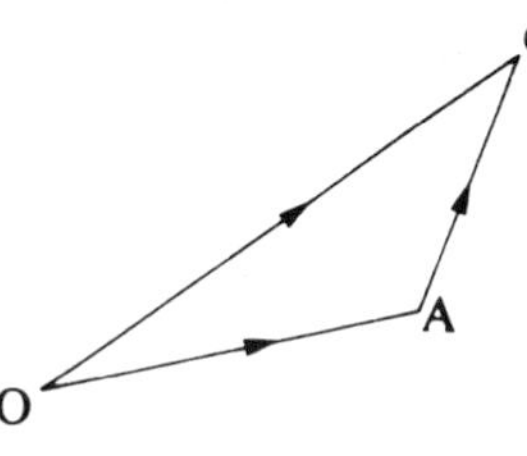

Figure 1.16

$$\begin{aligned} \overrightarrow{OC} &= \overrightarrow{OA} + \overrightarrow{AC} \\ &= (3\boldsymbol{j} + 2\boldsymbol{k}) + (-2\boldsymbol{i} + 4\boldsymbol{k}) \\ &= -2\boldsymbol{i} + 3\boldsymbol{j} + 6\boldsymbol{k} \end{aligned}$$

Hence point C is located at (−2, 3, 6) m.

Similarly $\overrightarrow{OC} = \overrightarrow{OB} + \overrightarrow{BC}$

so that

$$\begin{aligned} \overrightarrow{BC} &= \overrightarrow{OC} - \overrightarrow{OB} \\ &= (-2\boldsymbol{i} + 3\boldsymbol{j} + 6\boldsymbol{k}) - (3\boldsymbol{i} + 4\boldsymbol{j} + 5\boldsymbol{k}) \\ &= (-5\boldsymbol{i} - 1\boldsymbol{j} + 1\boldsymbol{k})\text{ m} \end{aligned}$$

Example 1.3

Points A, B and P are located at (2, 2, −4) m, (5, 7, −1) m and (3, 4, 5) m respectively. Determine the scalar component of the vector $\overrightarrow{OP}$ in the direction B to A and the vector component parallel to the line AB.

Solution To determine the component of a given vector in a particular direction, we first obtain the unit vector for the direction and then form the dot product between the unit vector and the given vector. This gives the magnitude of the component, otherwise known as the *scalar* component.

The vector $\overrightarrow{BA}$ is determined from the relationship

$$\overrightarrow{OB} + \overrightarrow{BA} = \overrightarrow{OA}$$

thus

$$\begin{aligned} \overrightarrow{BA} &= \overrightarrow{OA} - \overrightarrow{OB} \\ &= (2\boldsymbol{i} + 2\boldsymbol{j} - 4\boldsymbol{k}) - (5\boldsymbol{i} + 7\boldsymbol{j} - 1\boldsymbol{k}) \\ &= -(3\boldsymbol{i} + 5\boldsymbol{j} + 3\boldsymbol{k})\text{ m} \end{aligned}$$

The length of the vector $\overrightarrow{BA}$ is given by

$$\text{BA} = |\overrightarrow{AB}| = \surd(3^2 + 5^2 + 3^2) = \surd 43\text{ m}$$

and the unit vector

$$e = \frac{\overrightarrow{BA}}{BA} = \frac{-(3i+5j+3k)}{\sqrt{43}}$$

The required scalar component is

$$\begin{aligned}\overrightarrow{OP}\cdot e &= (3i+4j+5k)\\ &\quad\cdot(-3i-5j-3k)/\sqrt{43}\\ &= -(3\times3+4\times5+5\times3)/\sqrt{43}\\ &= -6.17 \text{ m}\end{aligned}$$

The minus sign indicates that the component of OP (taking the direction from O to P as positive) parallel to BA is opposite in sense to the direction from B to A.

If we wish to represent the component of OP in the specified direction as a vector, we multiply the scalar component by the unit vector for the specified direction. Thus

$$\begin{aligned}&-(3i+5j+3k)(-6.17)/\sqrt{43}\\ &\qquad = (2.82i+4.70j+2.82k) \text{ m}\end{aligned}$$

Example 1.4

See Fig. 1.17. Points C and D are located at (1, 2, 4) m and (2, −1, 1) m respectively. Determine the length of DC and the angle COD, where O is the origin of the co-ordinates.

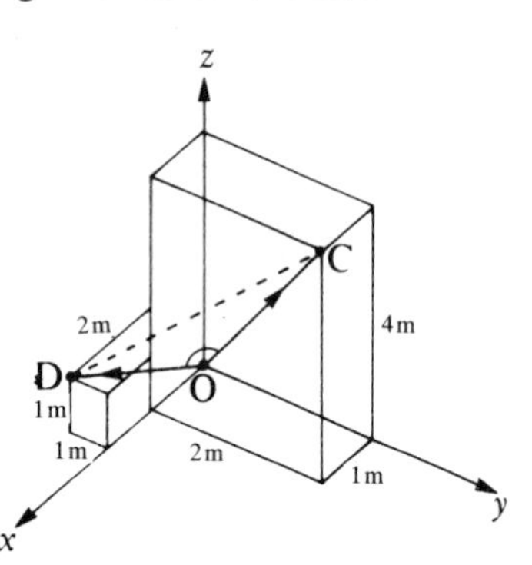

Figure 1.17

Solution If we first obtain an expression for CD in vector form, then the modulus of this vector will be the required length.

From the rule for vector addition, $\overrightarrow{OC}+\overrightarrow{CD}=\overrightarrow{OD}$, so that

$$\begin{aligned}\overrightarrow{CD} &= \overrightarrow{OD}-\overrightarrow{OC}\\ &= (2i-1j+1k)-(1i+2j+4k)\\ &= (1i-3j-3k) \text{ m}\end{aligned}$$

and $|\overrightarrow{CD}| = \sqrt{[1^2+(-3)^2+(-3)^2]} = \sqrt{19}$
$= 4.36$ m

The scalar or dot product involves the angle between two vectors and we can use the property of this product to determine this angle. By definition of the scalar product,

$$\overrightarrow{OC}\cdot\overrightarrow{OD} = (OC)(OD)\cos(\angle COD)$$

therefore $\cos(\angle COD)$

$$\begin{aligned}&= \frac{\overrightarrow{OC}\cdot\overrightarrow{OD}}{(OC)(OD)}\\ &= \frac{(1i+2j+4k)\cdot(2i-1j+1k)}{\sqrt{[1^2+2^2+4^2]}\sqrt{[2^2+(-1)^2+1^2]}}\\ &= \frac{1\times2+2(-1)+4\times1}{\sqrt{21}\ \sqrt{6}} = \frac{4}{\sqrt{126}}\\ &= 0.3563\, m\end{aligned}$$

and $\angle COD = 69.12°$

As a check, we can determine $\angle COD$ from the cosine rule:

$$\begin{aligned}\cos(\angle COD) &= \frac{OC^2+OD^2-CD^2}{2(OC)(OD)}\\ &= \frac{6+21-19}{2\sqrt{6}\ \sqrt{21}}\\ &= \frac{4}{\sqrt{126}} \quad \text{as before.}\end{aligned}$$

Problems

1.1 A position vector is given by OP = $(3i+2j+1k)$ m. Determine its unit vector.

1.2 A line PQ has a length of 6 m and a direction given by the unit vector $\frac{2}{3}i+\frac{2}{3}j+\frac{1}{3}k$. Write PQ as a vector.

1.3 Point A is at (1, 2, 3) m and the position vector of point B, relative to A, is $(6i+3k)$ m. Determine the position of B relative to the origin of the co-ordinate system.

1.4 Determine the unit vector for the line joining points C and D, in the sense of C to D, where C is at point (0, 3, −2) m and D is at (5, 5, 0) m.

1.5 Point A is located at (5, 6, 7) m and point B at (2, 2, 6) m. Determine the position vector (a) from A to B and (b) from B to A.

1.6 P is located at point (0, 3, 2) m and Q at point (3, 2, 1). Determine the position vector from P to Q and its unit vector.

1.7 A is at the point (1, 1, 2) m. The position of point B relative to A is $(2i+3j+4k)$ m and that of point C relative to B is $(-3\boldsymbol{i}-2\boldsymbol{j}+2\boldsymbol{k})$ m. Determine the location of C.

1.8 The dimensions of a room at 6 m × 5 m × 4 m, as shown in Fig. 1.18. A cable is suspended from the point P in the ceiling and a lamp L at the end of the cable is 1.2 m vertically below P.

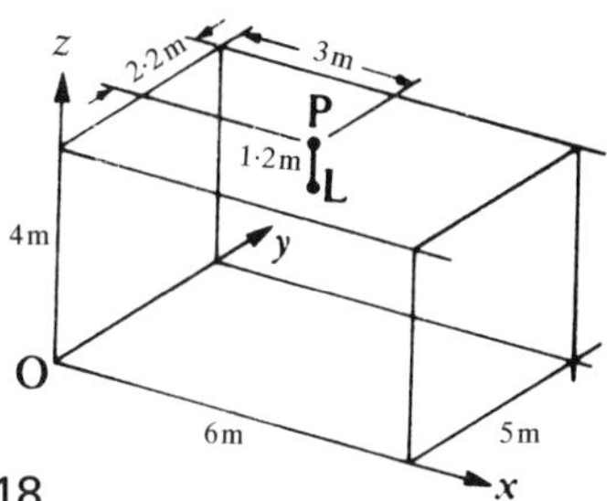

Figure 1.18

Determine the Cartesian and cylindrical co-ordinates of the lamp L relative to the x-, y-, z-axes and also find expressions for the corresponding cylindrical unit vectors $\boldsymbol{e}_R$, $\boldsymbol{e}_\theta$ and $\boldsymbol{e}_z$ in terms of $\boldsymbol{i}$, $\boldsymbol{j}$ and $\boldsymbol{k}$ (see Fig. 1.19).

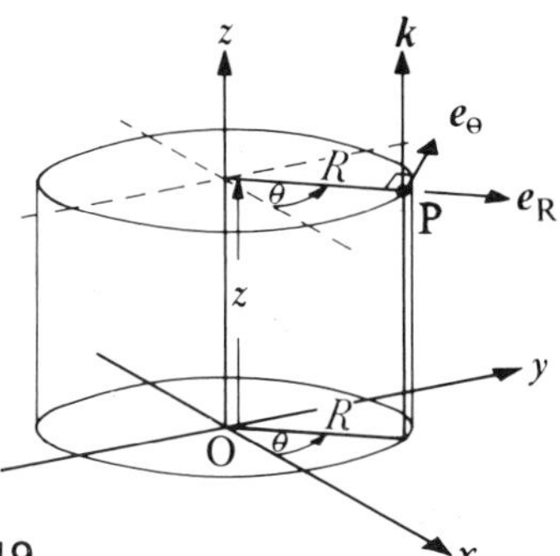

Figure 1.19

1.9 Show that the relationship between Cartesian and cylindrical co-ordinates is governed by the following equations (see Fig. 1.19):

$$x = R\cos\theta, \quad y = R\sin\theta,$$
$$R = (x^2+y^2)^{1/2}, \quad \theta = \arctan(y/x)$$

$$\boldsymbol{i} = \cos\theta\boldsymbol{e}_R - \sin\theta\boldsymbol{e}_\theta,$$
$$\boldsymbol{j} = \sin\theta\boldsymbol{e}_R + \cos\theta\boldsymbol{e}_\theta, \quad \boldsymbol{k} = \boldsymbol{e}_z$$

$$\boldsymbol{e}_R = \cos\theta\boldsymbol{i} + \sin\theta\boldsymbol{j},$$
$$\boldsymbol{e}_\theta = -\sin\theta\boldsymbol{i} + \cos\theta\boldsymbol{j}, \quad \boldsymbol{e}_z = \boldsymbol{k}$$

1.10 See Fig. 1.20. The location of an aircraft in spherical co-ordinates (r, θ, ϕ) relative to a radar installation is (20000 m, 33.7°, 12.5°). Determine the location in Cartesian and cylindrical co-ordinates.

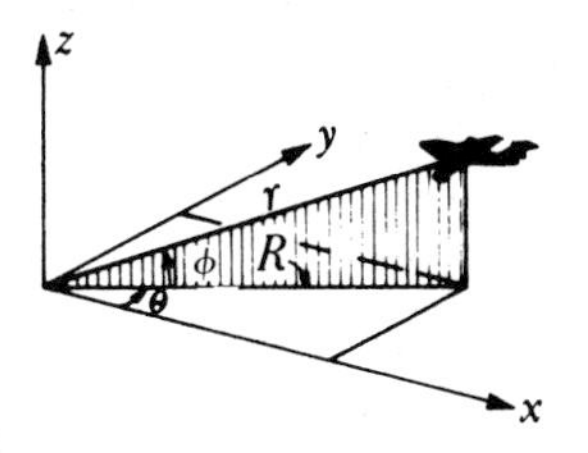

Figure 1.20

1.11 What are the angles between the line joining the origin O and a point at (2, −5, 6) m and the positive x-, y-, z-axes?

1.12 In problem 1.7, determine the angle ABC.

1.13 A vector is given by $(2\boldsymbol{i}+3\boldsymbol{j}+1\boldsymbol{k})$ m. What is the component of this vector (a) in the y-direction and (b) in a direction parallel to the line from A to B, where A is at point (1, 1, 0) m and B is at (3, 4, 5) m?

1.14 Find the perpendicular distances from the point (5, 6, 7) to each of the x-, y- and z-axes.

1.15 Points A, B and C are located at (1, 2, 1) m, (5, 6, 7) m and (−2, −5, 6) m respectively. Determine (a) the perpendicular distance from B to the line AC and (b) the angle BAC.

2

Kinematics of a particle in plane motion

2.1 Displacement, velocity and acceleration of a particle

A particle may be defined as a material object whose dimensions are of no consequence to the problem under consideration. For the purpose of describing the kinematics of such an object, the motion may be taken as being that of a representative point.

Displacement of a particle

If a particle occupies position A at time t_1 and at a later time t_2 it occupies a position B, then the displacement is the vector $\overrightarrow{AB}$ as shown in Fig. 2.1. In vector notation,

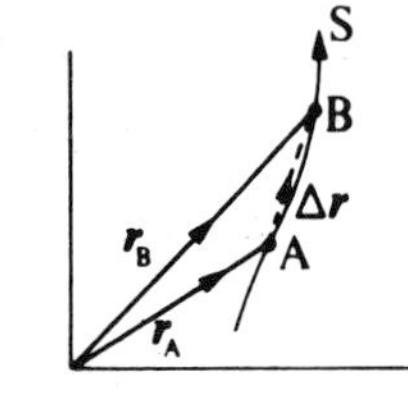

Figure 2.1

$$\boldsymbol{r}_\mathrm{B} = \boldsymbol{r}_\mathrm{A} + \Delta\boldsymbol{r}$$
$$\text{or} \quad \Delta\boldsymbol{r} = \boldsymbol{r}_\mathrm{B} - \boldsymbol{r}_\mathrm{A} \tag{2.1}$$

Here the symbol Δ signifies a finite difference.

If the time difference $\Delta t = t_2 - t_1$ is small, then $\lim_{\Delta t \to 0} |\Delta\boldsymbol{r}| = \mathrm{d}s$, an element of the path.

Velocity of a particle

The average velocity of a particle during the time interval Δt is defined to be

$$\boldsymbol{v}_\mathrm{average} = \frac{\Delta\boldsymbol{r}}{\Delta t}$$

This is a vector quantity in the direction of $\Delta\boldsymbol{r}$.

The instantaneous velocity is defined as

$$\boldsymbol{v} = \lim_{\Delta t \to 0}\left(\frac{\Delta\boldsymbol{r}}{\Delta t}\right) = \frac{\mathrm{d}\boldsymbol{r}}{\mathrm{d}t}$$

If $\boldsymbol{e}_\mathrm{t}$ is a unit vector tangential to the path, then as $\Delta t \to 0$, $\Delta\boldsymbol{r} \to \Delta s\boldsymbol{e}_\mathrm{t}$

$$\text{so} \quad \boldsymbol{v} = \lim_{\Delta t \to 0}\left(\frac{\Delta s}{\Delta t}\boldsymbol{e}_\mathrm{t}\right) = \frac{\mathrm{d}s}{\mathrm{d}t}\boldsymbol{e}_\mathrm{t} \tag{2.2}$$

The term $\mathrm{d}s/\mathrm{d}t$ is the rate of change of distance along the path and is a scalar quantity usually called speed.

Acceleration of a particle

The acceleration of a particle is defined (see Fig. 2.2) as

$$\boldsymbol{a} = \lim_{\Delta t \to 0}\left(\frac{\Delta\boldsymbol{v}}{\Delta t}\right) = \frac{\mathrm{d}\boldsymbol{v}}{\mathrm{d}t} = \frac{\mathrm{d}^2\boldsymbol{r}}{\mathrm{d}t^2} \tag{2.3}$$

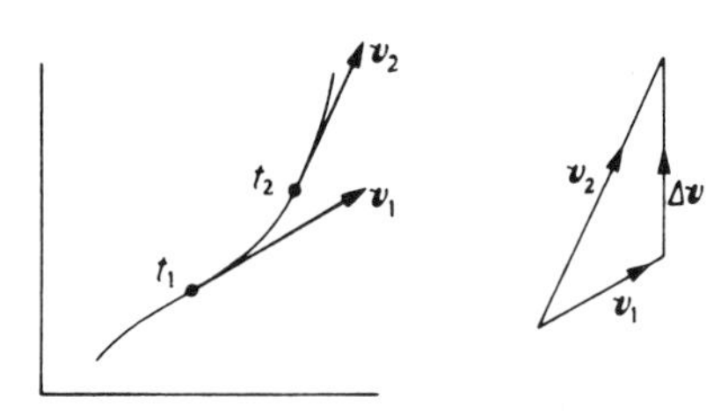

Figure 2.2

The direction of $\boldsymbol{a}$ is not obvious and will not be tangential to the path unless the path is straight.

Having defined velocity and acceleration in a quite general way, the components of these quantities for a particle confined to move in a plane can now be formulated.

It is useful to consider the ways in which a vector quantity may change with time, as this will help in understanding the full meaning of acceleration.

Since velocity is defined by both magnitude and direction, a variation in either quantity will constitute a change in the velocity vector.

If the velocity remains in a fixed direction, then the acceleration has a magnitude equal to the rate of change of speed and is directed in the same direction as the velocity, though not necessarily in the same sense.

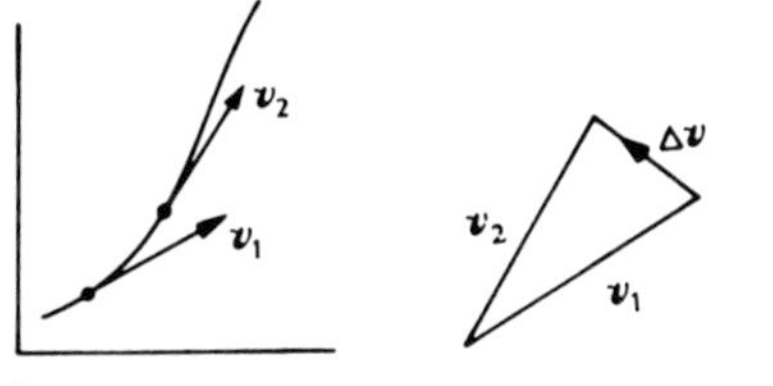

Figure 2.3

If the speed remains constant, then the acceleration is due solely to the change in direction of the velocity. For this case we can see that the vector diagram (Fig. 2.3) is an isosceles triangle. In the limit, for small changes in time, and hence small changes in direction, the change in velocity is normal to the velocity vector.

2.2 Cartesian co-ordinates

See Fig. 2.4.

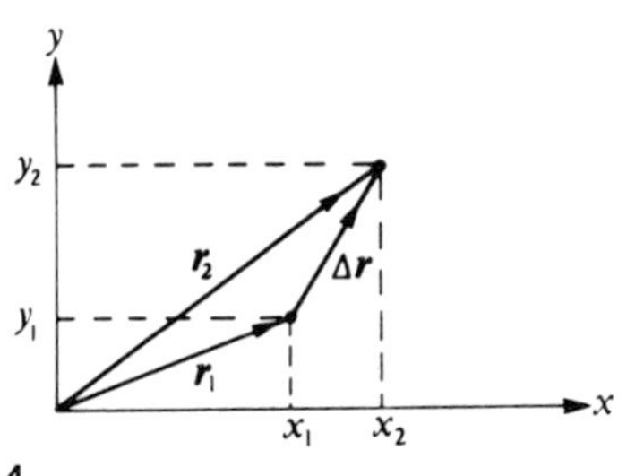

Figure 2.4

$$\Delta \boldsymbol{r} = (x_2 - x_1)\boldsymbol{i} + (y_2 - y_1)\boldsymbol{j}$$
$$= \Delta x \boldsymbol{i} + \Delta y \boldsymbol{j}$$

$$\frac{\Delta \boldsymbol{r}}{\Delta t} = \frac{\Delta x}{\Delta t}\boldsymbol{i} + \frac{\Delta y}{\Delta t}\boldsymbol{j}$$

$$\boldsymbol{v} = \lim_{\Delta t \to 0}\left(\frac{\Delta \boldsymbol{r}}{\Delta t}\right) = \frac{\mathrm{d}x}{\mathrm{d}t}\boldsymbol{i} + \frac{\mathrm{d}y}{\mathrm{d}t}\boldsymbol{j} \tag{2.4}$$

From Fig. 2.5 it is clear that

$$|\boldsymbol{v}| = \sqrt{(\dot{x}^2 + \dot{y}^2)} \tag{2.5}$$

where differentiation with respect to time is denoted by the use of a dot over the variable, i.e. $\mathrm{d}x/\mathrm{d}x = \dot{x}$.

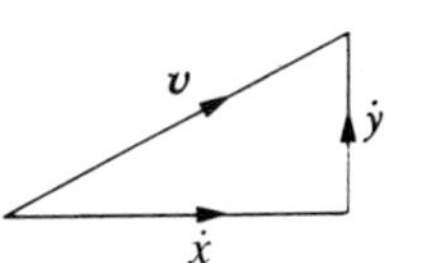

Figure 2.5

The acceleration is equally easy to derive. Since

$$\boldsymbol{v} = \dot{x}\boldsymbol{i} + \dot{y}\boldsymbol{j}$$

then

$$\boldsymbol{v} + \Delta \boldsymbol{v} = (\dot{x} + \Delta \dot{x})\boldsymbol{i} + (\dot{y} + \Delta \dot{y})\boldsymbol{j}$$

giving

$$\Delta \boldsymbol{v} = \Delta \dot{x}\boldsymbol{i} + \Delta \dot{y}\boldsymbol{j}.$$

$$\boldsymbol{a} = \lim_{\Delta t \to 0}\left(\frac{\Delta \boldsymbol{v}}{\Delta t}\right) = \lim_{\Delta t \to 0}\left(\frac{\Delta \dot{x}}{\Delta t}\boldsymbol{i} + \frac{\Delta \dot{y}}{\Delta t}\boldsymbol{j}\right)$$

$$\boldsymbol{a} = \frac{\mathrm{d}\dot{x}}{\mathrm{d}t}\boldsymbol{i} + \frac{\mathrm{d}\dot{y}}{\mathrm{d}t}\boldsymbol{j} \tag{2.6}$$

$$= \ddot{x}\boldsymbol{i} + \ddot{y}\boldsymbol{j}$$

and $|\boldsymbol{a}| = \sqrt{(\ddot{x}^2 + \ddot{y}^2)}$ (2.7)

Let us consider two simple cases and describe the motion in Cartesian co-ordinates.

i) *Motion in a straight line with constant acceleration*

Choosing the x-axis to coincide with the path of motion, we have

$$\ddot{x} = a$$

Intregration with respect to time gives

$$\int \ddot{x}\,\mathrm{d}t = \int(\mathrm{d}v/\mathrm{d}t)\,\mathrm{d}t = v = \int a\,\mathrm{d}t = at + C_1 \tag{2.8}$$

where C_1 is a constant depending on v when $t = 0$.

Integrating again,

$$\int v\,\mathrm{d}t = \int(\mathrm{d}x/\mathrm{d}t)\,\mathrm{d}t = x = \int(at + C_1)\,\mathrm{d}t$$
$$= \tfrac{1}{2}at^2 + C_1 t + C_2 \tag{2.9}$$

where C_2 is another constant depending on the value of x at $t = 0$.

ii) *Motion with constant speed along a circular path*

For the circular path shown in Fig. 2.6,

$$x^2 + y^2 = R^2 \tag{2.10}$$

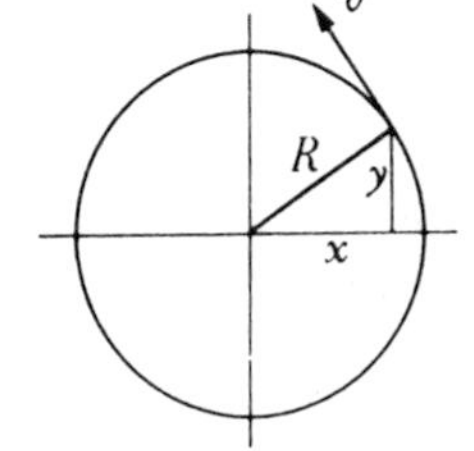

Figure 2.6

Differentiating twice with respect to time gives

$$2x\dot{x}+2y\dot{y}=0$$

and $\quad 2x\ddot{x}+2\dot{x}^2+2y\ddot{y}+2\dot{y}^2=0$

Since $2\dot{x}^2+2\dot{y}^2=2v^2$,

$$x\ddot{x}+y\ddot{y}=-v^2 \tag{2.11}$$

We see that, when $y=0$ and $x=R$,

$$\ddot{x}=-v^2/R$$

also, when $x=0$ and $y=R$,

$$\ddot{y}=-v^2/R$$

or, in general (Fig. 2.7), the component of acceleration resolved along the radius is

$$\begin{aligned} a_r &= \ddot{x}\cos\alpha+\ddot{y}\sin\alpha \\ &= \ddot{x}x/R+\ddot{y}y/R \end{aligned}$$

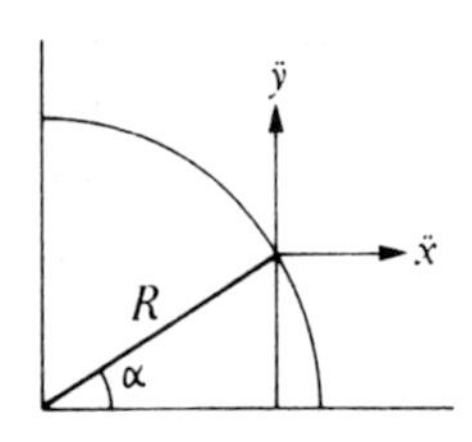

Figure 2.7

Using equation 2.11 we see that

$$a_r=-v^2/R$$

Resolving tangentially to the path,

$$\begin{aligned} a_t &= \ddot{y}\cos\alpha-\ddot{x}\sin\alpha \\ &= \ddot{y}x/R-\ddot{x}y/R \end{aligned}$$

Differentiating $\dot{x}^2+\dot{y}^2=v^2$ with respect to time, we have

$$2\dot{x}\ddot{x}+2\dot{y}\ddot{y}=0$$

hence

$$\ddot{y}/\ddot{x}=-\dot{x}/\dot{y}$$

and from the differentiation of equation 2.10 we have

$$\dot{y}/\dot{x}=-x/y$$

giving

$$\ddot{y}/\ddot{x}=-\dot{x}/\dot{y}=y/x$$

Thus we see that $a_t=0$.

This analysis should be contrasted with the more direct approach in terms of path and polar co-ordinates shown later in this chapter.

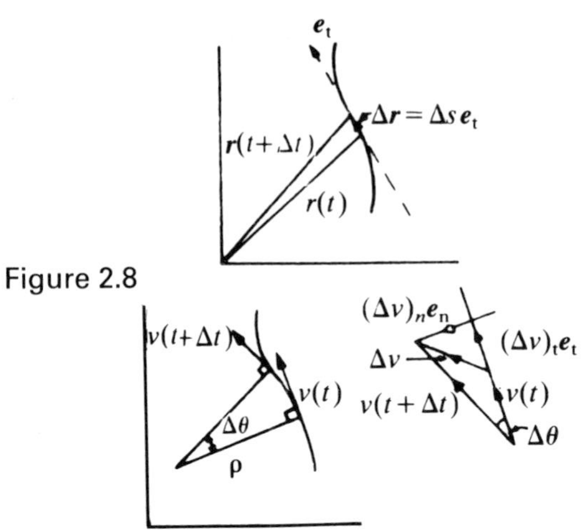

Figure 2.8

Figure 2.9

2.3 Path co-ordinates

The displacement $\Delta\boldsymbol{r}$ over a time interval Δt is shown in Fig. 2.8, where Δs is the elemental path length. Referring to Fig. 2.9, the direction of the path has changed by an angle $\Delta\theta$ and the speed has increased by Δv. Noting that the magnitude of $\boldsymbol{v}(t+\Delta t)$ is $(v+\Delta v)$, the change in velocity resolved along the original normal is

$$(v+\Delta v)\sin\Delta\theta$$

hence the acceleration in this direction is

$$a_n=\lim_{\Delta t\to 0}\left(\frac{(v+\Delta v)}{\Delta t}\right)\sin\Delta\theta$$

For small $\Delta\theta$, $\sin\Delta\theta\to\Delta\theta$; thus

$$a_n=\lim_{\Delta t\to 0}\left(\frac{v\Delta\theta}{\Delta t}+\frac{\Delta v\Delta\theta}{\Delta t}\right)=v\frac{d\theta}{dt}$$

and is directed towards the centre of curvature, i.e. in the direction of $\boldsymbol{e}_n$.

If ρ is the radius of curvature, then

$$ds=\rho\, d\theta$$

hence

$$\frac{ds}{dt}=\rho\frac{d\theta}{dt}$$

therefore

$$a_n=v\frac{1}{\rho}\frac{ds}{dt}=\frac{v^2}{\rho} \tag{2.12}$$

The change in velocity resolved tangentially to the path is

$$(v+\Delta v)\cos\Delta\theta-v$$

hence the acceleration along the path is

$$\lim_{\Delta t \to 0}\left(\frac{(v+\Delta v)\cos\Delta\theta - v}{\Delta t}\right) = \frac{dv}{dt} = a_t \quad (2.13)$$

Summarising, we have

$$\boldsymbol{v} = v\boldsymbol{e}_t = \frac{ds}{dt}\boldsymbol{e}_t \tag{2.14a}$$

$$\boldsymbol{a} = \frac{dv}{dt}\boldsymbol{e}_t + v\frac{d\theta}{dt}\boldsymbol{e}_n \tag{2.14b}$$

$$= \frac{d^2s}{dt^2}\boldsymbol{e}_t + \frac{v^2}{\rho}\boldsymbol{e}_n \tag{2.14c}$$

We will now reconsider the previous simple cases.

i) *Straight-line motion with constant acceleration*

$\boldsymbol{a} = a\boldsymbol{e}_t$ ($\boldsymbol{e}_t$ fixed in direction)

or $d^2s/dt^2 = a$ (2.15)

The solution is the same as before, with x replaced by s.

ii) *Motion in a circle at constant speed*

$\boldsymbol{a} = (v^2/\rho)\boldsymbol{e}_n$ (v and ρ are constant) (2.16)

2.4 Polar co-ordinates

Polar co-ordinates are a special case of cylindrical co-ordinates with $z = 0$, or of spherical co-ordinates with $\phi = 0$.

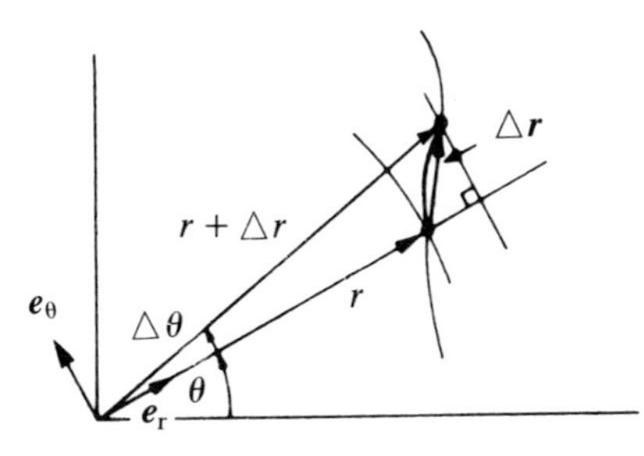

Figure 2.10

Referring to Fig. 2.10, it can be seen that

$$\Delta\boldsymbol{r} = [(r+\Delta r)\cos\Delta\theta - r]\boldsymbol{e}_r + (r+\Delta r)\sin\Delta\theta\boldsymbol{e}_\theta$$

hence the velocity is given by

$$\boldsymbol{v} = \lim_{\Delta t\to 0}\left(\frac{\Delta\boldsymbol{r}}{\Delta t}\right)$$

$$= \left(\frac{dr}{dt}\right)\boldsymbol{e}_r + r\left(\frac{d\theta}{dt}\right)\boldsymbol{e}_\theta = \dot{r}\boldsymbol{e}_r + r\dot{\theta}\boldsymbol{e}_\theta \tag{2.17}$$

Resolving the components of $\Delta\boldsymbol{v}$ along the $\boldsymbol{e}_r$ and $\boldsymbol{e}_\theta$ directions (Fig. 2.11) gives

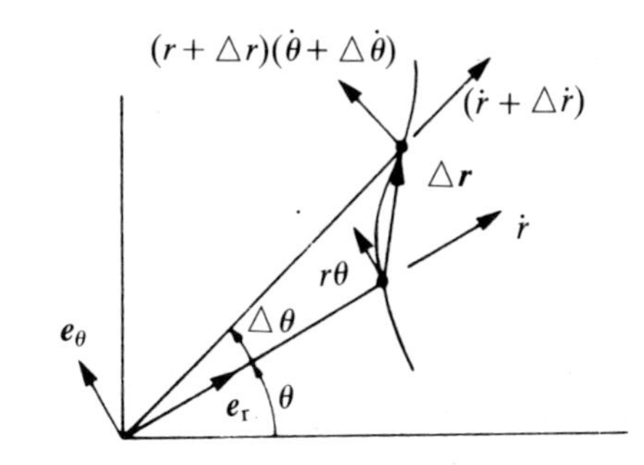

Figure 2.11

$$\begin{aligned}\Delta\dot{\boldsymbol{r}} = {} & [(\dot{r}+\Delta\dot{r})\cos\Delta\theta - (r+\Delta r) \\ & \times(\dot{\theta}+\Delta\dot{\theta})\sin\Delta\theta - \dot{r}]\boldsymbol{e}_r \\ & + [\dot{r}+\Delta\dot{r})\sin\Delta\theta + (r+\Delta r) \\ & \times(\dot{\theta}+\Delta\dot{\theta})\cos\Delta\theta - r\dot{\theta}]\boldsymbol{e}_\theta\end{aligned}$$

For small angles, $\sin\Delta\theta \to \Delta\theta$ and $\cos\Delta\theta \to 1$; thus

$$\boldsymbol{a} = \lim_{\Delta t\to 0}\left(\frac{\Delta\dot{\boldsymbol{r}}}{\Delta t}\right) = \left(\frac{d\dot{r}}{dt} - r\dot{\theta}\frac{d\theta}{dt}\right)\boldsymbol{e}_r + \left(\dot{r}\frac{d\theta}{dt} + r\frac{d\dot{\theta}}{dt} + \frac{dr}{dt}\dot{\theta}\right)\boldsymbol{e}_\theta$$

$$\boldsymbol{a} = (\ddot{r} - r\dot{\theta}^2)\boldsymbol{e}_r + (r\ddot{\theta} + 2\dot{r}\dot{\theta})\boldsymbol{e}_\theta \tag{2.18}$$

An alternative approach to deriving equations 2.17 and 2.18 is to proceed as follows.

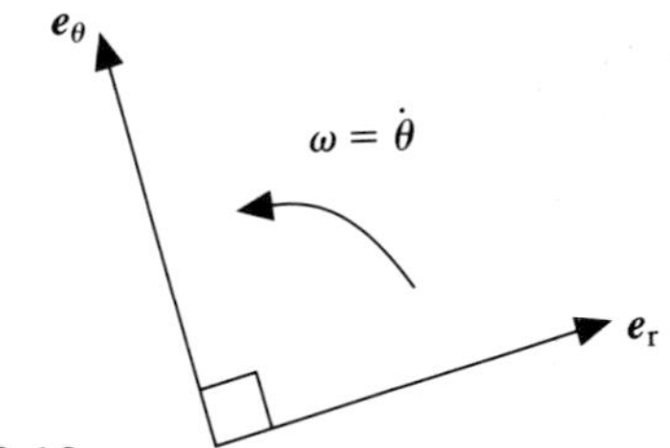

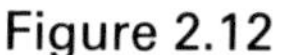

Figure 2.12

Consider the orthogonal unit vectors $\boldsymbol{e}_r$ and $\boldsymbol{e}_\theta$ which are rotating at an angular rate $\omega = \dot{\theta}$ as shown in Fig. 2.12. The derivative with respect to time of $\boldsymbol{e}_r$ is

$$\dot{\boldsymbol{e}}_r = \lim_{\Delta t\to 0}\left(\frac{\Delta\boldsymbol{e}_r}{\Delta t}\right)$$

where $\Delta\boldsymbol{e}_r$ is the change in $\boldsymbol{e}_r$ which occurs in the time interval Δt. During this interval $\boldsymbol{e}_r$ and $\boldsymbol{e}_\theta$

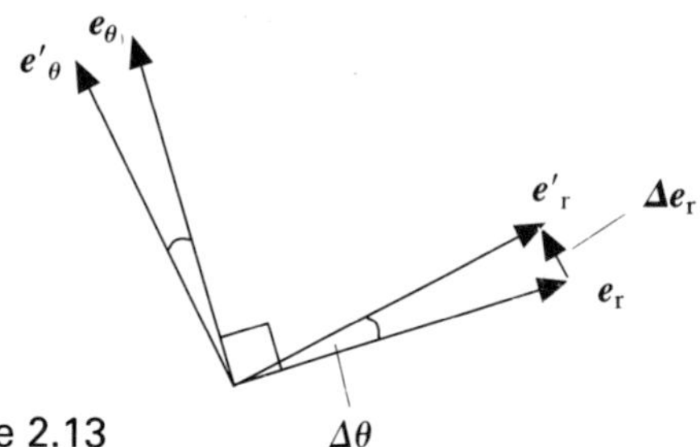

Figure 2.13

have rotated through the angle $\Delta\theta$, as shown in Fig. 2.13, so that they become the new unit vectors $\boldsymbol{e}'_r$ and $\boldsymbol{e}'_\theta$. The difference between $\boldsymbol{e}'_r$ and $\boldsymbol{e}_r$ is $\Delta\boldsymbol{e}_r = \boldsymbol{e}'_r - \boldsymbol{e}_r$. The magnitude of $\Delta\boldsymbol{e}_r$ for small $\Delta\theta$ is $1\times\Delta\theta$ since the magnitude of $\boldsymbol{e}_r$ is unity, by definition. For vanishingly small $\Delta\theta$, the vector $\Delta\boldsymbol{e}_r$ has the direction of $\boldsymbol{e}_\theta$,
hence

$$\dot{\boldsymbol{e}}_r = \lim_{\Delta t,\Delta\theta\to 0}\left(\frac{\Delta\boldsymbol{e}_r}{\Delta t}\right) = \lim_{\Delta t\to 0}\left(\frac{\Delta\theta\boldsymbol{e}_\theta}{\Delta t}\right) = \dot{\theta}\boldsymbol{e}_\theta \tag{2.19}$$

Similarly it can be shown that

$$\dot{\boldsymbol{e}}_\theta = -\dot{\theta}\boldsymbol{e}_r \tag{2.20}$$

The velocity $\boldsymbol{v}$ is the derivative with respect to time of the position vector $\boldsymbol{r} = r\boldsymbol{e}_r$. From the chain rule for differentiation we obtain

$$\boldsymbol{v} = \dot{\boldsymbol{r}} = \frac{\mathrm{d}}{\mathrm{d}t}(r\boldsymbol{e}_r) = \dot{r}\boldsymbol{e}_r + r\dot{\boldsymbol{e}}_r$$

$$= \dot{r}\boldsymbol{e}_r + r\dot{\theta}\boldsymbol{e}_\theta$$

from equation 2.19, which is the result previously obtained in equation (2.17).

The acceleration $\boldsymbol{a}$ can also be found from the chain rule, thus

$$\boldsymbol{a} = \dot{\boldsymbol{v}} = \frac{\mathrm{d}}{\mathrm{d}t}(\dot{r}\boldsymbol{e}_r + r\dot{\theta}\boldsymbol{e}_\theta)$$

$$= \ddot{r}\boldsymbol{e}_r + \dot{r}\dot{\boldsymbol{e}}_r + \dot{r}\dot{\theta}\boldsymbol{e}_\theta + r\ddot{\theta}\boldsymbol{e}_\theta + r\dot{\theta}\dot{\boldsymbol{e}}_\theta$$

Substituting from equations (2.19) and (2.20) we arrive at the result given in equation (2.18). (The differentiation of rotating vectors is dealt with more fully in Chapter 11).

As before we consider the two simple cases.

i) *Motion in a straight line*

$\theta = 0$ for all time

$$\boldsymbol{a} = \ddot{r}\boldsymbol{e}_r \tag{2.21}$$

ii) *Motion in a circle at constant speed*

$r =$ constant for all time

$\boldsymbol{v} = r\dot{\theta}\boldsymbol{e}_\theta$

Because r and v are constant, $\dot{\theta}$ is constant; so

$$\boldsymbol{a} = -r\dot{\theta}^2\boldsymbol{e}_r = -(v^2\text{V}/r)\boldsymbol{e}_r \tag{2.22}$$

We may also consider another simple example, that of a fly walking at a constant speed along a radial spoke of a wheel rotating at a constant speed. In this case

$$\boldsymbol{a} = [-r\dot{\theta}^2]\boldsymbol{e}_r + 2\dot{r}\dot{\theta}\boldsymbol{e}_\theta$$

so we see that there is a constant component of acceleration, $2\dot{r}\dot{\theta}$, at right angles to the spoke, independent of r. This component is often called the Coriolis component, after the French engineer Gustav-Gaspard Coriolis.

2.5 Relative motion

In this section we shall adopt the following notation:

$\boldsymbol{r}_{B/A}$ = position of B relative to A

$\dot{\boldsymbol{r}}_{B/A}$ = velocity of B relative to A, etc.

From Fig. 2.14,

$$\boldsymbol{r}_{B/O} = \boldsymbol{r}_{A/O} + \boldsymbol{r}_{B/A} \tag{2.23}$$

Differentiation with respect to time gives

$$\dot{\boldsymbol{r}}_{B/O} = \dot{\boldsymbol{r}}_{A/O} + \dot{\boldsymbol{r}}_{B/A} \tag{2.24}$$

and

$$\ddot{\boldsymbol{r}}_{B/O} = \ddot{\boldsymbol{r}}_{A/O} + \ddot{\boldsymbol{r}}_{B/A} \tag{2.25}$$

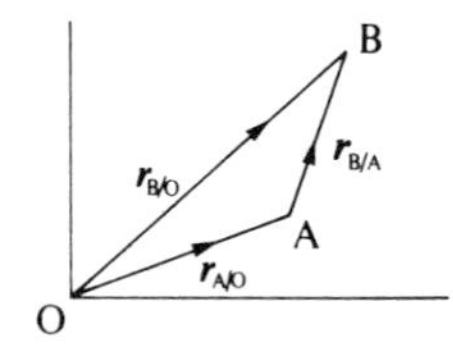

Figure 2.14

[The notation $\dot{\boldsymbol{r}}_B$ and $\ddot{\boldsymbol{r}}_B$ may be used in place of $\dot{\boldsymbol{r}}_{B/O}$ and $\ddot{\boldsymbol{r}}_{B/O}$ for velocity and acceleration relative to the reference axes.]

Consider now the case of a wheel radius r, centre A, moving so that A has rectilinear motion in the x-direction and the wheel is rotating at angular speed $\omega = \dot{\theta}$ (Fig. 2.15). The path traced out by a point B on the rim of the wheel is complex, but the velocity and acceleration of B may be easily obtained by use of equations 2.24 and 2.25.

Referring to Fig. 2.15,

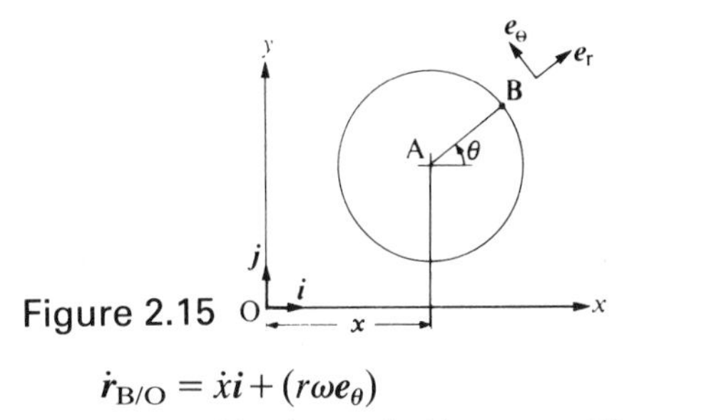

Figure 2.15

$$\dot{\boldsymbol{r}}_{\mathrm{B/O}} = \dot{x}\boldsymbol{i} + (r\omega\boldsymbol{e}_\theta)$$
$$= \dot{x}\boldsymbol{i} + (-r\omega\sin\theta\boldsymbol{i} + r\omega\cos\theta\boldsymbol{j}) \quad (2.26)$$

Similarly,

$$\ddot{\boldsymbol{r}}_{\mathrm{B/O}} = \ddot{x}\boldsymbol{i} + (-r\omega^2\boldsymbol{e}_\mathrm{r} + r\dot{\omega}\boldsymbol{e}_\theta)$$
$$= \ddot{x}\boldsymbol{i} - r\omega^2(\cos\theta\boldsymbol{i} + \sin\theta\boldsymbol{j})$$
$$+ r\dot{\omega}(-\sin\theta\boldsymbol{i} + \cos\theta\boldsymbol{j})$$
$$= (\ddot{x} - r\omega^2\cos\theta - r\dot{\omega}\sin\theta)\boldsymbol{i}$$
$$+ (-r\omega^2\sin\theta + r\dot{\omega}\cos\theta)\boldsymbol{j} \quad (2.27)$$

A special case of the above problem is that of rolling without slip. This implies that when $\theta = 3\pi/2$, $\dot{\boldsymbol{r}}_{\mathrm{B/O}} = 0$. Since

$$\dot{\boldsymbol{r}}_{\mathrm{B/O}} = (\dot{x} + r\omega)\boldsymbol{i} + 0\boldsymbol{j} = 0$$

then $\dot{x} = -r\omega$

Also,

$$\ddot{\boldsymbol{r}}_{\mathrm{B/O}} = (\ddot{x} + r\dot{\omega})\boldsymbol{i} + (r\omega^2)\boldsymbol{j}$$

but

$$\ddot{x} = -r\dot{\omega}$$

therefore

$$\ddot{\boldsymbol{r}}_{\mathrm{B/O}} = r\omega^2\boldsymbol{j}$$

Note that differentiating $\dot{\boldsymbol{r}}_{\mathrm{B/O}}(\theta = 3\pi/2)$ does *not* give $\ddot{\boldsymbol{r}}_{\mathrm{B/O}}(\theta = 3\pi/2)$: θ must be included as a variable of the differentiation.

2.6 One-dimensional motion

The description 'one-dimensional' is not to be taken as synonymous with 'linear', for, although linear motion is one-dimensional, not all one-dimensional motion is linear.

We have one-dimensional motion in path co-ordinates if we consider only displacement along the path; in polar co-ordinates we can consider only variations in angle, regarding the radius as constant. Let us consider a problem in path co-ordinates, Fig. 2.16, the location of P being determined by s measured along the path from some origin O. (This path could, of course, be a straight line.)

Speed is defined as $v = \mathrm{d}s/\mathrm{d}t$, and $\mathrm{d}v/\mathrm{d}t$ = rate of change of speed. This quantity is also the component of acceleration tangential to the path, but it is not the total acceleration.

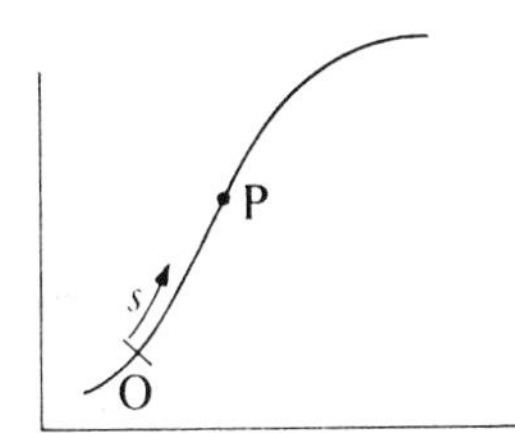

Figure 2.16

We may write

$$a_\mathrm{t} = \frac{\mathrm{d}v}{\mathrm{d}t} = \frac{\mathrm{d}s}{\mathrm{d}t}\frac{\mathrm{d}v}{\mathrm{d}s} = v\frac{\mathrm{d}v}{\mathrm{d}s}$$

Hence we have

$$a_\mathrm{t} = \frac{\mathrm{d}v}{\mathrm{d}t} = \frac{\mathrm{d}^2 s}{\mathrm{d}t^2} = v\frac{\mathrm{d}v}{\mathrm{d}s} \quad (2.28)$$

Most problems in one-dimensional kinematics involve converting data given in one set of variables to other data. As an example: given the way in which a component of acceleration varies with displacement, determine the variation of speed with time. In such problems the sketching of appropriate graphs is a useful aid to the solution.

2.7 Graphical methods

Speed–time graph (Fig. 2.17)

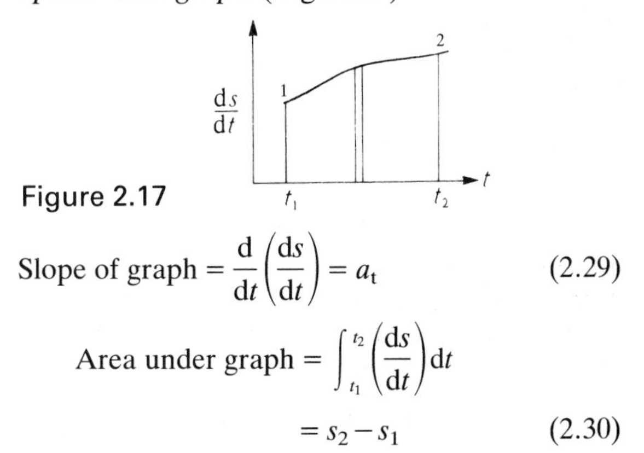

Figure 2.17

$$\text{Slope of graph} = \frac{\mathrm{d}}{\mathrm{d}t}\left(\frac{\mathrm{d}s}{\mathrm{d}t}\right) = a_\mathrm{t} \quad (2.29)$$

$$\text{Area under graph} = \int_{t_1}^{t_2}\left(\frac{\mathrm{d}s}{\mathrm{d}t}\right)\mathrm{d}t$$
$$= s_2 - s_1 \quad (2.30)$$

Hence, slope = rate of change of speed

and area = change of distance

If a_t is constant, then the graph is a straight line and

$$\text{area} = \tfrac{1}{2}(v_1 + v_2)(t_2 - t_1) = s_2 - s_1 \quad (2.31)$$

and slope $= a_t$

Distance–time graph (Fig. 2.18)

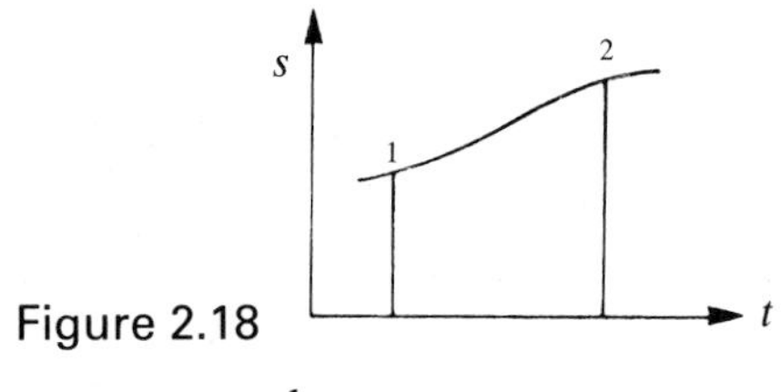

Figure 2.18

$$\text{Slope} = \frac{ds}{dt} = v \quad (2.32)$$

Rate-of-change-of-speed–time graph (Fig. 2.19)

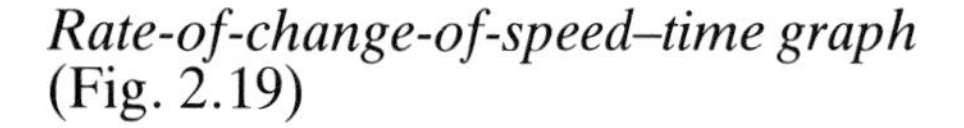

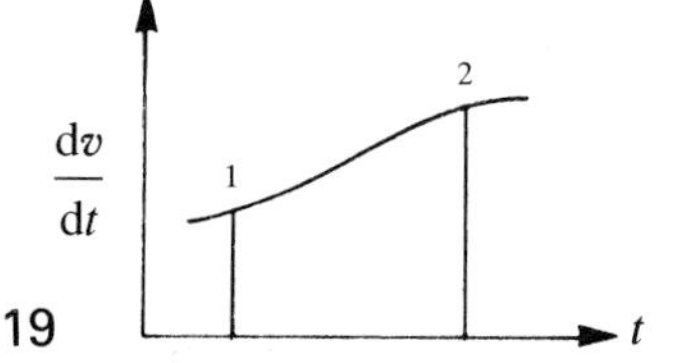

Figure 2.19

$$\text{Area} = \int_{t_1}^{t_2} \frac{dv}{dt}\, dt = v_2 - v_1 \quad (2.33)$$

If a_t is constant, then

$$\text{area} = a_t(t_2 - t_1) = v_2 - v_1 \quad (2.34)$$

Rate-of-change-of-speed–displacement graph (Fig. 2.20). Here we make use of equation (2.28).

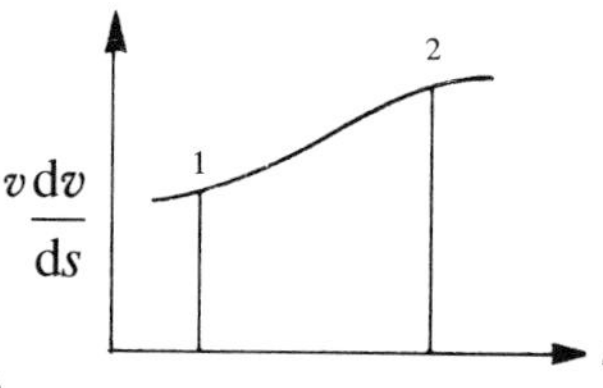

Figure 2.20

$$\text{Area} = \int_{s_1}^{s_2} v\frac{dv}{ds}\, ds = \tfrac{1}{2}v_2^2 - \tfrac{1}{2}v_1^2 \quad (2.35)$$

If a_t is constant, then

$$a_t(s_2 - s_1) = \tfrac{1}{2}v_2^2 - \tfrac{1}{2}v_1^2 \quad (2.36)$$

Inverse-speed–distance graph (Fig. 2.21)

$$\text{Area} = \int_{s_1}^{s_2} \frac{1}{v}\, ds = \int_{s_1}^{s_2} \frac{dt}{ds}\, ds = t_2 - t_1 \quad (2.37)$$

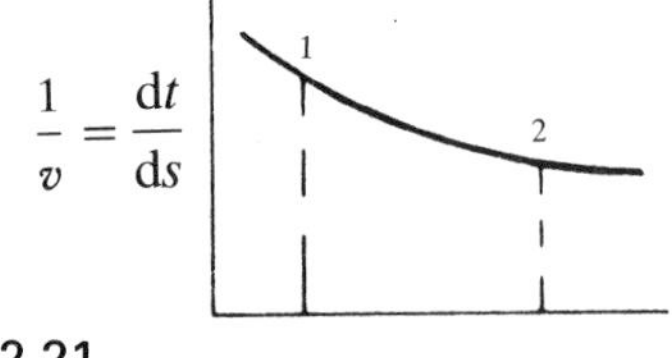

Figure 2.21

The advantages of sketching the graphs are many – even for cases of constant acceleration (see examples 2.2 and 2.3).

Discussion examples

Example 2.1

A point P moves along a path and its acceleration component tangential to the path has a *constant* magnitude a_{t0}. The distance moved along the path is s. At time $t = 0$, $s = 0$ and $v = v_0$. Show that (a) $v = v_0 + a_{t0}t$, (b) $s = v_0 t + \tfrac{1}{2}a_{t0}t^2$, (c) $v^2 = v_0^2 + 2a_{t0}s$ and (d) $s = \tfrac{1}{2}(v + v_0)t$.

Solution

a) Since $a_t = dv/dt$,

$$dv = a_{t0}\, dt$$

and $\int_{v_0}^{v} dv = a_{t0}\int_0^t dt$ since a_{t0} is constant.

Therefore $v - v_0 = a_{t0}t$

or $$v = v_0 + a_{t0}t \quad \text{(i)}$$

b) Since $v = ds/dt$,

$$\int_0^s ds = \int_0^t v\, dt = \int_0^t (v_0 + a_{t0}t)\, dt$$

$$s = v_0 t + \tfrac{1}{2}a_{t0}t^2 \quad \text{(ii)}$$

c) From (i), $t = (v - v_0)/a_{t0}$ and substituting for t in (ii) gives

$$v^2 = v_0^2 + 2a_{t0}s \quad \text{(iii)}$$

d) Also from (i), $a_{t0} = (v - v_0)/t$ and substituting for a_{t0} in (ii) gives

$$s = \tfrac{1}{2}(v_0 + v)t \quad \text{(iv)}$$

[As these equations for constant acceleration are often introduced before the case of variable acceleration has been discussed, it is a common mistake to try to apply them to problems dealing with variable acceleration. For such problems, however, the methods of section 2.7 should always be used (cf. example 2.3).]

Example 2.2

The variation with time of the tangential acceleration a_t of a vehicle is given in Fig. 2.22. At time $t = 0$ the speed is zero. Determine the speed when $t = t_3$.

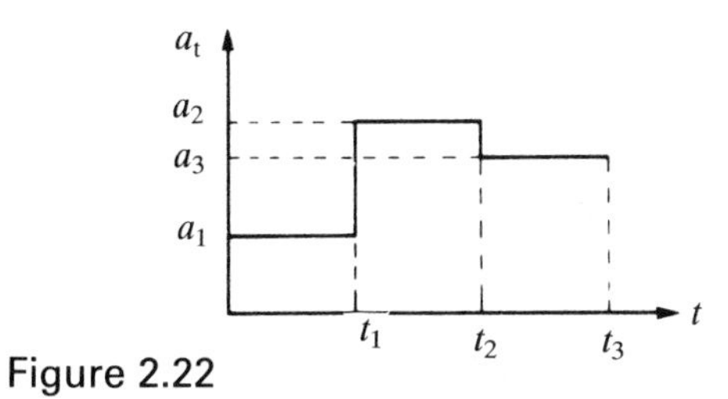

Figure 2.22

Solution Each portion of the graph represents constant acceleration and so we can use the appropriate formula (equation 2.34), $a_t(t_2 - t_1) = v_2 - v_1$, for each portion, using the final speed of one part as the initial speed of the next.

Time 0 to t_1:

$$v_1 - v_0 = a_1(t_1 - t_0), \quad v_1 = a_1 t_1$$

Time t_1 to t_2:

$$v_2 - v_1 = a_2(t_2 - t_1), \quad v_2 = a_2(t_2 - t_1) + v_1$$
$$= a_2(2 - t_1) + a_1 t_1$$

Time t_2 to t_3:

$$v_3 - v_2 = a_3(t_3 - t_2), \quad v_3 = a_3(t_3 - t_2) + v_2$$
$$v_3 = a_3(t_3 - t_2) + a_2(t_2 - t_1) + a_1 t_1$$

Alternatively we can dispense with the constant-acceleration formulae and obtain the same result more rapidly by noting that the speed change is equal to the area under the graph of tangential acceleration versus time (see equation 2.33), so that the speed at $t = t_3$ can be written down immediately.

Example 2.3

An accelerometer mounted in a vehicle measures the magnitude of the tangential acceleration a_t. At the same time the distance travelled, s, is recorded with the following results (see section 3.3):

$a_t/(\mathrm{m\,s^{-2}})$	s/m	$a_t/(\mathrm{m\,s^{-2}})$	s/m
1.2	0	−1.3	25
2.1	5	−0.8	30
2.6	10	0.1	35
2.1	15	0.9	40
0.4	20		

Given that the initial forward speed is 3.0 m/s and the acceleration varies smoothly with distance, find for $s = 40$ m (a) the speed and (b) the time taken.

Solution

a) We are given a_t in terms of s and require to find v, therefore we must use an expression relating these three parameters. The constant-acceleration formulae are of course not relevant here. The basic definition $a_t = dv/dt$ cannot be used directly and we must use the alternative form $a_t = v(dv/ds)$, equation 2.28, which relates the three required parameters. Integration gives

$$\int_{v_1}^{v_2} v\,dv = \int_{s_1}^{s_2} a_t\,ds$$

or $\frac{1}{2}(v_2^2 - v_1^2)$ is equal to the area under the graph of a_t versus s between $s = s_1$ and $s = s_2$, Fig. 2.23.

Letting $s_1 = 0$ and $s_2 = 40$ m, the area is found to be 32.0 $(\mathrm{m/s})^2$. This area can be determined by counting the squares under the graph, by the trapezium rule, by Simpson's rule, etc., depending on the order of accuracy required. (The trapezium rule and Simpson's rule are given in Appendix 3.)

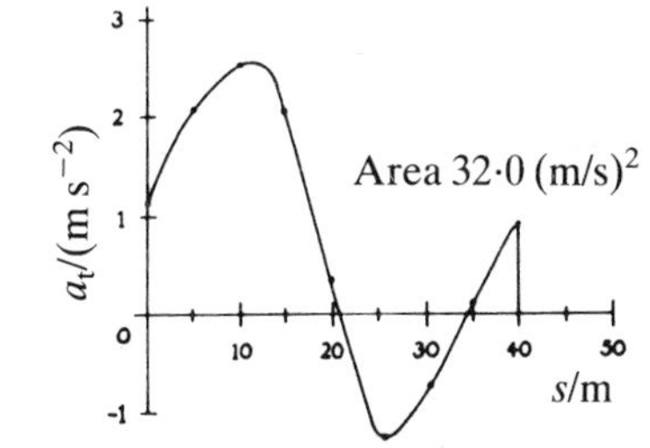

Figure 2.23

Thus $\frac{1}{2}(v_{40}^2 - 3^2) = 32$,

$$v_{40} = \sqrt{[2(32) + 3^2]} = 8.54 \text{ m/s}$$

b) Given a_t as a function of s, time cannot be found directly. We can, however, make use of the relationship $v = ds/dt$ in the form $dt = (1/v)ds$ provided we can first establish the relationship between v and s. To find values of v at various values of s, we can use repeated applications of the method of (a) above.

It is useful to set out the calculations in tabular form:

s/m	area/(m^2 s^{-2})	v/(m s^{-1}) $= [2(\text{area} + v_0^2)]^{1/2}$
0–5	8.4	5.08
0–10	20.2	7.03
0–15	32.6	8.61
0–20	39.4	9.37
0–25	36.8	9.09
0–30	31.2	8.45
0–35	29.0	8.91
0–40	32.0	8.54

Since $t_2 - t_1 = (1/v)\,ds$, the area under the graph of $1/v$ versus s will give the required time. Corresponding values are given below and are plotted in Fig. 2.24.

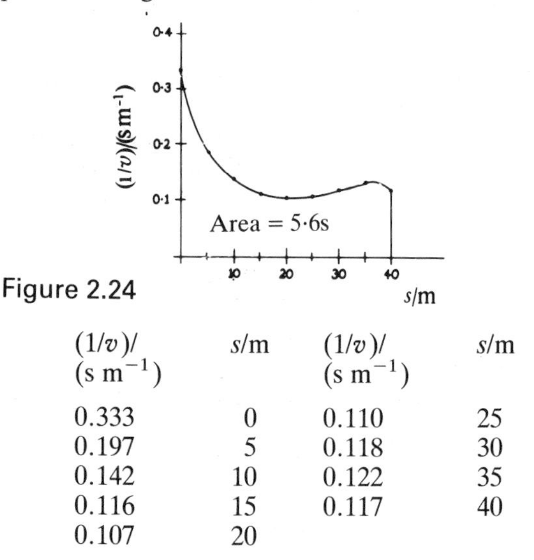

Figure 2.24

$(1/v)$/(s m^{-1})	s/m	$(1/v)$/(s m^{-1})	s/m
0.333	0	0.110	25
0.197	5	0.118	30
0.142	10	0.122	35
0.116	15	0.117	40
0.107	20		

The time taken is found to be approximately 5.6 s.

Example 2.4

At a particular instant, a point on a mechanism has a speed of 5.0 m/s and a tangential acceleration of magnitude 2.0 m/s^2. If the magnitude of the total acceleration is 3.0 m/s^2, what is the radius of curvature of the path being traced out by the point at this instant?

Solution Choice of co-ordinates is not difficult for this problem since radius of curvature is featured only in path co-ordinates. In these co-ordinates the total acceleration $\boldsymbol{a}$ (see Fig. 2.25) is given by

$$\boldsymbol{a} = a_t\boldsymbol{e}_t + a_n\boldsymbol{e}_n$$

$$= \ddot{s}\boldsymbol{e}_t + (v^2/\rho)\,\boldsymbol{e}_n \quad \text{(see equations 2.14)}$$

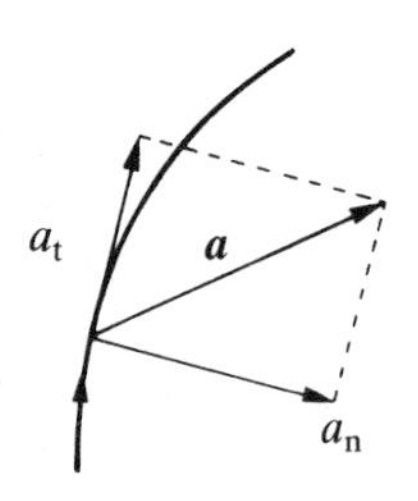

Figure 2.25

The magnitude of $\boldsymbol{a}$ is $\surd[\ddot{s}^2 + (v^2/\rho)^2]$ and substitution of the numerical values gives

$$3.0 = \surd[2^2 + (5^2/\rho)^2]$$

and $\rho = 11.18$ m

Example 2.5

See Fig. 2.26. The centre C of the wheel of radius 0.5 m has a constant velocity of 2.5 m/s to the right. The angular velocity of the wheel is constant and equal to 6 rad/s clockwise. Point P is at the bottom of the wheel and is in contact with a horizontal surface. Points Q and R are as shown in the figure.

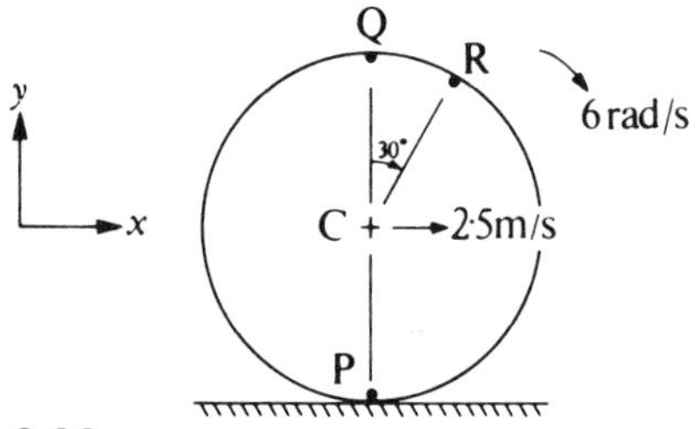

Figure 2.26

Determine (a) whether or not the wheel is slipping on the surface, (b) the velocities and accelerations of the points P and Q and (c) the velocity and acceleration of the point R.

Solution Usually the simplest way of dealing with the motion of a point on a wheel which is rotating and translating is to determine the motion of the wheel centre and add on the motion of the point relative to the centre. So for an arbitrary point A and centre C we can make use of

$$\boldsymbol{v}_A = \boldsymbol{v}_C + \boldsymbol{v}_{A/C} \quad \text{(see equation (2.24)}$$

and $\boldsymbol{a}_A = \boldsymbol{a}_C + \boldsymbol{a}_{A/C}$ (see equation 2.25)

a) If the wheel is not slipping then the velocity of point P must be the same as the velocity of the surface, namely zero.

From equation 2.17, the velocity of P relative to C is given by

$$\boldsymbol{v}_{P/C} = \dot{r}\boldsymbol{e}_r + r\dot{\theta}\boldsymbol{e}_\theta$$

where r is the length of the line CP and θ is the angle of the line CP measured from some datum in the plane of the motion. Since r has a constant value (0.5 m) then $\dot{r} = 0$ and $\boldsymbol{v}_{P/C}$ has no component in the direction of CP. The angular velocity of the line CP is $\dot{\theta}$ in the anticlockwise direction (since θ is defined as positive in this sense); thus $\dot{\theta} = -6$ rad/s, and [see Fig. 2.27(a)]

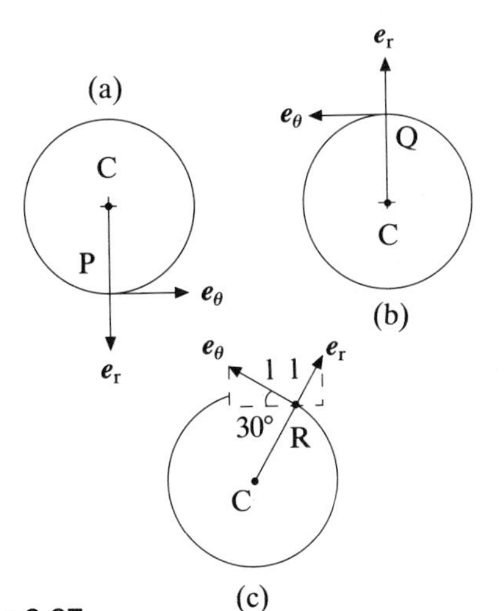

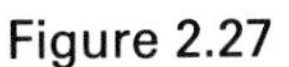
Figure 2.27

$$\boldsymbol{v}_{P/C} = v_\theta \boldsymbol{e}_\theta = r\dot{\theta}\boldsymbol{e}_\theta = 0.5(-6)\boldsymbol{i} = -3\boldsymbol{i} \text{ m/s}$$

The velocity of C is $\boldsymbol{v} = 2.5\boldsymbol{i}$ m/s and the total velocity of P is

$$\boldsymbol{v}_P = \boldsymbol{v}_C + \boldsymbol{v}_{P/C}$$
$$= 2.5\boldsymbol{i} - 3\boldsymbol{i} = -0.5\boldsymbol{i} \text{ m/s}$$

The wheel is therefore slipping.

b) See Fig. 2.27(b). For the radial line CQ we have $\boldsymbol{e}_r = \boldsymbol{j}$ and $\boldsymbol{e}_\theta = -\boldsymbol{i}$. The velocity of Q relative to C is

$$\boldsymbol{v}_{Q/C} = r\dot{\theta}\boldsymbol{e}_\theta = (-3)(-\boldsymbol{i}) = 3\boldsymbol{i} \text{ m/s}$$

so that

$$\boldsymbol{v}_Q = \boldsymbol{v}_C + \boldsymbol{v}_{Q/C} = 2.5\boldsymbol{i} + 3\boldsymbol{i} = 5.5\boldsymbol{i} \text{ m/s}$$

From equation 2.18, the acceleration of Q relative to C is given by

$$\boldsymbol{a}_{Q/C} = (\ddot{r} - r\dot{\theta}^2)\boldsymbol{e}_r + (r\ddot{\theta} + 2\dot{r}\dot{\theta})\boldsymbol{e}_\theta$$
$$= (-0.5)(-6)^2\boldsymbol{e}_r$$
$$= -18\boldsymbol{j} \text{ m/s}^2$$

The total acceleration of Q is

$$\boldsymbol{a}_Q = \boldsymbol{a}_C + \boldsymbol{a}_{Q/C}$$

but $\boldsymbol{v}_C$ is constant, and so $\boldsymbol{a}_C = 0$. Therefore

$$\boldsymbol{a}_Q = -18\boldsymbol{j} \text{ m/s}^2$$

c) See Fig. 2.27(c). For the radial line CR,

$$\boldsymbol{e}_r = \sin 30°\boldsymbol{i} + \cos 30°\boldsymbol{j} = \tfrac{1}{2}\boldsymbol{i} + \tfrac{1}{2}\sqrt{3}\boldsymbol{j}$$

and

$$\boldsymbol{e}_\theta = -\cos 30°\boldsymbol{i} + \sin 30°\boldsymbol{j} = -\tfrac{1}{2}\sqrt{3}\boldsymbol{i} + \tfrac{1}{2}\boldsymbol{j}$$
$$\boldsymbol{v}_{R/C} = r\dot{\theta}\boldsymbol{e}_\theta = 0.5(-6)(-\tfrac{1}{2}\sqrt{3}\boldsymbol{i} + \tfrac{1}{2}\boldsymbol{j})$$
$$= (2.6\boldsymbol{i} - 1.5\boldsymbol{j}) \text{ m/s}$$

and

$$\boldsymbol{v}_R = \boldsymbol{v}_C + \boldsymbol{v}_{R/C} = 2.5\boldsymbol{i} + 2.6\boldsymbol{i} - 1.5\boldsymbol{j}$$
$$= (5.1\boldsymbol{i} - 1.5\boldsymbol{j}) \text{ m/s}$$

The same result can be obtained from a velocity vector diagram, Fig. 2.28. Here $\boldsymbol{v}_C$ and $\boldsymbol{v}_{R/C}$ are drawn to some appropriate scale in the correct directions and are added graphically to give $\boldsymbol{v}_R$.

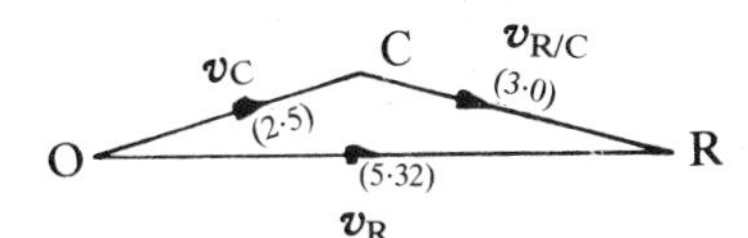

Figure 2.28

For the acceleration of R relative to C we have

$$\boldsymbol{a}_{R/C} = -r\dot{\theta}^2\boldsymbol{e}_r = -0.5(-6)^2(\tfrac{1}{2}\boldsymbol{i} + \tfrac{1}{2}\sqrt{3}\boldsymbol{j})$$
$$= (9\boldsymbol{i} + 15.6\boldsymbol{j}) \text{ m/s}^2$$

which is the total acceleration of R, since $\boldsymbol{a}_C = 0$.

Example 2.6

At the instant under consideration, the trolley T, Fig. 2.29, has a velocity of 4 m/s to the right and is decelerating at 2 m/s². The telescopic arm AB has a length of 1.5 m which is increasing at a constant rate of 2 m/s. At the same time, the arm has an anticlockwise angular velocity of 3 rad/s and a clockwise angular acceleration of 0.5 rad/s².

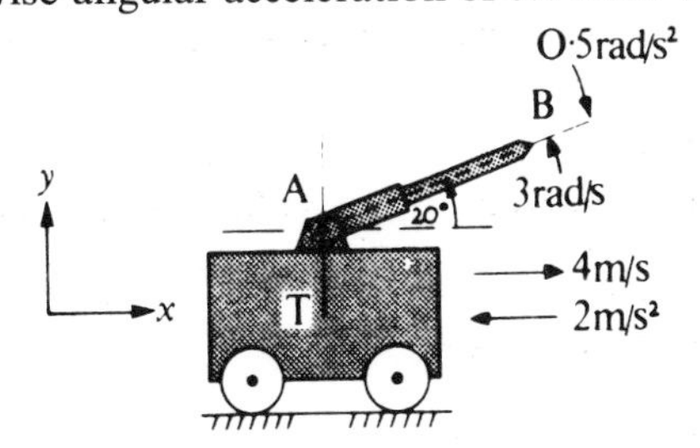

Figure 2.29

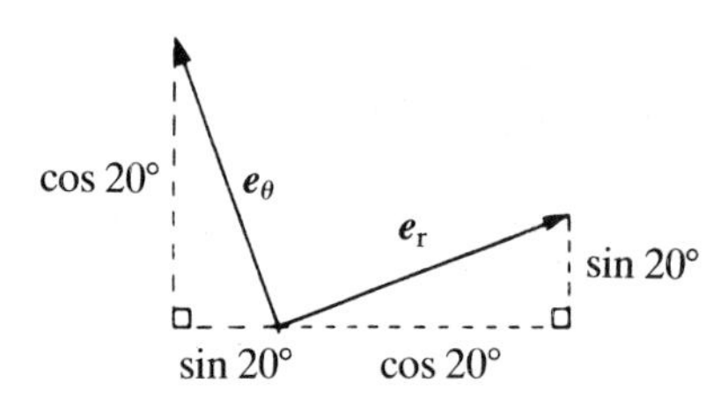

Figure 2.30

Determine for B (a) the velocity and speed, and (b) the acceleration and its magnitude. Give the vector quantities in terms of the unit vectors $\boldsymbol{i}$ and $\boldsymbol{j}$.

Solution Polar co-ordinates are again required, and we must first write down the expressions for $\boldsymbol{e}_r$ and $\boldsymbol{e}_\theta$ in terms of $\boldsymbol{i}$ and $\boldsymbol{j}$ (see Fig. 2.30).

$$\boldsymbol{e}_r = \cos 20°\boldsymbol{i} + \sin 20°\boldsymbol{j}$$
$$\boldsymbol{e}_\theta = -\sin 20°\boldsymbol{i} + \cos 20°\boldsymbol{j}$$

From equation 2.17,

$$\boldsymbol{v}_{B/A} = \dot{r}\boldsymbol{e}_r + r\dot{\theta}\boldsymbol{e}_\theta \quad \text{where } r = 1.5 \text{ and } \dot{r} = 2$$

thus $\boldsymbol{v}_{B/A} = 2(0.940\boldsymbol{i} + 0.342\boldsymbol{j})$
$$+ 1.5(3)(-0.342\boldsymbol{i} + 0.940\boldsymbol{j})$$
$$= 0.341\boldsymbol{i} + 4.91\boldsymbol{j} \text{ m/s}^2$$

From equation 2.24,

$$\boldsymbol{v}_B = \boldsymbol{v}_A + \boldsymbol{v}_{B/A}$$

thus
$$\boldsymbol{v}_B = 4\boldsymbol{i} + (0.34\boldsymbol{i} + 4.01\boldsymbol{j})$$
$$= (4.34\boldsymbol{i} + 4.91\boldsymbol{j}) \text{ m/s}^2$$

The speed of B is the magnitude of $\boldsymbol{v}_B$:

$$|\boldsymbol{v}_B| = \surd(4.34^2 + 4.91^2)$$
$$= 6.55 \text{ m/s}$$

The acceleration of B relative to A is, from equation 2.18,

$$\boldsymbol{a}_{B/A} = (\ddot{r} - r\dot{\theta}^2)\boldsymbol{e}_r + (r\ddot{\theta} + 2\dot{r}\dot{\theta})\boldsymbol{e}_\theta$$

and $\ddot{r} = 0$ since $\dot{r}$ is constant.

$$\boldsymbol{a}_{B/A} = -1.5(3)^2(0.940\boldsymbol{i} + 0.342\boldsymbol{j})$$
$$+ [(1.5)(-0.5) + 2(2)3]$$
$$\times (-0.342\boldsymbol{i} + 0.940\boldsymbol{j})$$
$$= -16.54\boldsymbol{i} + 5.96\boldsymbol{j} \text{ m/s}^2$$

From equation 2.25

$$\boldsymbol{a}_B = \boldsymbol{a}_A + \boldsymbol{a}_{B/A}$$
$$= -2\boldsymbol{i} + (-16.54\boldsymbol{i} + 5.96\boldsymbol{j})$$
$$= -18.54\boldsymbol{i} + 5.96\boldsymbol{j} \text{ m/s}^2$$

and the magnitude of the acceleration of B is

$$|\boldsymbol{a}_B| = [(-18.54)^2 + (5.96)^2]^{1/2} = 19.47 \text{ m/s}^2$$

A graphical solution is again appropriate, and somewhat quicker. For the velocity vector diagram we first draw, to scale, $\boldsymbol{v}_A$, the velocity of A, 4 m/s to the right (Fig. 2.31). The velocity of B relative to A, $\boldsymbol{v}_{B/A}$, having the components $\dot{r} = 2$ and $r\dot{\theta} = 1.5(3) = 4.5$ in the appropiate directions, is then added to $\boldsymbol{v}_A$ and the resultant is $\boldsymbol{v}_B$, which can be scaled from the figure.

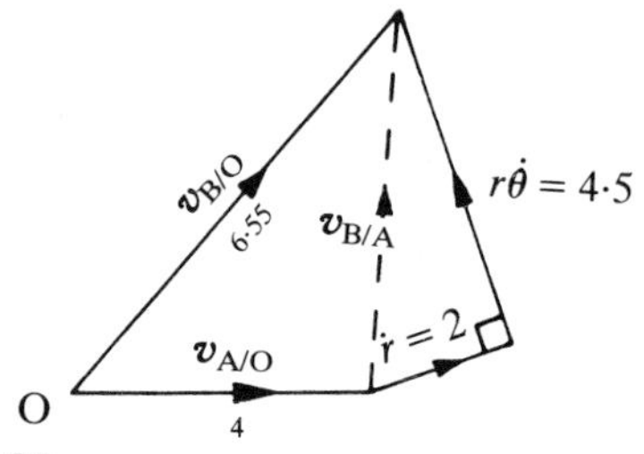

Figure 2.31

For the acceleration vector diagram of Fig. 2.32 we first draw a line to scale to represent the acceleration of A, $\boldsymbol{a}_A$. This is 2 m/s to the left. The acceleration of B relative to A, $\boldsymbol{a}_{B/A}$, is then added to $\boldsymbol{a}_A$. The components of $\boldsymbol{a}_{B/A}$ are $\ddot{r} - r\dot{\theta}^2 = 0 - 1.5(3)^2 = -13.5$ m/s^2 in the $\boldsymbol{e}_r$ direction and $r\ddot{\theta} + 2\dot{r}\dot{\theta} = 1.5(-0.5) + 2(2)3 = 11.25$ m/s^2 in the $\boldsymbol{e}_\theta$ direction. The acceleration of B, $\boldsymbol{a}_B$, can be scaled from the figure.

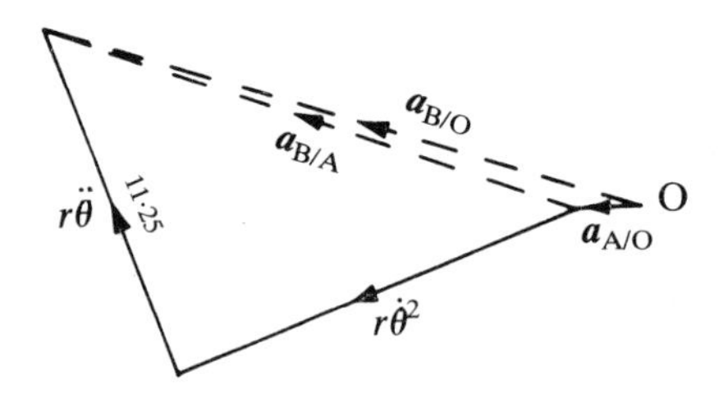

Figure 2.32

Example 2.7

A racing car B is being filmed from a camera mounted on car A which is travelling along a straight road at a constant speed of 72 km/h. The racing car is moving at a constant speed of 144 km/h along the circular track, centre O, which has a radius of 200 m. At the instant depicted in Fig. 2.33, A, B and O are co-linear.

Determine the angular velocity and the angular

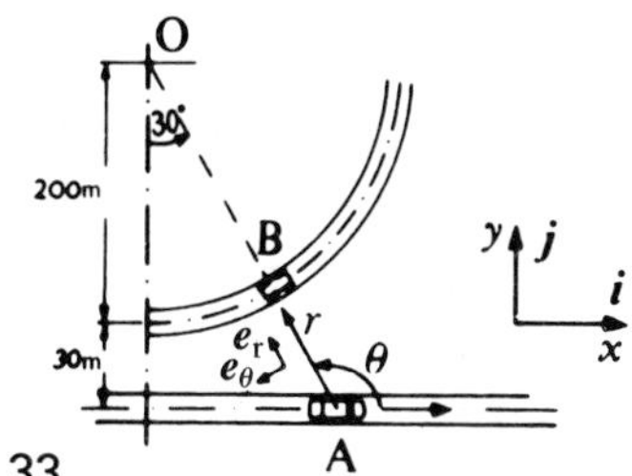

Figure 2.33

acceleration of the camera so that the image of B remains centrally positioned in the viewfinder.

Solution In order to find the required angular velocity and angular acceleration, we shall first need to determine the velocity and acceleration of B relative to A in the given polar co-ordinates and then make use of equations 2.17 and 2.18.

The velocity of B is perpendicular to the line AB, so that

$$\boldsymbol{v}_{\mathrm{B}} = 144\left(\frac{1000}{3600}\right)(-\boldsymbol{e}_\theta) = -40\boldsymbol{e}_\theta \text{ m/s}$$

The velocity of A is

$$\boldsymbol{v}_{\mathrm{A}} = 72\left(\frac{1000}{3600}\right)\boldsymbol{i} = 20\boldsymbol{i} \text{ m/s}$$

Resolving the unit vector $\boldsymbol{i}$ into the $\boldsymbol{e}_r$ and $\boldsymbol{e}_\theta$ directions we have

$$\boldsymbol{v}_{\mathrm{A}} = 20(-\cos 30°\,\boldsymbol{e}_\theta - \sin 30°\,\boldsymbol{e}_r)$$
$$= (-10\boldsymbol{e}_r - 17.32\boldsymbol{e}_\theta) \text{ m/s}$$

The velocity of B relative to A is

$$\boldsymbol{v}_{\mathrm{B/A}} = \boldsymbol{v}_{\mathrm{B}} - \boldsymbol{v}_{\mathrm{A}} = 10\boldsymbol{e}_r - 22.68\boldsymbol{e}_\theta \quad \text{(i)}$$

Also, from equation 2.17,

$$\boldsymbol{v}_{\mathrm{B/A}} = \dot{r}\boldsymbol{e}_r + r\dot{\theta}\boldsymbol{e}_\theta \quad \text{(ii)}$$

Comparing equations (i) and (ii) and noting from Fig. 2.33 that

$$r = (230/\cos 30°) - 200 = 65.58 \text{ m}$$

we find

$$\dot{r} = 10 \text{ m/s}$$

and the angular velocity of the camera is

$$\dot{\theta} = -22.58/65.58 = -0.346 \text{ rad/s}$$

The acceleration of B is most conveniently found from path co-ordinates (equations 2.14) and is

$$\boldsymbol{a}_{\mathrm{B}} = 0\boldsymbol{e}_\theta + \frac{40^2}{200}\boldsymbol{e}_r = 8\boldsymbol{e}_r$$

Since car A is travelling at a constant speed along a straight road,

$$\boldsymbol{a}_{\mathrm{A}} = 0$$

The acceleration of B relative to A is

$$\boldsymbol{a}_{\mathrm{B/A}} = \boldsymbol{a}_{\mathrm{B}} - \boldsymbol{a}_{\mathrm{A}} = 8\boldsymbol{e}_r \quad \text{(iii)}$$

Also, from equation 2.18,

$$\boldsymbol{a}_{\mathrm{B/A}} = (\ddot{r} - r\dot{\theta}^2)\boldsymbol{e}_r + (r\ddot{\theta} + 2\dot{r}\dot{\theta})\boldsymbol{e}_\theta \quad \text{(iv)}$$

Comparing equations (iii) and (iv) we see that

$$0 = r\ddot{\theta} + 2\dot{r}\dot{\theta} = 65.58\ddot{\theta} + 2(10)(-0.346)$$

hence the angular acceleration of the camera is

$$\ddot{\theta} = 20(0.346)/65.58 = 0.106 \text{ rad/s}^2$$

Problems

2.1 The position of a point, in metres, is given by $\boldsymbol{r} = (6t - 5t^2)\boldsymbol{i} + (7 + 8t^3)\boldsymbol{j}$, where t is the time in seconds. Determine the position, velocity and the acceleration of the point when $t = 3$ s.

2.2 The acceleration of a point P moving in a plane is given by $\boldsymbol{a} = 3t^2\boldsymbol{i} + (4t + 5)\boldsymbol{j}$ m/s^2, where t is the time in seconds. When $t = 2$, the position and velocity are respectively $(12\boldsymbol{i} + 26.333\boldsymbol{j})$ m and $(10\boldsymbol{i} + 21\boldsymbol{j})$ m/s. Determine the position and velocity at $t = 1$.

2.3 A point A is following a curved path and at a particular instant the radius of curvature of the path is 16 m. The speed of the point A is 8 m/s and its component of acceleration tangential to the path is 3 m/s^2. Determine the magnitude of the total acceleration.

2.4 A point P is following a circular path of radius 5 m at a constant speed of 10 m/s. When the point reaches the position shown in Fig. 2.34, determine its velocity and acceleration.

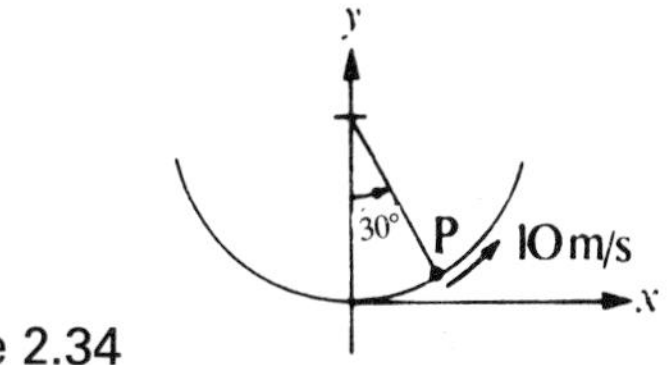

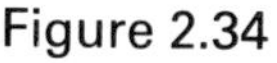
Figure 2.34

2.5 A ship A is steaming due north at 5 knots and another ship B is steaming north-west at 10 knots. Find the velocity of B relative to that of A. (1 knot = 1 nautical mile/h = 6082.66 ft/h = 0.515 m/s.)

2.6 A telescopic arm AB pivots about A in a vertical

plane and is extending at a constant rate of 1 m/s, the angular velocity of the arm remaining constant at 5 rad/s anticlockwise, Fig. 2.35. When the arm is at 30° to the horizontal, the length of the arm is 0.5 m. Determine the velocity and acceleration of B.

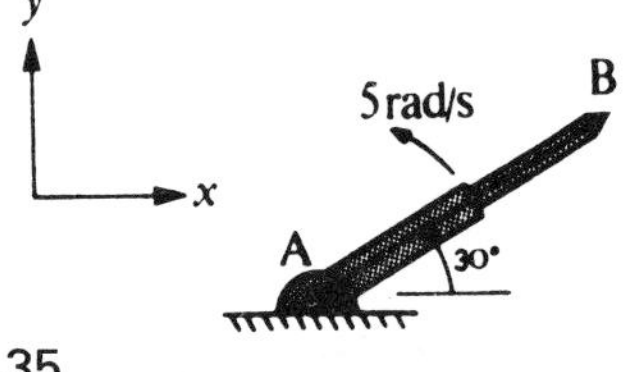

Figure 2.35

2.7 Repeat problem 2.6 assuming that the velocity of point A is $(7\boldsymbol{i}+2\boldsymbol{j})$ m/s and its acceleration is $(4\boldsymbol{i}+6\boldsymbol{j})$ m/s^2. Also determine for this case the speed of B and the magnitude of its acceleration.

2.8 For the mechanism shown in Fig. 2.36, determine the velocity of C relative to B and the velocity of C.

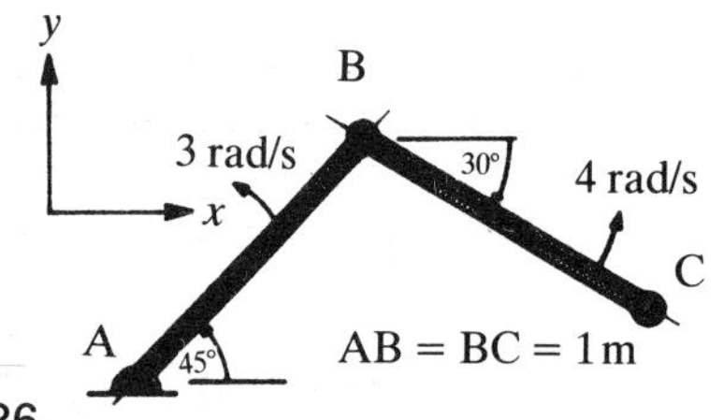

Figure 2.36

2.9 A point P moves along a straight line such that its acceleration is given by $a = (6s^2 + 3s + 2)$ m/s^2, where s is the distance moved in metres. When $s = 0$ its speed is zero. Find its speed when $s = 4$ m.

2.10 A point moves along a curved path and the forward speed v is recorded every second as given in the table below.

t/s	0	1	2	3	4	5	6
v/ms^{-1}	4.0	3.8	3.6	3.2	2.4	1.5	0.4

It can be assumed that the speed varies smoothly with time.

(a) Estimate the magnitude of the tangential acceleration at time $t = 3$ s and the distance travelled between $t = 0$ and $t = 6$ s.

(b) If, at $t = 3$ s, the magnitude of the total acceleration is 1.0 m/s^2, estimate the magnitude of the acceleration normal to the path and also the radius of curvature of the path.

2.11 The forward (tangential) acceleration a_t of the motion of a point is recorded at each metre of distance travelled, and the results are as follows.

s/m	0	1	2	3	4
a_t/ms^{-2}	2.0	2.1	2.5	2.9	3.5

At s = 4 m, the forward speed is 4.6 m/s.

Estimate

(a) the speed at s = 0 m, and

(b) the time taken to travel from s = 0 to s = 4 m.

Further problems involving variable acceleration are given in Chapter 3, problems 3.3, 3.4, 3.6, 3.12, 3.14, 3.15, 3.17, 3.18 and 3.19.

3
Kinetics of a particle in plane motion

3.1 Introduction

In the previous chapters we have studied the kinematics of a point moving in a plane; velocity and acceleration have been defined in various co-ordinate systems and for a variety of conditions. It is now necessary to consider the forces associated with the motion.

The concept of force is useful because it enables the branches of mechanical science to be brought together. For example, a knowledge of the force required to accelerate a vehicle makes it possible to decide on the size of the engine and transmission system suitable as regards both kinematics and strength; hence force acts as a 'currency' between thermodynamics or electro-technology or materials science.

3.2 Newton's laws of motion

Newton's laws define the concept of force in terms of the motion produced by the force if it acted alone – which is why we have yet to discuss statics.

We will first state the three laws in the form that is most common in current literature.

First law
Every body continues in a state of rest or of uniform rectilinear motion unless acted upon by a force.

Second law
The rate of change of momentum of a body is proportional to the force acting on the body and is in the direction of the force.

Third law
To each action (or force) there is an equal and opposite reaction.

The term 'momentum' is prominent in the formulation of the laws of mechanics and a formal definition is given below, together with a definition of mass. The reader concerned with the philosophical implications of the definitions of mass, length and time should consult a text on pure physics.

Momentum
Momentum is defined simply as the product of mass and velocity.

Mass
Mass is a measure of the quantity of matter in a body and it is regarded as constant. If two bodies are made from the same uniform material and have the same volume then their masses are equal.

The first law says that if a body changes its velocity then a force must have been applied. No mention is made of the frame of reference – whether a change in velocity occurs depends on the observer! This point will be considered in detail in section 3.6.

The second law establishes a relationship between the magnitude of the force and the rate of change of momentum:

$$\text{force} \propto \frac{\mathrm{d}}{\mathrm{d}t}(\text{momentum})$$

or $$\boldsymbol{F} \propto \frac{\mathrm{d}}{\mathrm{d}t}(m\boldsymbol{v}) = m\frac{\mathrm{d}\boldsymbol{v}}{\mathrm{d}t} = m\boldsymbol{a} \qquad (3.1)$$

when all points on the body have the same acceleration.

Equivalance of mass
If two objects made from different materials collide, then by Newton's third law they receive equal but opposite forces at any given time and it follows that the momentum gained by one body must be equal to that lost by the other.

If we conduct a simple collision experiment and measure the velocities of the bodies before and after impact, then we may obtain an expression for the ratio of their masses. Thus, equating the momentum before impact to that after impact,

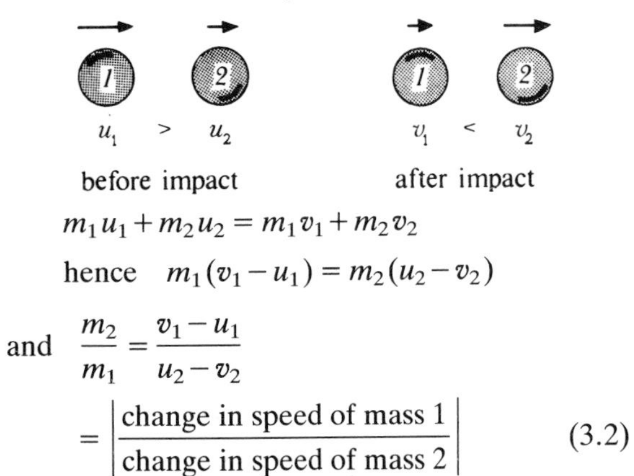

$$m_1u_1 + m_2u_2 = m_1v_1 + m_2v_2$$

hence $\quad m_1(v_1 - u_1) = m_2(u_2 - v_2)$

and $$\frac{m_2}{m_1} = \frac{v_1 - u_1}{u_2 - v_2}$$

$$= \left|\frac{\text{change in speed of mass 1}}{\text{change in speed of mass 2}}\right| \qquad (3.2)$$

Therefore Newton's laws provide, at least in principle, a means of measuring mass and also lead to the law of conservation of momentum (see section 8.3).

3.3 Units

At this stage it is convenient to consider the question of the units in terms of which the quantities encountered so far may be measured.

A statement defining the length of an object requires two parts: a number and a unit.

e.g. $L = n\text{m}$

where $\quad L$ = symbol signifying length,

n = pure number,

m = a unit, such as metre.

If other units are used, such as feet, then the length of the same object is

$$L = p\ \text{ft}$$

where p = a pure number

and ft = feet.

It is given that

1 ft = 0.3048 m exactly (see Appendix 2)

therefore $\quad \dfrac{\text{ft}}{\text{m}} = 0.3048$

(read as 1 ft ÷ 1 m, not as ft per m)

hence $\quad L = p\ \text{ft} = p\left(\dfrac{\text{ft}}{\text{m}}\right)\text{m} = p\ 0.3048\ \text{m}$

Note that in this treatment the symbol representing the unit is considered as a simple algebraic quantity. This approach simplifies the conversion from one system of units to another.

When plotting a graph of length against time, for example, the axes should be labelled as shown in Fig. 3.1, since pure numbers are being plotted (see Appendix 2, reference 3).

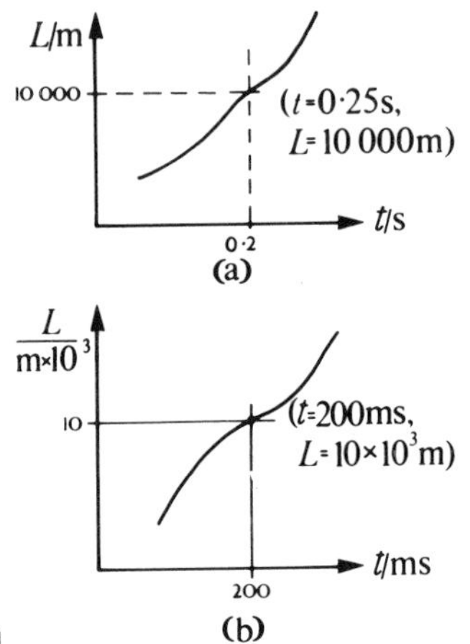

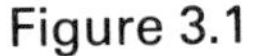
Figure 3.1

Time The unit for time is the second, symbol s, so that time $t = q$ s, where q is a pure number.

Mass The unit for mass is the kilogram, symbol kg, in the SI and the pound, symbol lb, in the 'British' absolute system.

It is given that

1 lb = 0.453 592 37 kg exactly (see Appendix 2)

or $\quad \dfrac{\text{lb}}{\text{kg}} \approx 0.454 = \dfrac{\text{1 lb}}{\text{1 kg}}$

Derived units

Velocity $\boldsymbol{v} = \text{d}\boldsymbol{r}/\text{d}t$, so that, in SI units, the magnitude of the velocity is

$$v = \frac{n\ \text{m}}{q\ \text{s}} = \left(\frac{n}{q}\right)\text{m/s}$$

hence the unit for speed is m/s (metres per second) and similarly the unit for acceleration is m/s^2 (metres per second^2).

The dimensions of these derived units are said to be

$$\text{length (time)}^{-1}$$

and

length $(\text{time})^{-2}$ respectively.

Force The unit for force is chosen so that when applying Newton's second law the constant of proportionality is unity. From the second law,

$$\text{force} \propto \frac{\mathrm{d}}{\mathrm{d}t}\left[(\text{mass})(\text{velocity})\right]$$

$$= \text{mass}\left[\frac{\mathrm{d}}{\mathrm{d}t}(\text{velocity})\right]$$

$$\text{force} \propto (\text{mass})(\text{acceleration})$$

Using consistent units,

$$\boldsymbol{F} = m\boldsymbol{a}$$

that is, if the numerical values of mass and acceleration are unity then the numerical value of the force is also unity. In the SI, in which the basic units are kg, m and s, the unit of force is the newton, N, so that

$$(p\ \mathrm{N}) = (q\ \mathrm{kg})(r\ \mathrm{m/s^2})$$

where p, q and r are pure numbers.

By definition, the numerical relationship is

$$p = qr$$

and the units are related by

$$\mathrm{N} = \mathrm{kg}\frac{\mathrm{m}}{\mathrm{s}^2}$$

We say that the 'dimensions' of the unit of force are kg m s^{-2} when expressed in terms of the basic units.

A list of SI units appears in Appendix 2.

3.4 Types of force

The nature of force is complex, so it is best to consider force as a concept useful in studying mechanics. It plays a role in mechanics similar to that of money in trade in that it enables us to relate a phenomenon in one discipline to one in another discipline. For example, in the simple case of a spring and a mass (Fig. 3.2) the results of Newton's second law and Hooke's law may be combined.

From Hooke's law*,

$$F = kx \quad (k = \text{constant})$$

and from Newton's second law, taking vectors acting to the right as positive

$$-F = m\frac{\mathrm{d}^2x}{\mathrm{d}t^2}$$

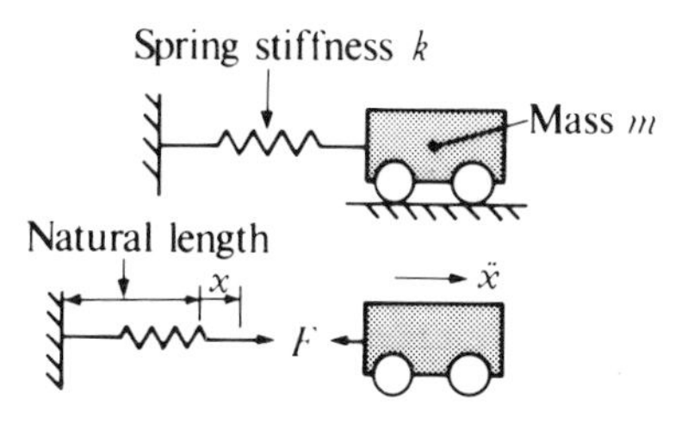

Figure 3.2

Eliminating F between the two equations gives

$$kx + m\frac{\mathrm{d}^2x}{\mathrm{d}t^2} = 0 \tag{3.3}$$

which is a differential equation relating displacement x to time t.

Note that force is used to represent the effect that one body has on the other. Let us now consider the definition of force. We need a formal definition to avoid ambiguity since not all writers mean the same thing when they use the term 'force'.

Definition of force

Force is the action of *one body upon another* which produces, when acting alone, a change in the motion of a body. (Newton's law gives the means of quantifying this force.)

It is convenient to group forces into two classes: (a) long-range forces and (b) short-range forces. Long-range forces are gravitational, electrostatic and magnetic forces and are also known as body forces. Short-range forces are the forces due to contact of two bodies. It might be argued that the latter are only special cases of the former, but in mechanical applications the distinction remains clear.

The forces of contact are often sub-divided into normal forces – i.e. normal to the tangent plane of contact – and tangential, shear or friction forces which are parallel to the plane of contact.

Dry friction

The friction force between two dry unlubricated surfaces is a quantity which depends on a large number of factors, but consideration of an ideal

* Hooke's law states that any deformation produced by a given loading system is proportional to the magnitude of the loading. A body obeying Hooke's law is said to be linearly elastic.

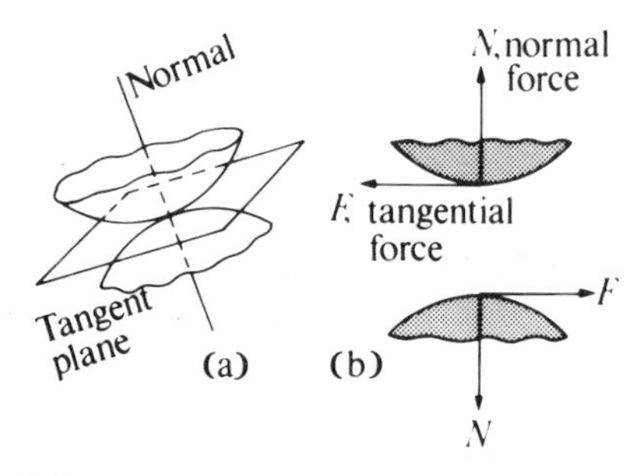

Figure 3.3

case known as Coulomb friction is often regarded as adequate. In this case the friction force is assumed to take any value up to a maximum or limiting value. This limiting value is considered to be proportional to the normal contact force between the two surfaces,

i.e. $F = \mu N$ (see Fig. 3.3) (3.4)

where μ is called the cofficient of limiting friction.

In practice μ is found to vary with sliding speed and often drops markedly as soon as sliding occurs.

So far we have considered the contact forces acting at a point, although they are most likely to be distributed over a finite area, A. The intensity of normal loading defined by

$$\lim_{\Delta A \to 0} \frac{\Delta P}{\Delta A} = \frac{dP}{dA} \quad (3.5)$$

is called 'pressure' or 'normal stress'.

It is conventional to speak of 'pressure' when dealing with fluids and 'stress' when dealing with solids.

3.5 Gravitation

Isaac Newton was also responsible for formulating the law of gravitation, which is expressed by

$$F = \frac{Gm_1 m_2}{d^2} \quad (3.6)$$

where F is the force of attraction between two bodies of masses m_1 and m_2 separated by a distance d; G is the universal gravitational constant and has a value

$$G = (6.670 \pm 0.005) \times 10^{-11}\ \text{m}^3\ \text{s}^{-2}\ \text{kg}^{-1}$$

The mass of the Earth is taken to be 5.98×10^{24} kg and its mean radius is 6.368×10^6 m. From equation 3.6, the force acting on 1 kg mass at the surface of the Earth is

$$F = \frac{6.670 \times 10^{-11} \times 5.98 \times 10^{24} \times 1}{(6.368 \times 10^6)^2}$$

$$= 9.8361\ \text{N}$$

If this force acts alone on a unit mass, it follows that the acceleration produced will be

$$9.8361\ \text{m/s}^2$$

This quantity is often called the acceleration due to gravity and is given the symbol g; thus

$$\text{gravitational force} = \text{mass} \times g$$

We prefer to regard g as the gravitational field intensity measured in N/kg.

The declared standard values of g is

$$g_n = 9.806\,65\ \text{m/s}^2 \text{ or N/kg}$$

This differs from the value calculated because the Earth is not a perfect sphere and also because the measured value is affected by the Earth's rotation.

Weight

The weight W of a body is usually defined as the force on the body due to gravity (mg); however it is normally interpreted as 'the force equal and opposite to that required to maintain a body at rest in a chosen frame of reference', that is relative to the surface of the Earth or relative to a freely orbiting spacecraft in the sense of 'weightlessness'. The difference between the two definitions on the Earth's surface is only 0.4 per cent.

Using the latter definition, $W = \text{mass} \times g'$, where g' is the apparent field intensity.

Unless otherwise stated, the value of g is taken to be g_n.

3.6 Frames of reference

In the previous section the term 'frame of reference' was used. It is clear that in the interpretation of Newton's first law we must have some reference frame from which to measure the velocity. For most elementary problems we consider the surface of the Earth to be a suitable frame, although we know that such a frame is rotating relative to the stars and is moving around the sun.

Intuitively, we would guess that a frame having no acceleration relative to the sun and not rotating relative to the stars would be the best possible. Let us regard such a frame as 'inertial' or 'Galilean'. It follows that any other frame moving with constant velocity relative to our original inertial frame will also be an inertial frame, since Newton's laws will be equally applicable. This is because force depends on rate of change of velocity, which will be the same when measured in either frame.

If we cannot observe the entire universe, how can we be sure that we have an inertial frame? The simple answer is that we cannot. Consider conducting experiments in a lift, with no means of observing the outside world. Assume that the lift is accelerating downwards, in which case we have no means of telling whether the force of gravity has reduced or the lift is accelerating – even the use of the property of light travelling in straight lines would not help. Such considerations as these led Einstein towards the general theory of relativity.

If we now consider experimenting on a rotating platform, we have the choice of assuming that the platform is rotating or, if this is denied, of inventing extra forces in order to explain the observed phenomena and preserve Newton's laws.

3.7 Systems of particles

So far we have either considered only a single particle or tacitly assumed that there is a representative point whose motion may be described. However, any real object is an assembly of basic particles constrained by internal forces and acted upon by outside bodies and surface forces.

Let us consider a collection of n particles of mass m_i and position $\boldsymbol{r}_i$. The force acting on any typical particle may be due (a) to external body forces, (b) to internal forces of one particle on another, or (c) if the particle is at the surface, then a contact force is possible.

For the ith particle (Fig. 3.4),

$$\boldsymbol{F}_i + \sum_j \boldsymbol{f}_{ij} = m_i \ddot{\boldsymbol{r}}_i \tag{3.7}$$

where $\boldsymbol{f}_{ij}$ is the force on particle i due to particle j. Thus

external force + sum of internal forces
= mass × acceleration

Note that

$$\sum_j \boldsymbol{f}_{ij} = \boldsymbol{f}_{ia} + \boldsymbol{f}_{ib} + \ldots + \boldsymbol{f}_{ij} + \ldots + \boldsymbol{f}_{in} \tag{3.8}$$

By Newton's third law,

$$\boldsymbol{f}_{ij} = -\boldsymbol{f}_{ji} \tag{3.9}$$

and in most cases they are collinear. (Some cases exist in electromagnetic theory where the equal and opposite forces are not collinear; Newton's law is then said to exist in its weak form.)

If we now sum all equations of the form of equation 3.7, we obtain

$$\sum_n \boldsymbol{F}_i + \sum_n \left(\sum_{j=1}^{j=n} \boldsymbol{f}_{ij}\right) = \sum_n m_i \ddot{\boldsymbol{r}}_i \tag{3.10}$$

The double summation is in fact quite simple, since for every $\boldsymbol{f}_{ij}$ there is an $\boldsymbol{f}_{ji}$ such that $\boldsymbol{f}_{ij} + \boldsymbol{f}_{ji} = \boldsymbol{0}$.

Hence we obtain

$$\sum_n \boldsymbol{F}_i = \sum_n m_i \ddot{\boldsymbol{r}}_i \tag{3.11}$$

3.8 Centre of mass

The centre of mass (c.m.) of a body is defined by the equation

$$\sum m_i \boldsymbol{r}_i = \left(\sum m_i\right) \boldsymbol{r}_G = M \boldsymbol{r}_G \tag{3.12}$$

where M is the total mass of the body and $\boldsymbol{r}_G$ is the position of the c.m. as shown in Fig. 3.5.

In scalar form,

$$\sum m_i x_i = M x_G; \quad \sum m_i y_i = M y_G; \quad \sum m_i z_i = M z_G \tag{3.13}$$

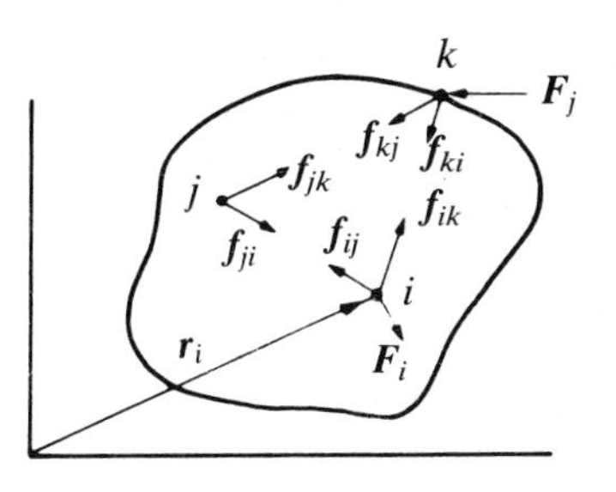

Figure 3.4

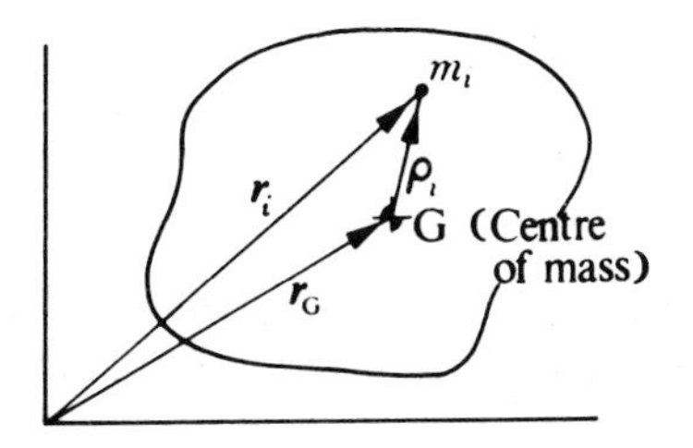

Figure 3.5

An alternative description may be obtained by writing

$$\boldsymbol{r}_i = \boldsymbol{r}_{\mathrm{G}} + \boldsymbol{\rho}_i$$

and substituting into equation 3.12 gives

$$\sum m_i(\boldsymbol{r}_{\mathrm{G}} + \boldsymbol{\rho}_i) = (\sum m_i)\boldsymbol{r}_{\mathrm{G}}$$

hence $\sum m_i \boldsymbol{\rho}_i = 0$ (3.14)

Differentiating equation 3.12 with respect to time gives

$$\sum m_i \dot{\boldsymbol{r}}_i = M\dot{\boldsymbol{r}}_{\mathrm{G}} \quad \text{and} \quad \sum m_i \ddot{\boldsymbol{r}}_i = M\ddot{\boldsymbol{r}}_{\mathrm{G}} \qquad (3.12\text{a})$$

Similarly, from equation 3.14,

$$\sum m_i \dot{\boldsymbol{\rho}}_i = \boldsymbol{0} \quad \text{and} \quad \sum m_i \ddot{\boldsymbol{\rho}}_i = \boldsymbol{0} \qquad (3.14\text{a})$$

As an example of locating the centre of mass for a body with a continuous uniform distribution of matter, we shall consider the half cylinder shown in Fig. 3.6.

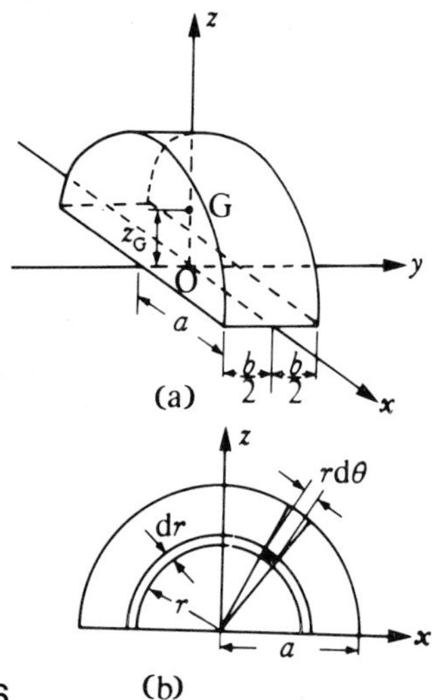

Figure 3.6

By symmetry, the centre of mass must lie on the z-axis.

The mass of the element with density ρ is

$$\rho b(r\,\mathrm{d}\theta)\,\mathrm{d}r$$

and its mass moment relative to the xy plane $(\sum m_i z_i)$ is

$$\rho b(r\,\mathrm{d}\theta)\,\mathrm{d}r\,r\sin\theta$$

For an elemental cylinder of thickness $\mathrm{d}r$ and radius r, the moment of the mass is

$$\int_{\theta=0}^{\theta=\pi} \rho b r^2 \,\mathrm{d}r \sin\theta\,\mathrm{d}\theta = \rho b r^2(-\cos\theta)\,\mathrm{d}r \Big|_{\theta=0}^{\theta=\pi}$$

$$= 2\rho b r^2\,\mathrm{d}r$$

and for the whole body the moment of mass is

$$\int_{r=0}^{r=a} 2\rho b r^2\,\mathrm{d}r = \left[2\rho b\left(\tfrac{1}{3}r^3\right)\right]_{r=0}^{r=a} = \tfrac{2}{3}\rho b a^3$$

The mass of the whole body is $\rho\frac{1}{2}\pi a^2 b$ so, by the definition of the centre of mass,

$$z_{\mathrm{G}} = \frac{\sum mz}{\sum m} = \frac{\iint \rho b r^2 \sin\theta\,\mathrm{d}r\,\mathrm{d}\theta}{\iint \rho b\,\mathrm{d}r\,r\,\mathrm{d}\theta}$$

$$= \frac{\frac{2}{3}\rho b a^3}{\frac{1}{2}\rho b \pi a^2} = \frac{4a}{3\pi}$$

In terms of the c.m., equation 3.11 becomes

$$\sum \boldsymbol{F}_i = M\ddot{\boldsymbol{r}}_{\mathrm{G}} \qquad (3.15)$$

This equation states that the vector sum of the external forces acting on a particular set of particles equals the total mass times the acceleration of the centre of mass, irrespective of the individual motion of the separate particles. This equation is equally applicable to any system of particles, whether they are rigidly connected or otherwise.

3.9 Free-body diagrams

The idea of a free-body diagram (f.b.d.) is central to the methods of solving problems in mechanics, and its importance cannot be overstated.

If we are to be able to use equation 3.15 properly, then we must show clearly *all* the forces acting on any bodies, or collection of bodies, and to do this we must remove all other bodies from the diagram and replace their actions by forces. As an example, consider a rear-wheel-drive car towing a trailer (Fig. 3.7(a)) – the f.b.d. for the car is shown in Fig. 3.7(b). We will assume that W and F are known.

Because the earth has been removed from the diagram, we must introduce the contact forces between the tyres and the road (here we have made an engineering assumption that the tangential force at the front wheel is small). Also, we have the sum of all the gravitational attractions, $\sum m_i g = W$, acting at a point G, the centre of gravity of the body. It can be shown that, for a uniform field, the centre of gravity and the centre of mass are coincident points. Removing the trailer exposes the force on the towing bar, shown as two components for convenience. As the path of the vehicle is a straight line, $\ddot{y}_{\mathrm{G}} = 0$ and $\ddot{x}_{\mathrm{G}} = a$, as yet unknown.

Equation 3.15 gives

$$\sum F_x = M\ddot{x}_G$$

and $\sum F_y = M\ddot{y}_G$

Resolving the forces gives

$$F - P - W\sin\alpha = M\ddot{x}_G = Ma \qquad (3.16)$$

and $Q - W\cos\alpha + R_1 + R_2 = M\ddot{y}_G = 0 \qquad (3.17)$

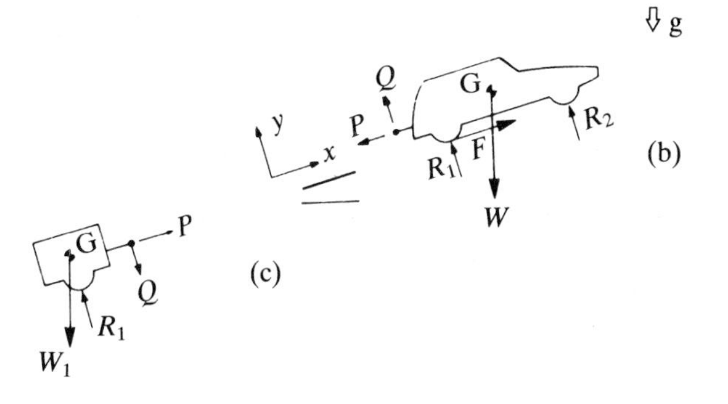

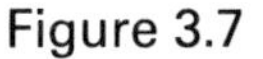

Figure 3.7

If we now draw a free-body diagram for the trailer, another useful equation may be derived. Note that on the free-body diagram, Fig. 3.7(c), P and Q are drawn equal and opposite to the P and Q on the car, as required by Newton's third law.

Resolving parallel to the road,

$$P - W_t\sin\alpha = M_t a \qquad (3.18)$$

Adding equations 3.16 and 3.18, we have

$$F - (W + W_t)\sin\alpha = (M + M_t)a$$

Hence a is determined so that using equation 3.16 the force P can be found.

3.10 Simple harmonic motion

As an example of one-dimensional motion we shall consider a special type of motion which is very common in physics. The motion is that due to forces such that the acceleration is proportional to the displacement from some equilibrium or rest position and is always directed towards that position. In mathematical terms,

$$\ddot{x} \propto -x$$

We have seen in section 3.4 that for a simple mass-and-spring system

$$m\ddot{x} + kx = 0$$

i.e. $$\ddot{x} = -\frac{k}{m}x \qquad (3.19)$$

A first integral can be obtained by writing

$$\ddot{x} = v\frac{dv}{dx} = -\frac{k}{m}x$$

hence $$\int v\,dv = \int -\frac{k}{m}x\,dx$$

and $$\frac{v^2}{2} = -\frac{k}{m}\frac{x^2}{2} + \frac{C^2}{2}$$ where C is a constant

Thus $$v = \frac{dx}{dt} = \sqrt{\left(C^2 - \frac{k}{m}x^2\right)} \qquad (3.20)$$

Now a second integral involves a substitution – that is, some guesswork – so let us guess that $x = A\sin\omega t$, A and ω being constants. Substituting in equation 3.19 gives

$$(-\omega^2)A\sin\omega t = -\frac{k}{m}(A\sin\omega t)$$

therefore $$\omega^2 = \frac{k}{m}$$

The same result would have been achieved had the substitution $x = B\cos\omega t$ been made, hence we conclude that the general solution of equation 3.19 is

$$x = A\sin\omega t + B\cos\omega t \quad \text{where } \omega = \sqrt{(k/m)} \qquad (3.21)$$

The velocity at time t is

$$v = \frac{dx}{dt} = \omega A\cos\omega t - \omega B\sin\omega t$$

The values of A and B depend on the initial conditions. If, when $t = 0$, $x = x_0$ and $v = v_0$ then

$$x_0 = B \quad \text{and} \quad v_0 = \omega A$$

This leads to

$$x = \frac{v_0}{\omega}\sin\omega t + x_0\cos\omega t, \quad \omega = \sqrt{(k/m)}$$

or alternatively

$$x = X\sin(\omega t + \phi)$$

where $X = \sqrt{[(v_0/\omega)^2 + x_0^2]}$ and is called the amplitude and $\phi = \arctan(x_0\omega/v_0)$ and is called a phase angle.

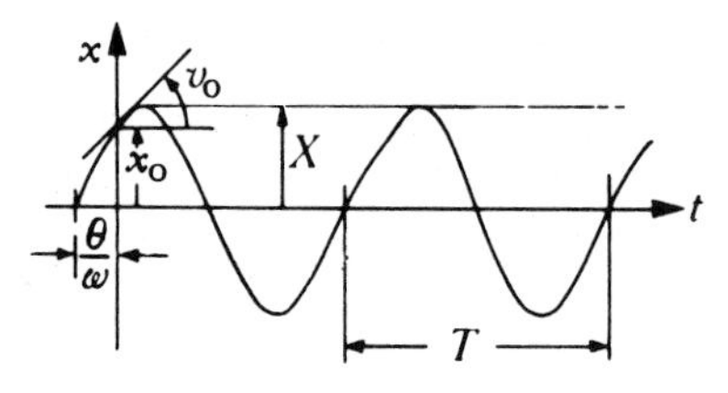

Figure 3.8

A graph of x against t is shown in Fig. 3.8.

The function of x is seen to repeat exactly after a time interval of T called the periodic time. We know that the sine function repeats when its argument has increased by 2π, therefore if time increases by $2\pi/\omega$ this must be equal to the periodic time. Hence

$$2\pi/\omega = T \qquad (3.22)$$

The inverse of the periodic time is the frequency, ν. If the periodic time is measured in seconds then the frequency will be measured in cycles per second or, in SI units, hertz (Hz) – where 1 hertz = 1 cycle per second.

$$\text{Therefore} \quad \text{frequency } \nu = \frac{1}{T} = \frac{\omega}{2\pi} \qquad (3.23)$$

Referring to Fig. 3.9, it is seen that the projection of the line OA which is rotating at an angular velocity ω rad/s produces simple harmonic motion.

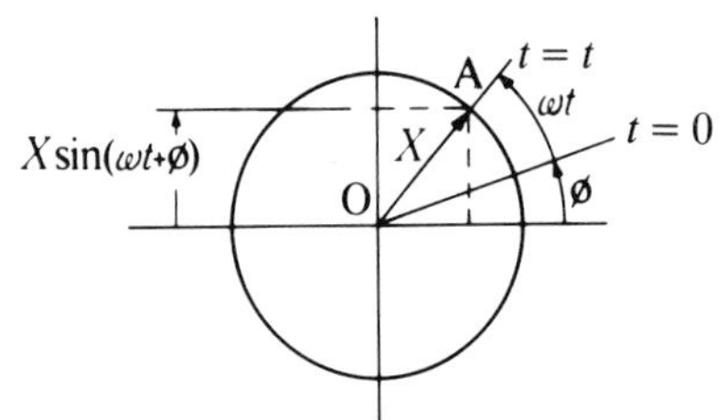

Figure 3.9

For the previously mentioned reason, ω is called the circular frequency (or angular frequency or pulsatance).

3.11 Impulse and momentum

Equation 3.15 may be written as

$$\sum \boldsymbol{F} = M\frac{d\boldsymbol{v}_G}{dt}$$

and integrated to give

$$\sum\int_1^2 \boldsymbol{F}\,dt = \int_1^2 M\,d\boldsymbol{v}_G = M\boldsymbol{v}_2 - M\boldsymbol{v}_1 \qquad (3.24)$$

The integral $\sum\int_1^2 \boldsymbol{F}\,dt$ is called the *impulse* and is usually given the symbol $\boldsymbol{J}$. (Note that impulse is a vector quantity.)

Hence impulse = change in momentum

or

$$\boldsymbol{J} = \Delta(M\boldsymbol{v}_G) \qquad (3.25)$$

This equation may be used directly if a force–time history is available as shown in Fig. 3.10. In this case the area under the curve is the impulse and may be equated to the change in momentum.

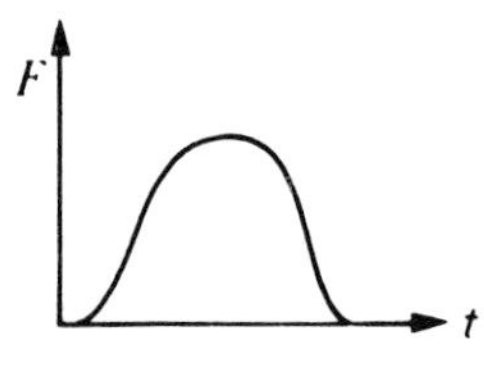

Figure 3.10

In collision problems, the impulse–momentum relationship is used in conjunction with Newton's third law. By this law, the force of contact on one body during collision is equal and opposite to that on the other, and so the impulse received by one body will be equal and opposite to that received by the other. It follows that the momentum received by one body will be equal to that lost by the other.

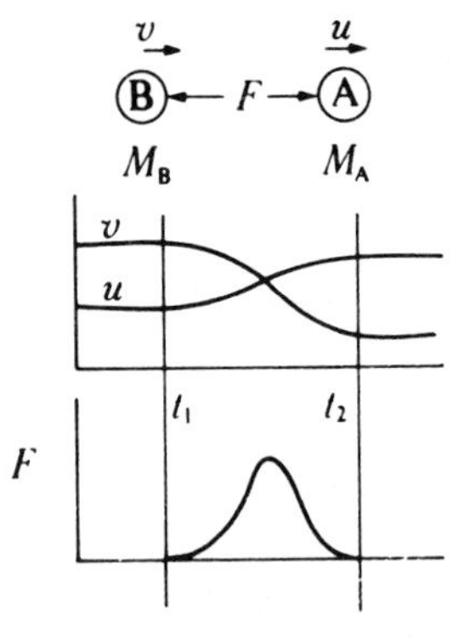

Figure 3.11

Consider the co-linear impact of two spheres A and B, as shown in Fig. 3.11.

For mass A,

$$\int_{t_1}^{t_2} F\,dt = J = M_A(u_2 - u_1)$$

For mass B,

$$\int_{t_1}^{t_2} -F\,dt = -J = M_B(v_2 - v_1)$$

Adding these two equations,

$$0 = M_A u_2 + M_B v_2 - (M_A u_1 + M_B v_1)$$

or $$(M_A u_1 + M_B v_1) = (M_A u_2 + M_B v_2) \qquad (3.26)$$

thus, momentum before impact = momentum after impact. A fuller treatment is given in Chapter 8.

3.12 Work and kinetic energy

It is also possible to integrate equation 3.15 with respect to distance. In this case we rewrite the equation for a particle, $\boldsymbol{F} = m\,\mathrm{d}\boldsymbol{v}/\mathrm{d}t$, in its component forms

$$F_x = mv_x\frac{\mathrm{d}v_x}{\mathrm{d}x}, \quad F_y = mv_y\frac{\mathrm{d}v_y}{\mathrm{d}y}, \quad F_z = mv_z\frac{\mathrm{d}v_z}{\mathrm{d}z}$$

Each equation may now be integrated, to give

$$\int F_x \mathrm{d}x = \int mv_x \mathrm{d}v_x = \tfrac{1}{2}mv_x^2 + \text{constant}$$
$$\int F_y \mathrm{d}y = \int mv_y \mathrm{d}v_y = \tfrac{1}{2}mv_y^2 + \text{constant}$$
$$\int F_z \mathrm{d}z = \int mv_z \mathrm{d}v_z = \tfrac{1}{2}mv_z^2 + \text{constant}$$

Adding these three equations,

$$\int(F_x \mathrm{d}x + F_y \mathrm{d}y + F_z \mathrm{d}z) = \tfrac{1}{2}mv^2 + \text{constant} \qquad (3.27)$$

where $$v^2 = \boldsymbol{v}\cdot\boldsymbol{v} = v_x^2 + v_y^2 + v_z^2 \qquad (3.28)$$

The integral on the left-hand side is seen to be equivalent to $\int \boldsymbol{F}\cdot\mathrm{d}\boldsymbol{s}$ since

$$\int \boldsymbol{F}\cdot\mathrm{d}\boldsymbol{s} = \int (F_x\boldsymbol{i} + F_y\boldsymbol{j} + F_z\boldsymbol{k})\cdot(\mathrm{d}x\boldsymbol{i} + \mathrm{d}y\boldsymbol{j} + \mathrm{d}z\boldsymbol{k})$$
$$= \int(F_x \mathrm{d}x + F_y \mathrm{d}y + F_z \mathrm{d}z)$$

Hence $\int \boldsymbol{F}\cdot\mathrm{d}\boldsymbol{s} = \tfrac{1}{2}mv^2 + \text{constant}$

or $$\int_1^2 \boldsymbol{F}\cdot\mathrm{d}\boldsymbol{s} = \tfrac{1}{2}mv_2^2 - \tfrac{1}{2}mv_1^2 \qquad (3.29)$$

The term $\int \boldsymbol{F}\cdot\mathrm{d}\boldsymbol{s}$ is defined as the work done by the force $\boldsymbol{F}$ when acting on a particle moving along a given path. The definition shows that only the component of force acting along the path does work on the particle.

The term $\tfrac{1}{2}mv^2$ is called the *kinetic energy* of the particle; hence equation 3.29 reads

work done = change in kinetic energy

The dimensions of work are those of (force) × (distance), so in SI units the dimensions are N m = J (joules).

The dimensions of kinetic energy are

$$\text{kg m}^2/\text{s}^2 = (\text{kg m/s}^2)\text{ m} = \text{N m} = \text{J}$$

Equation 3.29 was derived by integrating the equation of motion for a particle and thus it is not possible to include other forms of energy (thermal, rotational, etc.) in this development. Chapter 7 gives a fuller treatment of energy methods.

Note that work and energy are scalar quantities.

3.13 Power

Power is defined as the rate at which work is being performed; therefore

$$\text{power} = \frac{\mathrm{d}}{\mathrm{d}t}(\text{work})$$
$$= \frac{\mathrm{d}}{\mathrm{d}t}(\int \boldsymbol{F}\cdot\mathrm{d}\boldsymbol{s}) = \frac{\mathrm{d}}{\mathrm{d}t}\left[\int\left(\boldsymbol{F}\cdot\frac{\mathrm{d}\boldsymbol{s}}{\mathrm{d}t}\right)\mathrm{d}t\right]$$
$$= \boldsymbol{F}\cdot\frac{\mathrm{d}\boldsymbol{s}}{\mathrm{d}t} = \boldsymbol{F}\cdot\boldsymbol{v} \qquad (3.30)$$

Since for a particle

work = kinetic energy + constant

then $$\text{power} = \frac{\mathrm{d}}{\mathrm{d}t}(\text{k.e.}) = \frac{\mathrm{d}}{\mathrm{d}t}(\tfrac{1}{2}m\boldsymbol{v}\cdot\boldsymbol{v})$$

thus $$\boldsymbol{F}\cdot\boldsymbol{v} = m\boldsymbol{v}\cdot\frac{\mathrm{d}\boldsymbol{v}}{\mathrm{d}t}$$
$$= m\boldsymbol{a}\cdot\boldsymbol{v}$$

The dimensions of power are

N m/s = J/s = W (watts)

Discussion examples

Example 3.1

Figure 3.12 shows two small bodies which collide. The masses of the bodies are $m_A = 3m$ and $m_B = m$. Before impact, A is stationary and B has a velocity $\boldsymbol{u}_B$ in the direction shown. After impact the velocities are $\boldsymbol{v}_A$ and $\boldsymbol{v}_B$ as shown.

Assuming that external forces have a negligible effect, determine in terms of u_B the speeds v_A and v_B.

Solution There is no change in momentum in the absence of external forces (section 3.2). Equating the initial and final momenta gives

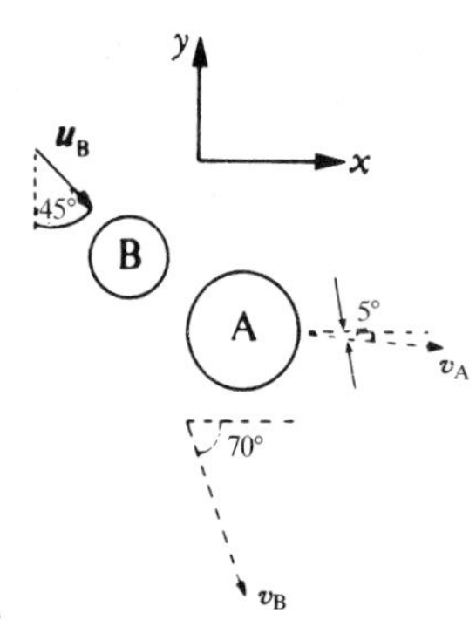

Figure 3.12

$$m_B \boldsymbol{u}_B = m_A \boldsymbol{v}_A + m_B \boldsymbol{v}_B \qquad \text{(i)}$$

For the x-components of equation (i),

$$m u_B \cos 45° = 3m v_A \cos 5° + m v_B \cos 70° \qquad \text{(ii)}$$

and, for the y-components,

$$m u_B \sin 45° = 3m v_A \sin 5° + m v_B \sin 70° \qquad \text{(iii)}$$

Solving for v_A and v_B from equations (ii) and (iii), we find

$$v_A = 0.155 u_B$$

and $v_B = 0.709 u_B$

Example 3.2

A force $\boldsymbol{R} = (3t\boldsymbol{i} + 0.4t^2\boldsymbol{j})$ N is applied to a particle of mass 0.1 kg which can move freely in a gravitational field of intensity 2.36 N/kg. The gravitational force acts in the $(-\boldsymbol{j})$-direction and t is the time in seconds.

At time $t = 0$ the velocity of the particle is $(700\boldsymbol{i} + 200\boldsymbol{j})$ m/s. Determine its velocity when $t = 2.0$ s.

Solution The free-body diagram (Fig. 3.13) for the particle shows the force $\boldsymbol{R}$ and the weight $\boldsymbol{W}$ acting on it.

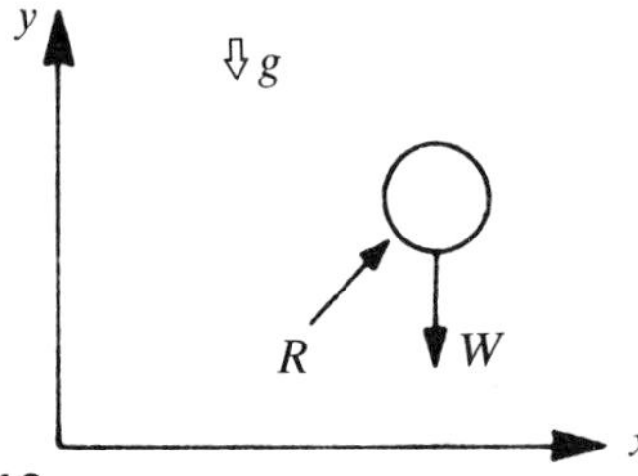

Figure 3.13

$$\boldsymbol{W} = W(-\boldsymbol{j}) = -mg\boldsymbol{j}$$

From Newton's second law (equation 3.1),

$$\sum \boldsymbol{F} = m\boldsymbol{a}$$

where $\boldsymbol{a}$ is the acceleration of the body. Thus

$$\boldsymbol{R} + \boldsymbol{W} = m\boldsymbol{a}$$

$$(3t\boldsymbol{i} + 0.4t\boldsymbol{j}^2) - (0.1)(2.36)\boldsymbol{j} = (0.1)\boldsymbol{a}$$

$$\boldsymbol{a} = [30t\boldsymbol{i} + (2.36 + 4t^2)\boldsymbol{j}] \text{ m/s} \qquad \text{(i)}$$

To find the velocity we shall have to integrate equation (i). Now

$$\boldsymbol{a} = \mathrm{d}\boldsymbol{v}/\mathrm{d}t$$

and

$$\int_{v_1}^{v_2} \mathrm{d}\boldsymbol{v} = \int_{t_1}^{t_2} \boldsymbol{a}\,\mathrm{d}t$$

$$\boldsymbol{v}_2 - \boldsymbol{v}_1 = \int_0^2 [30t\boldsymbol{i} + (2.36 + 4t^2)\boldsymbol{j}]\,\mathrm{d}t$$

Noting that $\boldsymbol{i}$ and $\boldsymbol{j}$ are fixed unit vectors, we obtain

$$\boldsymbol{v}_2 - (700\boldsymbol{i} + 200\boldsymbol{j}) = \frac{30(2)^2}{2}\boldsymbol{i} + \left(2.36(2) + \frac{4(2)^3}{3}\right)\boldsymbol{j}$$

$$\boldsymbol{v}_2 = (760\boldsymbol{i} + 215.4\boldsymbol{j}) \text{ m/s}$$

Example 3.3

A box of mass m is being lowered by means of a rope ABCD which passes over a fixed cylinder, the angle of embrace being α as shown in Fig. 3.14. The stretch in the rope and its mass can both be neglected.

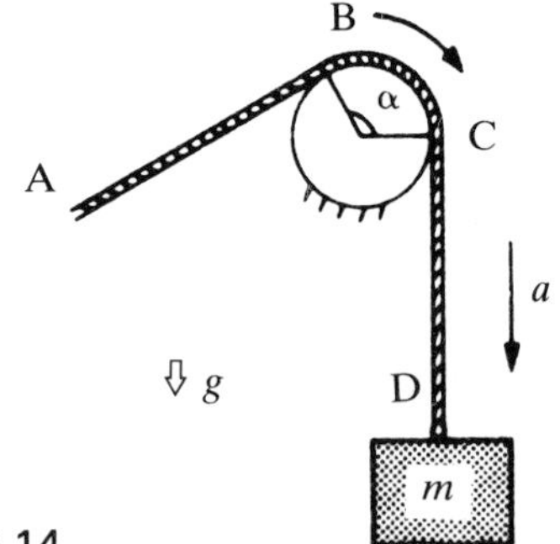

Figure 3.14

If the coefficient of friction between the rope and the cylinder is μ, show that the tensions in the rope at B and C are governed by the relationship

$$T_C/T_B = e^{\mu\alpha}$$

If the downward acceleration of the box is a, determine the tension T_B.

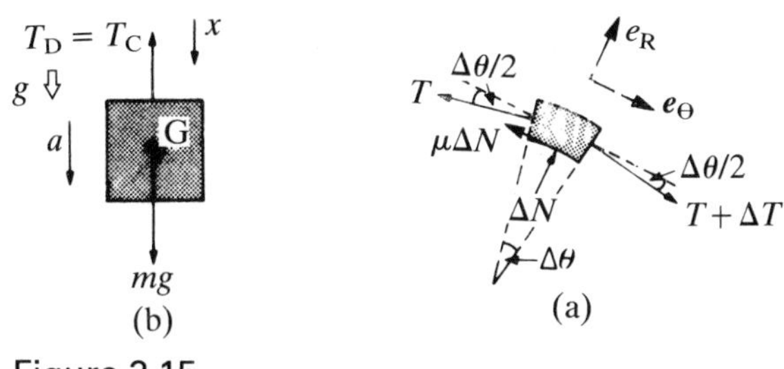

Figure 3.15

Solution Figure 3.15(a) is the free-body diagram for an element of the rope in contact with the cylinder which subtends the small angle $\Delta\theta$ at the centre of curvature. The change in tension across the element is shown by the forces T and $T+\Delta T$. The contact force with the cylinder has been resolved into components in the $\boldsymbol{e}_R$ and $\boldsymbol{e}_\theta$ directions. Since slip is occurring, the component in the $\boldsymbol{e}_\theta$ direction is μ times that in the $\boldsymbol{e}_R$ direction (equation 3.4).

Since $\Delta\theta/2$ is small we can replace $\cos(\Delta\theta/2)$ by unity and write

$$[\sum F_\theta = ma_\theta]$$

$$T+\mu\Delta N-(T+\Delta T)=0 \qquad \text{(i)}$$

$$\Delta T = \mu\Delta N$$

For the radial direction we can replace $\sin(\Delta\theta/2)$ by $\Delta\theta/2$ and write

$$[\sum F_R = ma_R]$$

$$\Delta N-(2T+\Delta T)\Delta\theta/2=0$$

and, neglecting the term of second order of smallness,

$$T\Delta\theta = \Delta N \qquad \text{(ii)}$$

Hence, eliminating ΔN,

$$\Delta T/T = \mu\Delta\theta \qquad \text{(iii)}$$

In the limit as $\Delta\theta$ approaches zero, equation (iii) becomes

$$\mathrm{d}T/T = \mu\,\mathrm{d}\theta$$

and $$\int_{T_B}^{T_C} \mathrm{d}T/T = \mu\int_0^\alpha \mathrm{d}\theta$$

assuming μ is constant. Thus

$$\ln(T_C/T_B) = \mu\alpha$$

$$T_C/T_B = \mathrm{e}^{\mu\alpha} \qquad \text{(iv)}$$

This is a well-known relationship. Note that the shape of the cylinder need not be circular.

Since the mass of the rope is negligible, there is no change in tension between C and D, as a free-body diagram and equation of motion would confirm.

From the f.b.d. for the box (Fig. 3.15(b)),

$$[\sum F_x = m\ddot{x}_G]$$

$$mg - T_C = ma$$

$$T_C = m(g-a) \qquad \text{(v)}$$

Combining equations (v) and (iv),

$$T_B = m(g-a)\mathrm{e}^{-\mu\alpha}$$

Example 3.4

The trolley with telescopic arm of example 2.6 is reproduced in Fig. 3.16. The arm carries a body of concentrated mass 3.0 kg. Determine the force $\boldsymbol{R}$ exterted by the arm on the body for the position shown.

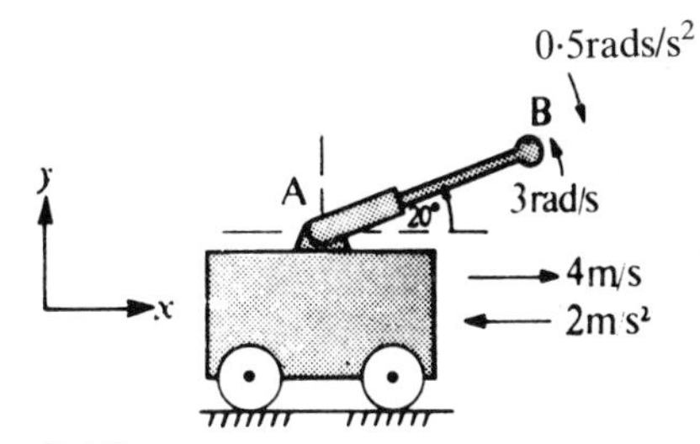

Figure 3.16

Solution The free-body diagram (Fig. 3.17) for B discloses only two forces: the weight $\boldsymbol{W}$ and the required force $\boldsymbol{R}$. From equation 3.1 (Newton's second law),

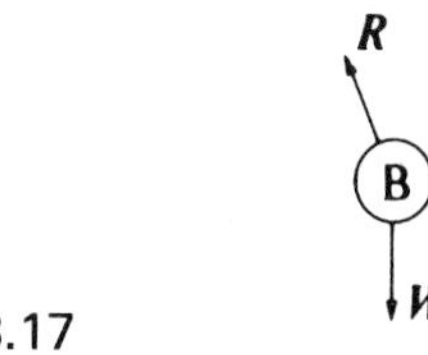

Figure 3.17

$$\sum \boldsymbol{F} = m\boldsymbol{a}_B$$

$$\boldsymbol{R}+\boldsymbol{W} = m\boldsymbol{a}_B$$

Since

$$\boldsymbol{W} = mg(-\boldsymbol{j})$$

and $\boldsymbol{a}_B$ (from example 2.6) is given by

$$\boldsymbol{a}_B = (-18.54\boldsymbol{i}+5.96\boldsymbol{j})\ \text{m/s}^2$$

then $$\boldsymbol{R} = (3.0)(-18.54\boldsymbol{i}+5.96\boldsymbol{j}) - (3.0)(9.81)(-\boldsymbol{j}) = (-55.62\boldsymbol{i}+47.31\boldsymbol{j})\ \text{N}$$

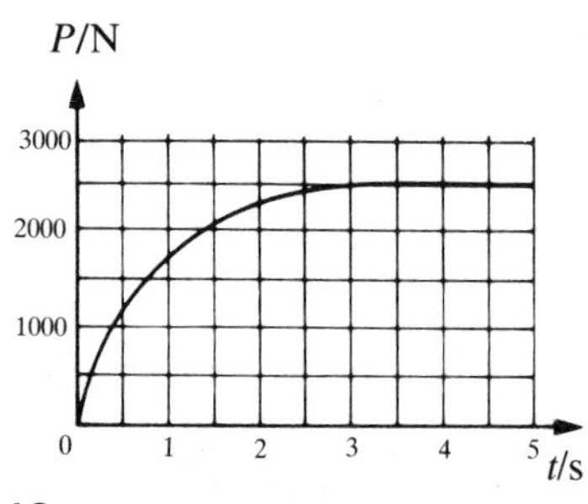

Figure 3.18

Example 3.5

See Fig. 3.18. An unmanned spacecraft having a mass of 1100 kg is to lift off vertically from the surface of the Moon where the value of g may be taken to be 1.62 N/kg. At time $t = 0$ the rocket motors are ignited and the variation of the thrust P of the motors with time t is shown.

Neglecting the mass of fuel burnt, determine the velocity when $t = 5$ s.

Solution The free-body diagram is shown in Fig. 3.19. Since there is no air resistance on the Moon, the only forces acting on the spacecraft when in flight are its weight $W = mg$ and the thrust P.

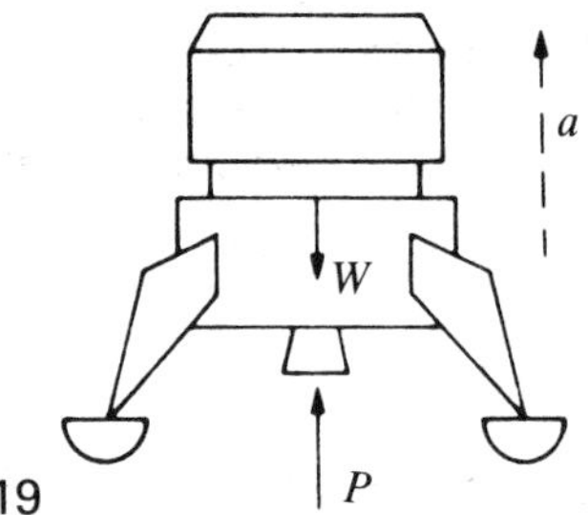

Figure 3.19

The equation of motion is thus

$$P - mg = ma \qquad \text{(i)}$$

We shall have to integrate equation (i) to determine the required velocity. For the small time interval considered, we can neglect the variation of g with height so that the weight W has the constant value of $W = 1100(1.62) = 1782$ N.

Writing $a = \mathrm{d}v/\mathrm{d}t$ and integrating equation (i) we have

$$\int_{t_1}^{t_2} P\,\mathrm{d}t - \int_{t_1}^{t_2} 1782\,\mathrm{d}t = \int_{v_1}^{v_2} 1100\,\mathrm{d}v$$

P is not known as an analytic function of t and so a numerical method must be used to evaluate the first integral. This is equivalent to measuring the area under the P–t curve. Thus

$$[\text{area under } P\text{–}t \text{ curve from } t_1 \text{ to } t_2] - 1782[t_2 - t_1] = 1100[v_2 - v_1] \qquad \text{(ii)}$$

We require the velocity v_2 at $t = 5$ s and know that when $t = 0$, $v = 0$. However, it would not be correct to substitute the values $t_1 = 0$, $t_2 = 5$, $v_1 = 0$ in the above relationship. Equation (i) applies only when the spacecraft is in flight. When the motors are first ignited the upthrust is less than the weight and contact forces will exist between the spacecraft and the surface of the Moon. The spacecraft remains in equilibrium until the thrust P exceeds the weight $W = 1782$ N, at which instant the contact forces disappear and equation (i) applies.

From Fig. 3.18 we note that P attains the value of 1782 N at time $t = 1.1$ s approximately. Using this value for t_1 in relationship (ii) with $v_1 = 0$ gives

$$[\text{area under } P\text{–}t \text{ curve from } t = 1.1 \text{ s to } t = 5 \text{ s}] - 1782[5 - 1.1] = 1100 v_2.$$

The required area is found to be 9180 N s approximately and hence

$$v_2 = 2.19 \text{ m/s}$$

The motion of rocket-powered vehicles is considered in more detail in Chapter 8.

Example 3.6

The loaded cage of a vertical hoist has a total mass of 500 kg. It is raised through a height of 130 m by a rope. The initial upwards acceleration of the cage is 1.65 m/s^2 and this remains constant until a speed of 10 m/s is reached. This speed remains constant until, during the final stage of the motion, the cage has a constant retardation which brings it to rest. The total time taken is 16.7 s.

Calculate (a) the tension in the rope at each stage, (b) the total work done by the tensile force on the cage and (c) the maximum power required.

Solution

a) The times and distances for each of the three stages of the motion can be found by writing simultaneous equations for constant acceleration for each stage and laboriously solving them. A more direct solution can be found by noting that the distance travelled is simply the area under the velocity–time graph, Fig. 3.20.

The time t_{AB} from A to B is found from

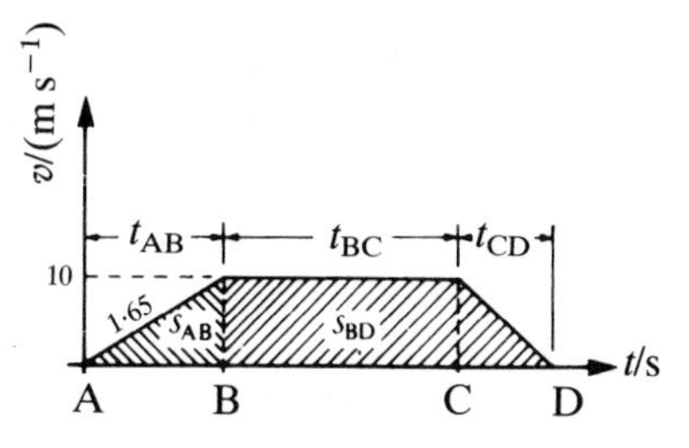

Figure 3.20

$$\frac{10}{t_{AB}} = 1.65$$

$$t_{AB} = 6.061 \text{ s}$$

From the area under the graph,

$$s_{AB} = \tfrac{1}{2}(6.061)10 = 30.31 \text{ m}$$

The remaining time, t_{BD}, is

$$t_{BD} = 16.7 - 6.061 = 10.64 \text{ s}$$

and the remaining distance, s, is

$$s_{BD} = 130 - 30.31 = 99.69 \text{ m}$$

Also from the area under the graph,

$$s_{BD} = \left(\frac{t_{BC} + t_{BD}}{2}\right)10$$

and hence

$$t_{BC} = 9.30 \text{ s}$$

The distance $\quad s_{BC} = (\text{velocity})(\text{time})$

$$= (10)(9.30) = 93.0$$

and $\quad s_{CD} = 130 - 30.31 - 93.0 = 6.69 \text{ m}$

The time from C to D is

$$t_{CD} = 16.7 - 6.061 - 9.30 = 1.339 \text{ s}$$

Finally, the constant acceleration between C and D is

$$a_{CD} = -10/1.339 = -7.468 \text{ m/s}^2$$

The forces acting on the cage (Fig. 3.21) are T (due to the tension in the rope) and $W = mg$.

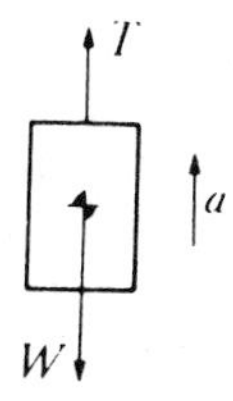

Figure 3.21

$$T - WT = ma$$
$$= m(g + a)$$

The different tensions are

$$T_{AB} = 500(9.81 + 1.65) \quad = 5730 \text{ N}$$
$$T_{BC} = 500(9.81 + 0) \quad = 4905 \text{ N}$$

and $\quad T_{CD} = 500(9.81 - 7.47) \quad = 1170 \text{ N}$

b) Denoting the upward displacement by y, the work done by the tensile force T (see section 3.12) is

$$\int_{s_1}^{s_2} \boldsymbol{F} \cdot \mathrm{d}\boldsymbol{s} = \int_{x_1}^{x_2} F_x \,\mathrm{d}x$$
$$= \int_0^{30.31} T_{AB}\,\mathrm{d}x + \int_0^{93} T_{BC}\,\mathrm{d}x$$
$$+ \int_0^{6.69} T_{CD}\,\mathrm{d}x$$
$$= 5730(30.3) + 4905(93)$$
$$+ 1170(6.69)$$
$$= 637600 \text{ N m}$$
$$= 637.6 \text{ kN m}$$

[It will be seen from the techniques of Chapter 7 that this final result could have been obtained simply by multiplying the weight of the cage by the total vertical distance travelled.]

c) The power required to lift the cage (see equation 3.30) is

$$P = \boldsymbol{T} \cdot \boldsymbol{v} = Tv$$

since the tension and the velocity are in the same direction. The power clearly has a maximum value just at the end of the first stage of motion. Thus

$$P_{max} = (5730)(10) = 57300 \text{ W}$$
$$= 57.3 \text{ kW}$$

Problems

3.1 Two bodies A and B collide and coalesce. The masses of the bodies are $m_A = 1$ kg and $m_B = 2$ kg. The velocities before impact were $\boldsymbol{v}_A = (15\boldsymbol{i} + 30\boldsymbol{j})$ m/s and $\boldsymbol{v}_B = (-20\boldsymbol{i} - 10\boldsymbol{j})$ m/s. Determine their velocity after the impact, assuming that only the impact forces are significant.

3.2 A railway truck A of mass 3000 kg is given a velocity of 4.0 m/s at the top of a 1 in 100 incline which is 50 m long. Neglecting all frictional resistance, determine the speed at the bottom of the incline.

Just beyond the bottom of the incline, truck A collides with a stationary truck B of mass 4000 kg and the two trucks become coupled automatically. Determine the speed of the trucks after the collision.

3.3 A body of mass m is initially at rest. Forces whose resultant is $\boldsymbol{R} = R\boldsymbol{i}$ and then applied to the body.

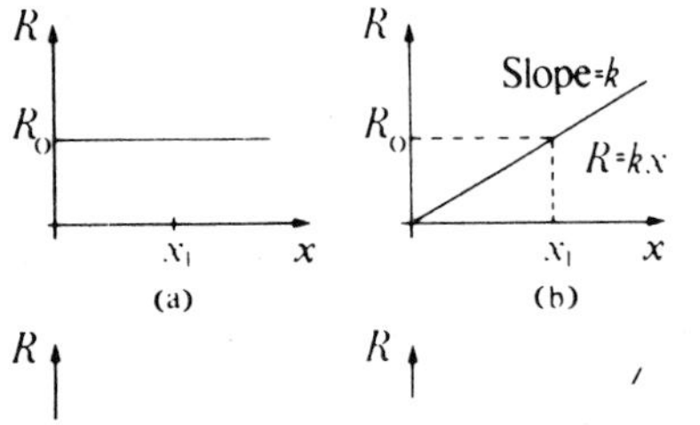

Figure 3.22

For the cases indicated in Fig. 3.22, which show the variation of the modulus R of the resultant with displacement x, find the velocity when $x = x_1$.

3.4 A resultant force $\boldsymbol{R} = R_x\boldsymbol{i} + R_y\boldsymbol{j}$ acts on a body of mass 0.5 kg. $R_x = (10 + 3t^2)$N and $R_y = (2t^3)$N, where t is the time in seconds. At time $t = 0$, the velocity $\boldsymbol{v}_G$ of the centre of mass G is $(5\boldsymbol{i} + 3\boldsymbol{j})$ m/s. Find $\boldsymbol{v}_G$ when $t = 2$ s.

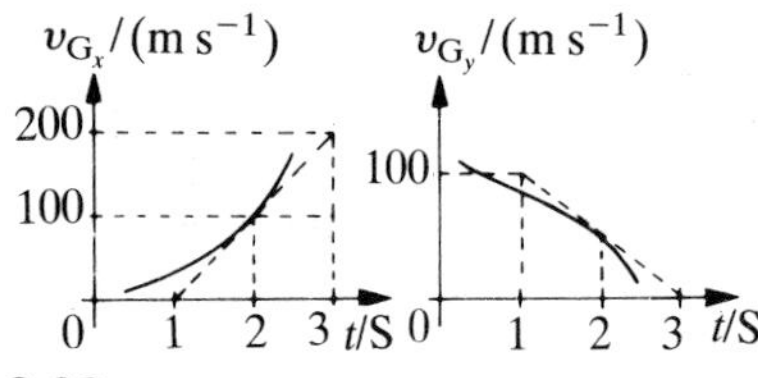

Figure 3.23

3.5 A link AB of a mechanism moves in the xy-plane. The mass of the link is 3.2 kg and the velocity components v_{Gx} and v_{Gy} of the centre of mass G are shown in Fig. 3.23. Determine the resultant force acting on the link when $t = 2$ s.

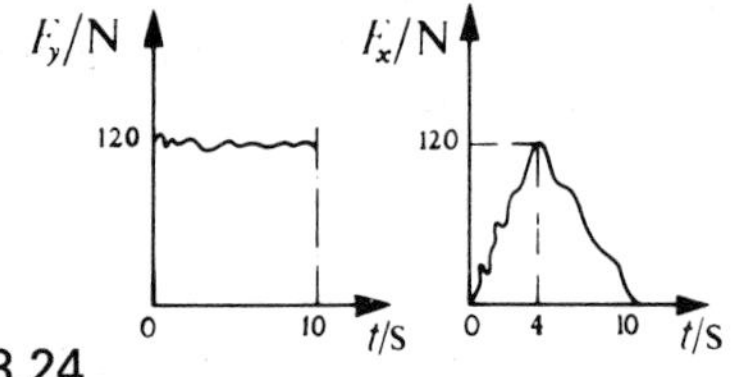

Figure 3.24

3.6 A small military projectile is launched from rest by a rocket motor whose thrust components F_x and F_y vary with time of flight as shown in Fig. 3.24. The vertically upwards direction is $+y$ and the value of g may be taken as 10 N/kg.

The mass of the projectile is 10 kg and is assumed to remain essentially constant. If air resistance is neglected, estimate (a) the magnitude of the velocity of the projectile after 10 s, and (b) the distance travelled by the projectile in the x-direction during this time.

3.7 The coefficient of friction between a box and a straight delivery chute is 0.5. The box is placed on the chute and is then released. Establish whether or not motion takes place and, if it does, the acceleration down the chute if its angle of inclination to the horizontal has the following values: (a) 20°, (b) 30°, (c) 40°.

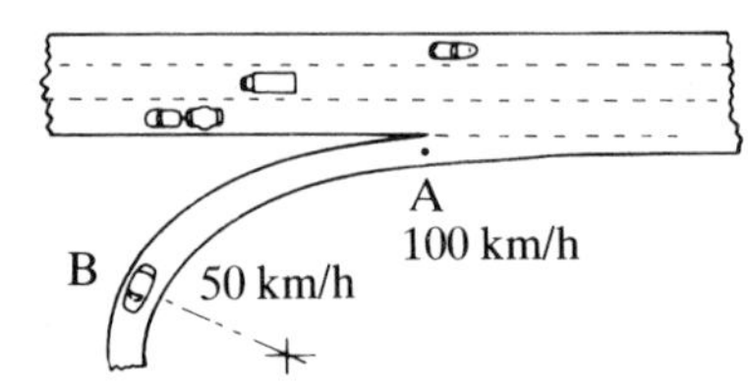

Figure 3.25

3.8 A car leaves a motorway at point A with a speed of 100 km/h and slows down at a uniform rate. Five seconds later, as it passes B, its speed is 50 km/h (Fig. 3.25). The radius of curvature of the exit road at B is 110 m. The mass of the car is 1500 kg.

Find (a) the acceleration of the car at B and (b) the total force exerted by the car on the road at B.

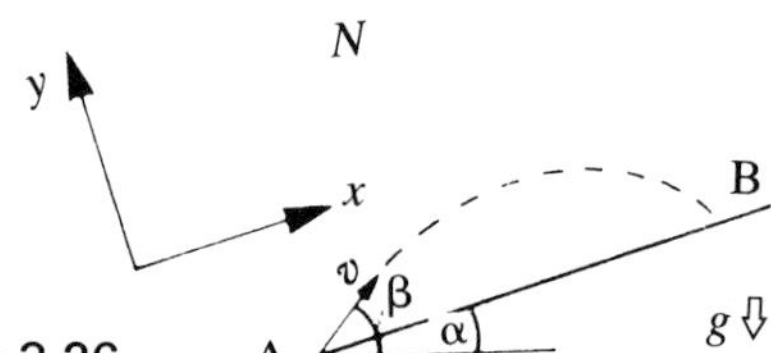

Figure 3.26

3.9 A missile is launched from point A (Fig. 3.26) with a velocity $\boldsymbol{v}$ inclined at an angle β to the horizontal and strikes the plane inclined at α to the horizontal at B. Show that

$$\mathrm{AB} = \frac{2v^2 \sin\gamma}{g\cos\alpha}[\cos\gamma - \tan\alpha \sin\gamma]$$

where $\gamma = \beta - \alpha$. Neglect air resistance.

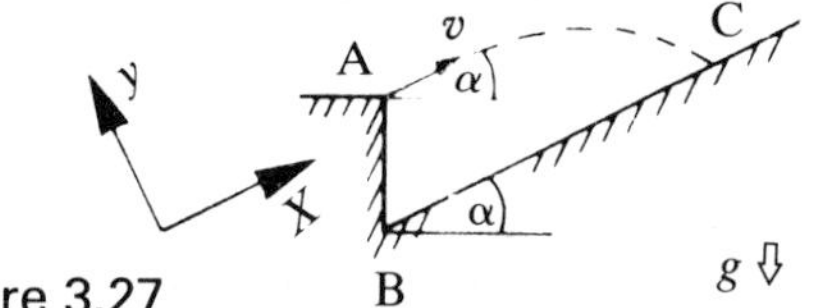

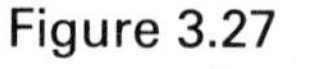

Figure 3.27

3.10 For the missile launched with velocity v for the configuration shown in Fig. 3.27, show that the distance BC does not depend on the angle α if air resistance can be neglected.

3.11 Figure 3.28 shows a block A on a horizontal surface and the coefficient of friction between block and surface is 0.7. Body B is connected to block A by a cord passing over a light pulley with negligible friction. The mass of A is 2 kg and that of B is 1.6 kg. Draw free-body diagrams for A and B to establish that, if the

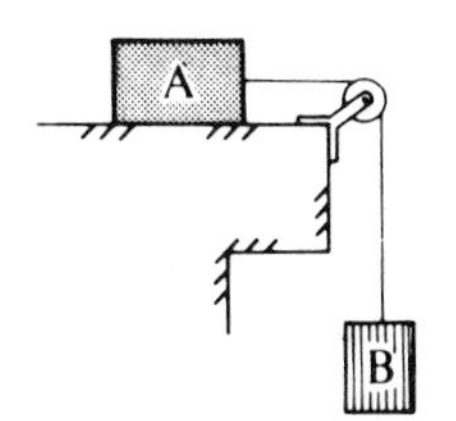

Figure 3.28

system is released from rest, motion takes place, and find the tension in the cord. Neglect the stretch in the cord and its mass.

3.12 A power boat whose mass is 2000 kg is heading towards a mooring buoy at a steady speed of 10 m/s. The combined water and air resistance of the hull varies with speed as shown in Fig. 3.29.

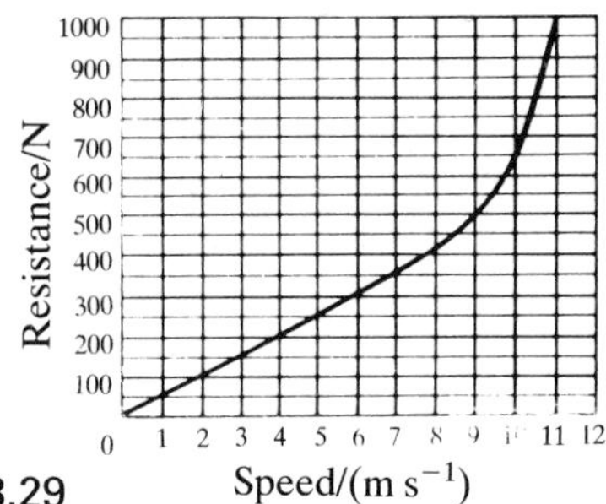

Figure 3.29

The approach to the buoy is then in two stages. Engine output is first reduced so that the thrust is halved. After a further period at the steady lower speed, the engine is shut down completely. What is the deceleration immediately following the first reduction, and what steady speed is achieved during this stage?

Calculate also the distance from the buoy at which final shut-down should occur, for the boat to come to rest at the buoy without further manoeuvring.

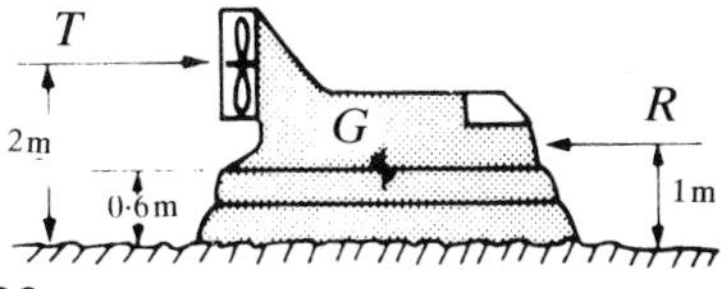

Figure 3.30

3.13 The hovercraft illustrated in Fig. 3.30 has a total mass of 600 kg with a centre of mass at G. The propulsion unit produces a thrust T on the craft of 900 N which gives a top speed of 120 km/h in still air.

Assuming that the air resistance R is proportional to the square of the air speed, and that the tangential force between the craft and the ground is negligible, determine the acceleration of the craft when $T = 900$ N and the speed through still air is 50 km/h.

3.14 A saloon motor car with driver has a mass of 700 kg. Wind-tunnel tests are used to predict how D, the total resistance to motion on a level road, varies with forward speed v. Engine and transmission tests are used to predict how T, the maximum tangential force obtainable between the road and the driving wheels, also varies with forward speed v.

The results are given below:

v/(km h^{-1})	D/newton	T/newton
18	325	1500
36	350	2000
54	390	2200
72	500	2100
90	650	1900
108	850	1600
126	1150	1300

Estimate the minimum time in which the car can accelerate forwards from 18 km/h to 126 km/h on a level road under conditions similar to those simulated in the tests.

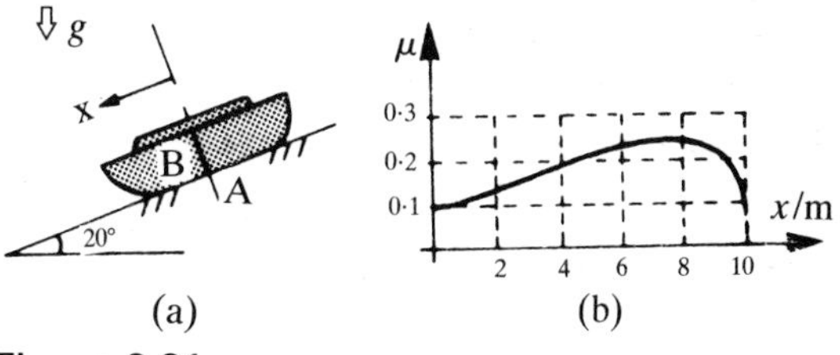

Figure 3.31

3.15 See Fig. 3.31(a). The lifeboat B is travelling down the incline and as it passes point A ($x = 0$) its speed is 3.0 m/s. The coefficient of friction μ varies with x as shown at (b). Estimate the speed of the lifeboat when it has travelled 10 m past A.

3.16 Car A is being driven along a main highway at a steady speed of 25 m/s towards a junction. Car B is being driven at a steady speed of 20 m/s towards the same junction along a straight road up an incline of 10° to the horizontal. At a particular instant car A is at 200 m from the junction and car B is at 135 m. A few seconds later the driver of car B observes car A and applies his brakes immediately, causing all four wheels to skid. His car just stops at the junction as car A passes through.

Determine the coefficient of friction between the tyres of car B and the road.

3.17 A road test is carried out on a sports car on a level road on a windless day, and the car is driven in such a way as to achieve the maximum possible acceleration through the gears. Results from the test are plotted in Fig. 3.32.

Estimate the following: (a) the time taken to travel the first 0.4 km of the test, (b) the maximum gradient the car can ascend in still air at a steady speed of 110 km/h in third gear and (c) the magnitude of the maximum possible acceleration for straight-line motion in still air at 160 km/h in fourth gear when the car is descending a gradient of 1 in 20.

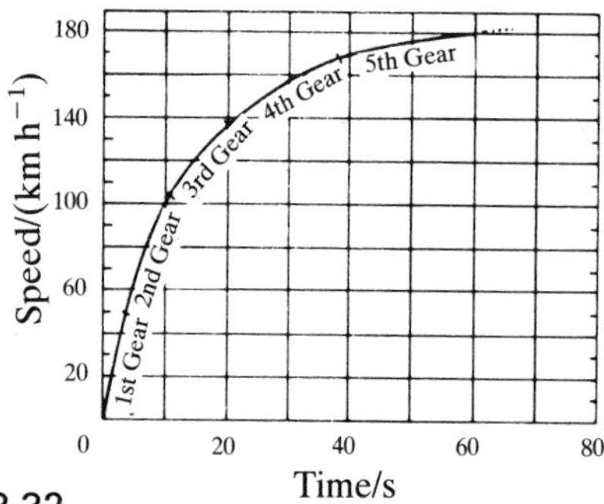

Figure 3.32

3.18 In the test rig for an ejector seat (Fig. 3.33(a)) the seat containing a dummy has a total mass of 500 kg. It is propelled up the launching guide by a rocket which develops a thrust whose magnitude T varies as shown by the graph in Fig. 3.33(b).

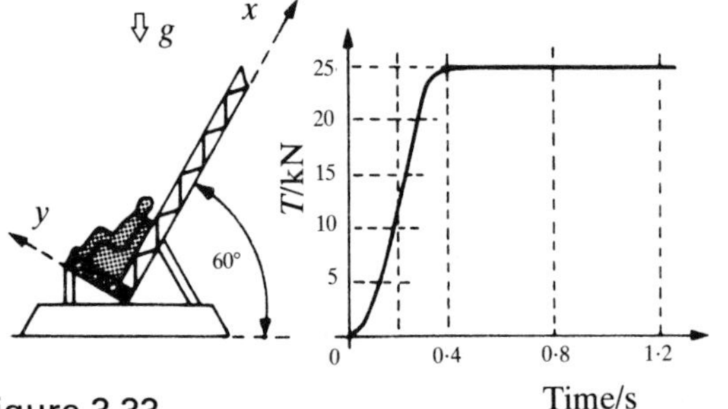

Figure 3.33

a) Find the maximum acceleration to which the dummy is subjected.
b) Estimate the velocity of the dummy 1.2 s after the rocket is fired.
Air resistance and friction in the guides may be neglected. The mass of the fuel burned is small and may also be neglected.

3.19 A small jet aircraft has a mass of 7350 kg and a jet thrust of 50 kN. During take-off, resistance to motion is equal to the sum of the aerodynamic drag force and the rolling resistance of the wheels; the latter force is equal to 0.02 times the normal reaction. Lift forces also act on the aircraft, and both the lift and drag vary with the horizontal velocity of the aircraft in the manner shown in Fig. 3.34.

Estimate the minimum length of runway required for take-off. Assume that the jet thrust and drag are always horizontal and that the lift forces act only vertically during take-off. The entire aircraft can be considered to be a rigid body in translation with wheels of negligible mass.

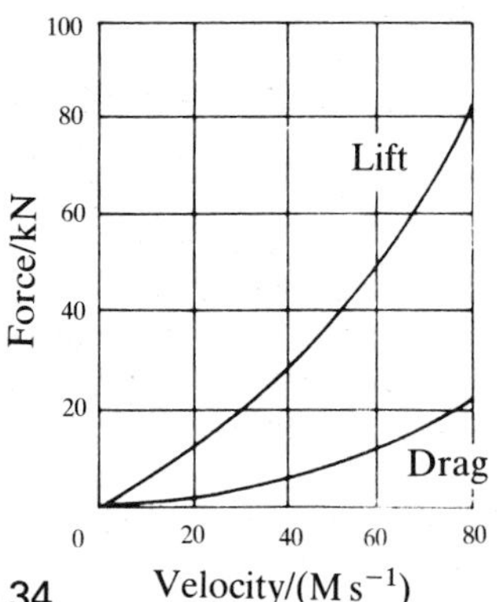

Figure 3.34

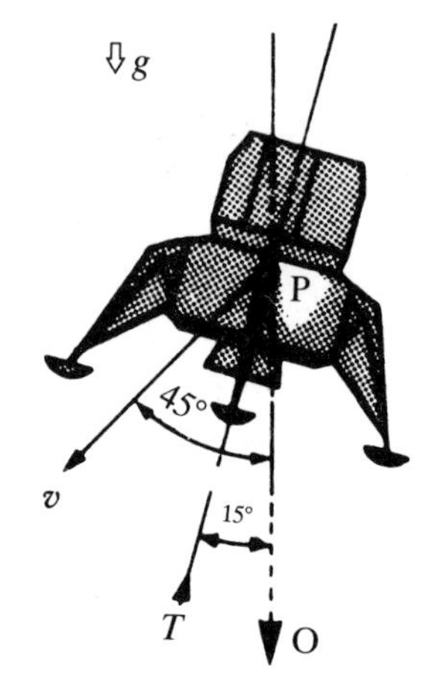

Figure 3.35

3.20 See Fig. 3.35. A lunar module P of mass 15 000 kg is approaching the Moon, which has a surface gravity of 1.62 N/kg and a mean radius of 1738 km. The centre of the Moon is taken as an origin O. When the module is 80 km from the surface, its velocity $\boldsymbol{v}$ is inclined at 45° from the line OP. The reverse thrust $\boldsymbol{T}$ of the descent engine then has a magnitude of 44.5 kN and is inclined at 15° to the line OP.

Determine (a) the magnitude of the component of the acceleration of the module along the radial line OP and (b) the magnitude of the component of the acceleration of the module normal to its flight path.

4
Force systems and equilibrium

4.1 Addition of forces

Force has been defined in Chapter 3 in the context of Newton's laws of motion. The action of a single force has been quantified by the changes it produces in the motion of a particle.

As we have shown that force is a vector quantity, any two forces acting at a point may be replaced by a resultant force $\boldsymbol{R}$ (Fig. 4.1). If a third force is now introduced, this may be added to $\boldsymbol{R}$ in just the same way as $\boldsymbol{F}_2$ was added to $\boldsymbol{F}_1$ (Fig. 4.2).

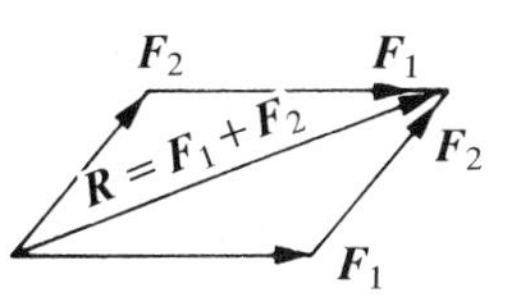

Figure 4.1

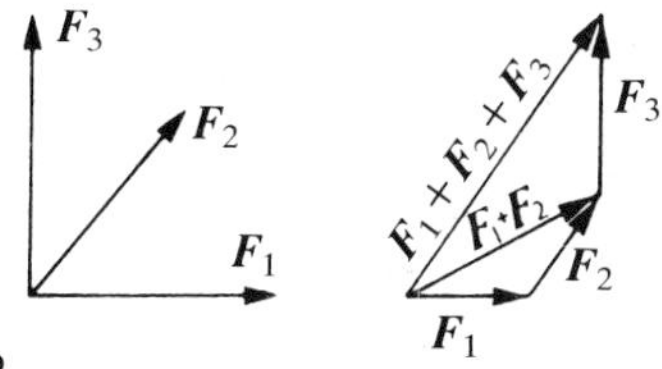

Figure 4.2

It is obvious that the position of the point of application is important. Consider two forces, equal in magnitude and direction, acting on different points on a body as shown in Fig. 4.3; their effects are clearly not the same. If, now, the force at A is applied at C, the overall effect is not altered; however, the internal effects will be different.

We conclude that the overall effect is governed by the line of action of the force and not by any specific point on that line. The difference between $\boldsymbol{F}_{1A}$ and $\boldsymbol{F}_{1B}$ is characterised by the separation d of the lines of action.

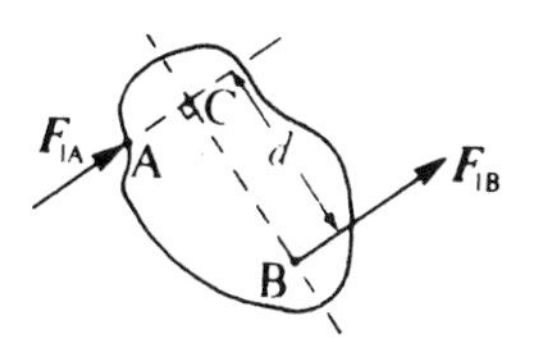

Figure 4.3

4.2 Moment of force

By definition, the magnitude of the moment of $\boldsymbol{F}$ (Fig. 4.4) about O is Fd.

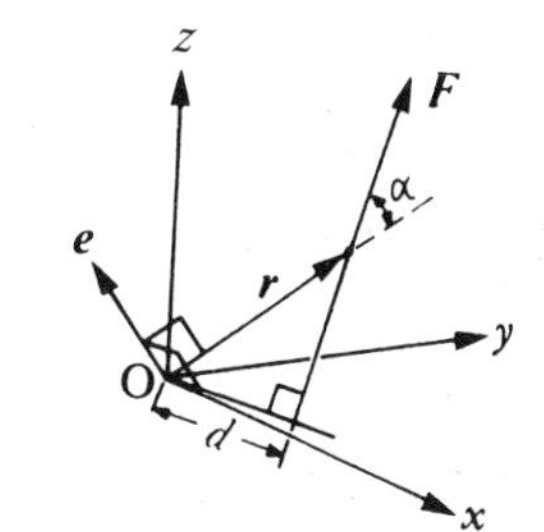

Figure 4.4

Clearly any force $\boldsymbol{F}$ acting tangentially to a sphere, radius d, gives a moment of the same magnitude, but the effect is uniquely defined if we associate with the magnitude a direction perpendicular to the plane containing $\boldsymbol{F}$ and $\boldsymbol{r}$ in a sense given by the right-hand screw rule. The moment of a force may therefore be regarded as a vector with a magnitude $Fd = Fr\sin\alpha$ and in a direction $\boldsymbol{e}$ as defined in Fig. 4.4. Hence we may write the moment of the force $\boldsymbol{F}$ about the point O as

$$\boldsymbol{M}_{\mathrm{O}} = Fr\sin\alpha\boldsymbol{e} = Fd\boldsymbol{e} \tag{4.1}$$

4.3 Vector product of two vectors

The vector or cross product of two vectors $\boldsymbol{A}$ and $\boldsymbol{B}$ is written $\boldsymbol{A}\times\boldsymbol{B}$ and is defined to have a magnitude $|\boldsymbol{A}||\boldsymbol{B}|\sin\alpha$, where α is the angle between the two vectors. The direction of $\boldsymbol{A}\times\boldsymbol{B}$ is given by the right-hand screw rule as shown in Fig. 4.5. Note that the vector product is not commutative since by definition $\boldsymbol{B}\times\boldsymbol{A} = -\boldsymbol{A}\times\boldsymbol{B}$, see Fig. 4.5.

If $\boldsymbol{A}$ and $\boldsymbol{B}$ are expressed in terms of their Cartesian components,

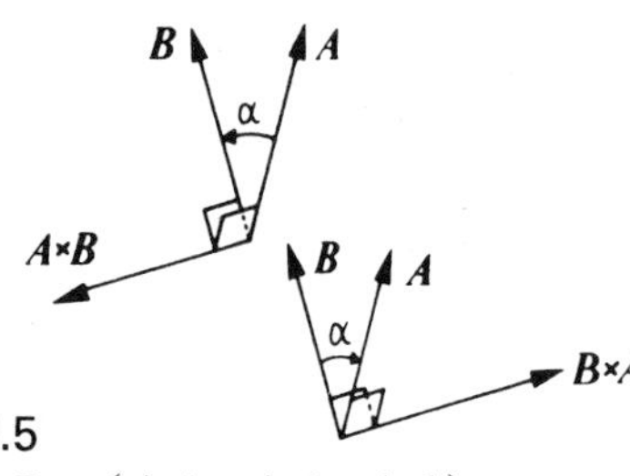

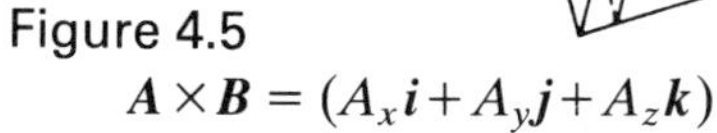

Figure 4.5

$$A \times B = (A_x i + A_y j + A_z k) \times (B_x i + B_y j + B_z k)$$

We must first consider the vector product of orthogonal unit vectors. By inspection,

$$i \times j = k = -j \times i$$
$$j \times k = i = -k \times j$$
$$k \times i = j = -i \times k$$

also $i \times i = j \times j = k \times k = 0$

hence

$$\begin{aligned} A \times B &= A_x B_y i \times j + A_x B_z i \times k \\ &\quad + A_y B_x j \times i + A_y B_z j \times k \\ &\quad + A_z B_x k \times i + A_z B_y k \times j \\ &= A_x B_y k - A_x B_z j \\ &\quad - A_y B_x k + A_y B_z i \\ &\quad + A_z B_x j - A_z B_y i \\ &= (A_y B_z - A_z B_y) i + (A_z B_x - A_x B_z) j \\ &\quad + (A_x B_y - A_y B_x) k \end{aligned} \tag{4.2}$$

This result is summarised by the following determinant:

$$A \times B = \begin{vmatrix} i & j & k \\ A_x & A_y & A_z \\ B_x & B_y & B_z \end{vmatrix} \tag{4.3}$$

From Fig. 4.4 we note that by the definition of the vector product of two vectors, equation 4.1 may be written as

$$M_O = r \times F = rF \sin \alpha e$$

4.4 Moments of components of a force

Consider two forces F_1 and F_2 whose resultant is R acting at point A (Fig. 4.6). The moment about O is

$$\begin{aligned} M_O &= r \times R \\ &= r \times (F_1 + F_2) \\ &= r \times F_1 + r \times F_2 \end{aligned} \tag{4.4}$$

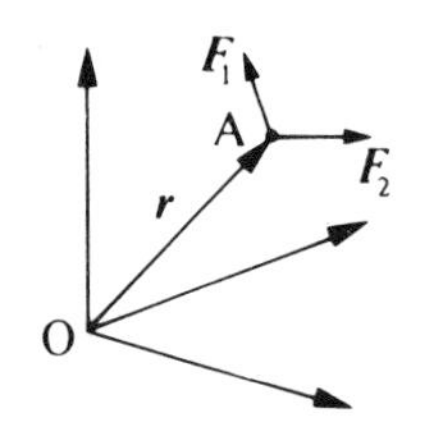

Figure 4.6

i.e. the moment of the resultant of F_1 and F_2 is equal to the vector sum of the moments of the components.

If a force F is replaced by its Cartesian components then the moment about O is, by inspection of Fig. 4.7,

$$M_O = (yF_z - zF_y) i + (zF_x - xF_z) j + (xF_y - yF_x) k \tag{4.5}$$

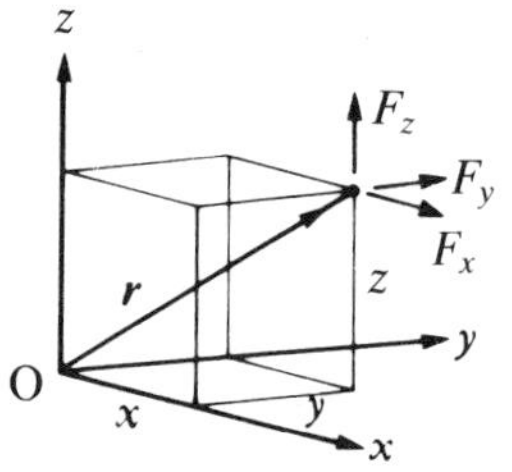

Figure 4.7

and this is seen to be the same as the vector-algebra definition

$$M_O = r \times F = (xi + yj + zk) \times (F_x i + F_y j + F_z k)$$

$$= \begin{vmatrix} i & j & k \\ x & y & z \\ F_x & F_y & F_z \end{vmatrix} \tag{4.6}$$

4.5 Couple

A couple is defined as a system of two non-collinear forces equal in magnitude but opposite in direction, i.e. in Fig. 4.8 $F_1 = -F_2$.

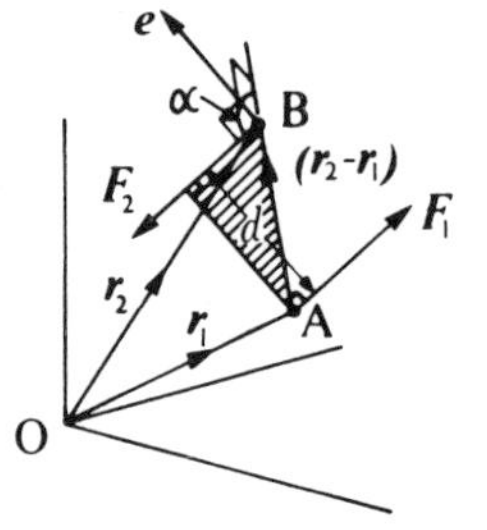

Figure 4.8

The moment of the couple about O is

$$\begin{aligned} M_O &= r_2 \times F_2 + r_1 \times F_1 \\ &= r_2 \times F_2 + r_1 \times (-F_2) \\ &= (r_2 - r_1) \times F_2 \\ &= (r_2 - r_1)(\sin\alpha) F_2 e = dF_2 e \end{aligned} \quad (4.7)$$

where e is normal to the plane containing d and F.

We see from this result that the moment of a couple is *independent of the origin* O and its magnitude is equal to the magnitude of one of the forces times the shortest distance between their lines of action.

It should also be noted that the resultant force of a couple is zero and the moment of a couple is often referred to as its torque.

4.6 Distributed forces

In most cases we have regarded forces as being applied at a point, but in practice this single force is the resultant of a distributed-force system which may be considered to be many small forces closely spaced.

Consider a small plane surface of area δA acted upon by a force having normal component δF_z and tangential components δF_x and δF_y (Fig. 4.9).

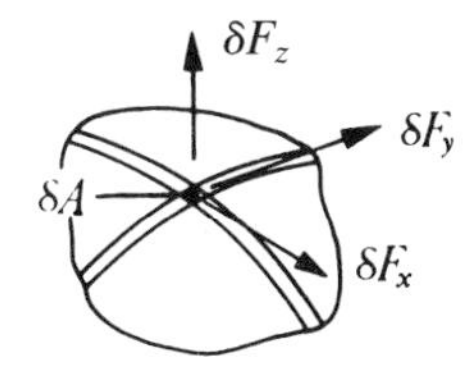

Figure 4.9

The intensity of loading normal to the surface is defined as

$$\lim_{\delta A \to 0}\left[\frac{\delta F_z}{\delta A}\right] = \sigma \quad \text{or} \quad p \quad (4.8)$$

When dealing with forces in solid mechanics this is referred to as a normal stress, σ, and in fluid mechanics it is called the pressure, p.

The terms $\lim_{\delta A \to 0}[\delta F_x/\delta A]$ and $\lim_{\delta A \to 0}[\delta F_y/\delta A]$ are called shear stresses (τ).

Equivalent co-planar force systems

It is sometimes convenient to replace a single force by an equal force along some different line of action together with a couple, where the moment of the couple C is $d \times F$, as shown in Fig. 4.10.

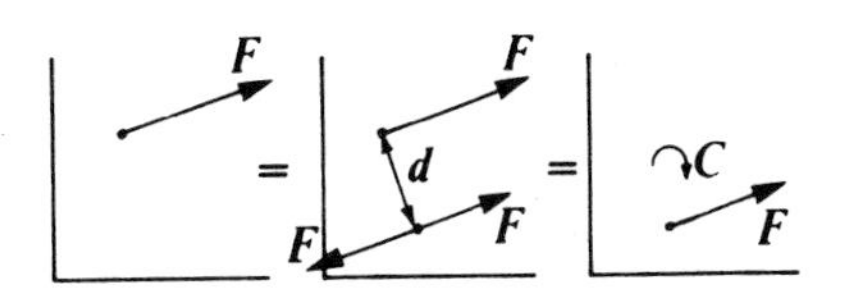

Figure 4.10

In general, two systems of forces are said to be equivalent, or equipollent, if they have the same resultant and the same moment about any arbitrary point.

It follows directly that any set of co-planar forces may be replaced by a single force plus a couple. The value of the couple depends on the line of action chosen for the resultant, but the value of the resultant is, of course, constant.

Since a single force may be replaced by a force plus a couple, the converse is also true provided that the resultant is not zero. Hence we may make the following statement:

> Any system of co-planar forces may be replaced by a single force, or a force plus a couple. If the resultant is zero, then the system may be replaced by a couple.

4.7 Equivalent force system in three dimensions

For a general system of forces, the resultant is

$$R = \sum F_i \quad (4.9)$$

and the moment about some arbitrary origin O is

$$M_O = \sum r_i \times F_i \quad (4.10)$$

We may therefore replace the system by a single force plus a couple. It is not generally possible to replace this system by a single force; however, it is possible to simplify this system to a single force plus a parallel couple.

In Fig. 4.11 the couple C may be replaced by two components, one parallel to R and one perpendicular to R.

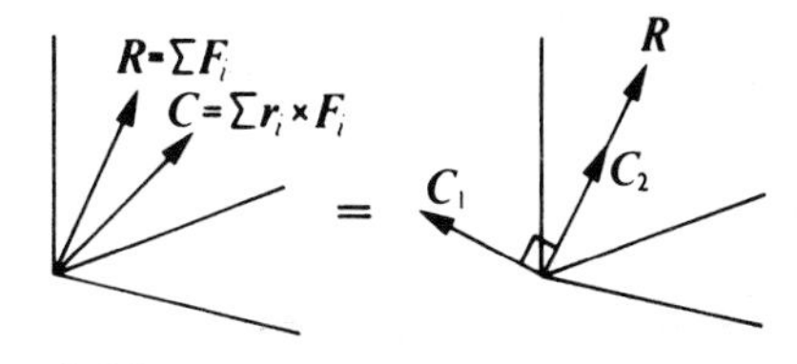

Figure 4.11

By moving the line of application of R in the plane normal to C_1 by a distance $|C_1|/|R|$, the system is now replaced by a single force R, the resultant, and a single co-axial couple C_2 called the wrench.

4.8 Equilibrium

There are two slightly different definitions of the concept of equilibrium; they are

i) a body is said to be in equilibrium when it is at rest (or, since velocity is relative, this implies that *all* points have zero acceleration);
ii) a system of forces is said to be in equilibrium when the resultant force is zero and the moment about any arbitrary point is zero.

Condition (i) implies (ii), but (ii) does not imply (i). For example the external forces acting on a flywheel rotating at constant speed satisfies (ii) but not (i); also a spring being compressed by equal and opposite forces satisfies (ii) but not always (i).

4.9 Co-planar force system

If we have a set of forces acting in the xy-plane then the condition for equilibrium of the system of forces is simply

$$\sum \boldsymbol{F} = \boldsymbol{0} \tag{4.11}$$

and $\sum \boldsymbol{M}_{\mathrm{O}} = 0$ (4.12)

or, in scalar form,

$$\left.\begin{aligned} \sum F_x &= 0 \\ \sum F_y &= 0 \\ \sum M_{\mathrm{O}z} &= 0. \end{aligned}\right\} \tag{4.13}$$

For a force system in the xy-plane $M_{\mathrm{O}x} = M_{\mathrm{O}y} = 0$, so that $M_{\mathrm{O}z}$ is often replaced by M_{O} without ambiguity.

If the set consists of three forces and no couples then, from Fig. 4.12,

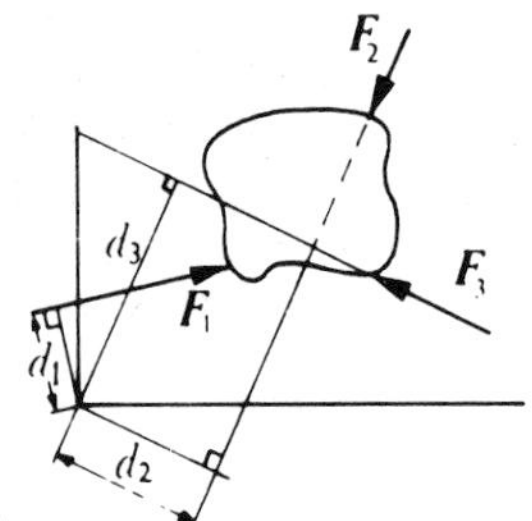

Figure 4.12

$$F_{1x} + F_{2x} + F_{3x} = 0 \tag{4.14a}$$

$$F_{1y} + F_{2y} + F_{3y} = 0 \tag{4.14b}$$

$$-F_1 d_1 - F_2 d_2 + F_3 d_3 = 0 \tag{4.14c}$$

As there are three independent equations there must be exactly three unknowns in the problem. For example, if $\boldsymbol{F}_1$ and $\boldsymbol{F}_2$ are known then the unknowns are F_{3x}, F_{3y} and d_3. Equations 4.14(a) and (b) determine F_{3x} and F_{3y}, therefore F_3 is now known in magnitude and direction. Equation 4.14(c) is used to find d_3.

The solution to this problem may also be found graphically by drawing a force diagram to some convenient scale, as shown in Fig. 4.13 to find $\boldsymbol{F}_3$. Since the resultant must be zero, $\boldsymbol{F}_3$ is the vector required to close the figure. The line of action can easily be found. Because the moment about any arbitrary point must be zero, it must be zero about the point of intersection of the lines of action of $\boldsymbol{F}_1$ and $\boldsymbol{F}_2$; hence $\boldsymbol{F}_3$ must pass through this same point, i.e. the three forces must be concurrent.

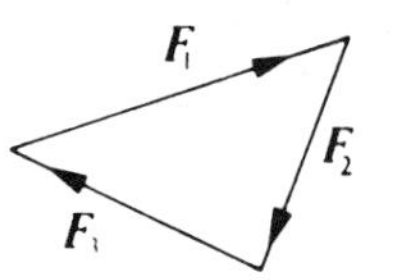

Figure 4.13

A simple plane frame will serve to examine some of the typical applications of equilibrium. Once again the concept of the free-body diagram is of great importance.

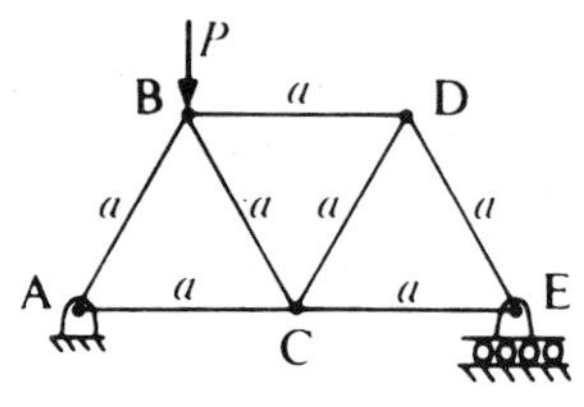

Figure 4.14

Example The structure shown in Fig. 4.14 is constructed of bars which are assumed to be connected by frictionless pins at their ends.

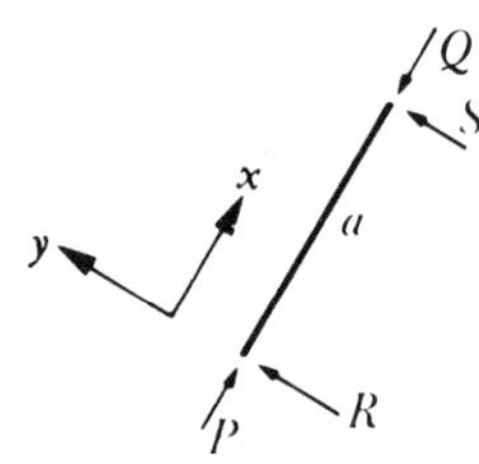

Figure 4.15

A free-body diagram of a typical bar (Fig. 4.15) soon shows that the loading is purely axial in all bars provided that the loads are applied only at the joints.

Equilibrium equations are

$$\sum F_x = 0 \quad \therefore P - Q = 0$$
$$\sum F_y = 0 \quad \therefore R + S = 0$$
$$\sum \text{moments about O} = 0 \quad \therefore Sa = 0$$

therefore $S = 0$, $R = 0$ and $P = Q$.

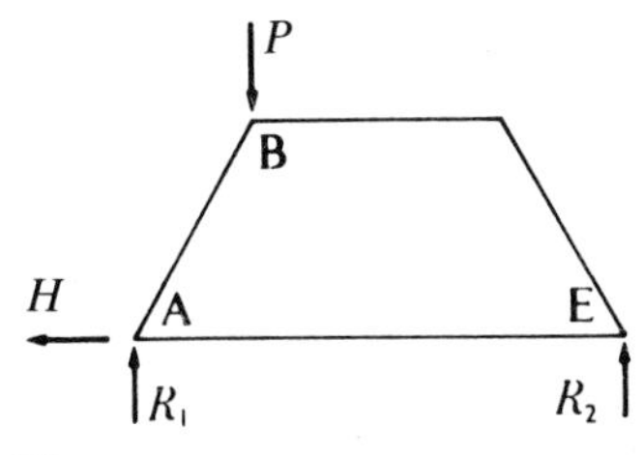

Figure 4.16

The free-body diagram for the complete structure is shown in Fig. 4.16. For equilibrium the equations are

$$-H = 0$$
$$R_1 + R_2 + P = 0$$

and $R_2(2a) + P(\frac{1}{2}a) = 0$

hence $R = \frac{1}{4}P$
$R_1 = \frac{3}{4}P$
$H = 0$

Note that the form of support ensures that there are exactly three unknown reaction forces. The reactions in this case are said to be statically determinate. Too few supports would lead to the possibility of collapse and too many would mean that the reactions would depend on the elastic properties of the structure.

If the forces acting on individual members of the structure are required, then these forces must be exposed by suitably 'cutting' the structure and producing a new free-body diagram.

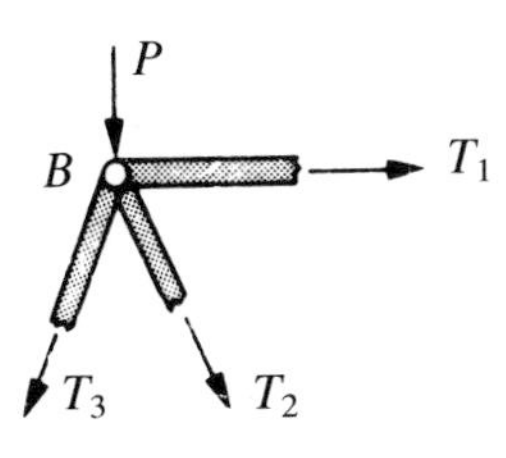

Figure 4.17

Consider the joint at B (Fig. 4.17). There are three unknowns but only two equations, since taking moments about B yields no information. However, resolving vertically,

$$P + T_2 \cos 30° + T_3 \cos 30° = 0$$

and horizontally

$$T_1 + T_2 \cos 60° - T_3 \cos 60° = 0$$

Consider the joint at A (Fig. 4.18) :

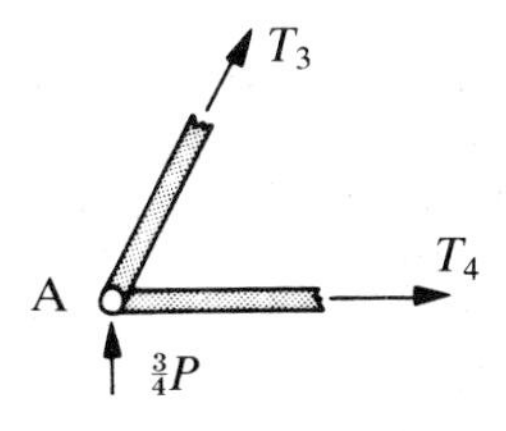

Figure 4.18

$$\tfrac{3}{4}P + T_3 \cos 30° = 0$$
$$T_3 \cos 60° + T_4 = 0$$

therefore $T_3 = -\frac{3}{4}P/\cos 30° = -P\sqrt{3}/2$

and $T_4 = -\frac{1}{2}T_3 = P\sqrt{3}/4$

Hence we can solve for T_2 and T_1 above.

Because the structure itself is statically determinate, each joint may be considered in turn.

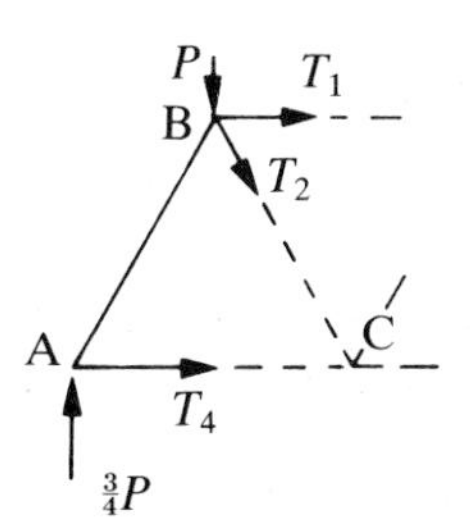

Figure 4.19

It is not always necessary to solve for all the internal forces in order to determine just one particular force. For example, if the load carried by member BC is required, then the free-body diagram (Fig. 4.19) will expose the force in that member. Resolving forces in the vertical direction gives

$$\tfrac{3}{4}P - P - T_2 \cos 30° = 0$$

hence $T_2 = -P/(2\sqrt{3})$

Similarly, moments about C will yield T_1.

It is seen that the key to the problem is drawing the right free-body diagrams, and practice is the only sure way to gain the essential experience.

4.10 Equilibrium in three dimensions

In three-dimensional problems there are six scalar equations, namely

$$\left.\begin{array}{ll}\sum F_x = 0 & \sum M_{Ox} = 0 \\ \sum F_y = 0 & \sum M_{Oy} = 0 \\ \sum F_z = 0 & \sum M_{Oz} = 0\end{array}\right\} \quad (4.15)$$

The basic problem is still the same, except that the geometry is more involved and simple plane force diagrams cannot be drawn. It is now that the benefits of vector algebra can be seen: as an example consider the problem shown in Fig. 4.20.

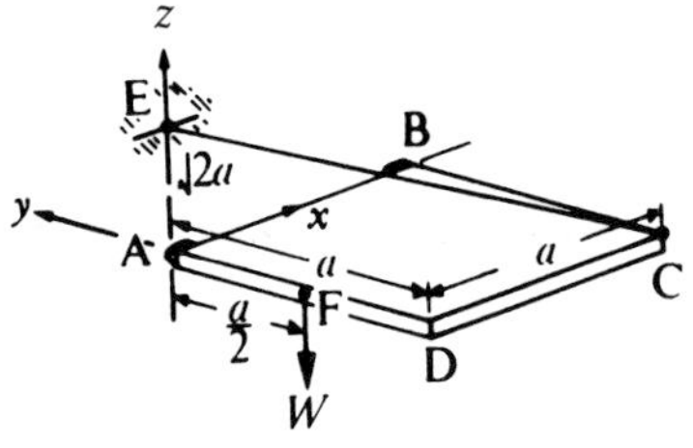

Figure 4.20

A flap ABCD carries a load $\boldsymbol{W}$ at F and is hinged about the x-axis on two hinges, one at A and one at B. Only the hinge at A can resist a load in the x-direction. A cable EC supports the door in a horizontal plane.

Free-body diagram

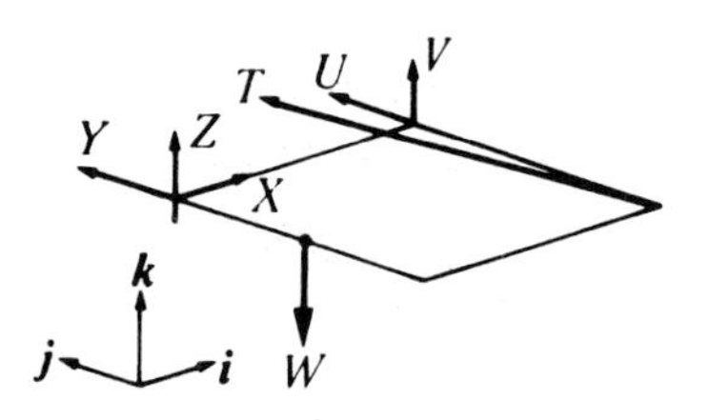

Figure 4.21

From Fig. 4.21, we see that there are six unknowns X, Y, Z, are the components of force $\boldsymbol{F}_A$ and U, V are the components of $\boldsymbol{F}_B$.$\boldsymbol{T}$. is the tension in the cable and may be written as

$$\boldsymbol{T} = T\boldsymbol{e}$$

where $\boldsymbol{e}$ is the unit vector in direction CE.

$$\boldsymbol{e} = \overrightarrow{CE}/|\overrightarrow{CE}| = \frac{-a\boldsymbol{i} + a\boldsymbol{j} + a\sqrt{2}\boldsymbol{k}}{\sqrt{(a^2 + a^2 + 2a^2)}}$$

$$= -\tfrac{1}{2}\boldsymbol{i} + \tfrac{1}{2}\boldsymbol{j} + \boldsymbol{k}/\sqrt{2}$$

so that $\boldsymbol{T} = T(-0.5\boldsymbol{i} + 0.5\boldsymbol{j} + 0.707\boldsymbol{k})$

Taking moments about A gives

$$\boldsymbol{M}_A = (-\tfrac{1}{2}a\boldsymbol{j}) \times (-W)\boldsymbol{k} + a\boldsymbol{i} \times (U\boldsymbol{j} + V\boldsymbol{k}) + (a\boldsymbol{i} - a\boldsymbol{j}) \times \boldsymbol{T}$$

The component of $\boldsymbol{M}_A$ along the x-axis is given by

$$\begin{aligned}\boldsymbol{M}_A \cdot \boldsymbol{i} &= -\tfrac{1}{2}a\boldsymbol{j} \times (-W)\boldsymbol{k} \cdot \boldsymbol{i} \\ &\quad - a\boldsymbol{j} \times T(0.5\boldsymbol{j} + 0.707\boldsymbol{k}) \cdot \boldsymbol{i} \\ &= \tfrac{1}{2}Wa - aT(0.707) = 0\end{aligned}$$

therefore $T = W/1.414$

If V is required, then $\boldsymbol{M}_A \cdot \boldsymbol{e}$ will exclude $\boldsymbol{T}$ since, by definition, $\boldsymbol{T}$ is parallel to $\boldsymbol{e}$ and will therefore have no moment about any axis parallel to $\boldsymbol{e}$.

$$\begin{aligned}\boldsymbol{M}_A \cdot \boldsymbol{e} &= (-\tfrac{1}{2}a\boldsymbol{j}) \times (-W)\boldsymbol{k} \cdot (-0.5\boldsymbol{i} + 0.5\boldsymbol{j} + 0.707\boldsymbol{k}) \\ &\quad + a\boldsymbol{i} \times (U\boldsymbol{j} + V\boldsymbol{k}) \cdot (-0.5\boldsymbol{i} + 0.5\boldsymbol{j} + 0.707\boldsymbol{k}) \\ &= -\tfrac{1}{4}aW + a(U0.707 - V0.5) = 0\end{aligned}$$

after some manipulation (see next section).

Moments about a vertical axis through A give, by inspection, $U = 0$; hence $V = -\tfrac{1}{2}W$. (This result can also be obtained by considering moments about the line AC.)

4.11 Triple scalar product

In the previous section use was made of the triple scalar product, the properties of which are now discussed.

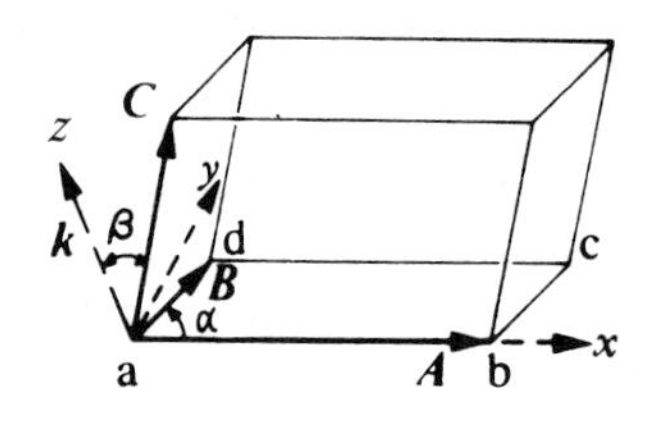

Figure 4.22

From section 4.3, $\boldsymbol{A} \times \boldsymbol{B} = AB\sin\alpha\boldsymbol{k}$.

$AB\sin\alpha$ is the area of the parallelogram *abcd* in Fig. 4.22; therefore

$$\begin{aligned}\boldsymbol{A} \times \boldsymbol{B} \cdot \boldsymbol{C} &= AB\sin\alpha\boldsymbol{k} \cdot \boldsymbol{C} \\ &= (AB\sin\alpha)(C\cos\beta) \\ &= \text{the volume of the parallelepiped}\end{aligned}$$

By symmetry, the volume is also $\boldsymbol{B} \times \boldsymbol{C} \cdot \boldsymbol{A} = \boldsymbol{C} \times \boldsymbol{A} \cdot \boldsymbol{B}$ and, because the dot product is commutative, the volume is $\boldsymbol{A} \cdot \boldsymbol{B} \times \boldsymbol{C} = \boldsymbol{B} \cdot \boldsymbol{C} \times \boldsymbol{A} = \boldsymbol{C} \cdot \boldsymbol{A} \times \boldsymbol{B}$, so we see that the position of the dot and the cross may be interchanged. It is also clear that if any two vectors are parallel then the product is zero.

It is easily shown that the triple scalar product

$$A \times B \cdot C = A \cdot B \times C = \begin{vmatrix} A_x & A_y & A_z \\ B_x & B_y & B_z \\ C_x & C_y & C_z \end{vmatrix} \qquad (4.16)$$

The sign of the result is unaltered provided the same cyclic order is preserved: reverse cyclic order introduces a change of sign.

4.12 Internal forces

At any internal surface of a body the sum of the forces acting over the surface will have a resultant $\boldsymbol{R}$ and the moment of all the elemental forces about the centroid is equivalent to a couple $\boldsymbol{C}$ (see Fig. 4.23). If the z-axis coincides with the axis of

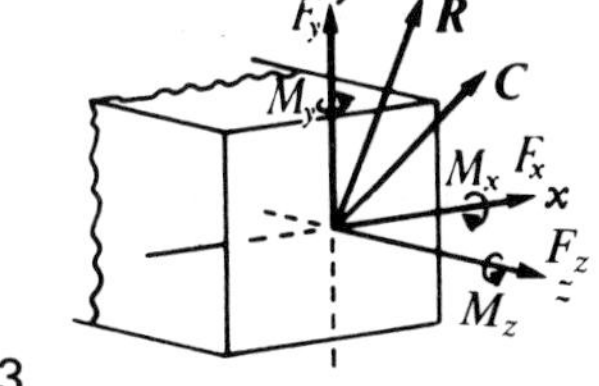

Figure 4.23

the body, then the components of $\boldsymbol{R}$ and $\boldsymbol{C}$ have the following descriptions:

$F_x = F_y =$ shear force
$F_z =$ axial force (tension or compression)
$M_x = M_y =$ bending moment

and $M_z =$ twisting moment

In these definitions the forces and couples are being applied *to* the surface; in some analyses the sign is defined in terms of the deformation produced, e.g. a positive M_x is one producing a positive curvature in the yz-plane.

4.13 Fluid statics

Pressure at a point. A fluid is characterised by the property of taking the shape of any vessel into which it is placed – this means that a fluid will not sustain a shear stress when static. Consider an element of fluid, density ρ, as shown in Fig. 4.24.

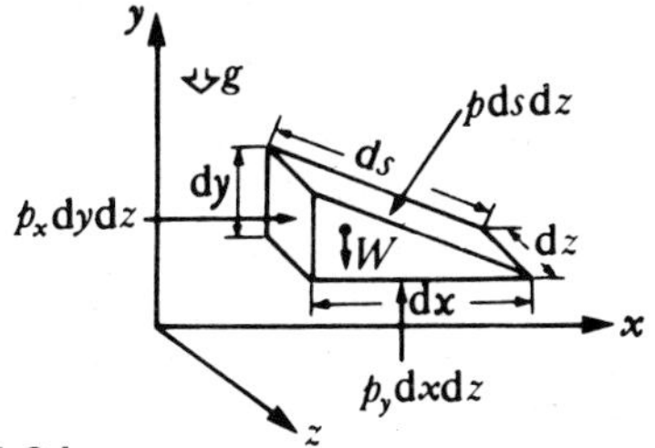

Figure 4.24

The weight of the element is $\rho g \frac{1}{2}(\mathrm{d}x\,\mathrm{d}y\,\mathrm{d}z)$.

Resolving forces in the vertical direction,

$$p_y\,\mathrm{d}x\,\mathrm{d}z - p\,\mathrm{d}z\,\mathrm{d}s\frac{\mathrm{d}x}{\mathrm{d}s} - \tfrac{1}{2}\rho g\,(\mathrm{d}x\,\mathrm{d}y\,\mathrm{d}z) = 0$$

hence $(p_y - p)\,\mathrm{d}x\,\mathrm{d}z - \frac{1}{2}\rho g\,\mathrm{d}x\,\mathrm{d}y\,\mathrm{d}z = 0$

therefore in the limit, as the volume tends to zero, $(p_y - p) \to 0$.

Resolving forces horizontally,

$$p_x\,\mathrm{d}y\,\mathrm{d}z - p\,\mathrm{d}z\,\mathrm{d}s\frac{\mathrm{d}y}{\mathrm{d}s} = 0$$

hence $p_x - p = 0$

We see that, for an infinitesimally small volume,

$$p_x = p_y = p \qquad (4.17)$$

that is, the pressure is the same in all directions at any point in a fluid at rest.

Pressure at a depth in a liquid. Since the fluid within any volume is in equilibrium, the vector sum of all the forces acting on the surface of that volume must be equal to the weight of fluid within that volume.

Resolving vertically for the forces shown in Fig. 4.25,

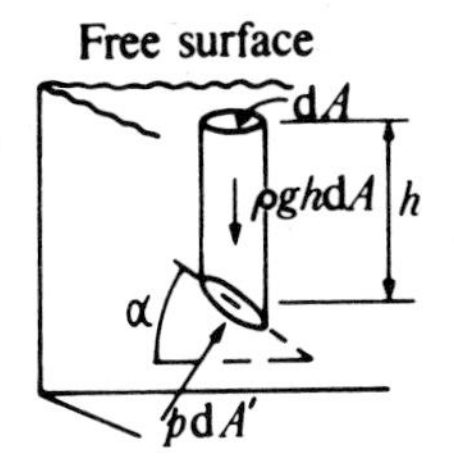

Figure 4.25

$$p\,\mathrm{d}A'\cos\alpha - \rho g\,\mathrm{d}A h = 0$$

but $\mathrm{d}A'\cos\alpha = \mathrm{d}A$

$$\text{hence} \quad p = \rho g h \qquad (4.18)$$

Force on a plane submerged surface. In Fig. 4.26,

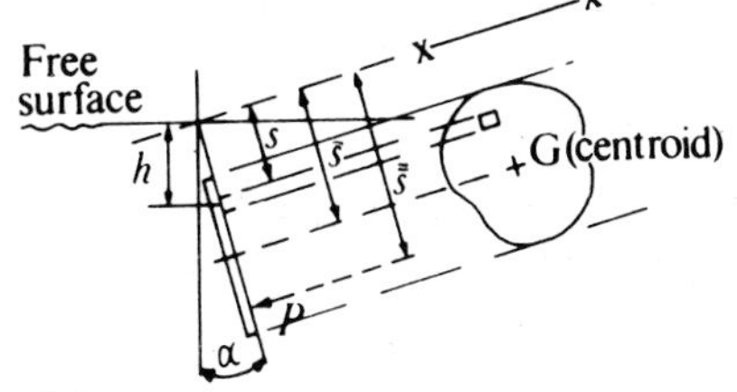

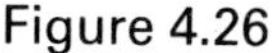

Figure 4.26

force on the elemental area $= \rho g (s\cos\alpha)\,\mathrm{d}A$

thus total force $= \int \rho g (s\cos\alpha)\,\mathrm{d}A$

$= \rho g \int s\,\mathrm{d}A \cos\alpha$

The integral $\int s\,\mathrm{d}A$ is the first moment of area about XX and is equal to $A\bar{s}$, where $\bar{s}$, by definition, is the location of the centroid of the area.

Therefore the total force acting on the area is given by

$$P = A\rho g\bar{s}\cos\alpha$$
$$= \text{area} \times (\text{pressure at the centroid}) \quad (4.19)$$

The moment of the force about XX is

$$P\bar{\bar{s}} = \int \rho g (s\cos\alpha) s\,\mathrm{d}A$$
$$= \rho g \cos\alpha \int s^2 \mathrm{d}A$$

The position of the centre of pressure is denoted by $\bar{\bar{s}}$. The integral $\int s^2 \mathrm{d}A$ is defined to be the second moment of area about XX and is denoted by $I_{XX} = Ak_{XX}{}^2$. The term k_{XX} is called the radius of gyration and is analogous to the term used in the description of the moment of inertia of a thin lamina (see Chapter 6). Thus

$$P\bar{\bar{s}} = \rho g \cos\alpha A k_{XX}{}^2$$

giving
$$\bar{\bar{s}} = \frac{\rho g \cos\alpha A k_{XX}{}^2}{A\rho g\bar{s}\cos\alpha}$$
$$= k_{XX}{}^2/\bar{s} \quad (4.20)$$

4.14 Buoyancy

Consider a region of a fluid, density ρ, at rest bounded by a surface S as shown in Fig. 4.27. It is clear that the vector sum of the surface forces due to static pressure must be equal and opposite to the gravitational force. The weight of the fluid is ρVg where V is the volume of the region so this must be equal to the value of the upthrust and this force acts at the centroid of the region.

If the body floats then it will displace its own weight of fluid so the upthrust will act through the centroid of the submerged volume, also known as the centre of buoyancy.

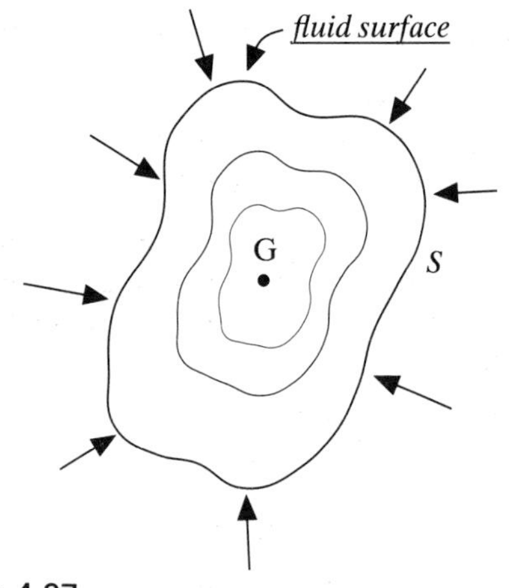

Figure 4.27

4.15 Stability of floating bodies

Figure 4.28 shows a vessel floating in the normal position. If the centre of gravity, G, of the vessel is below the centre of buoyancy the configuration is clearly stable. But if the centre of gravity is uppermost then we must investigate the conditions for stability.

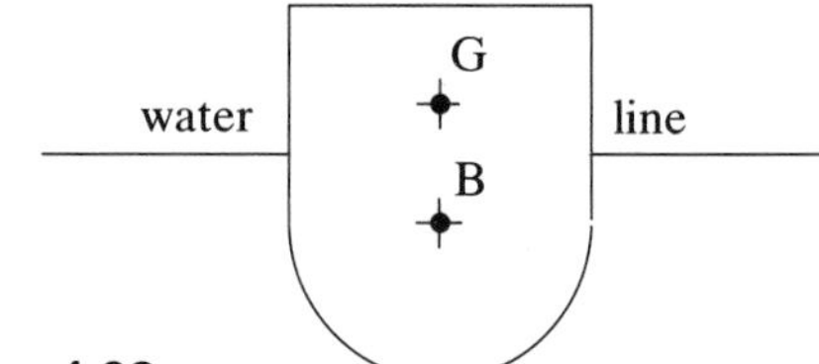

Figure 4.28

If the vessel is rotated by a small angle θ as shown in Fig. 4.29 then the upthrust will be an amount equal to mg through the original centre of buoyancy B plus a couple due to the gain in buoyancy on the low side and the loss of buoyancy on the high side.

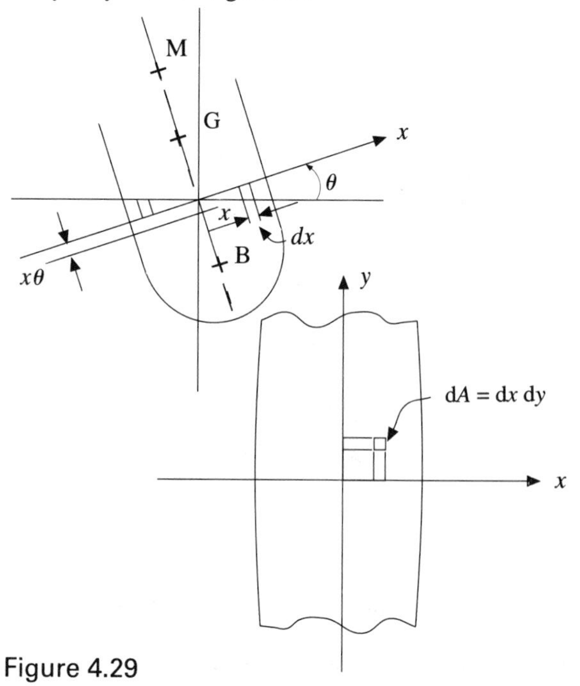

Figure 4.29

The moment due to an elemental area $\mathrm{d}x\,\mathrm{d}y$ is $\rho g x\theta x\,\mathrm{d}x\,\mathrm{d}y$ so writing $\mathrm{d}A = \mathrm{d}x\,\mathrm{d}y$ the total couple

$$= \int \rho g\theta x^2 \mathrm{d}A = \rho g\theta \int x^2 \mathrm{d}A$$

where the integral is taken over the cross-section area at the water-line. This integral is, by definition, the second moment of area and will be denoted by I, so the couple $C = \rho g\theta I$.

The original upthrust $mg = \rho V g$, where V is the displacement of the vessel, and acts through the original centre of buoyancy B. This force plus the couple may be combined into a single force acting through a point M. The position of M can be found by the fact that the moment about M of the hydro-static forces has to be zero.

Thus $\rho V g \overline{\mathrm{BM}}\,\theta - C \quad = 0$

or $\rho V g \overline{\mathrm{BM}}\,\theta \quad = \rho g\theta I$

giving $\overline{\mathrm{BM}} \quad = I/V \qquad (4.21)$

The height of M above G is known as the metacentric height and the point M as the metacentre.

Thus the metacentric height $\overline{\mathrm{GM}} = \overline{\mathrm{BM}} - \overline{\mathrm{BG}}$

$$= \mathrm{I/V} - \overline{\mathrm{BG}}. \qquad (4.22)$$

Clearly for stability the metacentric height must be positive.

Discussion examples

Example 4.1

In the simple structure shown in Fig. 4.30, links AB and BC are pinned together at B and to supports at A and C. Neglect the effects of gravity and determine the forces in the pins in terms of the applied load P.

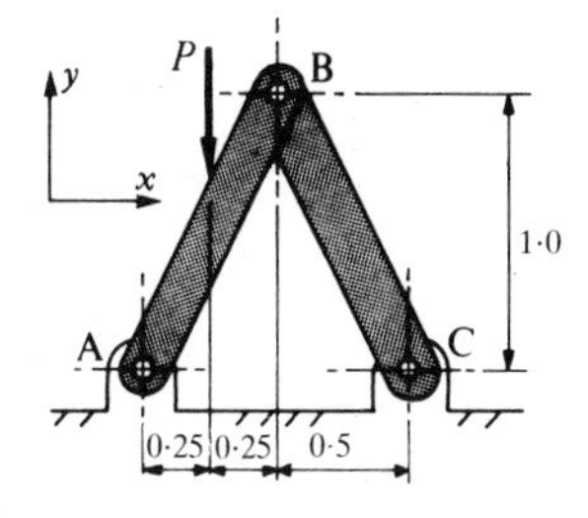

Figure 4.30

Solution One way of solving this problem is to resolve the forces into x- and y-components and write force and moment equations for each link.

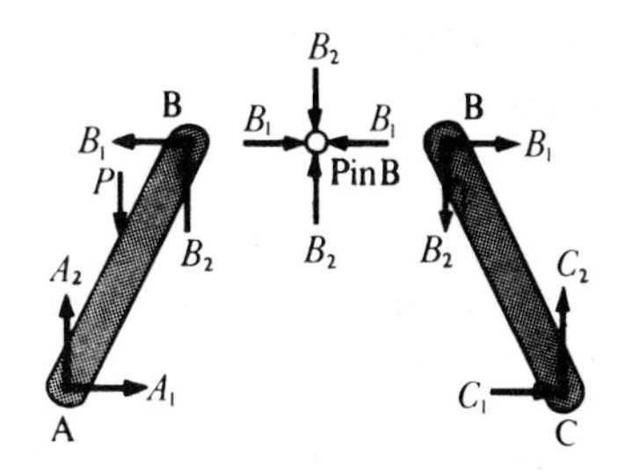

Figure 4.31

Free-body diagrams for AB and BC are shown in Fig. 4.31 with the force components at B on AB equal and opposite to those at B on BC (Newton's third law), the directions of the forces otherwise being chosen arbitrarily.

Since each link is in equilibrium, we can use equations 4.12: $\sum F_x = 0$, $\sum F_y = 0$ and $\sum M_{\mathrm{O}} = 0$ where the subscript O indicates an arbitrary axis perpendicular to the xy-plane.

For link AB:

$$\sum F_x = \mathrm{A}_1 - B_1 = 0 \qquad \text{(i)}$$

$$\sum F_y = \mathrm{A}_2 + B_2 - P = 0 \qquad \text{(ii)}$$

$$\sum M_{\mathrm{A}} = (0.5)B_2 - (0.25)P + (1)B_1 = 0 \qquad \text{(iii)}$$

For link BC:

$$\sum F_x = \mathrm{B}_1 + C_1 = 0 \qquad \text{(iv)}$$

$$\sum F_y = -\mathrm{B}_2 + C_2 = 0 \qquad \text{(v)}$$

$$\sum M_{\mathrm{C}} = -(1)B_1 + (0.5)B_2 = 0 \qquad \text{(vi)}$$

There are six independent equations, with six unknown, which can be laboriously solved to give $A_1 = \frac{1}{8}P$, $A_2 = \frac{3}{4}P$, $B_1 = \frac{1}{8}P$, $P_2 = \frac{1}{4}P$, $C_1 = -\frac{1}{8}P$, $C_2 = \frac{1}{4}P$.

The magnitudes of the forces in the pins are

$$F_{\mathrm{A}} = [(\tfrac{1}{8})^2 + (\tfrac{3}{4})^2]^{1/2}P \quad = 0.7603P$$

$$F_{\mathrm{B}} = [(\tfrac{1}{8})^2 + (\tfrac{1}{4})^2]^{1/2}P \quad = 0.2795P$$

$$F_{\mathrm{C}} = [(-\tfrac{1}{8})^2 + (\tfrac{1}{4})^2]^{1/2}P = 0.2795P$$

If we draw a single free-body diagram (Fig. 4.32) for the two connected links, the forces at B become internal forces and do not appear on the diagram.

$$\sum F_x = A_1 - C_1 = 0 \qquad \text{(vii)}$$

$$\sum F_y = A_2 - P + C_2 = 0 \qquad \text{(viii)}$$

$$\sum M_{\mathrm{A}} = (1)C_2 - (0.25)P = 0 \qquad \text{(ix)}$$

Equations (viii) and (ix) give

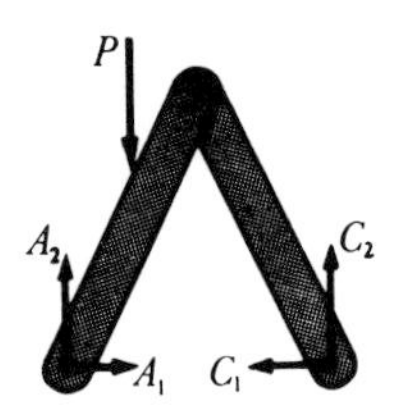

Figure 4.32

$$A_2 = \tfrac{3}{4}P \quad \text{and} \quad C_2 = \tfrac{1}{4}P$$

and these results enable equations (i) and (vi) to be solved more easily.

These solutions have made no use of the special relationships governing connected bodies in equilibrium where some of the bodies have only two forces or only three forces acting on them. For the former case, the forces must be equal, opposite and collinear, otherwise there would be a couple acting. Link BC is such a body and has one force at B, the other at C. The directions of these forces must therefore lie along BC. The free-body diagrams for the separate and connected links may now be drawn as shown in Fig. 4.33.

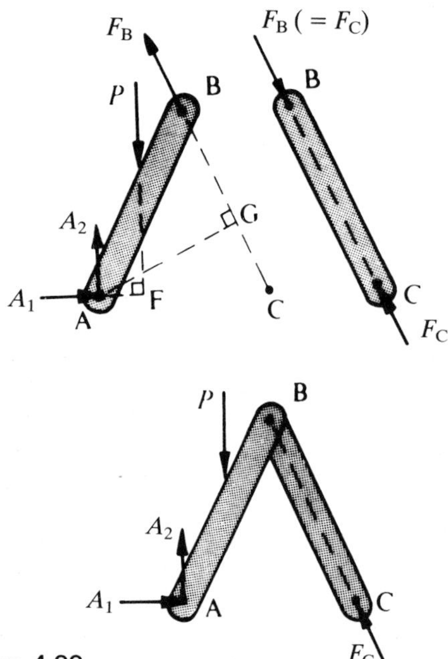

Figure 4.33

We can see immediately that the single equation $\sum M_A = 0$ for either AB alone or for the connected links will give the forces at B and C. For link AB:

$$\sum M_A = -(AF)P + (AG)F_B = 0$$

The length AG is found to be 0.894, therefore

$$-(0.25)P + (0.894)F_B = 0$$

$$F_B = 0.279P = F_C$$

We can then use, for example, the two equations $\sum F_x = 0$, $\sum F_y = 0$ to find A_1 and A_2. Thus we have reduced the number of equilibrium equations written to three.

We still have not made use of the relationship that, if a body is in equilibrium under the action of three non-parallel forces, those forces must be concurrent (section 4.9). Link AB is such a body. We know the direction of the force at B from the properties of the two-force link BC.

The force at B intersects the force P at point X on Fig. 4.34(a), and so the force at A also passes through this point. If we draw the linkage to scale to determine the location of X then we can draw the force triangle for forces $\boldsymbol{P}$, $\boldsymbol{F}_A$ and $\boldsymbol{F}_B$ acting on AB and determine $\boldsymbol{F}_A$ and $\boldsymbol{F}_B$ in magnitude and direction (Fig. 4.34(b)).

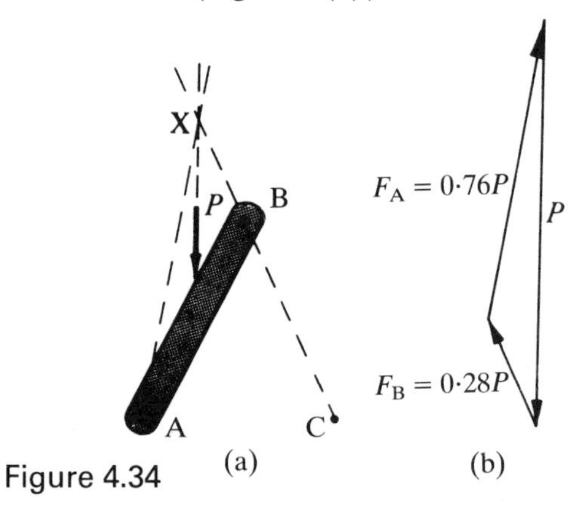

Figure 4.34

The solution to this problem has now been reduced to the drawing of a single force triangle.

Where use is made of the special properties of two- or three-force links in equilibrium, the technique is known as a semi-graphical method. This technique is employed in the outline solution of the next problem.

Example 4.2

Figure 4.35, which is drawn to scale, shows a spring-assisted hinge mechanism for a motor-car bonnet. Two such mechanisms are attached to the bonnet and are symmetrically disposed about the fore-and-aft centre line of the bonnet. Each mechanism consists of the cranked links ABCD

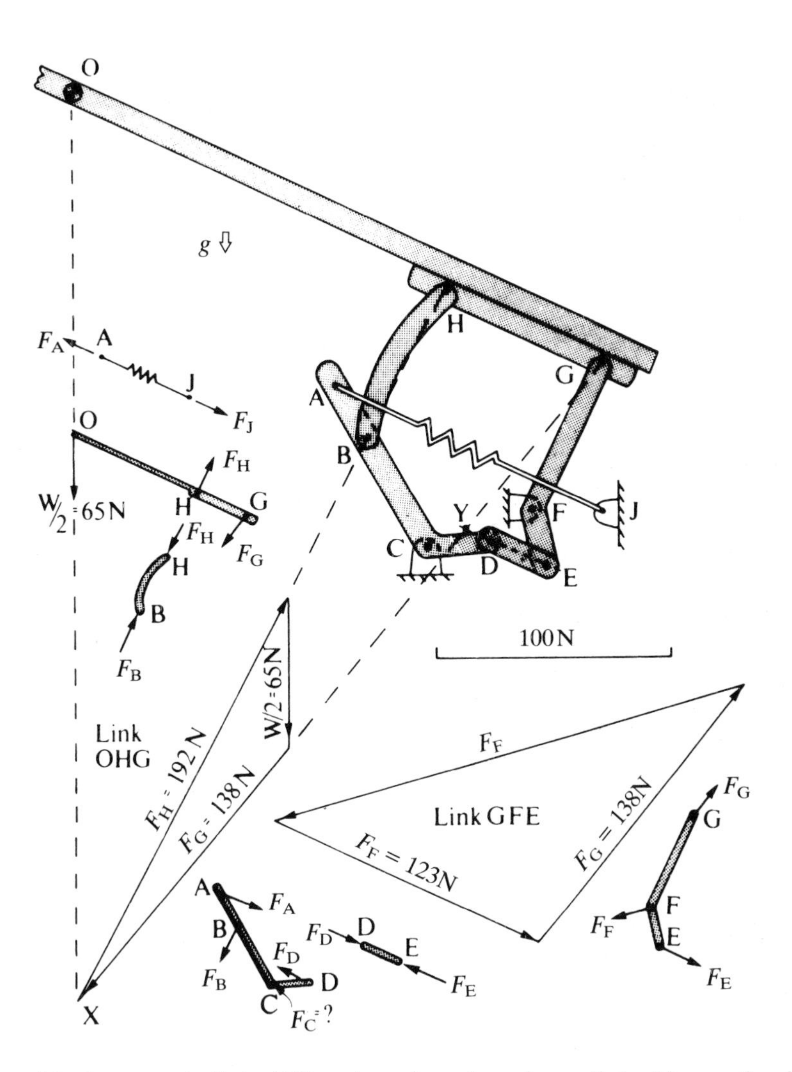

Figure 4.35

and EFG together with the two-pin links BH and DE, and spring assistance is provided by the spring AJ. The bonnet weighs 130 N, its centre of gravity is at O, and it is attached to pins at H and G.

The system is in equilibrium in the position shown. Neglecting frictional effects and the weights of all the members other than that of the bonnet, (a) find the magnitudes of the forces at pins B, E and G; (b) find the force in the springs and (c) state whether the springs are in tension or compression.

Solution Each mechanism supports half the weight of the bonnet, so that the effective vertical load at O for one mechanism is 130/2 = 65 N.

The two-force links are BH, DE and AJ, and the three-force links are OHG and EFG. ABCD is a four-force link. Hence the forces on OHG meet at the point X and those on EFG therefore meet at Y. The downward force $W = 65$ N at O is known and we can thus draw to scale the force triangle for OHG and determine the magnitudes and directions of forces F_H and F_G acting on this link. The force at G on EFG is equal and opposite to that on OHG and we can now draw the force triangle for GFE. Consideration of the free-body diagrams for BH and DE shows that the forces at B and D on link ABCD are equal and opposite to those at H on OHG and E on EFG respectively.

The magnitude and direction of the force at C is as yet unknown, but if we take moments about C for link ABCD, measuring the moment arms of the forces F_A, F_B and F_D directly from the figure, F_A can be determined. (The result is found to be $F_A = 222$ N.) Alternatively we could replace the known forces F_B and F_D on ABCD by their resultant and thus convert ABCD to a three-force link, leaving F_A to be found from a force triangle, and coincidently revealing F_C.

The free-body diagram for the spring shows it to be in tension.

The required results are thus: (a) 192 N, 123 N, 138 N; (b) 222 N; (c) tension.

Example 4.3

The mechanism shown in Fig. 4.36 is in equilibrium. Link AB is light and the heavy link BC weighs 480 N, its mass centre G being midway between B and C. Friction at the pins A and C is negligible. The limiting friction couple Q_B in the hinge at B is 10 N m. Pin C can slide horizontally and the horizontal force P is just sufficient to prevent the collapse of the linkage.

Find the value of P.

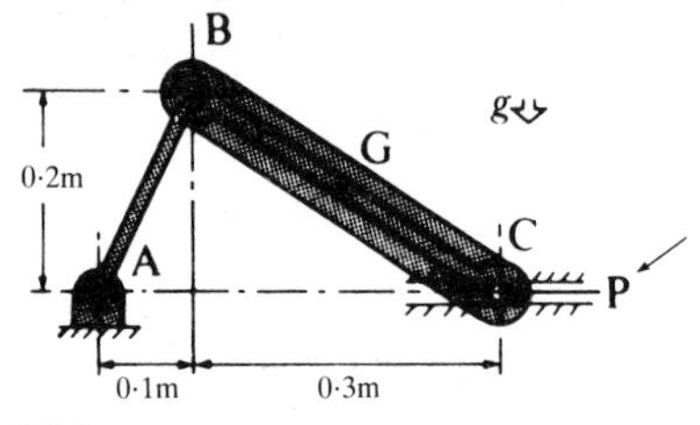

Figure 4.36

Solution If the line of action of an unknown force is known, but not the sense, problems can often be solved by assuming one of the two possible directions for the force and allowing the sign to 'take care of itself' in the ensuing mathematical process. This is not the case for a friction force where slip is occurring or is about to take place. A similar point is applicable for the case of a friction couple.

In the present problem the solution cannot proceed until the directions of Q_B applied to each link have been determined. Since the linkage is on the point of collapse, BC is about to rotate anticlockwise relative to AB. Thus BC imparts an anticlockwise couple $Q_B = 10$ N m to AB. Similarly, or as a consequence of Newton's third law, AB imparts a clockwise couple to BC.

The separate free-body diagrams for AB and BC are shown in Fig. 4.37. The directions of the weight W and of the couples at B are known. If any of the other arbitrarily chosen directions for the remaining forces were to be reversed, the result would be unaffected.

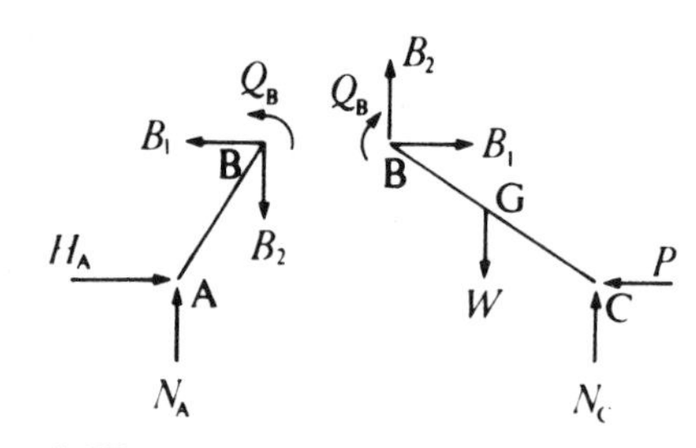

Figure 4.37

The value of P could be found by writing one moment and two force equations for each link, resulting in six equations with six unknowns. A quicker solution can be obtained as follows. If we take moments about B for link BC only we shall obtain a relationship between P and the unknown force N_C. This force can be found by taking moments about A for the two connected links.

For BC only $[\sum M_B = 0]$:

$$N_C(0.3) - W(0.15) - P(0.2) - Q_B = 0$$

$$N_C(0.3) - 480(0.15) - P(0.2) - 10 = 0 \qquad \text{(i)}$$

For the whole system $[\sum M_A = 0]$:

$$N_C(0.4) - W(0.25) = 0$$

$$N_C = 480(0.25)/0.4$$

Substituting the value of N_C in equation (i), we find

$$P = 40 \text{ N}$$

This problem is considered again in Chapter 7.

Example 4.4

Figure 4.38 shows a machine part in equilibrium. The part is the cranked rod ABC, where angle ABC = 120°. The rod is in contact with other machine parts (not shown) at B and C and these cause the forces $\boldsymbol{F}_B$ and $\boldsymbol{F}_C$ and the couple $\boldsymbol{C}_C$ to be applied to the rod. $\boldsymbol{F}_B = (-30\boldsymbol{k})$ N, $\boldsymbol{F}_C = (-15\boldsymbol{i} - 15\boldsymbol{j} - 10\boldsymbol{k})$ N and $\boldsymbol{C}_C = (-3\boldsymbol{i} + 5\boldsymbol{k})$ N m.

Determine the force and couple at A applied by the rod to the support. Also determine the direct

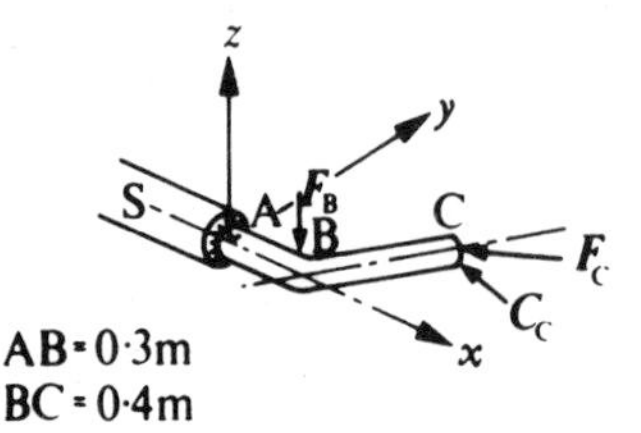

Figure 4.38

force, shear force, twisting moment and bending moment in a plane transverse section of the rod at A.

Solution The free-body diagram (Fig. 4.39) shows $\boldsymbol{F}_B$, $\boldsymbol{F}_C$ and $\boldsymbol{C}_C$, together with the force $\boldsymbol{F}_A$ and the couple $\boldsymbol{C}_A$ applied to the arm by the support S to maintain equilibrium.

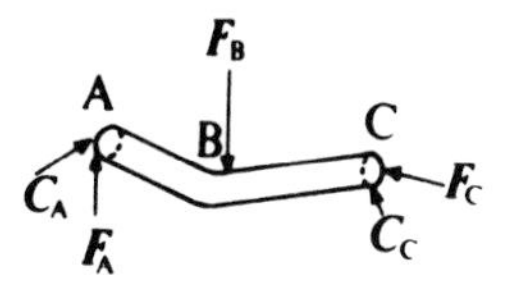

Figure 4.39

The force $\boldsymbol{F}_A$ can be found directly from the equation $\sum \boldsymbol{F} = 0$. Thus

$$\boldsymbol{F}_A + \boldsymbol{F}_B + \boldsymbol{F}_C = 0$$

$$\boldsymbol{F}_A = -(-30\boldsymbol{k}) - (-15\boldsymbol{i} - 15\boldsymbol{j} - 10\boldsymbol{k})$$

$$= (15\boldsymbol{i} + 15\boldsymbol{j} + 40\boldsymbol{k})\ \text{N}$$

The couple $\boldsymbol{C}_A$ can be found by taking moments about any point, but any error that may have been made in the determination of $\boldsymbol{F}_A$ will not be carried forward if the chosen point is A.

The position vectors $\boldsymbol{r}_{B/A}$ and $\boldsymbol{r}_{C/A}$ will be required for the moment equation.

$$\boldsymbol{r}_{B/A} = \overrightarrow{AB} = (\text{AB})\boldsymbol{i} = (0.3\boldsymbol{i})\ \text{m}$$

$$\boldsymbol{r}_{C/A} = \overrightarrow{AB} + \overrightarrow{BC}$$

$$= \boldsymbol{r}_{B/A} + \boldsymbol{r}_{C/B}$$

$$= 0.3\boldsymbol{i} + (0.4\cos 60°\boldsymbol{i} + 0.4\sin 60°\boldsymbol{j})$$

$$= (0.5\boldsymbol{i} + 0.346\boldsymbol{j})\ \text{m}$$

$[\sum \boldsymbol{M}_A = 0]$

$$\boldsymbol{C}_A + (\boldsymbol{r}_{B/A} \times \boldsymbol{F}_B) + (\boldsymbol{r}_{C/A} \times \boldsymbol{F}_C) + \boldsymbol{C}_C = 0$$

$$\boldsymbol{C}_A = -[(\boldsymbol{r}_{B/A} \times \boldsymbol{F}_B) + (\boldsymbol{r}_{C/A} \times \boldsymbol{F}_c) + \boldsymbol{C}_C] \qquad \text{(i)}$$

$$\text{Now } \boldsymbol{r}_{B/A} \times \boldsymbol{F}_b = \begin{vmatrix} \boldsymbol{i} & \boldsymbol{j} & \boldsymbol{k} \\ 0.3 & 0 & 0 \\ 0 & 0 & -30 \end{vmatrix}$$

$$= -\boldsymbol{j}(-9) = (9\boldsymbol{j})\ \text{N m}$$

$$\boldsymbol{r}_{C/A} \times \boldsymbol{F}_C = \begin{vmatrix} \boldsymbol{i} & \boldsymbol{j} & \boldsymbol{k} \\ 0.5 & 0.346 & 0 \\ -15 & -15 & -10 \end{vmatrix}$$

$$= \boldsymbol{i}(-3.46) - \boldsymbol{j}(-5) + \boldsymbol{k}(-7.5 + 5.19)$$

$$= (-3.46\boldsymbol{i} + 5\boldsymbol{j} - 2.31\boldsymbol{k})\ \text{N m}$$

and $\qquad \boldsymbol{C}_C = (-3\boldsymbol{i} + 5\boldsymbol{k})\ \text{N m}$

Substituting in equation (i) we find

$$\boldsymbol{C}_A = (6.46\boldsymbol{i} - 14\boldsymbol{j} - 2.69\boldsymbol{k})\ \text{N m}$$

$\boldsymbol{F}_A$ and $\boldsymbol{C}_A$ are the force and couple acting on ABC. The force and couple acting on the support S are $-\boldsymbol{F}_A$ and $-\boldsymbol{C}_A$.

See section 4.12. For the transverse plane section at A the *direct force*, F_d, is the component of $\boldsymbol{F}_A$ which is parallel to the axis of the rod (Fig. 4.40). The unit vector $\boldsymbol{e}$ for this axis for the present case is $\boldsymbol{i}$, and the component in this direction is

$$F_d = \boldsymbol{F}_A \cdot \boldsymbol{e} = (15\boldsymbol{i} + 15\boldsymbol{j} + 40\boldsymbol{k}) \cdot \boldsymbol{i} = 15\ \text{N}$$

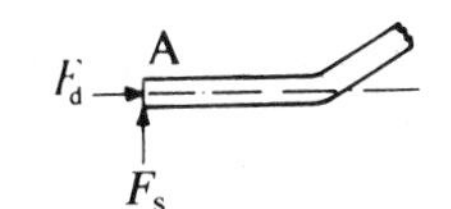

Figure 4.40

The *shear force*, F_s, is the component of $\boldsymbol{F}_A$ which is perpendicular to the rod axis. In general this is most readily found by Pythagoras's theorem, since $F_s = [F_A{}^2 - F_d{}^2]^{1/2}$. Thus

$$F_s = [(15^2 + 15^2 + 40^2) - 15^2]^{1/2} = 42.72\ \text{N}$$

Alternatively we can find F_s from

$$F_a = |\boldsymbol{F}_A \times \boldsymbol{e}| = |(15\boldsymbol{i} + 15\boldsymbol{j} + 40\boldsymbol{k}\boldsymbol{j}) \times \boldsymbol{i}|$$

$$= |-15\boldsymbol{k} + 40\boldsymbol{k}|$$

$$= [15^2 + 40^2]^{1/2}$$

$$= 42.72\ \text{N}$$

The *twisting moment*, C_t, is the component of $\boldsymbol{C}_A$ which is parallel to the rod axis.

$$C_t = \boldsymbol{C}_A \cdot \boldsymbol{e} = (6.46\boldsymbol{i} - 14\boldsymbol{j} - 2.69\boldsymbol{k}) \cdot \boldsymbol{i}$$

$$= 6.46\ \text{N m}$$

Finally the *bending moment*, C_b, is the component of $\boldsymbol{C}_A$ perpendicular to the rod axis.

$$C_b = [C_A{}^2 - C_t{}^2]^{1/2}$$

$$= [6.46^2 + 14^2 + 2.69^2 - 6.46^2]^{1/2}$$

$$= 14.26\ \text{N m}$$

(Alternatively, we can find C_b from $\boldsymbol{C}_b = \boldsymbol{e} \times (\boldsymbol{C}_A \times \boldsymbol{e})$.

Problems

4.1 A system of pinned links and sliders in equilibrium is shown in Fig. 4.41. Link CE is a uniform bar of mass 10 kg. The weights of all the other parts and the effects of friction can be neglected. Equilibrium is maintained by the application of a couple $\boldsymbol{Q}$ to link AB.

Determine, by a semi-graphical method or other-

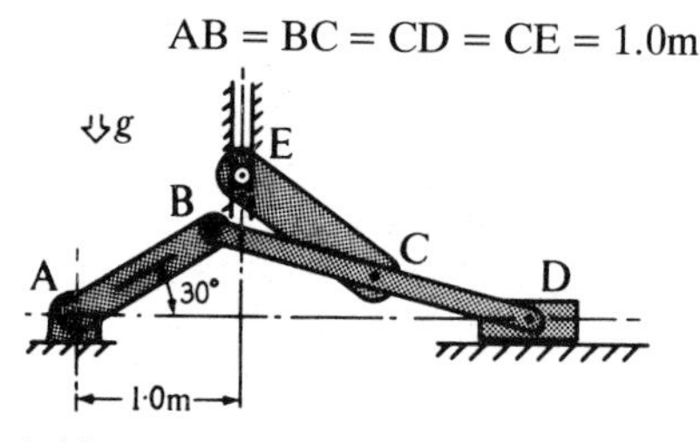

Figure 4.41

wise, the magnitude and sense of the couple $\boldsymbol{Q}$. Take g to be 10 N/kg.

4.2 Figure 4.42 shows a plane pin-jointed structure which is drawn to scale. An anticlockwise couple $\boldsymbol{Q}$ is applied to the cranked link ABC, causing a force of 40 N in link BD.

Determine the magnitude of the forces in pins F and G. State whether EF is in tension or compression and find the magnitude of $\boldsymbol{Q}$.

4.3 The pin-jointed plane structure shown in Fig. 4.43 is constructed from the light rigid links ABC, CDE, EF, DG, BH and AJ. The figure is drawn to scale.

If the magnitude of the force on the pin at C is not to exceed 10 kN, determine by means of a semi-graphical method, or otherwise, the maximum permissible value of the load $\boldsymbol{P}$. For this value of $\boldsymbol{P}$, find the magnitude of the forces acting at A and B.

4.4 The gantry illustrated in Fig. 4.44 lowers a 300 kg load with an acceleration of 2 m/s^2. The masses of the links of the gantry itself and the inertia of the pulley at D may be neglected.

Find, assuming that the joints of the gantry are pinned, (a) the resultant force on the vertical post AE at A and (b) the force in the member AC (state whether compression or tension).

4.5 Figure 4.45 is drawn to scale and shows a brake for a winding gear consisting of a drum D, of diameter 1.2 m, and a flexible belt AB, whose angle of embrace

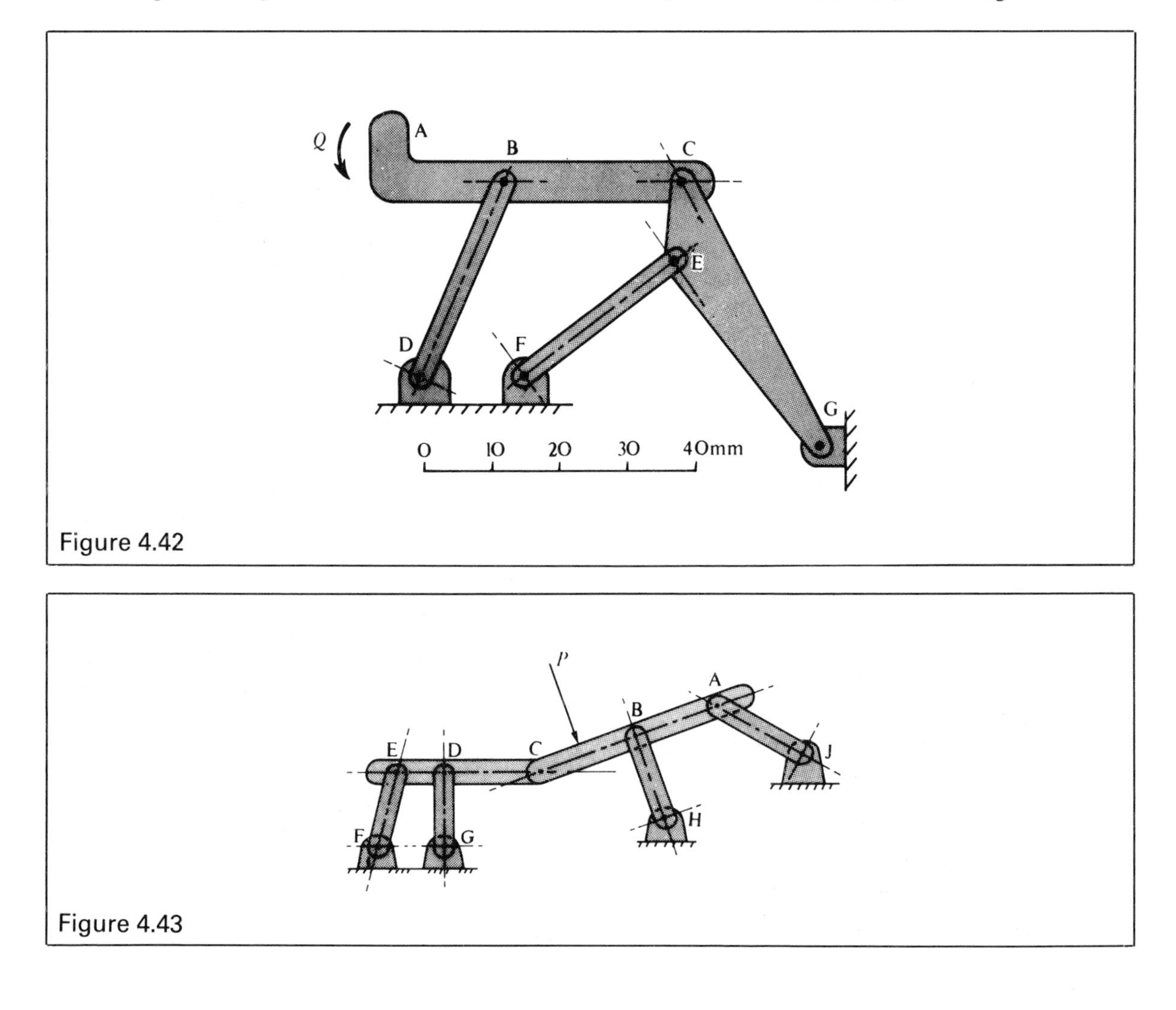

Figure 4.42

Figure 4.43

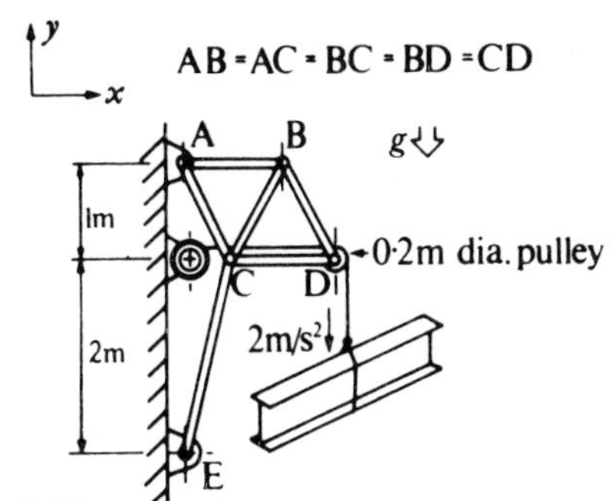

Figure 4.44

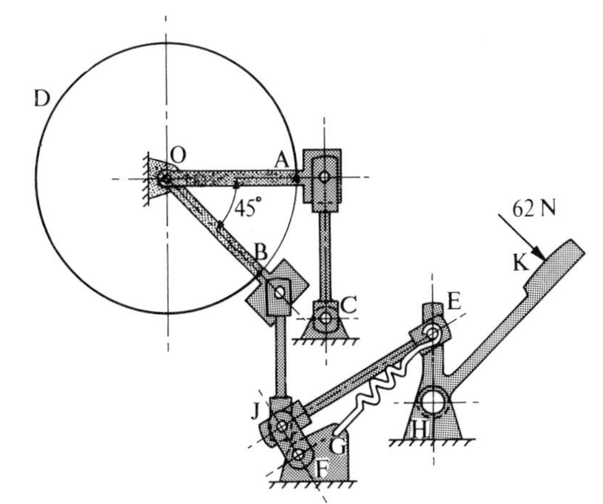

Figure 4.45

is 315°. The coefficient of friction between the belt and pulley is 0.2 (see example 3.3).

The belt is put in tension by applying a force 62 N to the pedal at K in a direction perpendicular to KH. The crank EHK is pivoted to link EJ at E. The links EJ, BJ, and FJ are pivoted to each other at J, and the end A of the belt is anchored to the link AO. The ends of the spring are secured at G and E; in the position shown the spring tension is 50 N.

Find, for the position shown, neglecting gravity, (a) the magnitude and direction of the force on the crank EHK at H, and (b) the braking torque applied to the drum D which is rotating anticlockwise. All the pivots of the linkage may be assumed to be frictionless.

4.6 The mechanism shown in Fig. 4.46 consists of two light links AB and AC. The system is in equilibrium under the action of the vertical load of 240 N at B. Limiting friction in the hinge at B is 5.0 N m and friction elsewhere can be neglected.

Show that, if P is increased to more than 115 N, the mechanism will be set into motion.

4.7 A light rod ABC lies in the xy-plane and is fixed to a support at A. The location of A, B and C are (0, 0, 0) m , (1, 0, 0) m and (1, 1, 0) m respectively and sections AB and BC are straight. Forces of $(-20\boldsymbol{k})$ N and

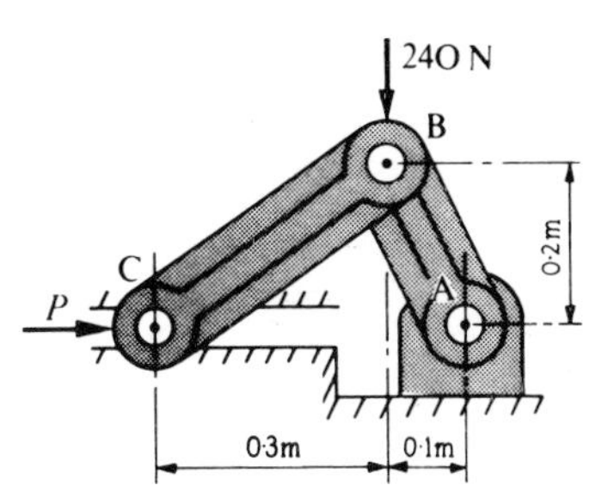

Figure 4.46

$(-10\boldsymbol{k})$ N are applied to B and C respectively. What force and couple are applied to the wall?

4.8 The bent rod ABD in Fig. 4.47 has a length AB in the y-direction and a length BD inclined at 30° to the xy-plane and parallel to the xz-plane. It is fixed to a wall at A and is subjected to forces $\boldsymbol{F}_B$, $\boldsymbol{F}_D$ and couples $\boldsymbol{C}_B$, $\boldsymbol{C}_D$ applied as shown.

$$\boldsymbol{F}_B = -500\boldsymbol{k} \text{ N}$$
$$\boldsymbol{F}_D = 300\boldsymbol{i} + 200\boldsymbol{j} - 300\boldsymbol{k} \text{ N}$$
$$\boldsymbol{C}_B = -70\boldsymbol{i} + 20\boldsymbol{j} \text{ N m}$$
$$\boldsymbol{C}_D = 60\boldsymbol{i} - 10\boldsymbol{j} - 100\boldsymbol{k} \text{ N m}$$

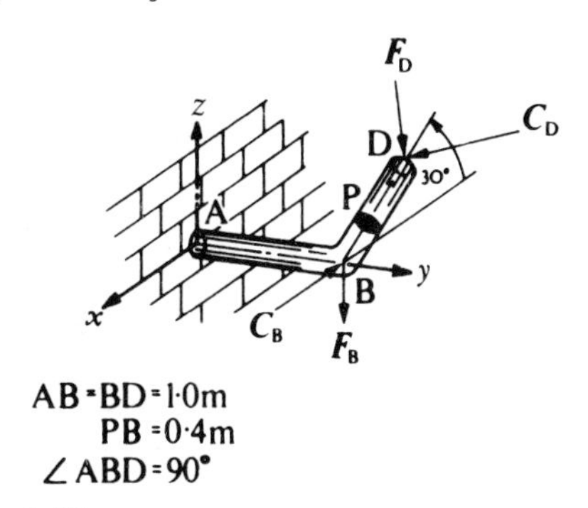

Figure 4.47

Determine the magnitudes of the shear force and of the twisting and bending couples acting in a normal section at P, neglecting the weight of the rod.

4.9 A screen is fastened rigidly to the free end P of a heavy pole clamped to a wall at O in the oblique position shown in Fig. 4.48. The screen has a mass of 50 kg and its mass centre is at G. The pole has a uniform cross-section over its length and a mass per unit length of 5 kg/m.

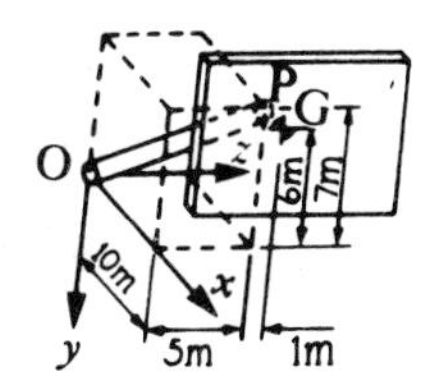

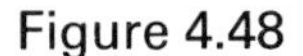

Figure 4.48

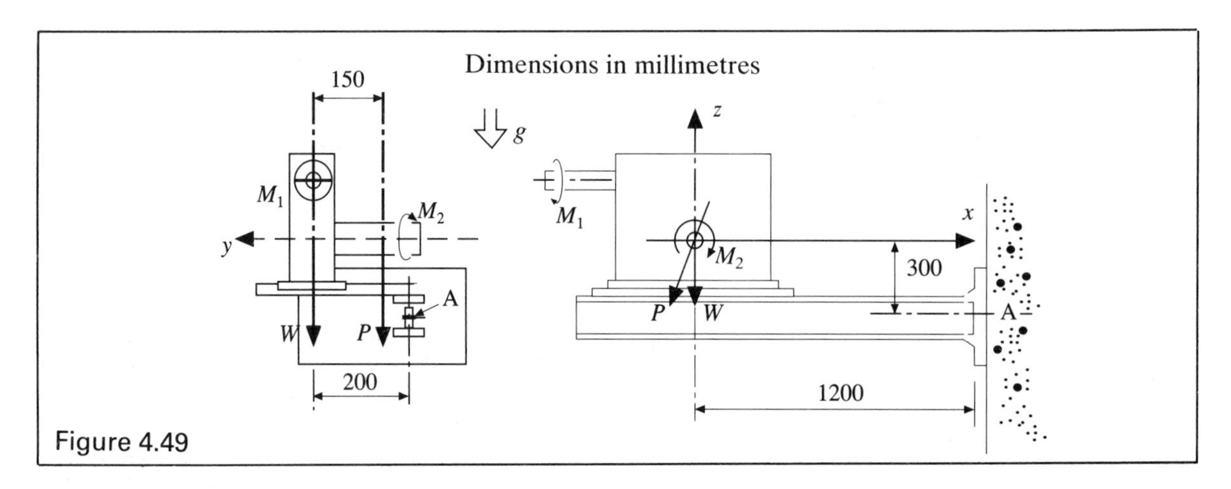

Figure 4.49

Estimate the magnitudes of the twisting couple and of the bending couple acting on a cross-section normal to the axis of the pole at O. Determine the direction of the axis of this bending couple. Assume $g = 10$ N/kg.

4.10 A system of forces can be reduced to a single force $(46\boldsymbol{i}+20\boldsymbol{j}+30\boldsymbol{k})$ N, whose line of action passes through a point O, and a couple $(50\boldsymbol{i}-43\boldsymbol{j}-48\boldsymbol{k})$ N m.
a) Show that the force and couple can be replaced by a single force.
b) Express this force in vector form.
c) Find the co-ordinates, relative to O, of a point on the line of action of the single force.

4.11 Figure 4.49 shows a reduction-gear unit and base which is secured to a cantilever I-beam. The input and output shafts are subject to couples $\boldsymbol{M}_1$ and $\boldsymbol{M}_2$ respectively. The unit is also subject to forces $\boldsymbol{P}$ and $\boldsymbol{W}$ acting as indicated in the figure.

$$\boldsymbol{M}_1 = +\ 60\boldsymbol{i} \text{ N m} \quad \boldsymbol{P} = (-500\boldsymbol{i}-800\boldsymbol{k}) \text{ N}$$
$$\boldsymbol{M}_2 = +250\boldsymbol{j} \text{ N m} \quad \boldsymbol{W} = -700\boldsymbol{k} \text{ N}$$

Calculate the following forces and moments acting on the I-beam at A: (a) the axial force, (b) the shear force, (c) the twisting moment and (d) the bending moment.

4.12 The mass of the motor and pulley assembly shown in Fig. 4.50 is 40 kg, and its mass centre is located at $(600\boldsymbol{j} + 1500\boldsymbol{k})$ mm relative to x-, y-, z-axes of origin O at the base of the pillar as shown. The 100 mm diameter pulley is at $200\boldsymbol{i}$ mm relative to the mass centre, and its plane of rotation is parallel to the yz-plane. The tension T_1 in the horizontal run of the belt is 12 N, and T_2 is 20 N.

Determine (a) the resultant force $\boldsymbol{F}$ at the centre of the pulley and the accompanying couple $\boldsymbol{M}$, due to the belt tensions, (b) the bending moment, twisting moment, axial force and shear force at the base of the pillar, point O, caused by the belt tensions.

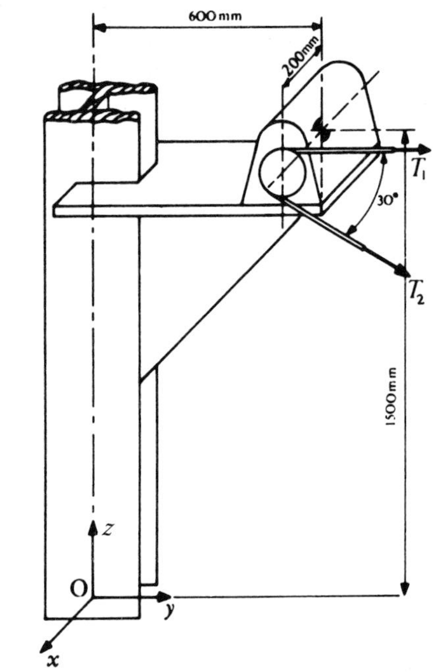

Figure 4.50

4.13 The root fixing of an aircraft wing is shown in Fig. 4.51. It consists of three lugs, A, B and C, each of which can support a force only in its own plane; for example, lug C cannot support a load in the x-direction.

Determine the load carried by each lug due to the given equivalent aerodynamic loading.

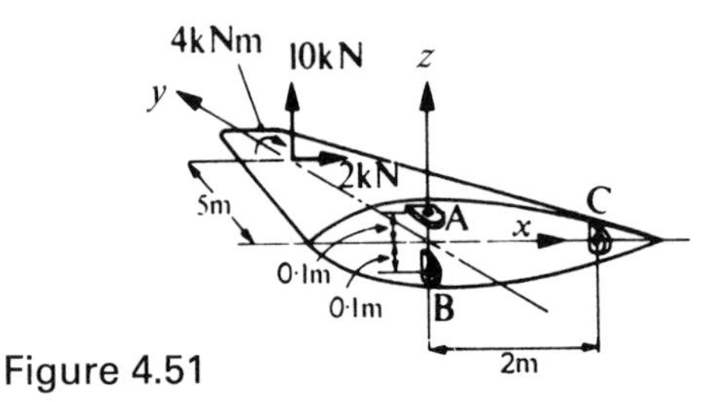

Figure 4.51

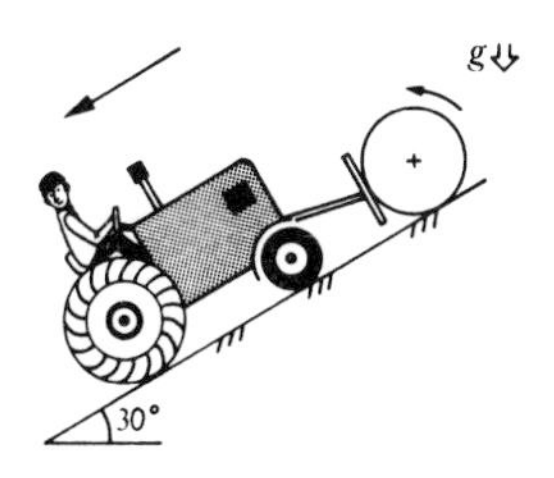

Figure 4.52

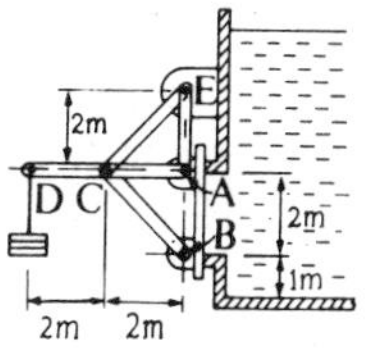

4.14 In Fig. 4.52 the cylinder of mass 1200 kg is rolling down the incline as the tractor reverses at constant speed. Determine the normal force that the blade of the tractor exerts on the cylinder. The coefficient of friction between the cylinder and the blade of the tractor is 0.4 and that between the cylinder and the ground is 0.45. Confirm that slipping does not take place between the cylinder and the ground.

4.15 Figure 4.53 shows a tank with a rectangular opening 2 m deep and 1.5 m wide, sealed by a flat plate AB which is attached to a pin-jointed frame freely hinged at E. The level of water in the tank may be limited by suspending suitable masses from the end D of the rigid link ACD.

Figure 4.53

Calculate the mass required at D to limit the total depth of water in the tank to 5 m and the resultant force in member BC. The mass of the plate and links and friction at the pins may be neglected. The density of water is 1000 kg/m^3.

4.16 A buoy is constructed from a hollow sphere of radius R and a mast which passes through the centre of the sphere. The buoy is weighted so that the waterline is R/4 below the centre of the sphere and the centre of gravity is R/3 below the centre.

Show using equation 4.21 that the metacentre is at the centre of the sphere and therefore the metacentric height is R/3.

5
Kinematics of a rigid body in plane motion

5.1 Introduction

A rigid body is defined as being a system of particles in which the distance between *any* two particles is fixed in magnitude. The number of co-ordinates required to determine the position and orientation of a body in plane motion is three: the system is said to have three degrees of freedom.

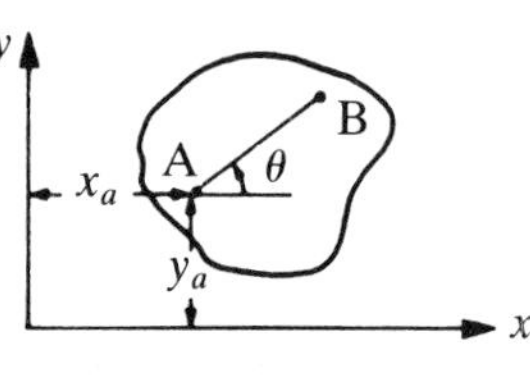

Figure 5.1

One way of describing the position of the body is shown in Fig. 5.1. The position of some representative point such as A and the angle which a line AB makes with the x-axis are three possible co-ordinates.

5.2 Types of motion

The simplest type of motion is that of rectilinear translation, in which θ remains constant and A moves in a straight line (Fig. 5.2). It follows that all particles move in lines parallel to the path of A; thus the velocity and acceleration of all points are identical.

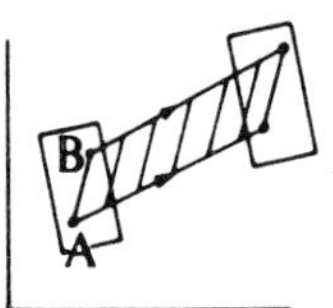

Figure 5.2

This is still true if A is describing a curved path, since if $\theta =$ constant all paths are identical in shape but displaced from each other. This motion is called curvilinear translation (Fig. 5.3).

If the angle θ changes during translation, then this motion is described as general plane motion (Fig. 5.4). In Fig. 5.4 the body has rotated by an angle $(\theta_2 - \theta_1)$. If this change takes place between time t_1 and t_2 then the average angular speed is

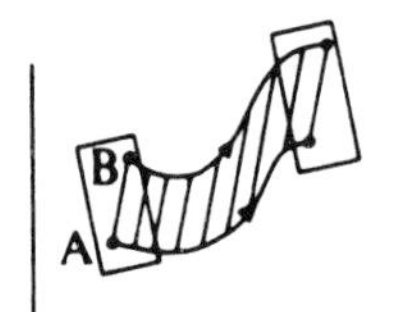

Figure 5.3

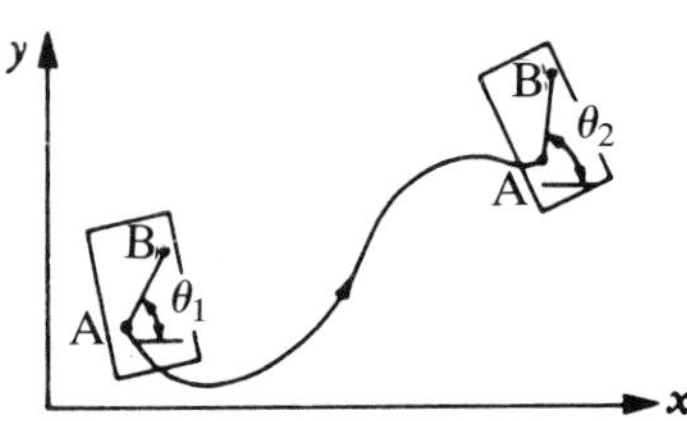

Figure 5.4

$$\frac{\theta_2 - \theta_1}{t_2 - t_1} = \omega_{\text{average}}$$

As $(t_2 - t_1) \to 0$, $(\theta_2 - \theta_1) \to 0$, and the angular speed is defined as

$$\omega = \lim_{\Delta t \to 0} \frac{\Delta\theta}{\Delta t} = \frac{\mathrm{d}\theta}{\mathrm{d}t} \tag{5.1}$$

The angular velocity vector is defined as having a magnitude equal to the angular speed and a direction perpendicular to the plane of rotation, the positive sense being given by the usual right-hand screw rule. In the present case,

$$\boldsymbol{\omega} = \omega \boldsymbol{k} \tag{5.2}$$

It should be noted that infinitessimal rotations and angular velocity are vector quantities, whereas finite angular displacement is *not.*

A very important point to note is that the angular speed is not affected by the translation, therefore we do not have to specify any point in the plane about which rotation is supposed to be taking place.

5.3 Relative motion between two points on a rigid body

The definition of the vector product of two vectors has been already introduced in Chapter 4 in connection with the moment of a vector; the same definition is useful in expressing the relative velocity between two points on a rigid body due to rotation.

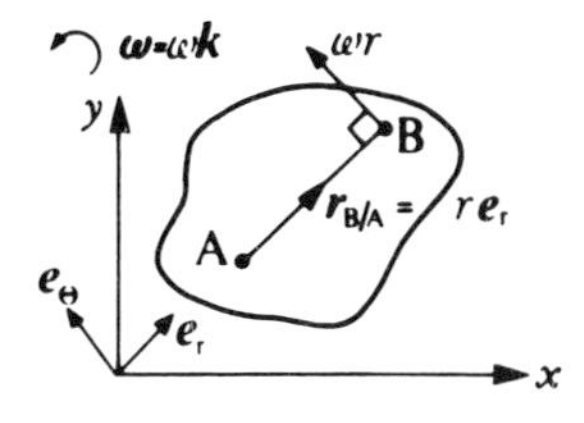

Figure 5.5

Referring to Fig. 5.5, it is seen that $\boldsymbol{v}_{\mathrm{B/A}}$ has a magnitude of $v_{\mathrm{B/A}} = \omega r$ and is in the $\boldsymbol{e}_\theta$ direction. From the definition of the vector product of two vectors given in section 4.3,

$$\boldsymbol{\omega} \times \boldsymbol{r}_{\mathrm{B/A}} = \omega \boldsymbol{k} \times r\boldsymbol{e}_{\mathrm{r}} = \omega r(\boldsymbol{k} \times \boldsymbol{e}_{\mathrm{r}})$$
$$= \omega r \boldsymbol{e}_\theta = \boldsymbol{v}_{\mathrm{B/A}} \quad (5.3)$$

If A and B are not in the same xy-plane, so that

$$\boldsymbol{r}_{\mathrm{B/A}} = r\boldsymbol{e}_{\mathrm{R}} + z\boldsymbol{k}$$

then
$$\boldsymbol{v}_{\mathrm{B/A}} = \boldsymbol{\omega} \times \boldsymbol{r}_{\mathrm{B/A}}$$
$$= \omega \boldsymbol{k} \times (r\boldsymbol{e}_{\mathrm{R}} + z\boldsymbol{k})$$
$$= \omega r \boldsymbol{e}_\theta \quad \text{as in equation 5.3}$$

because $\boldsymbol{k} \times \boldsymbol{k} = 0$.

A complete description of the motion of the body is possible if the motion of point A and the angular motion of AB are specified. We have that

$$\dot{\boldsymbol{r}}_{\mathrm{B/O}} = \dot{\boldsymbol{r}}_{\mathrm{A/O}} + \dot{\boldsymbol{r}}_{\mathrm{B/A}}$$

and, from equation 2.17,

$$\dot{\boldsymbol{r}}_{\mathrm{B/0}} = \dot{\boldsymbol{r}}_{\mathrm{A/O}} + \omega r_{\mathrm{B/A}} \boldsymbol{e}_\theta \quad (5.4)$$

Similarly, from equation 2.18,

$$\ddot{\boldsymbol{r}}_{\mathrm{B/O}} = \ddot{\boldsymbol{r}}_{\mathrm{A/O}} + \ddot{\boldsymbol{r}}_{\mathrm{B/A}}$$
$$= \ddot{\boldsymbol{r}}_{\mathrm{A/O}} + (-\omega^2 r_{\mathrm{B/A}} \boldsymbol{e}_{\mathrm{r}} + r_{\mathrm{B/A}} \dot{\omega} \boldsymbol{e}_\theta) \quad (5.5)$$

A special case of general plane motion is that of rotation about a fixed point. In this type of motion, one point, say A, is permanently at rest, so that

$$\dot{\boldsymbol{r}}_{\mathrm{B/O}} = \dot{\boldsymbol{r}}_{\mathrm{B/A}} = r_{\mathrm{B/A}} \omega \boldsymbol{e}_\theta \quad (5.6)$$

and
$$\ddot{\boldsymbol{r}}_{\mathrm{B/O}} = \ddot{\boldsymbol{r}}_{\mathrm{B/A}} = -\omega^2 r_{\mathrm{B/A}} \boldsymbol{e}_{\mathrm{r}} + \dot{\omega} r_{\mathrm{B/A}} \boldsymbol{e}_\theta \quad (5.7)$$

The applications of the equations of this section to the graphical solution of plane mechanisms are described in the following sections.

5.4 Velocity diagrams

One very simple yet common mechanism is the four-bar chain, shown in Fig. 5.6. It is seen that if the motion of AB is given then the motion of the rest of the mechanism may be determined.

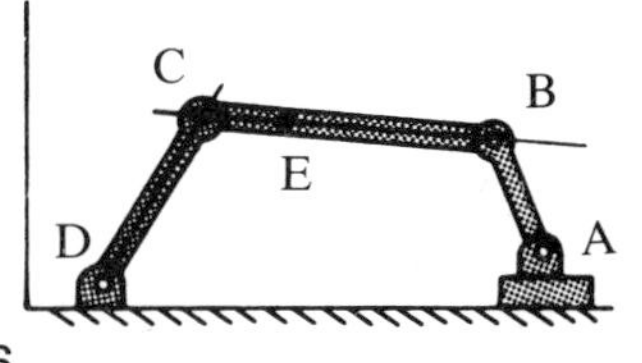

Figure 5.6

This problem can be solved analytically, but the solution is surprisingly lengthy and is best left to a computer to solve if a large number of positions of the mechanism are being examined. However, a simple solution may be found by using vector diagrams; this also has an advantage of giving considerable insight into the behaviour of mechanisms.

For the purpose of drawing velocity and acceleration vector diagrams it is helpful to define a convention for labelling. The convention is best illustrated by considering two particles P and Q in plane motion, Fig. 5.7.

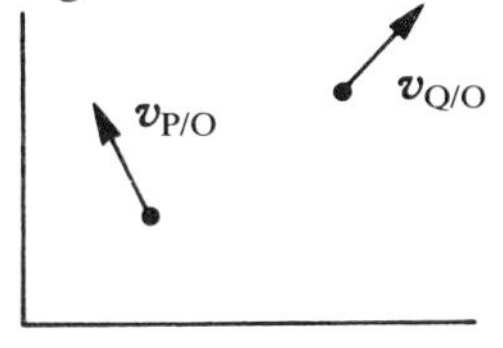

Figure 5.7

The velocity of P relative to O is written $\boldsymbol{v}_{\mathrm{P/O}}$. It is clear that $\boldsymbol{v}_{\mathrm{O/P}} = -\boldsymbol{v}_{\mathrm{P/O}}$, therefore if $\boldsymbol{v}_{\mathrm{P/O}}$ is represented by an arrow thus ↖ then $\boldsymbol{v}_{\mathrm{O/P}}$ is represented by ↘. This information can be concisely given by a single line $^{p}\searrow_{o}$. Similarly the line oq may be drawn. Thus, for Fig. 5.7 we may draw a vector diagram as shown in Fig. 5.8. The velocity of Q relative to P is then $_{p}\nearrow^{q}$ and the velocity of P relative to Q is $_{p}\swarrow^{q}$. This convention will be used throughout.

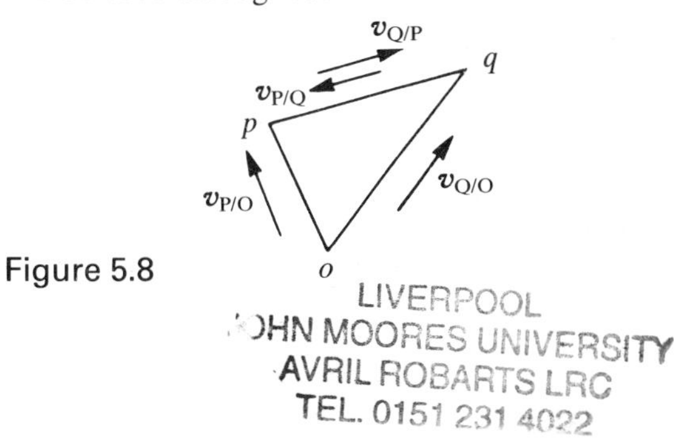

Figure 5.8

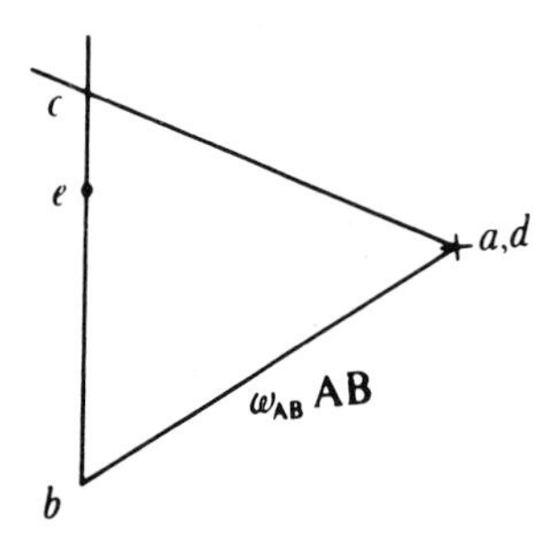

Figure 5.9

Returning to Fig. 5.6 and assuming that ω_{AB} is anticlockwise and of given magnitude, we can place the points *a*, *b* and *d* on the diagram (Fig. 5.9). Note that *a* and *d* are the same point as there is no relative velocity between A and D.

To construct the point *c* we must view the motion of C from two vantage points, namely D and B. Since DC is of fixed length, the only motion of C relative to D is perpendicular to DC; hence we draw *dc* perpendicular to DC. Similarly the velocity of C relative to B is perpendicular to CB; hence we draw *bc* perpendicular to BC. The intersection of these two lines locates *c*.

The angular velocity of CB is obtained from $v_{C/B}/CB$ (clockwise). The direction of rotation is determined by observing the sense of the velocity of C relative to B and remembering that the relative velocity is due only to the rotation of CB.

Note again that angular velocity is measured with respect to a plane and not to any particular point on the plane.

5.5 Instantaneous centre of rotation

Another graphical technique is the use of instantaneous centres of rotation. The axes of rotation of DC and AB are easily seen, but BC is in general plane motion and has no fixed centre of rotation. However, at any instant a point of zero velocity may be found by noting that the line joining the centre to a given point is perpendicular to the velocity of that point.

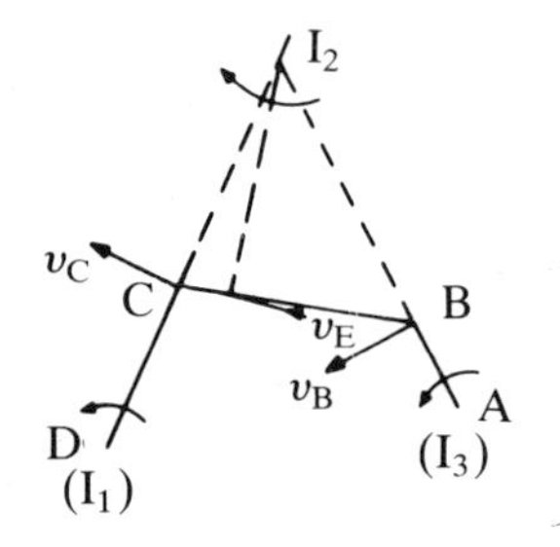

Figure 5.10

In Fig. 5.10 the instantaneous centre for member BC is found to be the intersection of AB and DC, since the velocity of B is perpendicular to AB and the velocity of C is perpendicular to CD.

If the velocity of B is known then

$$\omega_{BC} = \frac{v_B}{I_2B} = \frac{v_C}{I_2C} = \frac{v_E}{I_2E} \tag{5.8}$$

Each point on link CB is, instantaneously, rotating about I_2.

5.6 Velocity image

If the velocity diagram has been constructed for two points on a rigid body in plane motion, then the point on the velocity diagram for a third point on the link is found by constructing a triangle on the vector diagram similar to that on the space diagram. Hence in our previous example a point E situated at, say, one third of the length of BC from C will be represented on the velocity diagram by a point *e* such that $ce/cb = \frac{1}{3}$, as shown in Fig. 5.9.

More generally, see Fig. 5.11, since *ab* is perpendicular to AB, *ac* is perpendicular to AC and *bc* is perpendicular to BC, triangle *abc* is similar to triangle ABC.

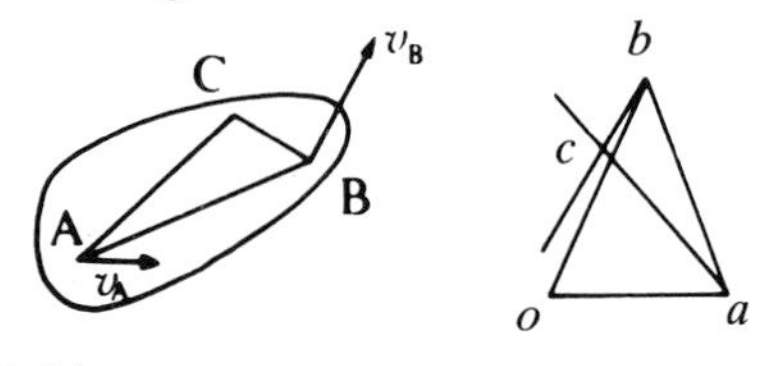

Figure 5.11

Problems with sliding joints

In the mechanism shown in Fig. 5.12, the block or slider B is free to move in a slot in member AO. In order to construct a velocity diagram as shown in Fig. 5.13, we designate a point B′ fixed on the link AO coincident in space with B. The velocity of B relative to C is perpendicular to CB, the velocity of B′ relative to O is perpendicular to OB′ and the velocity of B relative to B′ is parallel to the tangent of the slot at B.

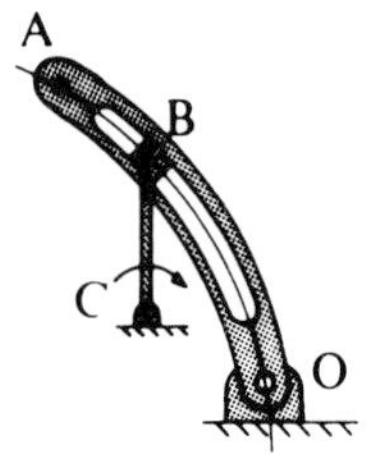

Figure 5.12

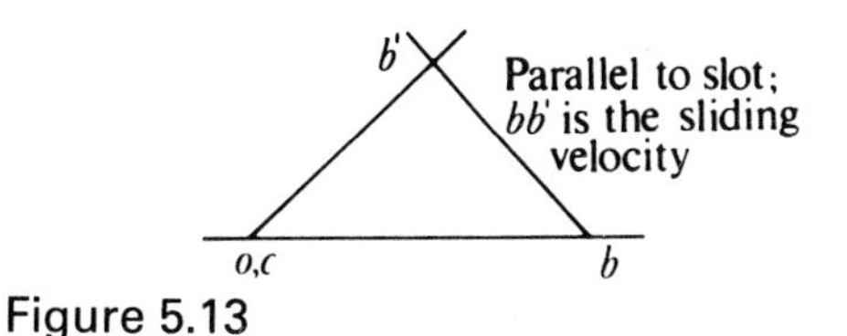

Figure 5.13

The two mechanisms used as examples, namely the four-bar chain and the slidercrank chain, employ just two methods of connection which are known as turning pairs and sliding pairs. It is remarkable how many mechanisms are constructed using just these simple arrangements.

5.7 Acceleration diagrams

Having constructed the velocity diagram, it is now possible to draw the relevant acceleration diagram. The relative acceleration between two points is shown in polar co-ordinates in Fig. 5.14.

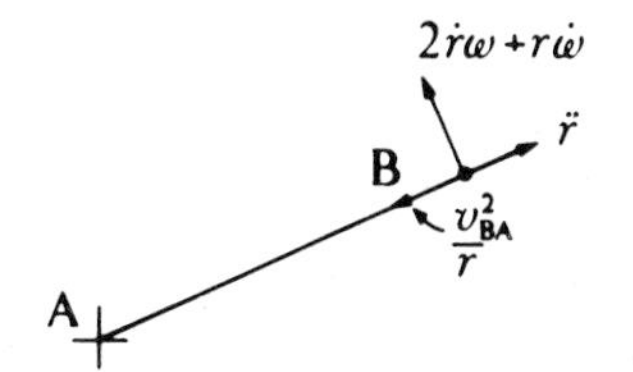

Figure 5.14

If AB is of fixed length, then only two components remain (see Fig. 5.15). One term depends on the angular velocity, which is known from the velocity diagram, and the other term depends on the angular acceleration, which is unknown in magnitude but is in a direction perpendicular to AB.

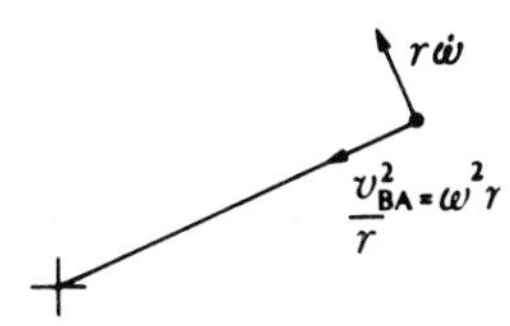

Figure 5.15

Referring to the four-bar chain shown in Fig. 5.6 and given the angular acceleration of link AB, the acceleration vector of B relative to A may be drawn (Fig. 5.16). Note carefully the directions of the accelerations: B is accelerating centripetally towards A.

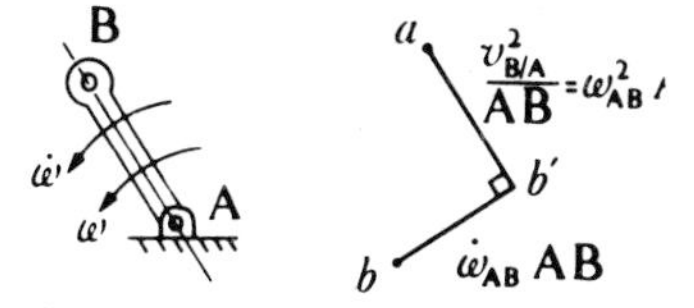

Figure 5.16

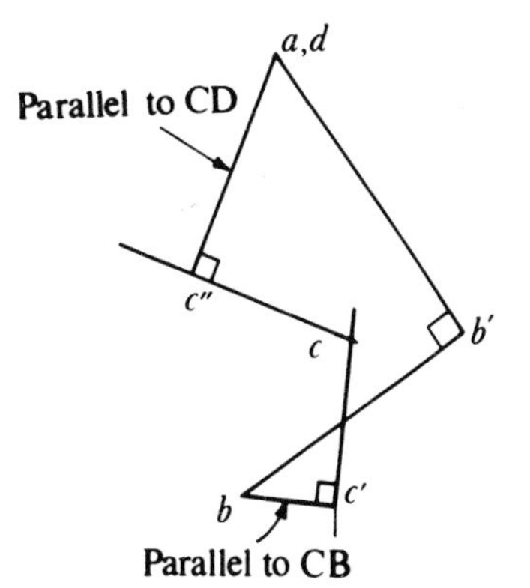

Figure 5.17

The complete acceleration diagram for the mechanism can now be constructed as shown in Fig. 5.17 (see also example 5.1). The acceleration of C is given by the line ac and the angular acceleration of CB is given by cc'/CB (clockwise), since $cc' = \dot{\omega}_{CB}$ CB.

5.8 Acceleration image

In the same way that the velocity of a point on a rigid body may be constructed once the velocities of any two other points are known, the acceleration can be found from the known accelerations of two other points.

$$\boldsymbol{a}_{C/A} = -\omega^2 r_1 \boldsymbol{e}_{r1} + \dot{\omega} r_1 \boldsymbol{e}_{\theta 1}$$
$$\boldsymbol{a}_{C/B} = -\omega^2 r_2 \boldsymbol{e}_{r2} + \dot{\omega} r_2 \boldsymbol{e}_{\theta 2}$$

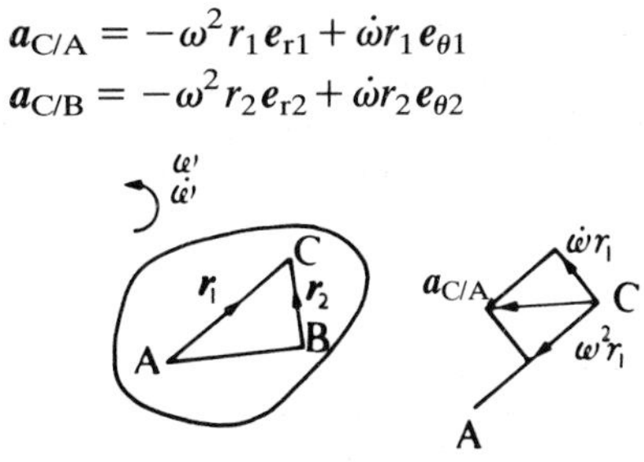

Figure 5.18

From Fig. 5.18, the angle between $\boldsymbol{a}_{C/A}$ and $\boldsymbol{r}_{A/C}$ is

$$\arctan\left(\frac{\dot{\omega} r_1}{\omega^2 r_1}\right) = \arctan\left(\frac{\dot{\omega}}{\omega^2}\right)$$

which is independent of r_1. The angle between $\boldsymbol{a}_{C/A}$ and $\boldsymbol{a}_{B/C}$ is therefore the same as the angle between $\boldsymbol{r}_1$ and $\boldsymbol{r}_2$; hence the triangle abc in the acceleration diagram is similar to triangle ABC.

5.9 Simple spur gears

When two spur gears, shown in Fig. 5.19, mesh together, the velocity ratio between the gears will be a ratio of integers if the axes of rotation are

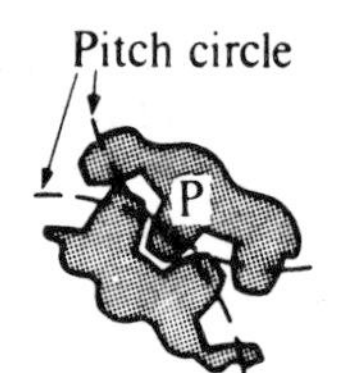

Figure 5.19

fixed. If the two wheels are to mesh then they must have the same circular pitch, that is the distance between successive teeth measured along the pitch circle must be the same for both wheels.

If T is the number of teeth on a wheel then the circular pitch p_C is $\pi D/T$, where D is the diameter of the pitch circle. The term 'diametral pitch' is still used and this is defined as $P = T/D$. Another quantity used is the module, $m = D/T$.

The number of teeth passing the pitch point in unit time is $2\pi\omega T$, so for two wheels A and B in mesh

$$|\omega_A T_A| = |\omega_B T_B| \tag{5.9}$$

or $$\frac{\omega_A}{\omega_B} = -\frac{D_B}{D_A} = -\frac{T_B}{T_A}$$

the minus sign indicating that the direction of rotation is reversed.

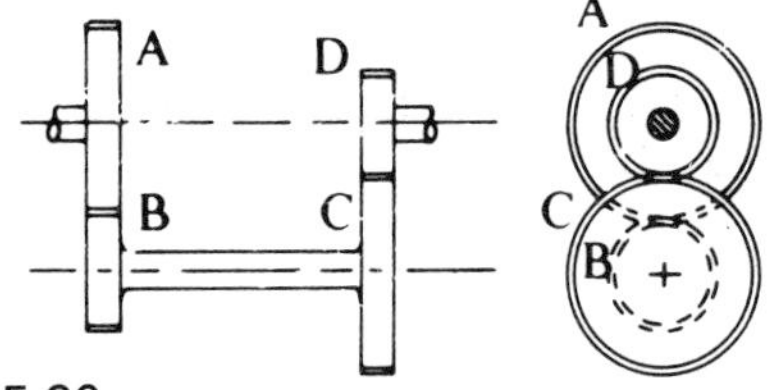

Figure 5.20

Figure 5.20 shows a compound gear train in which wheel B is rigidly connected to wheel C; thus $\omega_B = \omega_C$. The velocity ratio for the gear is

$$\frac{\omega_D}{\omega_A} = \frac{\omega_D}{\omega_C} \cdot \frac{\omega_B}{\omega_A}$$

$$= \left(-\frac{T_C}{T_D}\right)\left(-\frac{T_A}{T_B}\right) = \frac{T_C T_A}{T_D T_B} \tag{5.10}$$

5.10 Epicyclic motion

If the axle of a wheel is itself moving on a circular path, then the motion is said to be epicyclic.

Figure 5.21 shows the simplest type of epicyclic motion. If no slip occurs at P, the contact point, then the velocity of P is given as

$$v_{P/O1} = v_{O2/O1} + v_{P/O2}$$

hence $$\omega_A r_A = \omega_C(r_A + r_B) - \omega_B r_B$$

or $$\frac{\omega_A - \omega_C}{\omega_B - \omega_C} = -\frac{r_B}{r_A} \tag{5.11}$$

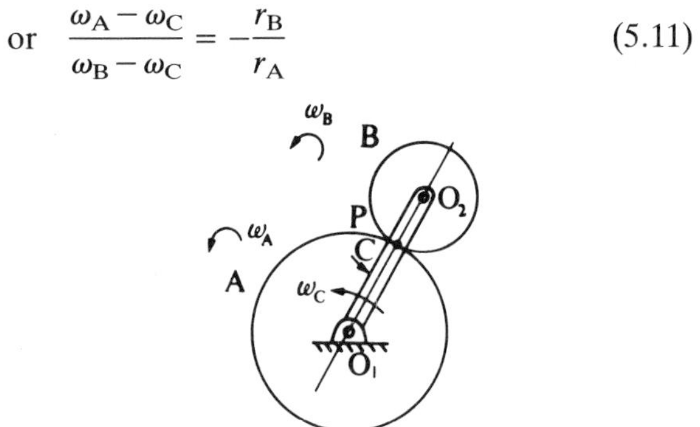

Figure 5.21

that is the motion relative to the arm or carrier is independent of the speed of the arm. For example, if $\omega_C = 0$ we have the case of a simple gear train where

$$\frac{\omega_A}{\omega_B} = -\frac{r_B}{r_A} \tag{5.12}$$

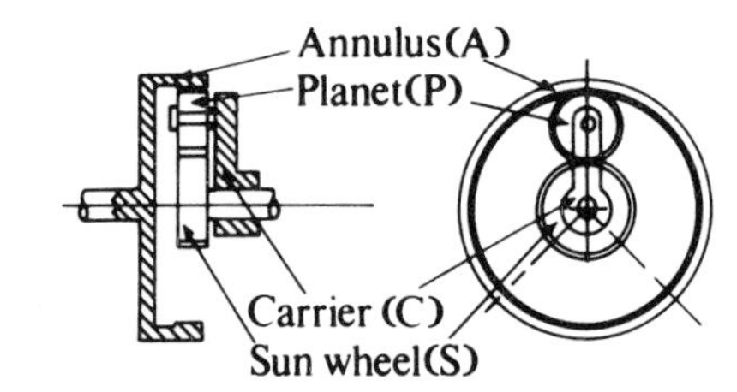

Figure 5.22

Figure 5.22 shows a typical arrangement for an epicyclic gear in which the planet is free to rotate on a bearing on the carrier, which is itself free to rotate about the central axis of the gear. If the carrier is fixed, the gear is a simple gear train so that the velocity ratio

$$\frac{\omega_A}{\omega_S} = \frac{\omega_A}{\omega_P}\frac{\omega_P}{\omega_S} = \left(\frac{T_P}{T_A}\right)\left(-\frac{T_S}{T_P}\right) = -\frac{T_S}{T_A}$$

Note that the direction of rotation of the annulus is the same as that of the planet, since the annulus is an internal gear. Also, we see that the number of teeth on the planet wheel does not affect the velocity ratio – in this case the planet is said to act an an idler.

If the carrier is not fixed, then the above velocity ratio is still valid provided the angular speeds are relative to the carrier; thus

$$\frac{\omega_A - \omega_C}{\omega_S - \omega_C} = -\frac{T_S}{T_A} \tag{5.13}$$

If two of the speeds are known then the third may be calculated. In practice it is common to fix one of the elements (i.e. sun, carrier or annulus) and use the other two elements as input and output. Thus we see that it is possible to obtain three different gear ratios from the same mechanism.

5.11 Compound epicyclic gears

In order to obtain a compact arrangement, and also to enable a gearbox to have a wider choice of selectable gear ratios, two epicyclic gears are often coupled together. The ways in which this coupling can occur are numerous so only two arrangements will be discussed. The two chosen are common in the automotive industry and between them form the basis of the majority of automatic gearboxes.

Simpson gear train

In the arrangement shown in Fig. 5.23(a), the two sun wheels are on a common shaft and the carrier of the first epicyclic drives the annulus of the second. This second annulus is the output whilst the input is either the sun wheel or the annulus of the first epicyclic.

This design, used in a General Motors 3-speed automatic transmission, provides three forward gears and a reverse gear. These are achieved as follows.

First gear employs the first annulus as input and locks the carrier of the second. Second gear again uses the first annulus as input but fixes the sun wheel shaft. Third is obtained by locking the first annulus and the sun wheel together so that the whole assembly rotates as a solid unit. Reverse gear again locks the second carrier, as for the first gear, but in this case the drive is via the sun wheel.

Figure 5.23(b) shows a practical layout with three clutches and one band brake which carry out the tasks of switching the drive shafts and locking the second carrier or the sun wheel shaft.

To engage first gear drive is applied to the forward clutch and the second carrier is fixed. In normal drive mode this is achieved by means of the one-way Sprag clutch. This prevents the carrier from rotating in the negative sense, relative to the drive shaft, but allows it to free-wheel in the positive sense. This means that no engine braking is provided during over-run. To provide engine braking the reverse/low clutch is engaged in the lock-down mode. For second gear the reverse/low clutch (if applied) is released and the intermediate band brake is applied, thus locking the sun wheel. For third gear the intermediate band is released and the direct clutch activated hence locking the whole gear to rotate in unison. For reverse gear the forward clutch is released, then the direct clutch and the reverse/low clutch are both engaged thus only the second epicyclic gear is in use.

The operation of the various clutches and band brakes is conventionally achieved by a hydraulic circuit which senses throttle position and road speed. The system is designed to change down at a lower speed than it changes up at a given throttle position to prevent hunting. Electronic control is now used to give more flexibility in changing parameters to optimise for economy or for performance.

To determine the gear ratios two equations of the same type as equation 5.13 are required and they are solved by applying the constraints dictated by the gear selected. A more convenient set of symbols will be used to represent rotational speed. We shall use the letter A to refer to the annulus, C for the carrier and S for the sun, also we shall use 1 to refer to the first simple epicyclic gear and 2 for the second. In this notation, for example, the speed of the second carrier will be referred to as C_2.

For the first epicyclic gear

$$\frac{S_1 - C_1}{A_1 - C_1} = -\frac{T_{A1}}{T_{S1}} = -R_1 \tag{5.14}$$

and for the second epicyclic gear

$$\frac{S_2 - C_2}{A_2 - C_2} = -\frac{T_{A2}}{T_{S2}} = -R_2 \tag{5.15}$$

Where R is the ratio of teeth on the annulus to teeth on the sun. In all cases $S_2 = S_1$ and $C_1 = A_2 = \omega_0$, the output.

With the first gear selected $C_2 = 0$ and $A_1 = \omega_i$, the input.

From equation 5.14 $\quad S_1 = -\omega_i \times R_1 + \omega_0(1 + R_1)$

and from equation 5.15 $\quad S_1 = -\omega_0 \times R_2$

Eliminating S_1 $\quad \omega_i = \dfrac{\omega_0(1 + R_1 + R_2)}{R_1}$

thus the first gear ratio $= \omega_i/\omega_0 = (1 + R_1 + R_2)/R_1$

With second gear selected $S_1 = 0$ and ω_i is still A_1.

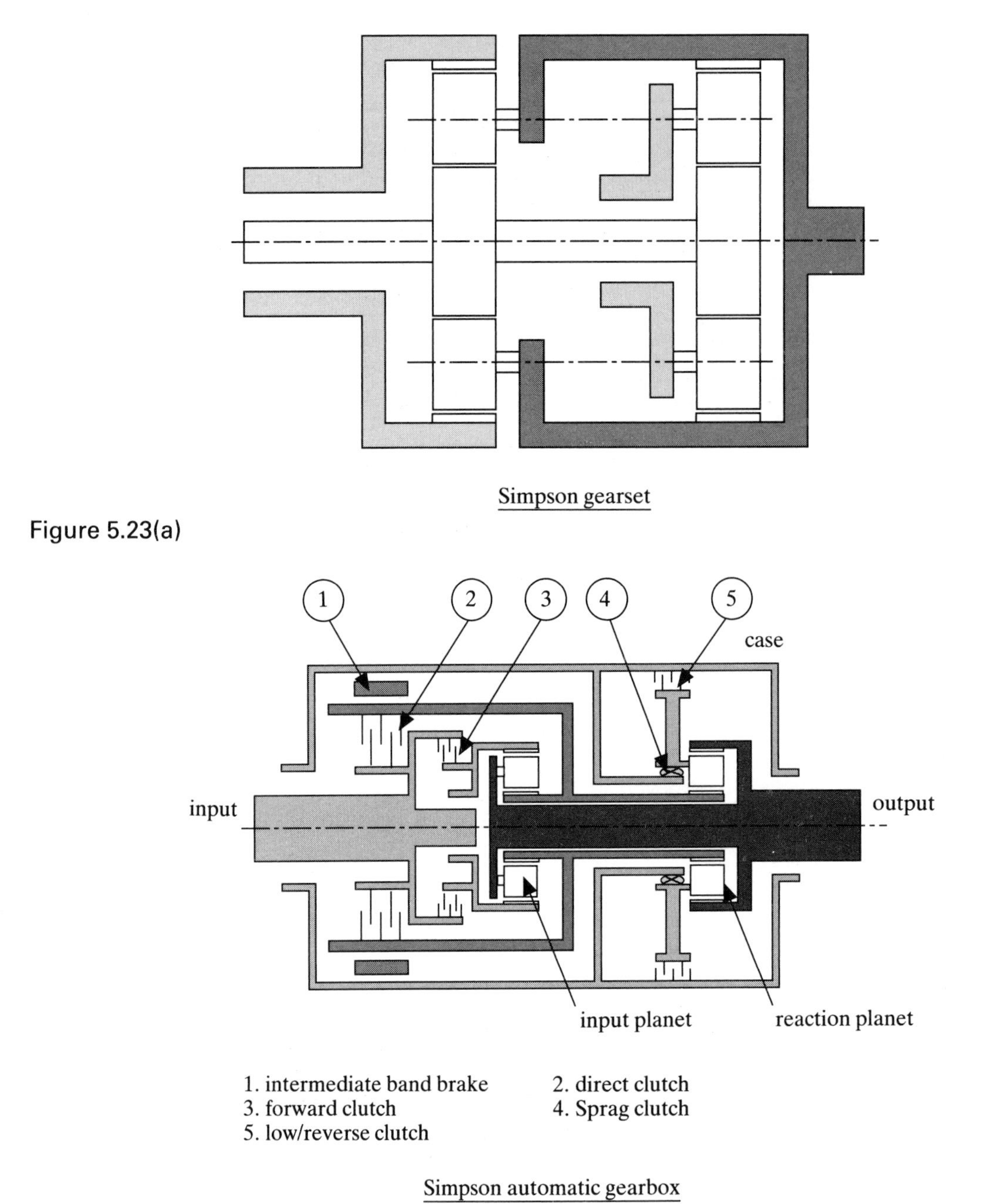

Figure 5.23(a)

Figure 5.23(b)

From equation 5.14 $0 = \omega_0(1+R_1) - \omega_i \times R_1$

thus the second gear ratio $\omega_i/\omega_0 = (1+R_1)/R_1$

The third gear is, of course, unity.

For the reverse gear $C_2 = 0$ and $\omega_i = S_1$ so from equation 5.17

$$\omega_i/\omega_0 = -R_2$$

Summarising we have

GEAR	GEAR RATIO
1st	$(1+R_1+R_2)/R_1$
2nd	$(1+R_1)/R_1$
3rd	1
Reverse	$-R_2$

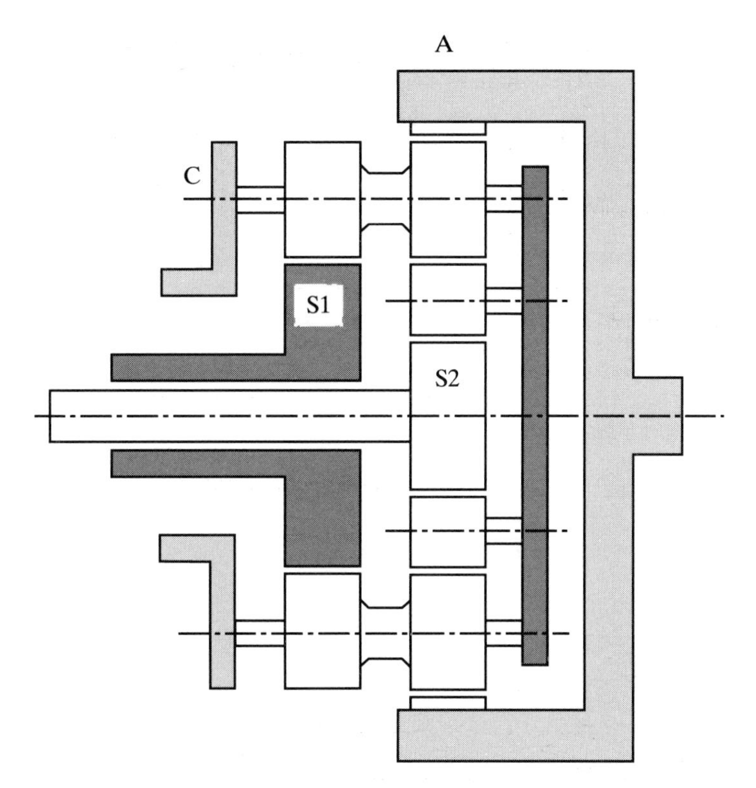

Figure 5.23(c)

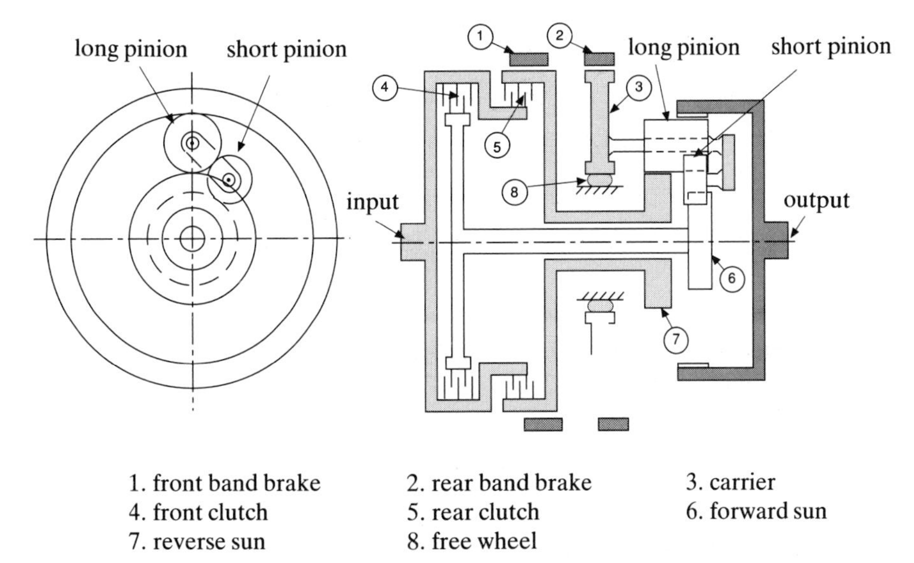

Figure 5.23(d)

Ravigneaux gearbox
The general arrangement of the Ravigneaux gear is shown in Fig. 5.23(c). This gear is used in the Borg Warner automatic transmission which is to be found in many Ford vehicles.

In this design there is a common planet carrier

and the annulus is rigidly connected to the output shaft. The second epicyclic has two planets to effect a change in the direction of rotation compared with a normal set. In the actual design, shown in Fig. 5.23(d), the first planet wheel doubles as the idler for the second epicyclic gear.

When first gear is selected, the front clutch provides the drive to the forward sun wheel and the common carrier is locked, either by the rear band brake in lock-down mode or by the free-wheel in normal drive. For second gear the drive is still to the forward sun wheel but the reverse sun wheel is fixed by means of the front band brake. For top gear drive both suns are driven by the drive shaft thereby causing the whole gear train to rotate as a unit. For the reverse gear the rear clutch applies the drive to the reverse sun wheel and the carrier is locked by the rear band brake.

For the first gear the input $\omega_i = S_2$ and $C_1 = C_2 = 0$, the output $\omega_0 = A_1 = A_2$. So, from equation 5.15,

$$S_2 = R_2 \times A$$

therefore $\omega_i/\omega_0 = S_2/A = R_2$

For second gear S_2 is the input but $S_1 = 0$

From equation 5.14 $0 = -A \times R_1 + (1 + R_1)C$ and from equation 5.15 $S_2 = R_2 \times A + C(1 - R_2)$

Elimination of C gives

$$S_2 = R_2 + A \times R_1 \times (1 - R_2)/(1 + R_1)$$

thus $\omega_i/\omega_0 = S_2/A = \dfrac{R_1 + R_2}{1 + R_1}$

The top gear ratio is again unity.

Reverse has $C = 0$ with input S_1 so from equation 5.14

$$S_1 = -R_1 \times A$$

giving the gear ratio

$$\omega_i/\omega_0 = S_1/A = -R_1.$$

Summarising we have

GEAR	GEAR RATIO
1st	R_2
2nd	$(R_1 + R_2)/(1 + R_1)$
3rd	1
Reverse	$-R_1$

Discussion examples

Example 5.1

The four-bar chain mechanism will now be analysed in greater detail. We shall consider the mechanism in the configuration shown in Fig. 5.24 and determine $\boldsymbol{v}_C$, $\boldsymbol{v}_E$, ω_2, ω_3, $\boldsymbol{a}_B$, $\boldsymbol{a}_C$, $\boldsymbol{a}_E$, $\dot{\omega}_2$ and $\dot{\omega}_3$, and the suffices 1, 2, 3 and 4 will refer throughout to links AB, BC, CD and DA respectively.

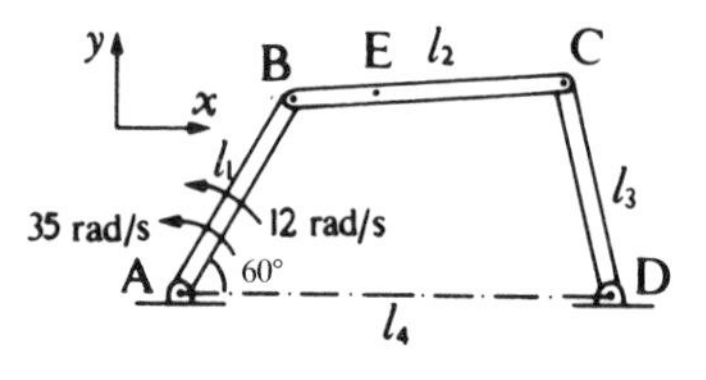

AB = l_1 = 150mm
BC = l_2 = 190mm
CD = l_3 = 150mm
DA = l_4 = 300mm
BE = 50mm

Figure 5.24

Velocities

In general, for any link PQ of length R and rotating with angular velocity ω (see Fig. 5.25(a)) we have, from equation 2.17,

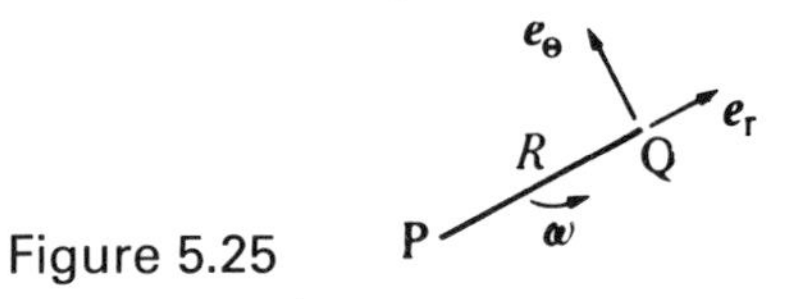

Figure 5.25

$$\boldsymbol{v}_{Q/P} = \dot{R}\boldsymbol{e}_r + R\omega\boldsymbol{e}_\theta$$

If PQ is of fixed length then $\dot{R} = 0$ and $\boldsymbol{v}_{Q/P}$ has a magnitude $R\omega$ and a direction perpendicular to the link and in a sense according the the direction of ω.

Velocity diagram (section 5.4). Since l_1 is constant, the magnitude of $\boldsymbol{v}_{B/A}$ is $\omega_1 l_1$ and its direction is perpendicular to AB in the sense indicated in Fig. 5.25(b), so we can draw to a suitable scale the vector $\overrightarrow{ab}$ which represents $\boldsymbol{v}_{B/A}$. The velocity of C is determined by considering the known directions of $\boldsymbol{v}_{C/B}$ and $\boldsymbol{v}_{C/D}$

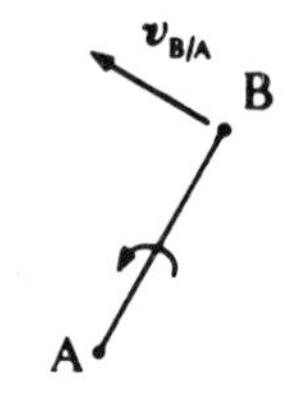

Link	Velocity	Direction	Sense	Magnitude (m/s)	Line
AB	$v_{B/A}$	⊥AB	↖	$(AB)\,\omega_1 = (0.15)12 = 1.8$	*ab*
BC	$v_{C/B}$	⊥BC	?	$(BC)\,\omega_2 = ?$	*bc*
CD	$v_{C/D}$	⊥CD	?	$(CD)\,\omega_3 = ?$	*cd*

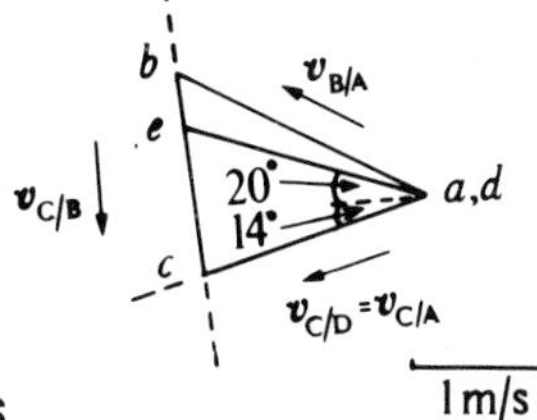

Figure 5.26

and by noting that (see equation 2.24)

$$v_{C/A} = v_{B/A} + v_{C/B} \qquad \text{(i)}$$

and $v_{C/A} = v_{C/D}$ since A and D each have zero velocity. There are sufficient data to draw the velocity triangle representing equation (i) (Fig. 5.26).

From this figure it can be seen that the location of point *c* on the velocity diagram is the intersection of a line drawn through *b* perpendicular to BC and a line drawn through *a*, *d* perpendicular to DC. By scaling we find that the magnitude of *dc* is 1.50 m/s and thus

$$v_{C/A} = v_C = 1.50 \text{ m/s} \qquad 14°$$

The magnitude of ω_2 is

$$\omega_2 = \frac{bc}{BC} = \frac{1.28}{0.19} = 6.7 \text{ rad/s}$$

To determine the direction, we note that $v_{C/B}$, the velocity of C relative to B is the sense from *b* to *c* (and that $v_{B/C}$ is in the opposite sense) so that BC is rotating clockwise (see Fig. 5.27). Thus

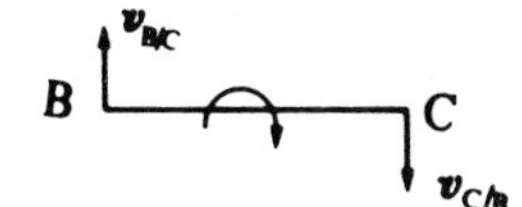

Figure 5.27

$$\omega_2 = -6.7\,\boldsymbol{k} \text{ rad/s}$$

where $\boldsymbol{k}$ is the unit vector coming out of the page.

The magnitude of $\boldsymbol{\omega}_3$ is

$$\boldsymbol{\omega}_3 = \frac{cd}{CD} = \frac{1.5}{0.15} = 10 \text{ rad/s}$$

and the direction is clearly anticlockwise, so that

$$\boldsymbol{\omega}_3 = 10\,\boldsymbol{k} \text{ rad/s}$$

From the concept of the velocity image we can find the position of *e* on *bc* from

$$\frac{be}{bc} = \frac{BE}{BC}$$

Thus

$$be = 1.28\left(\frac{50}{190}\right) = 0.337 \text{ m/s}$$

The magnitude of v_E is *ae* and this is found from the diagram to be 1.63 m/s. Thus

$$v_E = 1.63 \text{ m/s} \qquad 20°$$

Instantaneous centre (section 5.5). In Fig. 5.28, I, the instantaneous centre of rotation of BC, is at the intersection of AB and CD. The triangle IBC rotates instantaneously about I. From the known direction of v_B, the angular velocity of the triangle is clearly seen to be clockwise.

The magnitude of $\boldsymbol{\omega}_2$ is

$$\boldsymbol{\omega}_2 = \frac{v_B}{IB} = \frac{\omega_1(AB)}{IB} = \frac{12(0.15)}{0.27} = 6.7 \text{ rad/s}$$

and $\quad \boldsymbol{\omega}_2 = -6.7\,\boldsymbol{k}$ rad/s

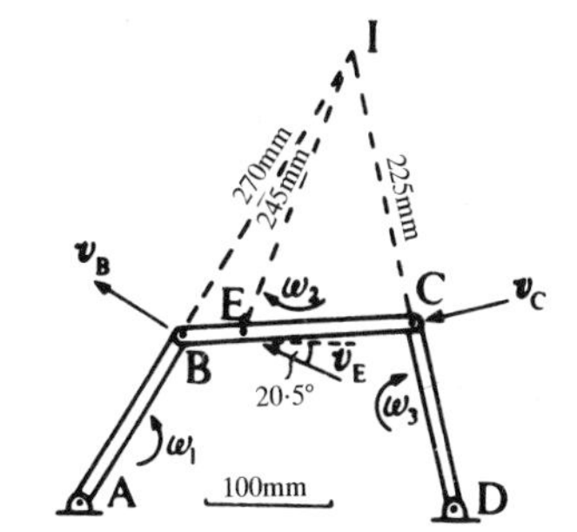

Figure 5.28

The magnitude of v_C is

$$v_C = \omega_2(IC) = 6.7(0.225) = 1.50 \text{ m/s}$$

and the sense is in the direction shown.

The magnitude of $\boldsymbol{\omega}_3$ is

$$\omega_3 = \frac{v_C}{CD} = \frac{1.47}{0.15} = 9.8 \text{ rad/s}$$

and the sense is clearly anticlockwise so that

$$\omega_3 = 9.8\boldsymbol{k} \text{ rad/s}$$

Point E lies on link BC so that the instantaneous centre for E is also I. The magnitude of $\boldsymbol{v}_E$ is

$$\boldsymbol{v}_E = \omega_2(IE) = 6.7(0.245) = 1.64 \text{ m/s}$$

and the sense is in the direction shown.

The discrepancies between the two methods are obviously due to inaccuracies in drawing.

Accelerations

For any link PQ of length R, angular velocity ω and angular acceleration $\dot{\omega}$ (see Fig. 5.29) we have, from equation 2.18,

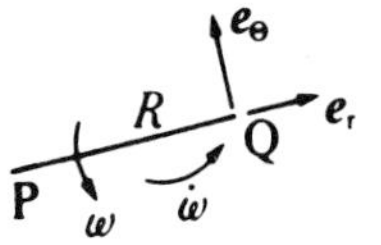

Figure 5.29

$$\boldsymbol{a}_{Q/P} = (\ddot{R} - R\omega^2)\boldsymbol{e}_r + (R\dot{\omega} + 2\dot{R}\omega)\boldsymbol{e}_\theta$$

If PQ is of fixed length then $\dot{R} = \ddot{R} = 0$ and $\boldsymbol{a}_{Q/P}$ has one component of magnitude $R\omega^2$ always in the sense of Q to P and another of magnitude $R\dot{\omega}$, perpendicular to PQ and directed according to the sense of $\dot{\boldsymbol{\omega}}$.

Acceleration diagram (section 5.7). See Fig. 5.30. The radial and normal components of $\boldsymbol{a}_{B/A}$ are both known, and summing these gives the total acceleration $\boldsymbol{a}_B$ since A is a fixed point ($\overrightarrow{ab'} + \overrightarrow{b'b} = \overrightarrow{ab}$ in the diagram). The radial component of $\boldsymbol{a}_{C/B}$ has a magnitude of $l_2\omega_2^2$ and is directed from C to B. The normal component of $\boldsymbol{a}_{C/B}$ is perpendicular to BC but is as yet unknown in magnitude or sense. Similar reasoning applies to $\boldsymbol{a}_{C/D}$. However we have enough data to locate point c on the acceleration diagram shown in Fig. 5.30.

The magnitudes and directions of $\boldsymbol{a}_B$ and $\boldsymbol{a}_C$ are taken directly from the diagram.

$$\boldsymbol{a}_B = \boldsymbol{a}_{B/A} = \overrightarrow{ab} = 22.0 \text{ m/s}^2 \qquad 46°$$

$$\boldsymbol{a}_C = \boldsymbol{a}_{C/D} = \overrightarrow{dc} = 31.6 \text{ m/s} \qquad 43°$$

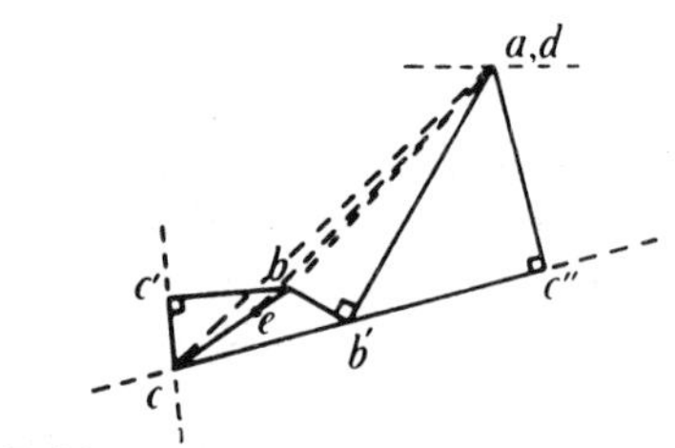

Figure 5.30

The magnitude of $\dot{\omega}_2$ is

$$\dot{\omega}_2 = \frac{c'c}{BC} = \frac{4.7}{0.19} = 24.7 \text{ rad/s}^2$$

To determine the sense of $\dot{\omega}_2$ we note that the normal component of $\boldsymbol{a}_{C/B}$ is $c'c$ in the sense of c' to c; thus BC has a clockwise angular acceleration.

$$\dot{\omega}_2 = -24.7\boldsymbol{k} \text{ rad/s}^2$$

Similarly we find that the magnitude of $\dot{\omega}_3$ is

$$\dot{\omega}_3 = \frac{c''c}{CD} = \frac{28}{0.15} = 187 \text{ rad/s}^2$$

and the sense is anticlockwise,

$$\dot{\omega}_3 = 187\boldsymbol{k} \text{ rad/s}^2$$

From the concept of the acceleration image we can find the position of e on bc from

$$\frac{be}{bc} = \frac{BE}{BC}$$

Thus

$$be = 0.99\left(\frac{50}{190}\right) = 0.260 \text{ m/s}^2$$

The magnitude and direction of $\boldsymbol{a}_E$ are taken from the diagram and we find

$$\boldsymbol{a}_E = \boldsymbol{a}_{E/A} = \overrightarrow{ae} = 24.2 \text{ m/s}^2 \qquad 45°$$

Link	*Acceleration*	*Direction*	*Sense*	*Magnitude*	*Line*
AB	$\boldsymbol{a}_{B/A}$ (radial)	$\parallel$AB	↙	$l_1\omega_1^2 = 0.15(12)^2 = 21.6$	ab'
	$\boldsymbol{a}_{B/A}$ (normal)	$\perp$AB	↖	$l_1\dot{\omega}_1 = 0.15(35) = 5.25$	$b'b$
BC	$\boldsymbol{a}_{C/B}$ (radial)	$\parallel$BC	↙	$l_2\omega_2^2 = 0.19(6.7)^2 = 8.53$	bc'
	$\boldsymbol{a}_{C/B}$ (normal)	$\perp$BC	?	$l_2\dot{\omega}_2 = ?$	$c'c$
CD	$\boldsymbol{a}_{C/D}$ (radial)	$\parallel$CD	↘	$l_3\omega_3^2 = 0.15(10)^2 = 15.0$	dc''
	$\boldsymbol{a}_{C/D}$ (normal)	$\perp$CD	?	$l_3\dot{\omega}_3 = ?$	$c''c$

Vector-algebra methods

Vector algebra can be used in the solution of mechanism problems. Such methods are a powerful tool in the solution of three-dimensional mechanism problems but usually take much longer than graphical methods for problems of plane mechanisms. They do, however, give a systematic approach which is amenable to computer programming.

An outline of a vector-algebra solution to the present problem is given below. Students who are following a course leading to the analysis of three-dimensional mechanisms should find this a useful introduction and are encouraged to try these techniques on a few simple plane mechanisms.

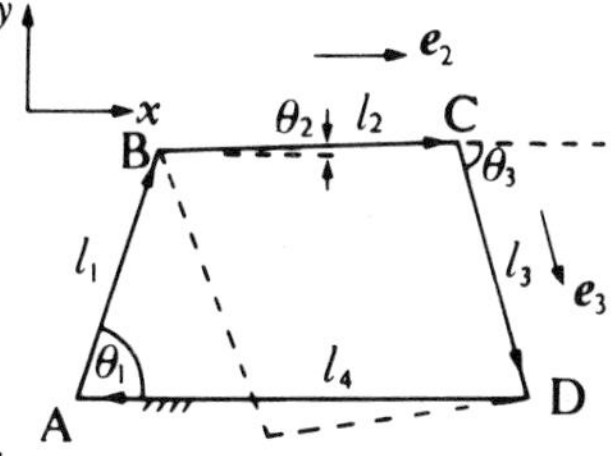

Figure 5.31

From Fig. 5.31 we note that

$$l_1 + l_2 + l_3 + l_4 = 0 \qquad \text{(ii)}$$

The vector $l_1 = l_1(\cos\theta_1 i + \sin\theta_1 j)$ is known and the vectors l_2 and l_3 can be determined by first evaluating angles θ_2 and θ_3 by the methods of normal trigonometry and then writing

$$l_2 = l_2(\cos\theta_2 i + \sin\theta_2 j)$$
$$l_3 = l_3(\cos\theta_3 i - \sin\theta_3 j)$$

Alternatively we can write

$$l_2 = l_2 e_2 = l_2(ai + bj)$$
$$l_3 = l_3 e_3 = l_3(ci + dj)$$

and determine the values of a, b, c and d. Noting that

$$d = \pm\sqrt{(1-c^2)} \qquad \text{(iii)}$$

and substituting in equation (ii) with $l_4 = -l_4 i$ and insertion of numerical values gives

$$0.190 e_2 = (0.225 - 0.180c)i - [0.1299 \pm 0.180\sqrt{(1-c^2)}]j$$

Taking the modulus of this equation eliminates e_2 and rearranging and squaring we find two values for c, each with two corresponding values of d from equation (iii). Only one of each pair of values of d is consistent with the links BC and CD joining at C, and one of the values of c corresponds with the mechanism being in the alternative position shown dotted in Fig. 5.31. The vector l_2 can then be found from equation (ii). The results are

$$l_1 = (0.0750i + 0.1299j)\ \text{m}$$
$$l_2 = (0.1893i + 0.1580j)\ \text{m}$$
$$l_3 = (0.0350i - 0.1457j)\ \text{m}$$

Now,

$$v_C = v_B + v_{C/B}$$

and, from equation 5.3,

$$v_C = \omega_1 \times l_1 + \omega_2 \times l_2$$

also

$$v_C = \omega_3 \times (-l_3) \qquad \text{(iv)}$$

Equating the two expressions for v_C,

$$\omega_1 \times l_1 + \omega_2 \times l_2 + \omega_3 \times l_3 = 0 \qquad \text{(v)}$$

Writing $\omega_1 = 12k$, $\omega_2 = \omega_2 k$ and $\omega_3 = \omega_3 k$, and carrying out the vector products in equation (v), gives

$$(-1.559 - 0.0158\omega_2 + 0.1457\omega_3)i + (0.9 + 0.1893\omega_2 + 0.03565\omega_3)j = 0$$

Equating the coefficients of i and j to zero and solving for ω_2 and ω_3, we find

$$\omega_2 = -6.634$$

and $\omega_3 = 9.980$

Using $v_B = \omega_1 \times l_1$ and equation (iv) leads to

$$v_B = -1.559i + 0.9j\ \text{m/s}$$
$$|v_B| = \sqrt{[(1.559)^2 + (0.9)^2]} = 1.800\ \text{m/s}$$

and

$$v_C = -(1.453i + 0.3558j)\ \text{m/s}$$
$$|v_C| = \sqrt{[(1.453)^2 + (0.3558)^2]} = 1.497\ \text{m/s}$$

A quicker way of finding v_C, if ω_2 is not required, is to note that since $v_{C/B}$ is perpendicular to BC, we can write

$$v_{C/B} \cdot l_2 = 0$$

or $(v_C - v_B) \cdot l_2 = 0$

v_B is known and writing from equation (iv)

$$v_C = \omega_3 k \times (-0.03501i + 0.1457j)$$

and carrying out the dot product we find

$\omega_3 = 9.98$ rad/s and hence v_C may be determined.

Differentiating equation (v) with respect to time,

$$\dot{\omega}_1 \times l_1 + \omega_1 \times \dot{l}_1 + \dot{\omega}_2 \times \dot{l}_2 + \omega_2 \times l_2 + \omega_3 \times l_3 + \omega_3 \times \dot{l}_3 = 0 \quad \text{(v)}$$

Note that the product of vectors can be differentiated in a manner similar to that for the product of scalars, see Appendix 1. Since l_1 is constant in magnitude then

$$\dot{l}_1 = \omega_1 \times l_1 \quad \text{since } \dot{l}_1 = v_{B/A} \quad \text{(see equation 5.3)}$$

and similarly for $\dot{l}_2$ and $\dot{l}_3$. Hence,

$$\dot{\omega}_1 \times l_1 + \omega_1 \times (\omega_1 \times l) + \dot{\omega}_2 \times l_2 + \omega_2 \times (\omega_2 \times l_2) + \dot{\omega}_3 \times l_3 + \omega_3 \times (\omega_3 \times l_3) = 0 \quad \text{(vi)}$$

Substituting the previously obtained values together with

$$\dot{\omega}_1 = 35k, \quad \dot{\omega}_2 = \omega_2 k \quad \text{and} \quad \dot{\omega}_3 = \omega_3 k$$

and carrying out the vector products, we find

$$\dot{\omega}_2 = -22.77 \text{ rad/s}^2 \quad \text{and} \quad \dot{\omega}_3 = 184.4 \text{ rad/s}^2$$

Differentiating equation (iv),

$$a_C = \dot{\omega}_3 \times (-l_3) + \omega_3 \times [\omega_3 \times (-l_3)]$$

Substituting the numerical values gives

$$a_C = -(23.32i + 21.09j) \text{ m/s}^2$$

and $a_C = 31.40$ m/s^2

The acceleration a_B can be found in a similar manner.

Example 5.2

In the mechanism shown in Fig. 5.32, FED is an offset slider-crank chain which is given an oscillatory motion by the rotation of crank AB. When B is vertically above A, the angular velocity of AB is 3.0 rad/s anticlockwise. Determine the corresponding velocity of slider D. All the lengths are given in mm.

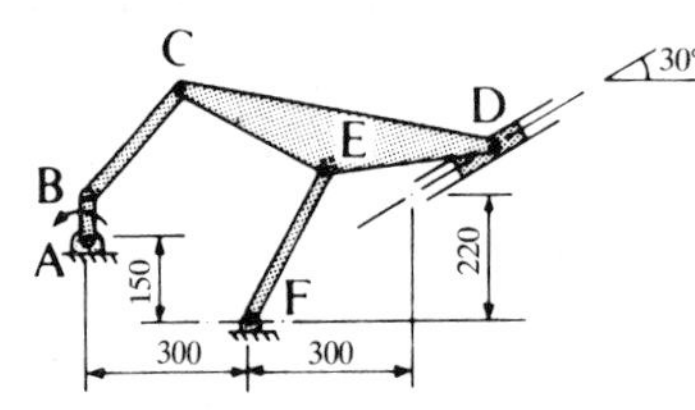

Figure 5.32 AB = 70, BC = 250, CD = 570
CE = ED = EF = 300

Solution As we attempt to draw the mechanism to scale, we are presented with an immediate difficulty. We know the location of B but we cannot readily determine the position of C. If we assume that B is fixed and D is not constrained by the slider, and we allow the four-bar chain BCEF to move, then the correct configuration is obtained when D coincides with the slider centre line. Thus we need a trial-and-error method to determine the correct positions. The difficulty in drawing the mechanism suggests that there will also be difficulties in drawing a velocity diagram, and this proves to be the case.

We know that the magnitude of v_B is

$$v_B = \omega_{AB}(AB) = 3(0.07) = 0.21 \text{ m/s}$$

and that the diagram is horizontally to the left, so we can draw $\overrightarrow{ab}$ on the velocity diagram to represent $v_{B/A} = v_B$.

We know the directions of $v_{C/B}$, $v_{C/E}$ and $v_{E/F}$, but more information is required before we can proceed (see Fig. 5.33).

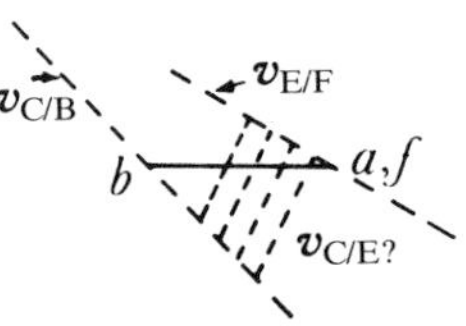

Figure 5.33

The instantaneous-centres method presents no difficulties once the mechanism has been drawn to scale. For the slider-crank chain FED the instantaneous centre is at I_1, the intersection of the lines perpendicular to the velocities at D and E. C has the same instantaneous centre since it is rigidly attached to DE, see Fig. 5.34.

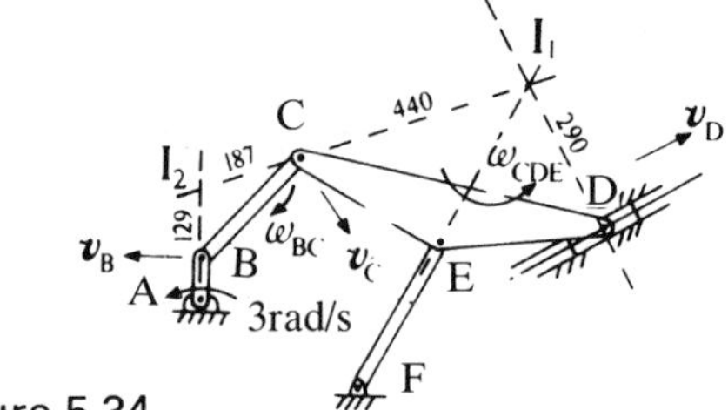

Figure 5.34

The velocity of C is perpendicular to I_1C and the velocity of B is perpendicular to AB. Thus the instantaneous centre for BC is at I_2.

For link BC:

$$\omega_{BC} = v_B/(I_2B), \qquad v_C = \omega_{BC}(I_2C)$$

For link CDE:

$$\omega_{CDE} = v_C/(I_1C), \quad v_D = \omega_{CDE}(I_1D)$$

Hence,

$$v_D = \frac{(I_1D)(I_2C)v_B}{(I_1C)(I_2B)}$$

$$= \frac{(0.29)(0.187)(0.21)}{(0.44)(0.129)} = 0.20 \text{ m/s}$$

Thus $\boldsymbol{v}_D = 0.20$ m/s 30°

This example shows the advantage of the instantaneous-centres method for certain mechanisms, but it should be noted that, where a slider moves in a link which is itself rotating, as in the next example, this method is not helpful. A solution is however possible by the velocity-diagram method. If, for instance, it is assumed that $\boldsymbol{v}_D$ is 1 m/s up the incline, then the velocity diagram can be constructed and the corresponding value of v_B determined. The diagram can then be rescaled to make $v_B = 0.21$ m/s and the correct value of v_D may be found. A solution by this method is left as an exercise for the reader.

Example 5.3

Figure 5.35 shows part of the essential kinematics of a linkage known as a quick-return mechanism. Crank AB rotates about A and slotted link CD rotates about C. Pin B on the end of AB engages in the slot of CD; AB $= r$ and AC $= l$. The angular velocity of AB is $\omega\boldsymbol{k}$ and its angular acceleration is $\dot{\omega}\boldsymbol{k}$.

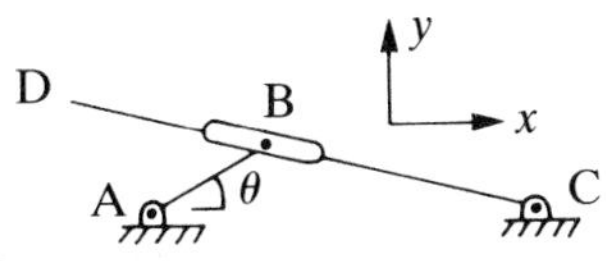

Figure 5.35

If $l/r = R$, show that the angular velocity of CD is

$$\boldsymbol{\omega}_{CD} = \frac{(1 - R\cos\theta)\omega}{1 + R^2 - 2R\cos\theta}\boldsymbol{k}$$

and that its angular acceleration is

$$\dot{\boldsymbol{\omega}}_{CD} = \left[\frac{R\sin\theta(R^2-1)\omega^2}{(1+R^2-2R\cos\theta)^2} + \frac{(1-R\cos\theta)\dot{\omega}}{1+R^2-2R\cos\theta}\right]\boldsymbol{k}$$

If $r = 50$ mm, $l = 140$ mm, $\omega = 1$ rad/s and $\dot{\omega} = 2$ rad/s^2, determine ω_{CD} and $\dot{\omega}_{CD}$ if $\theta = 30°$. Check this result from velocity and acceleration diagrams.

Solution See Fig. 5.36. If we obtain an expression for the angle ϕ in terms of r, l and θ, then differentiation will lead to the required results.

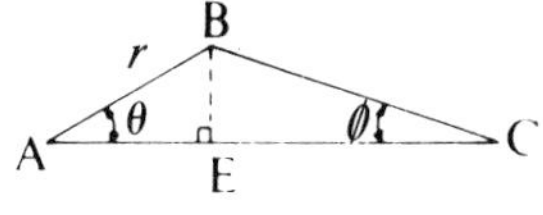

Figure 5.36

From the figure,

$$\tan\phi = \frac{BE}{EC} = \frac{r\sin\theta}{l - r\cos\theta} = \frac{\sin\theta}{R - \cos\theta} \qquad \text{(i)}$$

Differentiating with respect to time, using the quotient rule,

$$\sec^2\phi\dot{Q} = \frac{(R-\cos\theta)\cos\theta - \sin\theta(\sin\theta)}{(R-\cos\theta)^2}\omega$$

$$= \frac{R\cos\theta - 1}{(R-\cos\theta)^2}\omega \qquad \text{(ii)}$$

Combining equations (i) and (ii) and noting that

$$\sec^2\phi = 1 + \tan^2\phi$$

we find

$$\dot{\phi} = \frac{(R\cos\theta - 1)\omega}{1 + R^2 - 2R\cos\theta} \qquad \text{(iii)}$$

The angle ϕ is positive in the clockwise sense so that

$$\boldsymbol{\omega}_{CD} = -\dot{\phi}\boldsymbol{k}$$

hence the result.

Differentiating again and rearranging and collecting the terms, we find the appropriate expression for $\ddot{\phi}$ and

$$\dot{\boldsymbol{\omega}}_{CD} = -\ddot{\phi}\boldsymbol{k}$$

Substituting the numerical values, we find

$$\omega_{CD} = -0.3571\boldsymbol{k} \text{ rad/s}$$

and $\dot{\omega}_{CD} = -0.1127\boldsymbol{k}$ rad/s

The mechanism is drawn to scale in Fig. 5.37(a)

Velocity diagram (Fig. 5.37(b))

To draw the velocity diagram we let B_1 be a point fixed on CD which is momentarily coincident with

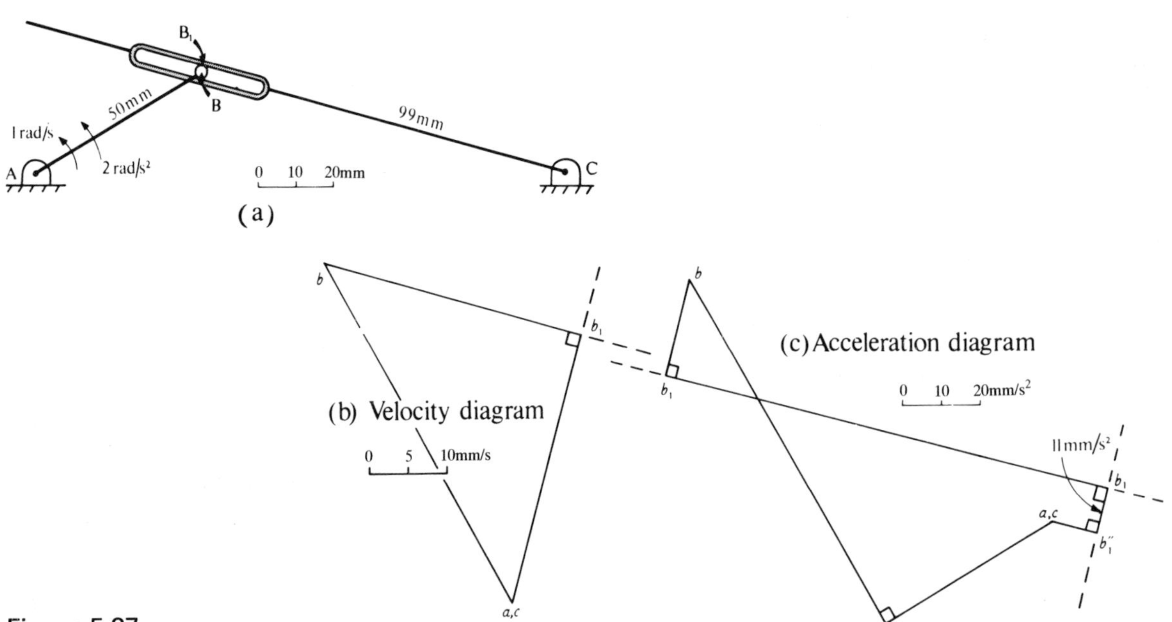

Figure 5.37

B. The velocity diagram will make use of the result

$$\boldsymbol{v}_{B1} = \boldsymbol{v}_B + \boldsymbol{v}_{B1/B}$$

and we note that $\boldsymbol{v}_{B1/B}$ is parallel to the slot.

Link	*Velocity*	*Direction*	*Sense*	*Magnitude* (m/s)	*Line*
AB	$\boldsymbol{v}_{B/A}$	$\perp$AB	↖	$(AB)\,\omega_{AB} = 50(1) = 50$	ab
—	$\boldsymbol{v}_{B1/B}$	$\parallel CB_1D$	?		bb_1
CB_1D	$\boldsymbol{v}_{B/C}$	$\perp CB_1D$	?	$(CB_1)\,\omega_{BC}$	cb_1

From the velocity diagram we note that $\boldsymbol{v}_{B1/B}$ has a magnitude of 35.25 mm/s and the sense is from b to b_1. The velocity $\boldsymbol{v}_{B1/C} = \boldsymbol{v}_{B1}$ has a magnitude of 35.0 mm/s, the sense being from c to b_1. The angular velocity ω_{CB1} thus has a magnitude of

$$\omega_{CB1} = \frac{v_{B1/C}}{CB_1} = \frac{cb_1}{CB_1} = \frac{35}{99} = 0.354 \text{ rad/s}$$

and the sense is clockwise.

An attempt to use the method of instantaneous centres would prove fruitless.

Acceleration diagram (Fig. 5.37(c))

A feature not encountered in the previous problem is the relative acceleration between coincident points such as B and B_1. Since the relative acceleration for any pair of points B and B_1 is

$$\boldsymbol{a}_{B1/B} = (\ddot{R} - R\omega^2)\boldsymbol{e}_r + (R\dot{\omega} + 2\dot{R}\omega)\boldsymbol{e}_\theta$$

then, if B and B_1 momentarily coincide so that $R = 0$,

$$\boldsymbol{a}_{B1/B} = \ddot{R}\boldsymbol{e}_r + 2\dot{R}\omega\boldsymbol{e}_\theta$$

We note that, for the general case where B and B_1 are not necessarily coincident, the line BB_1 always lies on the line CD so that it always has the same angular velocity as CD. Thus, in the above equation we use $\omega_{CD} = \omega_{CB1}$ for ω.

The term $2\dot{R}\omega\boldsymbol{e}_\theta$ is known as the Coriolis component of acceleration. Its magnitude can be determined by means of the velocity diagram but we need to determine the direction before we can complete the acceleration diagram.

We know from Chapter 2 that the direction depends on the directions of $\boldsymbol{v}_{B1/B}$ and ω_{BB1} ($=\omega_{CB1}$). It is convenient to note that the direction is the same as that obtained by rotating the vector $\boldsymbol{v}_{B1/B}$ through 90° in the sense of the angular velocity ω_{BB1} [note that this direction is that of $(\omega_{BB1} \times \boldsymbol{v}_{B1/B})$].

In the present case the direction of $\boldsymbol{v}_{B1/B} = bb_1$ is in the sense C to B_1 and the angular velocity of CB_1 is clockwise. The direction of the Coriolis component of $\boldsymbol{a}_{B1/B}$ is thus in the direction $\boldsymbol{e}_{Cor.}$ shown in Fig. 5.38. Similarly the Coriolis component of $\boldsymbol{a}_{B/B1}$ is in the opposite direction.

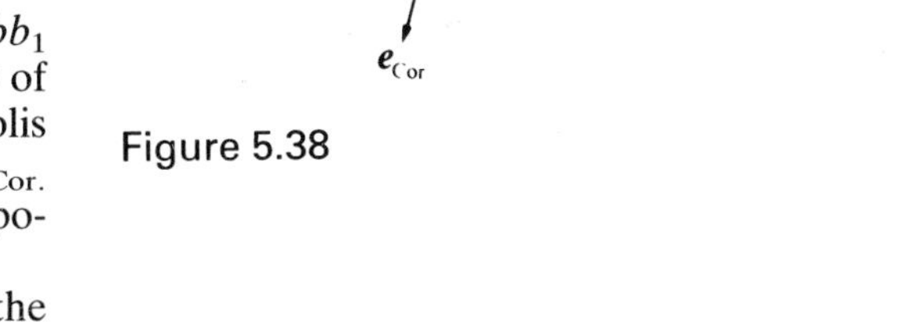

Figure 5.38

We can now proceed to draw the lines of the acceleration diagram in the order listed below.

Link	*Acceleration*	*Direction*	*Sense*	*Magnitude* (mm/s^2)	*Line*
AB	$\boldsymbol{a}_{B/A}$ (radial)	$\parallel$AB	↙	$(AB)\,\omega_{AB}^2 = 50(1)^2 = 50$	ab'
	$\boldsymbol{a}_{B/A}$ (normal)	$\perp$AB	↖	$(AB)\,\dot{\omega}_{AB} = 50(2) = 100$	$b'b$
—	$\boldsymbol{a}_{B1/B}$ (normal)	$\perp CB_1$	↙	$2(v_{B1/B})\,\omega_{CB1} = 2(35.25)(0.354) = 25.0$	bb_1'
	$\boldsymbol{a}_{B1/B}$ (radial)	$\parallel CB_1$	?	$\ddot{R}_{BB1}$	$b_1'b_1$
CB_1D	$\boldsymbol{a}_{B1/C}$ (radial)	$\parallel CB_1$	→	$(CB_1)\,\omega_{CB1}^2 = 99(0.354)^2 = 12.4$	cb_1''
	$\boldsymbol{a}_{B1/C}$ (normal)	$\perp CB_1$	?	$(CB_1)\,\dot{\omega}_{CB1}$	$b_1''b$

The component normal to CB_1 of $\boldsymbol{a}_{B1/C}$ is $b_1''b_1$ and the sense is from b_1'' to b_1. The magnitude of $\dot{\omega}_{CB1}$ is

$$\dot{\omega}_{CB1} = b_1''b_1/(CB_1) = 11/99 = 0.111 \text{ rad/s}^2$$

and the sense is clockwise.

Example 5.4

Figure 5.39 shows the main features of a simple two-speed epicyclic gearbox. The sun wheel S_1 is keyed to the input shaft I which is rotating at 1000 rev/min. The sun wheel S_2 is keyed to the annulus A_1. The planet carriers C_1 and C_2 are both keyed to the output shaft O. The numbers of teeth on the annulus and sun wheels are $T_{A1} = T_{A2} = 80$, $T_{S1} = 30$ and $T_{S2} = 28$.

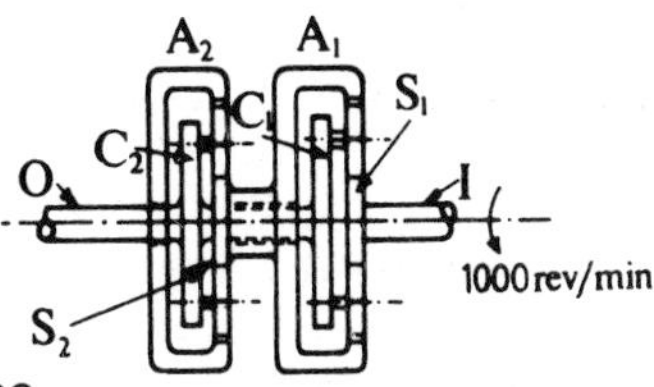

Figure 5.39

Determine the speeds of the output shaft O and the non-stationary annulus (a) when annulus A_1 is held fixed and (b) when annulus A_2 is held fixed.

Solution The effective radii of the wheels are proportional to the number of teeth. Writing down the relative angular-velocity equations, from equation 5.13, for the right-hand gear:

$$\frac{\omega_{S1/C1}}{\omega_{A1/C1}} = \frac{\omega_{S1} - \omega_{C1}}{\omega_{A1} - \omega_{C1}} = -\frac{T_{A1}}{T_{S1}}$$

$$\frac{1000 - \omega_{C1}}{\omega_{A1} - \omega_{C1}} = -\frac{80}{30} \qquad \text{(i)}$$

for the left-hand gear:

$$\frac{\omega_{S2/C2}}{\omega_{A2/C2}} = \frac{\omega_{S2} - \omega_{C2}}{\omega_{A2} - \omega_{C2}} = -\frac{T_{A2}}{T_{S2}}$$

Hence

$$\frac{\omega_{A1} - \omega_{C1}}{\omega_{A2} - \omega_{C1}} = -\frac{80}{28} \qquad \text{(ii)}$$

since $\omega_{S2} = \omega_{A1}$ and $\omega_{C2} = \omega_{C1}$.

a) Putting $\omega_{A1} = 0$ and solving the simultaneous equations in ω_{C1} and ω_{A2} gives

$$\omega_{C1} = 272.7 \text{ rev/min} = \omega_O$$

and $\omega_{A2} = 368.2$ rev/min

b) Putting $\omega_{A2} = 0$ and solving for ω_{C1} and ω_{A1} gives

$$\omega_{C1} = -151.1 \text{ rev/min} = \omega_O$$

and $\omega_{A1} = -582.7$ rev/min

The negative signs indicate that the rotation is in the opposite sense to that of the input shaft.

Example 5.5

A Simpson gear set of the type shown in Fig. 5.23(a) has been designed to have the following gear ratios. First gear 2.84, second gear 1.60 and third gear direct.

Determine the ratio of teeth on the annulus to those on the sun wheel for both the first and the second simple epicyclic gears which form the set.

Suggest practical values for the number of teeth on each wheel to give a good approximation to the desired ratios.

Solution From the example given in the text, section 5.11, we know that the second gear ratio depends only on the first simple epicyclic so

$$(1+R_1)/R_1 = 1.60$$

therefore

$$R_1 = 1/(1.60-1) = 1.67$$

Now using the expression for the first gear ratio

$$(1+R_1+R_2)/R_1 = 2.84$$

gives

$$R_2 = R_1(2.84) - 1 - R_1$$
$$= 1.67 \times 2.84 - 1 - 1.67 = 2.07$$

The reverse gear ratio is numerically equal to $R_2 = 2.07$.

The diameter of the sun wheel plus twice that of the planets must equal the diameter of the annulus. For a meshing gear train all gears will have the same diametral pitch, that is the ratio of the number of teeth to the diameter is constant. It now follows that the number of teeth on the sun wheel plus twice those of the planets will be equal to the number of teeth on the annulus. For manufacturing reasons we will assume that no wheel is to have fewer than 15 teeth. If we take the planet wheels of the first epicyclic gear to have 15 teeth then the number of teeth on the annulus

$$T_A = T_S + 2 \times 15$$

but $T_A/T_S = 1.67$

therefore $1.67 \times T_S = T_S + 30$

giving $T_S = 45$, to the nearest whole number

and $T_A = 1.67 \times T_S = 75$, to the nearest whole number.

These numbers satisfy the kinematic requirements but, because the number of teeth on the sun wheel are exactly three times the number on the planet, the same teeth will mesh every three revolutions of the planet relative to the carrier. The same ratio, to two places of decimals, can be achieved with $T_A = 85$, $T_P = 17$ and $T_s = 51$. Since T_P, the number of teeth on the planet, is a prime number even wear on the teeth will be assured.

We could start our design for the second simple epicyclic by taking the diameter of the annulus to be the same as the first gear so that, assuming the same diametral pitch, both annuli will have the same number of teeth, that is 85.

This means that $T_{S2} = 85/2.07 = 41$, to the nearest whole number. The actual ratio $85/41 = 2.07$ to two places of decimals. The number of teeth on the planet $= (85-41)/2 = 22$.

In this gear the number of teeth on the sun is a prime number and the number of teeth on the planet is $2\times$(prime number) thereby assuring even wear.

It is obvious that many other combinations of gear sizes are possible so there is no unique solution.

Example 5.6

A Ravigneaux gear as shown in Fig. 5.23d has gear wheels of the same diametral pitch. The number of teeth on the first (reverse) sun wheel is 32 and on the second (forward) is 28. The long pinion has 17 teeth and the short pinion has 16. Determine the gear ratios for the three forward gears and one reverse.

Solution The number of teeth on the annulus

$$T_A = T_{S1} + 2 \times T_{P(long)}$$
$$= 32 + 2 \times 17 = 66.$$

For the first simple epicyclic the ratio $T_A/T_{S1} = R_1$

$$= 66/32 = 2.06$$

For the second simple epicyclic $T_A/T_{S2} = R_2 = 66/28 = 2.36$.

From the summary for the gearbox, page 000

1st gear ratio $= 2.36$

2nd gear ratio $= (\mathbf{R_1} + \mathbf{R_2})/(1 + \mathbf{R_1}) = (2.06 + 2.36)/(1 + 2.06) = 1.44$

3rd gear ratio $= 1$

and reverse gear ratio $= -\mathbf{R_1} = -2.06$.

Problems

5.1 In the mechanism shown in Fig. 5.40, AB is rotating anticlockwise at 10 rad/s. When $\theta = 45°$, determine the angular velocity of link BDC and the velocities of C and D.

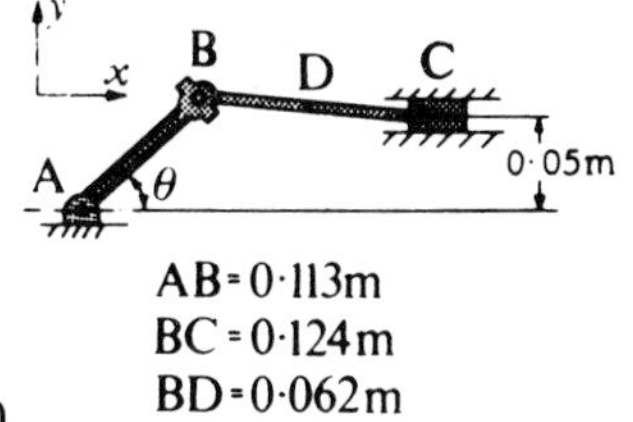

Figure 5.40

Solve this problem (a) by drawing a velocity diagram, (b) by the method of instantaneous centres and (c) analytically.

5.2 The device shown in Fig. 5.41 is for testing the resistance to wear between the material of a road Y and a 'shoe' X. The crank OA is rotating clockwise at 10 rad/s and the shoe is loaded such that contact is always maintained between the test surfaces and B lies on the line OD.

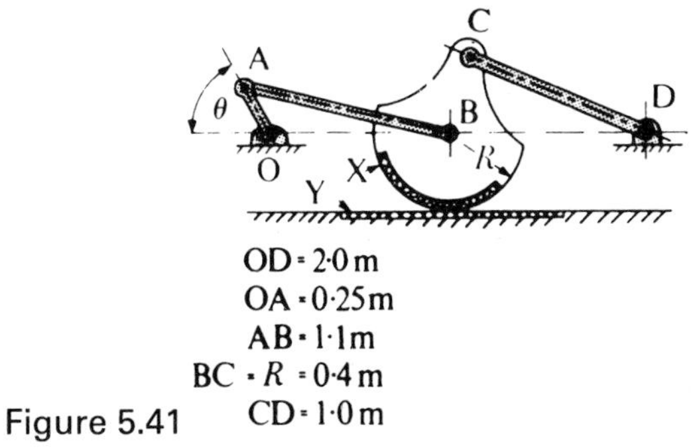

Figure 5.41

For the instant when the angle θ is 60°, determine (a) the rubbing speed between the two test materials and (b) the angular velocity of the shoe.

5.3 A flat-footed follower F slides in guides G and engages with cam C as shown in Fig. 5.42. The cam consists of a circular disc, centre A, radius r, rotating at constant speed ω about point O, and OA = e. The spring S maintains contact between the follower and the cam.

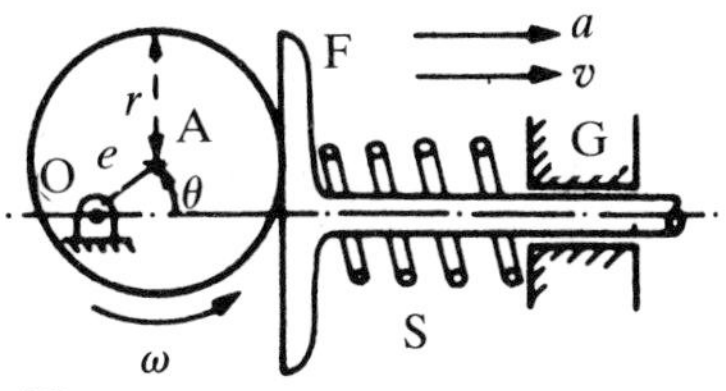

Figure 5.42

Find expressions for the velocity v and acceleration a of the follower.

5.4 Part of the control system for an engine is illustrated in Fig. 5.43. At the instant when the beam OA passes through the horizontal position, its angular speed ω is found to be 1.1 rad/s. The motion of the point A is transmitted through the push rod AB to the right-angled bell crank BCD. The cylindrical end at D is a sliding fit between the parallel faces of the collars fitted to the valve shaft EF.

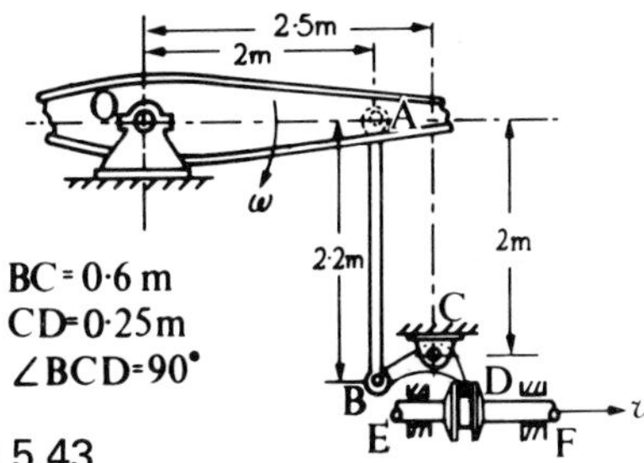

Figure 5.43

For the configuration shown, find (a) the angular velocity of BCD and (b) the linear velocity v of the shaft EF.

5.5 Figure 5.44 shows a four-slot Geneva mechanism which converts continuous rotation of a shaft with centre O_1 to intermittent rotation of a parallel shaft with centre O_2. Pin P rotates ar radius R about centre O_1, and engages with the slots of the Geneva wheel, centre O_2. The slots are tangential to the path of the pin at entry and exit.

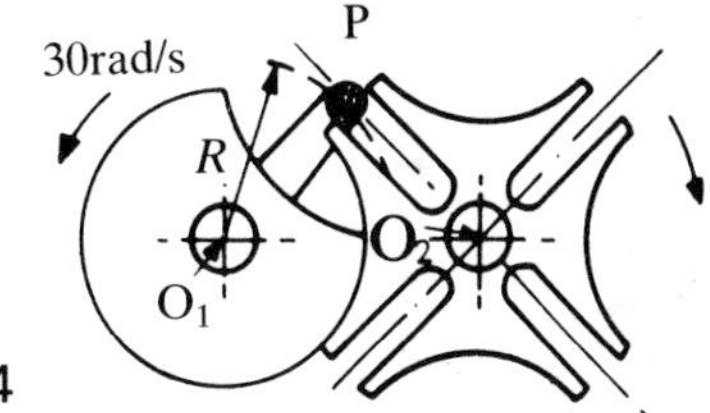

Figure 5.44

If crank O_1P rotates at a constant angular speed of 30 rad/s, determine the angular acceleration of the Geneva wheel just before the pin leaves a slot.

5.6 In the engine mechanism shown in Fig. 5.45, crank AB rotates at a constant angular velocity $\omega_O \mathbf{k}$. G is a point on the connecting rod BC such that BG = a, GC = b and $a + b = l$.

Show that

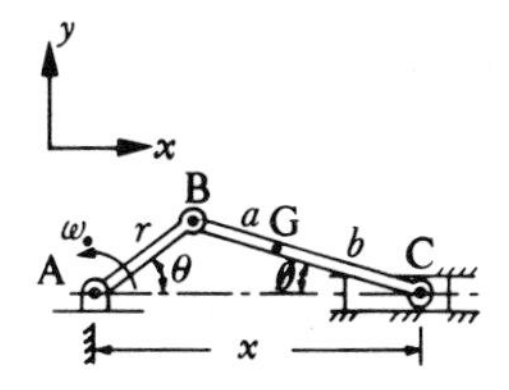

Figure 5.45

$$\omega_{BC} = -\left[\frac{r\cos\theta\,\omega_O}{l\cos\phi}\right]\boldsymbol{k}$$

$$\boldsymbol{v}_C = [-r\sin\theta\,\omega_O + l\sin\phi\,\omega_{BC}]\boldsymbol{i}$$

$$\boldsymbol{v}_G = [-r\sin\theta\,\omega_O + a\sin\phi\,\omega_{BC}]\boldsymbol{i} - [b\cos\phi\,\omega_{BC}]\boldsymbol{j}$$

$$\dot{\omega}_{BC} = \left[\sec\phi\left(\frac{r}{l}\sin\theta\,\omega_O^2 - \sin\phi\,\omega_{BC}^2\right)\right]\boldsymbol{k}$$

$$\boldsymbol{a}_C = -[r\cos\theta\,\omega_O^2 + l(\cos\phi\,\omega_{BC}^2 - \sin\phi\,\dot{\omega}_{BC})]\boldsymbol{i}$$

$$\boldsymbol{a}_G = -[r\cos\theta\,\omega_O^2 + a(\cos\phi\,\omega_{BC}^2 - \sin\phi\,\dot{\omega}_{BC})]\boldsymbol{i} - b[\sin\phi\,\omega_{BC}^2 + \cos\phi\,\dot{\omega}_{BC}]\boldsymbol{j}$$

where $\sin\phi = (r/l)\sin\theta$ and $\omega_{BC} = -\dot{\phi}$.

5.7 Figure 5.46 shows one of the cylinders C of a petrol engine. The crankshaft AB is rotating anticlockwise at a constant speed of 3000 rev/min about A. The piston E which slides in cylinder C is connected to the crankshaft by the connecting rod BD, and G is the mass centre of the connecting rod.

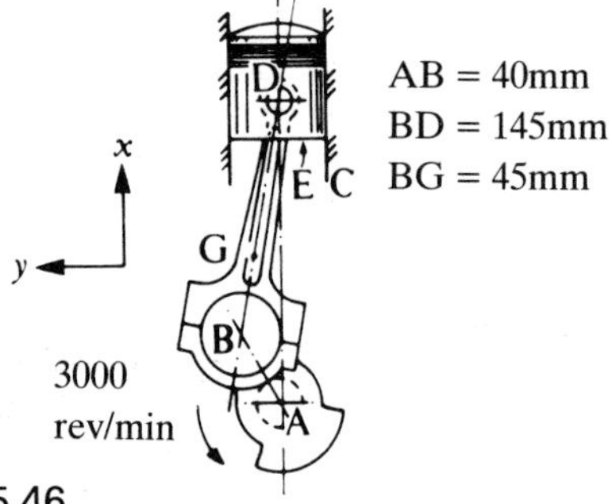

Figure 5.46

For angle DAB = 30°, determine (a) the velocities of E and G and the angular velocity of BD; (b) the accelerations of E and G and the angular acceleration of BD. Solve this problem graphically and check your results from the formulae of the previous question.

5.8 Figure 5.47 shows part of an opposed-piston diesel engine running at 2000 rev/min. Connecting rods AB and DE are connected to the flywheel at A and D respectively, the crank radius being 160 mm. Slider B is connected to piston C by link BC, and pistons C and E are in the same cylinder.

When angle BOA = 60°, find the velocities and accelerations of each piston.

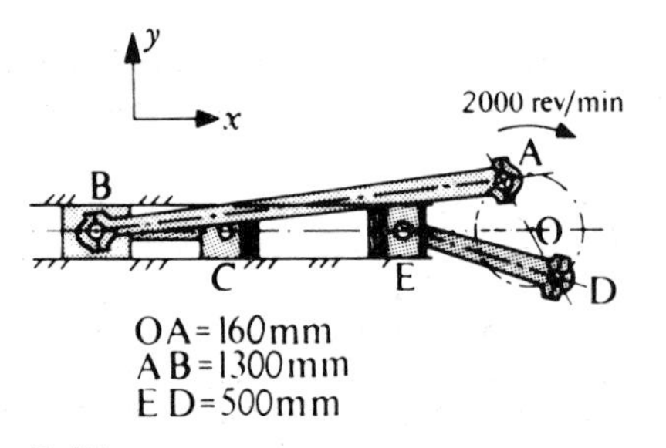

Figure 5.47

5.9 An 'up-and-over' mechanism for a garage door comprises two identical units of the type shown in Fig. 5.48, mounted one on each side of the door. Each unit consists of a trunnion block T which runs on two 0.1 m diameter rollers in a vertical guide, with the door carried on a pin at B. The link OA is pinned to the door at A and rotates about the fixed axis at O.

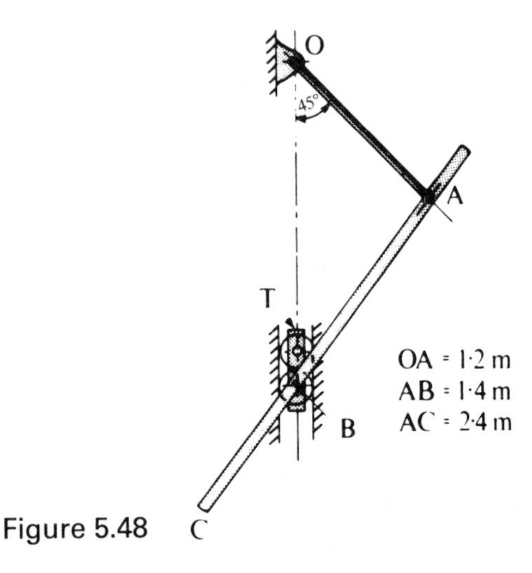

Figure 5.48 C

At the instant when the door is in the position shown, the trunnion block has an upward velocity of 0.75 m/s. For this position determine (a) the angular velocity of the link OA, (b) the velocity of the lower edge of the door at C and (c) the angular velocity of the trunnion block rollers, assuming no slip.

5.10 See Fig. 5.49. P is a representative water particle moving outward along the impeller blade of a 'centrifugal' pump. The radius of curvature ρ of the blades at the tip is 150 mm. The impeller has an angular velocity of 30 rad/s clockwise and an angular acceleration of 0.01 rad/s^2 in the same sense. At the blade tip the particle has, *relative to the impeller blade*, a tangential velocity of 15 m/s and a tangential acceleration of 10 m/s^2.

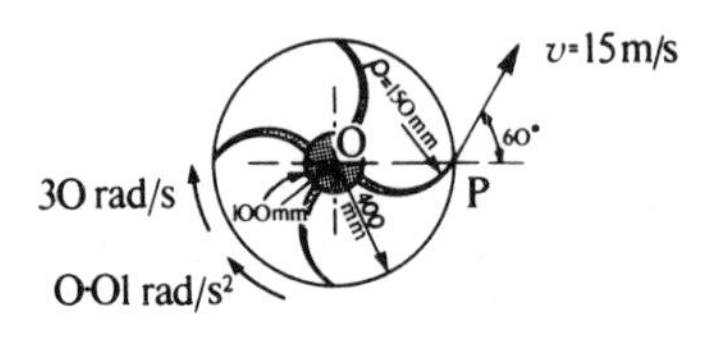

Figure 5.49

Find the total velocity and the total acceleration of the water particle P as it is on the point of leaving the blade. A semi-graphical method is suggested.

5.11 Figure 5.50 shows part of a shuttle drive mechanism for a sewing machine, the continuous rotation of crank AB at 60 rad/s causing an oscillatory motion of the shuttle drive shaft G. A, D and G are fixed centres and the lengths are all given in millimetres. The slotted link CDE which rotates about D is driven by the connecting bar BC and in turn drives the crank GF via the swivel block at F.

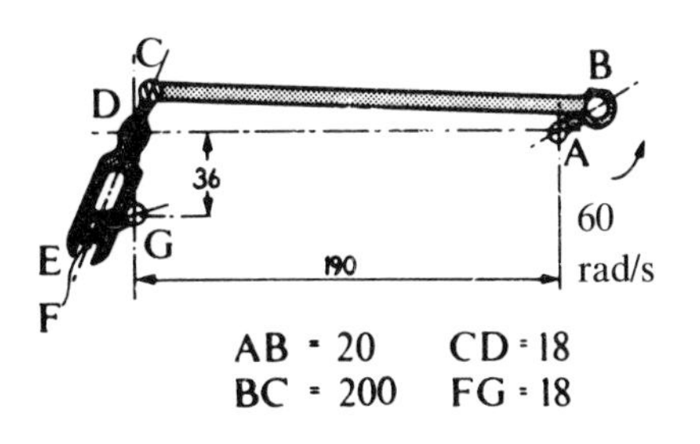

Figure 5.50

Find, for angle DAB = 150°, the angular velocity and angular acceleration of CDE and the sliding velocity of the block. Hence determine the angular velocity and angular acceleration of GF.

5.12 In the mechanism shown in Fig. 5.51, the crank OB rotates with uniform clockwise angular velocity of 1 rad/s. It drives link ABP whose end A is constrained to move vertically. The disc D rotates about the axis O; it is driven by a pin P, attached to ABP, which engages with the slot S. OB = 50 mm; AB = 90 mm; BP = 90 mm; angle AOB = 30°.

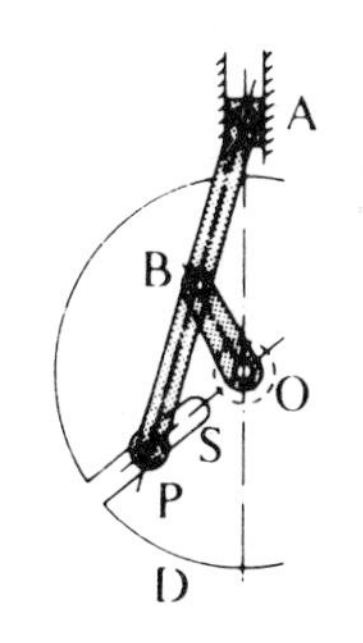

Figure 5.51

Construct the velocity and acceleration vector diagrams for the mechanism in this position, and from these find (a) the magnitude and sense of the angular velocity of the disc D and (b) the magnitude and sense of the angular acceleration of the disc D. (Suggested scales: 1 cm = 0.01 m/s, 1 cm = 0.01 m/s^2.)

5.13 Figure 5.52 shows the essential kinematics of a compound epicyclic gear designed to give a large speed reduction from the input shaft I to the output shaft O. Carrier C is keyed to the input shaft and carries a pin T on which the compounded planet wheels P_1 and P_2 are

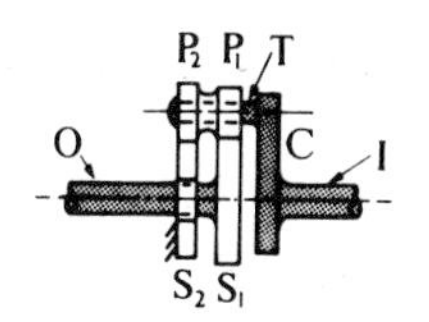

Figure 5.52

free to rotate. P_1 meshes with sun wheel S_1, which is keyed to O, and P_2 meshes with sun wheel S_2, which is fixed. The gear wheels all have teeth of the same pitch. The numbers of teeth are $N_{P1} = 20$, $N_{P2} = 21$, $N_{S1} = 70$.

Show that $N_{S2} = 69$ and that the speed ratio is $\omega_O/\omega_I = 9/147$.

5.14 The epicyclic gear shown in Fig. 5.53 consists of a sun wheel S which is fixed to the case, three compound planet wheels P_1–P_2 which are mounted on the carrier C, and an annulus A.

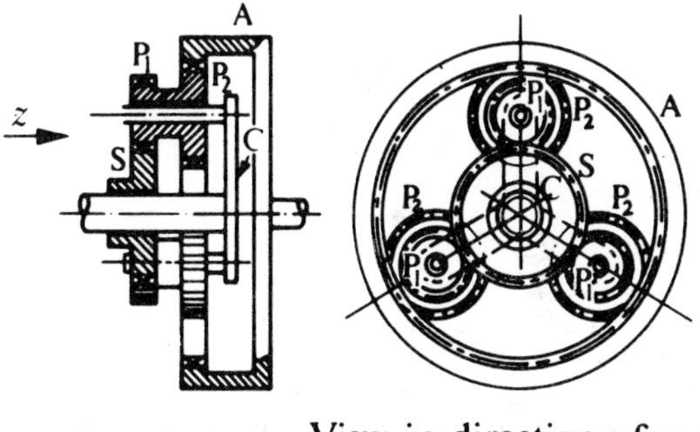

Figure 5.53

The number of teeth are as follows:

Gear	S	P_1	P_2
No. of teeth	40	20	30

The shaft attached to C has a speed of 15$\boldsymbol{k}$ rev/s; find the angular velocity of the output shaft attached to A.

5.15 Figure 5.54 illustrates the arrangement of an epicyclic gearbox. Wheel A is integral with the input shaft and drives the planet carrier C through the idler gear B. There is one compound planet DE. Wheel D meshes with wheel F, which is keyed to the output shaft, and wheel E meshes with the fixed gear G. All teeth are cut having a module of 4 mm.

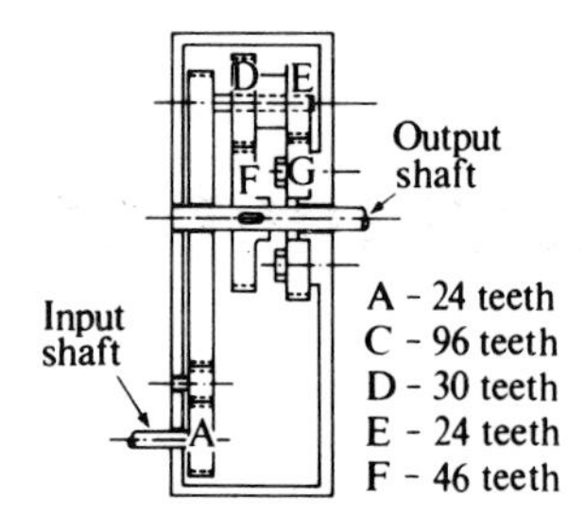

Figure 5.54

a) For the numbers of teeth given in Fig. 5.54, show that the number of teeth on wheel G is 52.

b) Determine the overall speed ratio of the gearbox.

6
Kinetics of a rigid body in plane motion

6.1 General plane motion

In this chapter we consider the motion of a rigid body in general plane motion, by which we mean that the centre of mass is moving in a plane and any rotation is about an *instantaneous* axis perpendicular to the plane.

In Chapter 3 it was shown that the resultant of the external forces on a body is equal to the product of the total mass and the acceleration of the centre of mass. We must now consider the effect of the positions of the lines of action of the applied forces, remembering that the acceleration of the centre of mass is the same whether or not the line of action of the resultant passes through the centre of mass.

Consider initially a group of particles in random motion. For a typical particle (see Fig. 6.1),

$$\sum_j \boldsymbol{f}_{ij} + \boldsymbol{F}_i = m_i \ddot{\boldsymbol{r}}_i \tag{6.1}$$

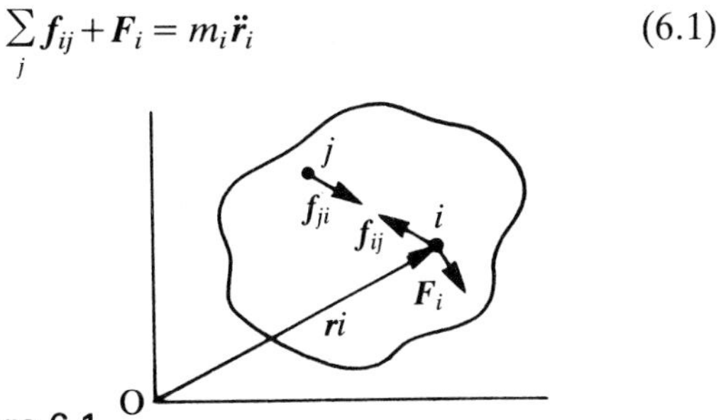

Figure 6.1

where $\boldsymbol{f}_{ij}$ is the force on particle i due to particle j and $\boldsymbol{F}_i$ is an externally applied force.

Taking the moments of the forces about O, we have

$$\boldsymbol{r}_i \times \sum_j \boldsymbol{f}_{ij} + \boldsymbol{r}_i \times \boldsymbol{F}_i = \boldsymbol{r}_i \times (m_i \ddot{\boldsymbol{r}}_i) \tag{6.2}$$

The total moment of the internal forces is zero, since the internal forces occur in pairs of collinear forces equal in magnitude but opposite in sense, and so summing over all the particles gives

$$\sum \boldsymbol{r}_i \times \boldsymbol{F}_i = \sum \boldsymbol{r}_i \times (m_i \ddot{\boldsymbol{r}}_i)$$

$$= \frac{\mathrm{d}}{\mathrm{d}t} \sum (\boldsymbol{r}_i \times m_i \dot{\boldsymbol{r}}_i) \tag{6.3}$$

or

moment of external forces
= $\sum$ moment of (mass × acceleration)
= moment of the rate of change of momentum
= rate of change of the moment of momentum

We may make use of the definition of the centre of mass and, by writing $\boldsymbol{r}_i = \boldsymbol{r}_G + \boldsymbol{\rho}_i$ (see Fig. 6.2), equation 6.3 becomes

$$\begin{aligned}\sum \boldsymbol{r}_i \times \boldsymbol{F}_i &= \sum \boldsymbol{r}_G \times m_i \ddot{\boldsymbol{r}}_G + \sum \boldsymbol{r}_G \times m_i \ddot{\boldsymbol{\rho}}_i \\ &\quad + \sum \boldsymbol{\rho}_i \times m_i \ddot{\boldsymbol{r}}_G + \sum \boldsymbol{\rho}_i \times m_i \ddot{\boldsymbol{\rho}}_i \\ &= \boldsymbol{r}_G \times M \ddot{\boldsymbol{r}}_G + \sum \boldsymbol{\rho}_i \times m_i \ddot{\boldsymbol{\rho}}_i\end{aligned} \tag{6.4}$$

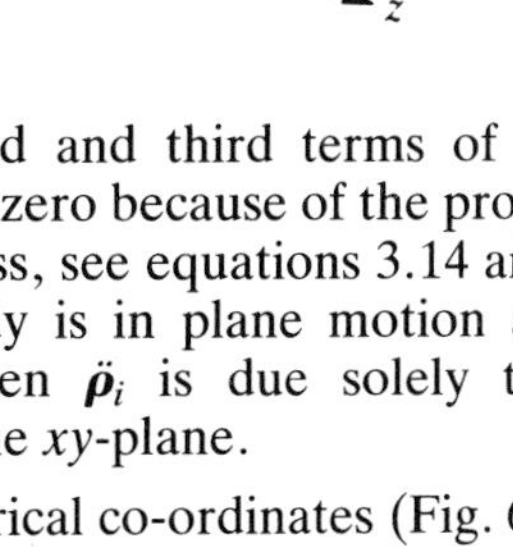

Figure 6.2

The second and third terms of the previous equation are zero because of the properties of the centre of mass, see equations 3.14 and 3.14(a).

If the body is in plane motion as previously specified, then $\ddot{\boldsymbol{\rho}}_i$ is due solely to rigid-body rotation in the xy-plane.

Using cylindrical co-ordinates (Fig. 6.3),

$$\boldsymbol{\rho}_i = R_{iG} \boldsymbol{e}_r + z_i \boldsymbol{k} \tag{6.5}$$

$$\ddot{\boldsymbol{\rho}}_i = -\omega^2 R_{iG} \boldsymbol{e}_R + \dot{\omega} R_{iG} \boldsymbol{e}_\theta \tag{6.6}$$

Considering moments about the Gz axis only,

$$M_G = \sum \boldsymbol{\rho}_i \times m_i \ddot{\boldsymbol{\rho}}_i \cdot \boldsymbol{k}$$

but $\boldsymbol{\rho}_i \times m_i \ddot{\boldsymbol{\rho}}_i \cdot \boldsymbol{k} = m_i \begin{vmatrix} R_{iG} & 0 & z_i \\ -\omega^2 R_{iG} & \dot{\omega} R_{iG} & 0 \\ 0 & 0 & 1 \end{vmatrix}$

(see equation 4.16)

$$= \dot{\omega} m_i R_{iG}{}^2$$

hence $\quad M_G = \dot{\omega} \sum m_i R_{iG}{}^2 \qquad (6.7)$

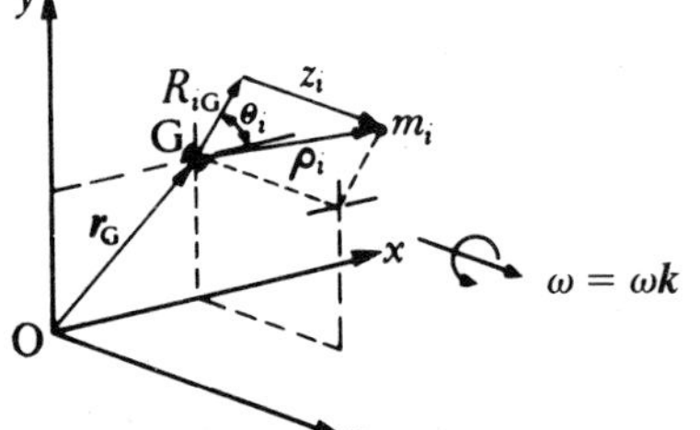

Figure 6.3

The term $\sum m_i R_{iG}{}^2$ is known as the moment of inertia of the body about an axis through G parallel to the z-axis and is given the symbol I_G. I_G is also written $Mk_G{}^2$, where M is the mass and k_G is called the radius of gyration.

Thus

$$M_G = I_G \dot{\omega}$$

and, since ω is fixed in direction and I_G is a constant for the rigid body,

$$M_G = \frac{d}{dt}(I_G \omega) \qquad (6.8)$$

For any rigid body in general plane motion we now have three important equations:

$$F_x = M\ddot{x}_G \qquad (6.9)$$

$$F_y = M\ddot{y}_G \qquad (6.10)$$

and $\quad M_G = I_G \dot{\omega} = Mk_G{}^2 \dot{\omega} \qquad (6.11)$

Should we choose to take moments about some point O other than the centre of mass then, from equation 6.4, *we must add a term equal to the moment of the total mass times the acceleration of the centre of mass* to the right-hand side of equation 6.11. Referring to Fig. 6.4,

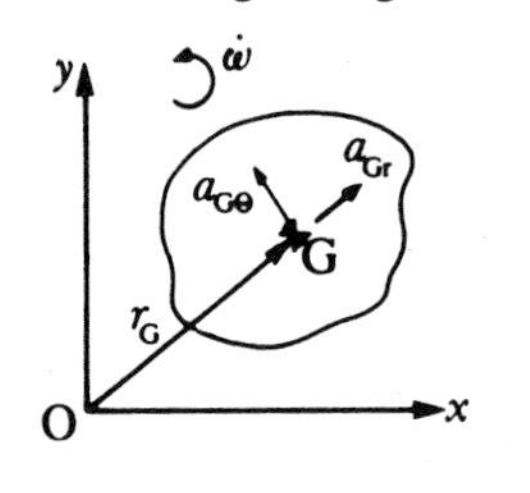

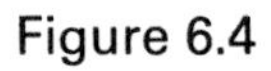

Figure 6.4

$$M_O = I_G \dot{\omega} + r_G M a_{G\theta} \qquad (6.12a)$$

It is sometimes convenient to use vector algebra here, and we note that the final term of equation 6.12a is the component of $(\boldsymbol{r}_G \times M\boldsymbol{a}_G)$ in the z-direction; thus

$$M_O = I_G \dot{\omega} + (\boldsymbol{r}_G \times M\boldsymbol{a}_G) \cdot \boldsymbol{k} \qquad (6.12b)$$

Notice that, even when $\dot{\omega} = 0$, M_O is not necessarily zero. This further emphasises the importance of the centre of mass, because by its use the kinetics of translation and of rotation may be treated separately.

6.2 Rotation about a fixed axis

A special but common case of general plane motion is rotation about a fixed axis. If the axis of rotation passes through O then the angular velocity of the line OG will be the same as that of the body, namely $\omega \boldsymbol{k}$. Referring to Fig. 6.5 and using equation 6.3 we see that the moment about Oz of the external forces is

$$M_O = \sum R_i (m R_i \dot{\omega})$$

$$= \dot{\omega} \sum m_i R_i^2 = \dot{\omega} I_O$$

where $I_O = \sum m_i R_i^2$ is defined to be the moment of inertia about Oz.

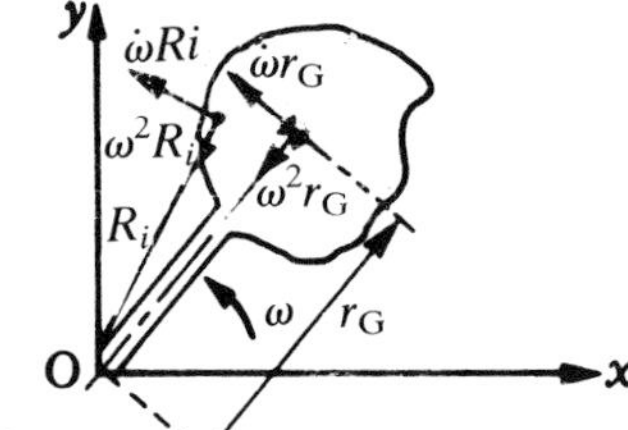

Figure 6.5

Also, from equation 6.12a

$$M_O = I_G \dot{\omega} + r_G M \dot{\omega} r_G$$

$$= (I_G + M r_G{}^2) \dot{\omega} \qquad (6.13a)$$

Combining the two expressions for M_O. We have

$$M_O = (I_G + M r_G{}^2) \dot{\omega} = I_O \dot{\omega} \qquad (6.13b)$$

6.3 Moment of inertia of a body about an axis

Parallel-axes theorem

The moment of inertia about the z-axis is defined to be

$$I_O = \sum m_i R_i^2 = \sum m_i (x_i{}^2 + y_i{}^2)$$

From Fig. 6.6,

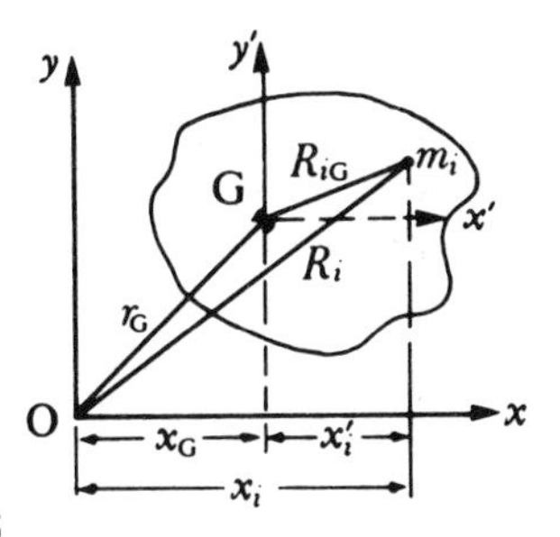

Figure 6.6

$$x_i = x_G + x_i' \quad \text{and} \quad y_i = y_G + y_i'$$

so $$x_i^2 + y_i^2 = (x_G^2 + y_G^2) + (x_i'^2 + y_i'^2) + 2x_G x_i' + 2y_G y_i'$$

$$= r_G^2 + R_{iG}^2 + 2x_G x_i' + 2y_G y_i'$$

By virtue of the properties of the c.m.,

$$x_G \sum m_i x_i' = 0$$

and $$y_G \sum m_i y_i' = 0$$

thus $$I_O = \sum m_i r_G^2 + \sum m_i R_{iG}^2$$

$$= M r_G^2 + I_G \tag{6.14}$$

$$= M(r_G^2 + k_G^2) = M k_O^2 \tag{6.15}$$

where k_O is the radius of gyration about the z-axis.

Perpendicular-axes theorem
Consider the thin lamina in the xy plane shown in Fig. 6.7.

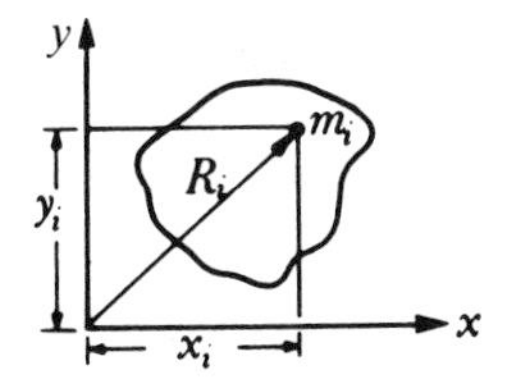

Figure 6.7

$$I_x = \sum m_i y_i^2$$

$$I_y = \sum m_i x_i^2$$

and $$I_z = \sum m_i R_i^2$$

$$= \sum m_i (x_i^2 - y_i^2)$$

$$= I_x + I_y \tag{6.16}$$

Moment of inertia of a right circular uniform cylinder
i) *Moment of inertia about the axis of the cylinder.* In Fig. 6.8, the mass of an elemental rod is $\rho L\,\mathrm{d}r(r\,\mathrm{d}\theta)$, where ρ is the density of the material.
Moment of inertia about the axis

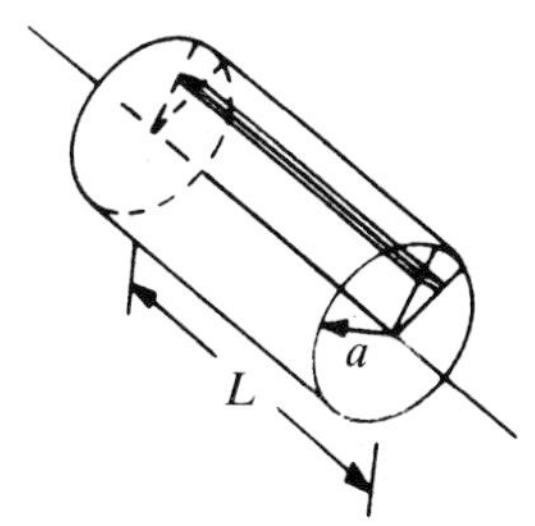

Figure 6.8

$$= (\rho L\,\mathrm{d}r\,r\,\mathrm{d}\theta)r^2$$

hence for the whole body

$$I_{Gz} = \int_0^{2\pi}\int_0^a \rho L r^3\,\mathrm{d}r\,\mathrm{d}\theta$$

$$= \int_0^a \rho L r^3\,\mathrm{d}r\,2\pi = \rho L 2\pi a^4/4 = \tfrac{1}{2}\pi\rho L a^4$$

The mass of the cylinder is $\rho\pi a^2 L$, therefore

$$I_{Gz} = Ma^2/2 = Mk_{Gz}^2$$

ii) *Moment of inertia about an end diameter.* For a circular lamina, relative to its own centre of mass (Fig. 6.9), $I_x = I_y$; hence, from the perpendicular-axes theorem,

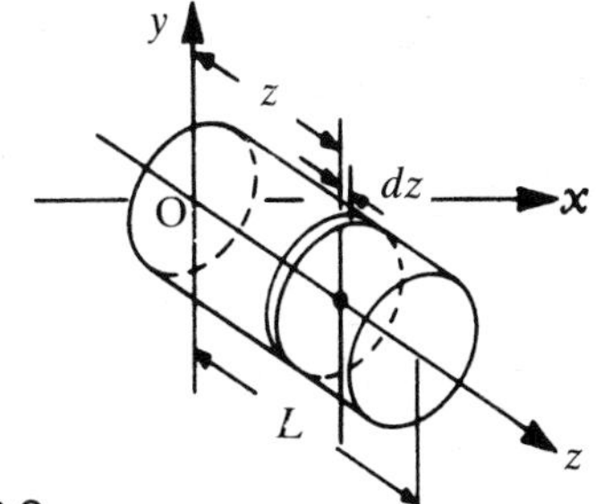

Figure 6.9

$$I_x = I_y = \tfrac{1}{2}I_z = \tfrac{1}{4}\pi\rho a^4\,\mathrm{d}z$$

The moment of inertia about the x-axis may be found through the parallel-axes theorem. Hence, for the lamina,

$$I_x = \tfrac{1}{4}\pi\rho a^4\,\mathrm{d}z + \rho\pi a^2\,\mathrm{d}z\,z^2$$

and integrating for the whole bar gives

$$I_x = \pi\rho a^2 \int_0^L (\tfrac{1}{4}a^2 + z^2)\,\mathrm{d}z$$

$$= (\rho\pi a^2 L)\left(\frac{a^2}{4} + \frac{L^2}{3}\right)$$

We may use the parallel-axes theorem to find the moment of inertia about a diameter through the centre of mass:

$$I_{Gx} = M\left(\frac{a^2}{4} + \frac{L^2}{3}\right) - M\left(\frac{L}{2}\right)^2$$

$$= M\left(\frac{a^2}{4} + \frac{L^2}{12}\right)$$

6.4 Application

As an example of the use of the preceding theory, consider the problem of the cable drum shown in Fig. 6.10.

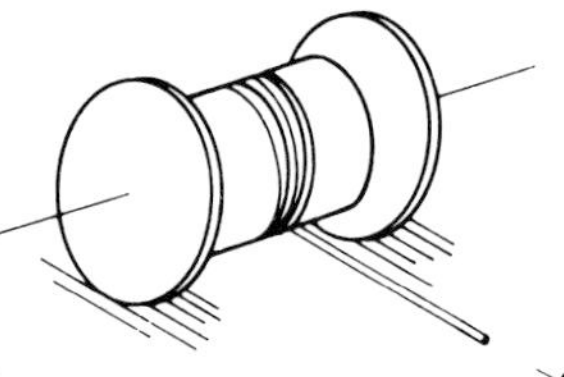

Figure 6.10

Let us assume that the drum has symmetry, that the cable is horizontal and that the friction between the ground and the drum is sufficient to prevent slip. If the tension in the cable is T, what is the acceleration of the drum and the direction of motion?

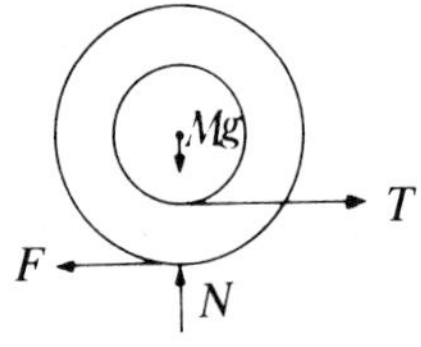

Figure 6.11

The first and important step is to draw the free-body diagram as shown in Fig. 6.11. The next step is to establish the kinematic constraints (see Fig. 6.12). In this case the condition of no slip at the ground gives

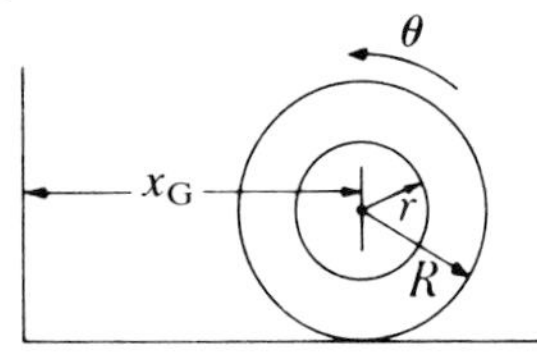

Figure 6.12

$$\dot{x}_G = -R\dot{\theta}, \; \dot{y}_G = 0$$

and $\ddot{x}_G = -R\ddot{\theta}, \quad \ddot{y}_G = 0$ (6.17)

We can now write the three equations of motion (see equations 6.9–6.11 and Fig. 6.11):

$$T - F = M\ddot{x}_G \tag{6.18}$$

$$N - Mg = 0 \tag{6.19}$$

$$Tr - FR = I_G\ddot{\theta} \tag{6.20}$$

Eliminating F and N leads to

$$T\left(1 - \frac{r}{R}\right) = M\ddot{x}_G - I_G\frac{\ddot{\theta}}{R} \tag{6.21}$$

and, since with no slip $\ddot{\theta} = -\ddot{x}_G/R$,

$$T(1 - r/R) = (M + I_G/R^2)\ddot{x}_G$$

hence
$$\ddot{x}_G = \frac{T(1 - r/R)}{(M + I_G/R^2)} \tag{6.22}$$

Since $R > r$, $\ddot{x}_G$ is positive and thus the drum will accelerate to the right. As the drum started from rest, it follows that the motion is directed to the right. An intuitive guess might well have produced the wrong result.

Discussion examples

Example 6.1

Figure 6.13 shows two pulleys, P_1 and P_2, connected by a belt. The effective radius of pulley P_2 is r and its axial moment of inertia is I. The system is initially at rest and the tension in the belt is T_0. The motor M which drives pulley P_1 is then started and it may be assumed that the average of the tensions T_{AB} and T_{CD} in sections AB and CD of the belt remains equal to T_0. Denoting the anticlockwise angular acceleration of pulley P_2 by α and the clockwise resisting couple on the same axle by Q, find expressions for T_{AB} and T_{CD}, neglecting the mass of the belt.

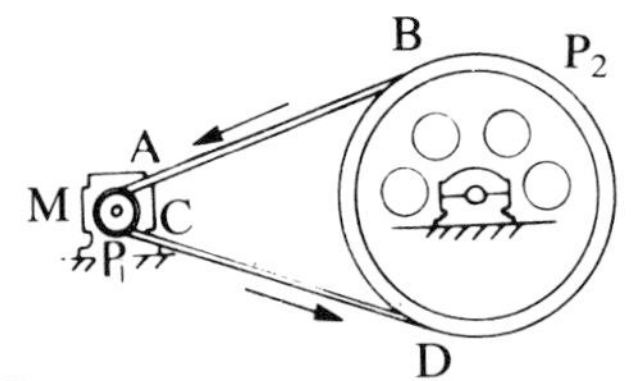

Figure 6.13

Solution The solutions of problems in this chapter start with a similar pattern to those of Chapter 3, first drawing the free-body diagram(s) and then writing down the appropriate equation(s) of motion.

In the present problem there are four forces and one couple acting on pulley P_2; these are shown in the free-body diagram (Fig. 6.14). T_{AB} and T_{CD} are the belt tensions and Q is the load couple mentioned above. R is the contact force at the axle and W is the weight; these two forces can be eliminated by taking moments about the pulley axle.

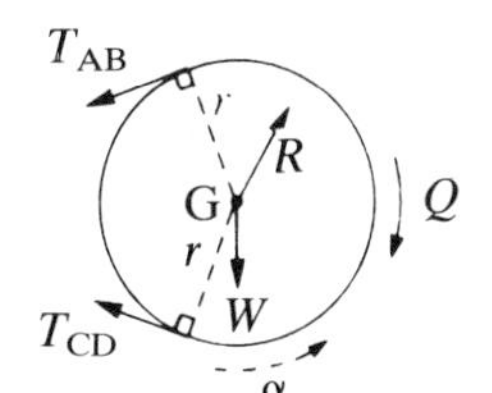

Figure 6.14

$$\sum M_G = I_G \dot{\omega} \quad \text{(equation 6.11)}$$

$$T_{AB}r - T_{CD}r - Q = I\alpha \qquad \text{(i)}$$

The average value of T_{AB} and T_{CD} is T_0, so that

$$\tfrac{1}{2}(T_{AB} + T_{CD}) = T_0 \qquad \text{(ii)}$$

Solving for T_{AB} and T_{CD} from equations (i) and (ii), we find

$$T_{AB} = T_0 + (I\alpha + Q)/(2r)$$

and $\quad T_{CD} = T_0 - (I\alpha + Q)/(2r)$

Example 6.2

The hoist shown in Fig. 6.15 consists of a winding drum D driven by an electric motor M. A pinion of N_M teeth on the motor shaft meshes with a gear wheel of N_D teeth attached to the drum. The effective radius of the drum is R. The total axial moment of inertia on the drum shaft is I_D and on the motor shaft is I_M. The rope wrapped round the drum carries a load of mass m at its lower end which is being raised with an acceleration a. Neglect bearing friction and the mass and stretch of the rope and determine the couple being applied to the rotor of the motor.

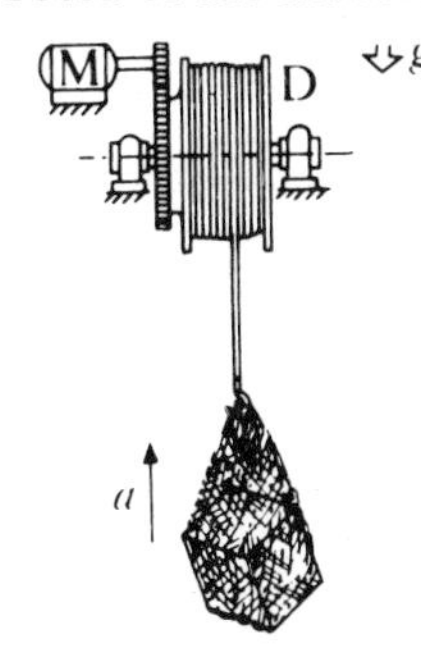

Figure 6.15

Solution The free-body diagrams for the motor, drum and load are shown in Fig. 6.16. Forces which pass through the axles of the motor and the drum will be eliminated by taking moments about the axles. The contact force between the teeth has been resolved into a tangential (F) and normal (N) component. The forces P_M and P_D are the axle contact forces and W_M and W_D are the weights. The tension T in the vertical portion of the cable does not vary since its mass is negligible. C_M is the required couple.

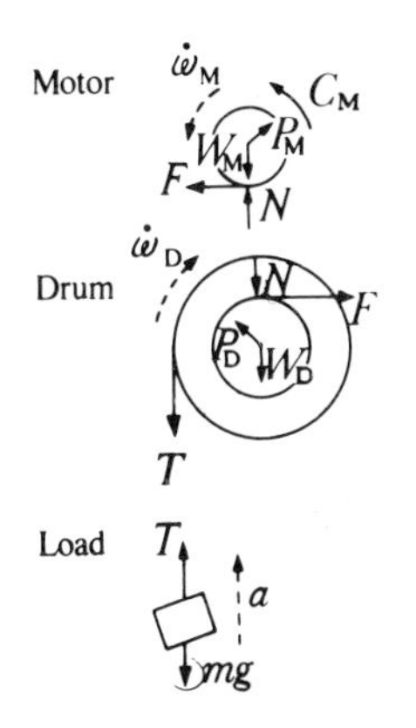

Figure 6.16

Taking moments about the axle of the motor, from equation 6.11,

$$C_M - Fr_M = I_M \dot{\omega}_M \qquad \text{(i)}$$

where r_M is the effective radius of the motor pinion.

Taking moments about the axle of the drum,

$$Fr_D - TR = I_D \dot{\omega}_D \qquad \text{(ii)}$$

where r_D is the effective radius of the drum gear wheel.

The force equation for the load is

$$T - mg = ma \qquad \text{(iii)}$$

The numbers of teeth on the pinion and wheel are proportional to their radii and hence

$$\frac{r_M}{r_D} = \frac{N_M}{N_D}$$

and it follows that

$$\frac{\omega_M}{\omega_D} = \frac{\dot{\omega}_M}{\dot{\omega}_D} = \frac{N_D}{N_M} \qquad \text{(iv)}$$

The final required relationship is

$$a = R\dot{\omega}_D \qquad \text{(v)}$$

since the rope does not stretch.

Combining equations (i) to (v), we find

$$C_M = \frac{N_M}{N_D}\left[\left\{I_D + I_M\left(\frac{N_D}{N_M}\right)^2 + mR^2\right\}\frac{a}{R} + Rmg\right]$$

This type of problem is readily solved by the energy methods described in the next chapter.

Example 6.3

Figure 6.17 shows an experimental vehicle powered by a jet engine whose thrust can be represented by the equivalent concentrated force P acting on the vehicle as shown. The vehicle is suspended from light wheels at A and B which run on the straight horizontal track. Friction at the wheels is negligible. The total mass of the vehicle is 4000 kg and the mass centre is at G.

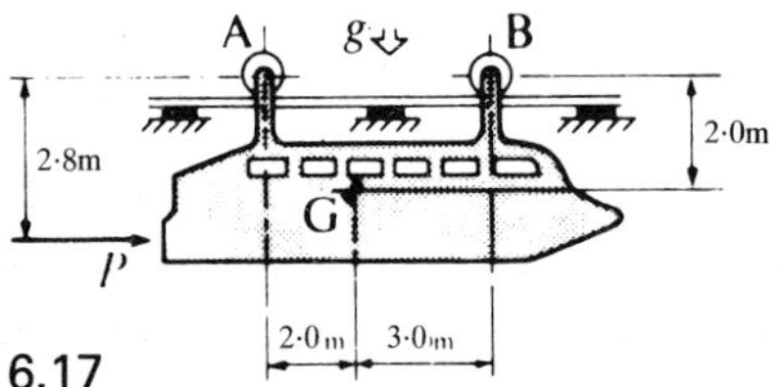

Figure 6.17

a) If wind resistance can be neglected, determine the maximum permissible value of P consistent with the wheel at B remaining in contact with the track. What would be the acceleration a of the vehicle for this value of P?

b) If wind resistance were taken into account, would the maximum permissible value of P consistent with the wheel at B remaining in contact with the track necessarily always exceed that obtained in (a) above? Give reasons for your answer.

Solution Let us first consider the motion of the wheels, whose mass is to be neglected. The right-hand side of any equation of motion for a body of negligible mass will be zero, and the equation will be the same as though the body were in equilibrium (Chapter 4). In the present case there are only two forces (and no couples) acting on a wheel: the contact force at the axle and the contact force with the track. These forces must therefore be equal, opposite and collinear. The contact points lie on a vertical line so that the forces are vertical (Fig. 6.18(a)).

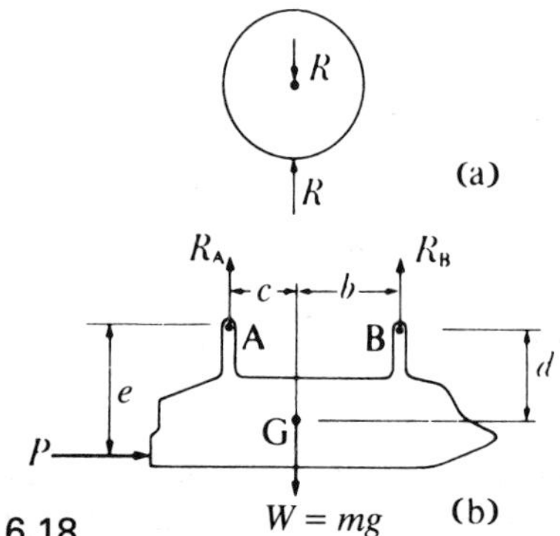

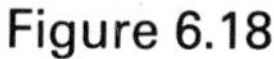

Figure 6.18

a) The free-body diagram for the vehicle has vertical forces at A and B together with the thrust P and the weight W, as shown in Fig. 6.18(b).

For the x-direction ($\sum F_x = m\ddot{x}_G$),

$$P = ma \qquad \text{(i)}$$

and, for the y-direction ($\sum F_y = m\ddot{y}_G$),

$$R_A + R_B - W = 0 \qquad \text{(ii)}$$

If we next take moments about G, ($\sum M_G = I_G\dot{\omega}$),

$$(e-d)P + bR_B - cR_A = 0 \qquad \text{(iii)}$$

We have assumed that there is no rotation ($\dot{\omega} = 0$). Denoting the required value of P by P_O we note that when $P = P_O$, $R_B = 0$ but P_O is just not sufficient to cause rotation. Eliminating R_A, we find

$$P_O = \frac{mgc}{e-d} \qquad \text{(iv)}$$

and, numerically,

$$P_O = \frac{4000(9.81)2}{2.8-2}$$

$$= 98100\text{N} = 98.1 \text{ kN}$$

The corresponding acceleration, a_O, from equation (i) is

$$a_O = P_O/m$$
$$= 98100/4000 = 24.53 \text{ m/s}^2$$

Since R_A is not required, we could have used a single equation for moments about A (and thus eliminated R_A) instead of equations (ii) and (iii). When taking moments about some general point O, the appropriate equation is

$$\sum M_o = I_G\dot{\omega} + r_G m a_{G\theta} \qquad \text{(equation 6.12a)}$$

or $\sum M_O = I_G\dot{\omega} + (\boldsymbol{r}_G \times m\boldsymbol{a}_G)\cdot\boldsymbol{k}$ (equation 6.12b)

If the second of these equations is used directly, the positive direction for moments is determined by the sign convention for the vector product.

In the present problem it is clear that the (anticlockwise) moment of ma_G about A is dma and it is unnecessary to carry out the vector products

$$(\boldsymbol{r}_G \times \boldsymbol{a}_G)\cdot\boldsymbol{k} = [(c\boldsymbol{i} - d\boldsymbol{j}) \times ma\boldsymbol{i}]\cdot\boldsymbol{k}$$
$$= dma\boldsymbol{k}\cdot\boldsymbol{k} = dma$$

Thus, from either equation 6.12a or 6.12b, taking moments about A and putting $R_B = 0$, $P = P_O$,

$$\sum M_A = eP_O - cmg = dma \qquad \text{(v)}$$

and substituting for a from equation (i) gives

$$P_O = mgc/(e-d) \quad \text{as before.}$$

b) The answer to this part of the question is 'not necessarily'. Suppose (see Fig. 6.19) that the resultant F of the wind resistance is horizontal and that the centre of pressure is a distance f below the axles. P_1 is the value of P that just makes $R_B = 0$ under these conditions. Equation (i) becomes

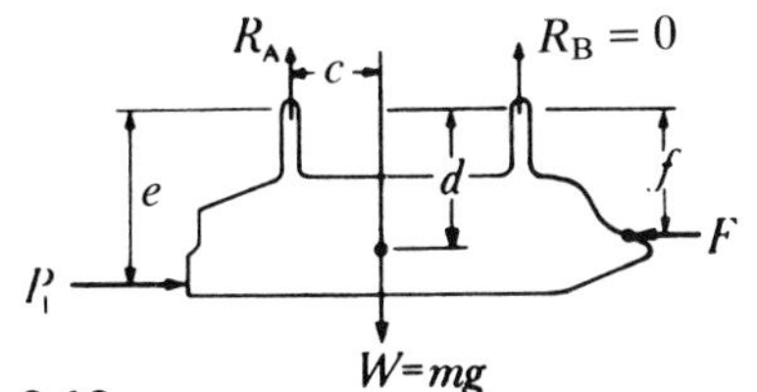

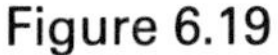
Figure 6.19

$$P_1 - F = ma$$

and equation (v) becomes

$$eP_1 - cmg - fF = mda$$

Eliminating a gives

$$P_1 = \frac{mgc + F(f-d)}{e-d} \qquad \text{(vi)}$$

Comparing equations (iv) and (vi) we see that P_1 is greater than P_O only if $f > d$.

Example 6.4

When predicting the maximum acceleration of a motorcycle, it is necessary to consider (a) the power available at a given speed, (b) the tendency of the front wheel to lift and (c) the tendency of the rear wheel to slip.

A motorcycle and rider are travelling over a horizontal road, the combined centre of mass being 0.7 m above the road surface and 0.8 m in front of the axle of the rear wheel (see Fig. 6.20). The wheelbase of the motorcycle is 1.4 m. If the coefficient of friction between the tyres and the road is 0.8, find the maximum possible acceleration, neglecting the resistance of the air and assuming that the acceleration is not limited by the power available. Neglect the mass of the wheels.

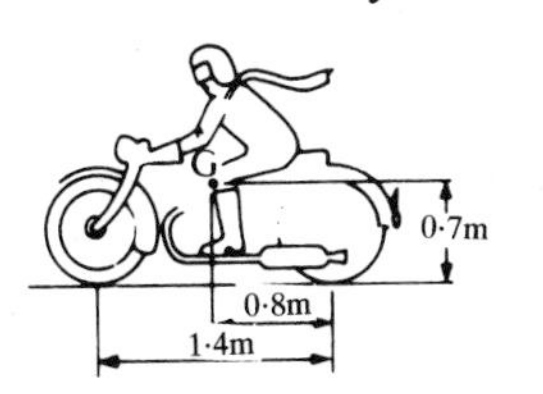

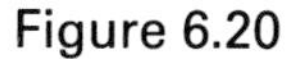
Figure 6.20

Solution If we make the assumption that the front wheel is on the point of lifting (i.e. zero force between front wheel and ground), the tangential component F_R and the normal component N_R of the contact force between the rear wheel and the road can be determined. If F_R is less than or equal to μN_R then the maximum acceleration is limited by front-wheel lift and our assumption was valid. If, on the other hand, F_R is found to be greater than μN_R our assumption was invalid since this is not possible. The problem must then be reworked assuming that slip is taking place at the rear wheel.

Front wheel on point of lifting (Fig. 6.21)
Taking moments about B, from equation 6.12a or 6.12b, replacing O by B,

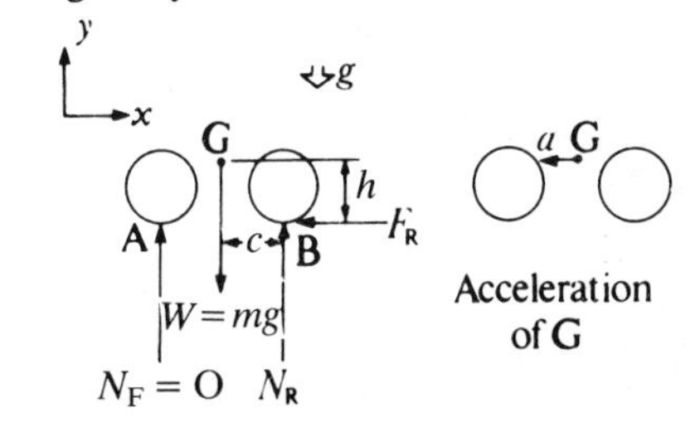

Figure 6.21

$$mgc = 0 + mah$$

$$a = gc/h$$

$$= 9.81(0.8)/0.7 = 11.21 \text{ m/s}^2$$

For the x-direction ($\sum F_x = m\ddot{x}_G$),

$$-F_R = m(-a)$$

$$F_R = m(11.21)$$

For the y-direction ($\sum F_y = m\ddot{y}_G$),

$$N_R - mg = 0$$

$$N_R = m(9.81)$$

The ratio $$\frac{F_R}{N_R} = \frac{11.21}{9.81} = 1.143$$

The ratio F/N cannot exceed the value of the coefficient of friction μ, which is 0.8, and so the original assumption is invalid. The maximum acceleration is therefore limited by rear-wheel slip.

Rear-wheel slip (Fig. 6.22)
Since the wheels are light, the contact force between the front wheel and the ground is vertical (see example 6.3) and we can replace F_R by μN_R.

For the x-direction $[\sum F_x = m\ddot{x}_G]$,

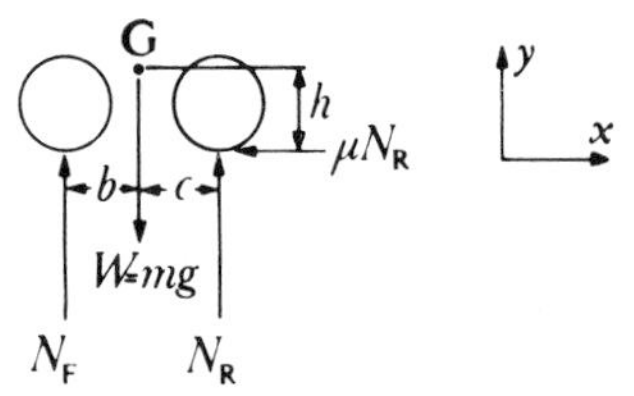

Figure 6.22

$$-\mu N_R = m(-a)$$

For the y-direction $[\sum F_y = m\ddot{y}_G]$,

$$N_F + N_R - mg = 0$$

Taking moments about G $[\sum M_G = I_G \dot{\omega}]$,

$$cN_R - bN_F - h\mu N_R = 0$$

Substituting numerical values and eliminating N_R and N_F we find that $a = 5.61\ \text{m/s}^2$.

Example 6.5

See Fig. 6.23(a). The slider B of mass m is constrained to move in vertical guides. A pin P fixed to the slider engages with the slot in link OA which rotates about O. The mass of the link is M and its moment of inertia about O is I_O. G is the mass centre of the link and OG $= a$. A spring of stiffness k restrains the motion of B and is unstrained when $\theta = 0$.

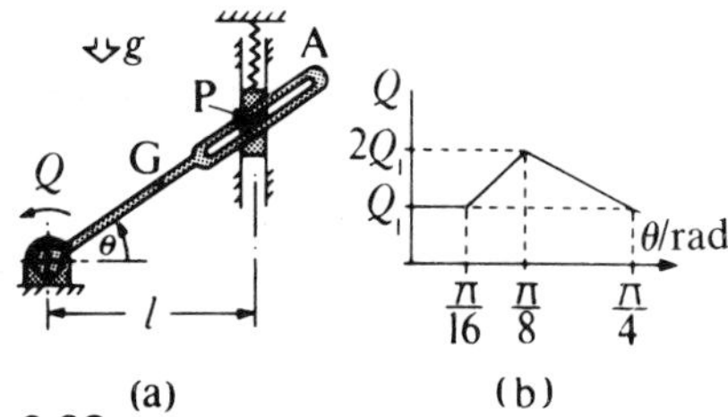

Figure 6.23

The system is released from rest at $\theta = 0$ under the action of the couple Q which is applied to link OA. The variation of Q with θ is shown in Fig. 6.23(b). Assuming that the couple is large enough to ensure that θ attains the value of 45°, determine the angular velocity ω of the link OA at this angle. Neglect friction.

Solution

Link OA. In the free-body diagram for the link OA (Fig. 6.24), S, the contact force with the pin P, is perpendicular to the link since friction is negligible. R is the contact force at the axis O.

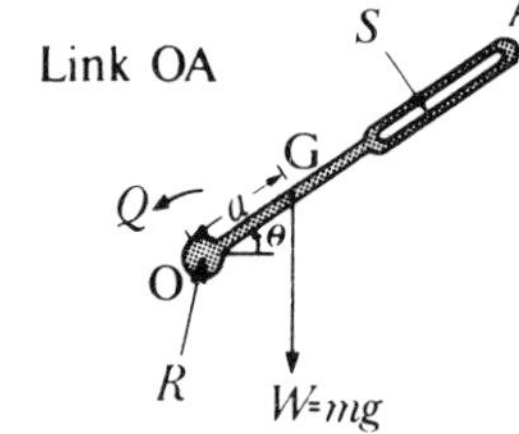

Figure 6.24

Since the link is rotating about a fixed axis, the appropriate moment equation is equation 6.13 and our aim is to replace $\dot{\omega}$ by $\omega\,\mathrm{d}\omega/\mathrm{d}\theta$ and to integrate the equation to find ω at the required value of θ.

$$[\sum M_O = I_O \dot{\omega}]$$

$$Q - Mga\cos\theta - Sl\sec\theta = I_O\,\omega\frac{\mathrm{d}\omega}{\mathrm{d}\theta} \qquad \text{(i)}$$

We need a suitable expression for S before ω at $\theta = \pi/4$ can be determined. Note that $\int_0^{\pi/4} Q\,\mathrm{d}\theta$ is simply the area under the graph of Q against θ.

Slider B. In the free-body diagram for the slider B (Fig. 6.25), S is the force on the pin P (equal and opposite to that on the link OA). Denoting the upward displacement of the block by y, the downward spring force on the slider is ky. N is the contact force of the guide on the slider.

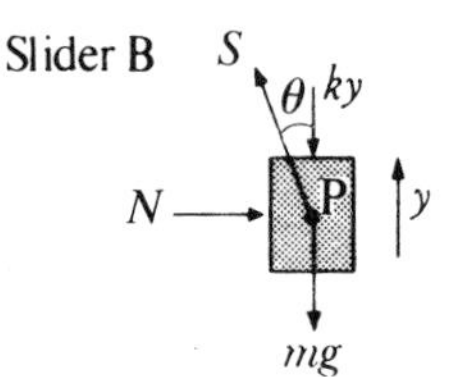

Figure 6.25

$$[\sum F_y = m\ddot{y}_G]$$

$$S\cos\theta - ky - mg = m\ddot{y} \qquad \text{(ii)}$$

From the geometry of the linkage, $y = l\tan\theta$ and hence

$$\ddot{y} = l\sec^2\theta(\dot{\omega} + 2\tan\theta\,\omega^2)$$

S is thus given by

$$S = \sec\theta\left[ml\sec^2\theta\left(\omega\frac{\mathrm{d}\omega}{\mathrm{d}\theta} + 2\tan\theta\,\omega^2\right) + kl\tan\theta + mg\right]$$

When this expression for S is substituted in equation (i), a cumbersome differential equation results. Since only the angular velocity of the link is required, we shall defer this problem to the next chapter, where it is readily solved by an energy method in Example 7.2.

Example 6.6

Figure 6.26 shows part of a mechanical flail which consists of links AB and BC pinned together at B. Link AB rotates at a constant anticlockwise angular velocity of 25 rad/s and, in the position shown, the instantaneous angular velocity of BC is 60 rad/s anticlockwise. The links are each made from uniform rod of mass 2 kg/m.

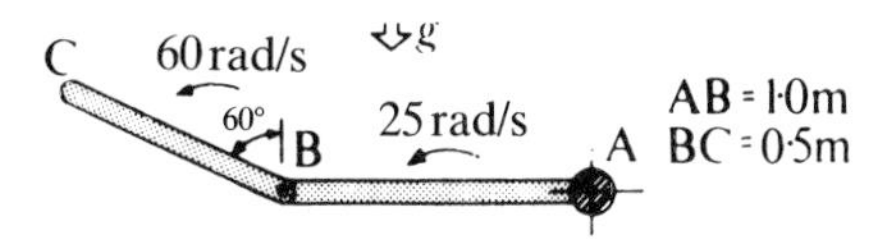

Figure 6.26

Determine the angular acceleration of BC and the bending moment in the rod AB at A.

Solution Figure 6.27 shows the separate free-body diagrams for AB and BC. Subscripts 1 and 2 relate to AB and BC respectively. R_B is the contact force at the pinned joint B. Since A is not pinned, there will be a force R_A and a couple Q acting there. The magnitude of Q is the required bending moment.

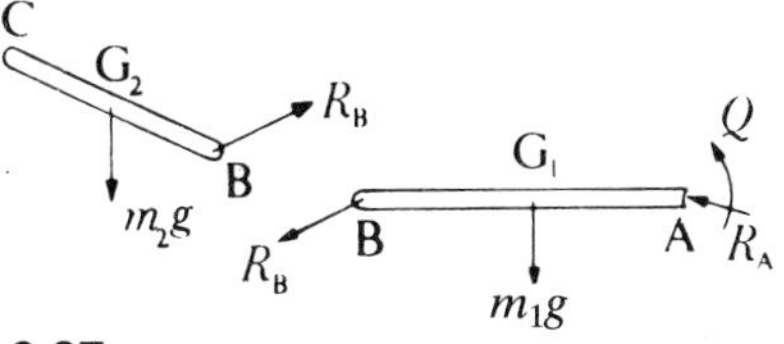

Figure 6.27

R_B can be found from equations of motion for link BC. If an equation for moments about A for link AB is then written, this will not contain R_A and Q can be found. A solution using this approach is left as an exercise for the reader, but a technique will be described below which does not involve the determination of R_B.

Just as equations of motion can be written for systems of particles, so they can be written for systems of rigid bodies. Suppose that n rigid bodies move in the xy-plane. If the force equations for the bodies are summed, we obtain

$$\sum \boldsymbol{F} = \sum_{i=1}^{n} m_i \boldsymbol{a}_{Gi} \qquad \text{(i)}$$

where $\sum \boldsymbol{F}$ is the sum of all the forces acting on the system. Summing the equations for moments about some point O, we obtain from equation 6.12b

$$\sum M_O = \sum_{i=1}^{n} I_{Gi}\dot{\omega}_i + \sum_{i=1}^{n} (\boldsymbol{r}_{Gi} \times m\boldsymbol{a}_{Gi}) \cdot \boldsymbol{k} \qquad \text{(ii)}$$

where $\sum M_O$ is the sum of all the moments acting on the system. These equations are often useful when two or more bodies are in contact, since the contact forces, appearing in equal and opposite pairs, do not appear in the equations.

Let us start the present problem in the usual way by using the free-body diagram for BC alone. The forces acting on the link are the weight m_2g and the contact force R_B (Fig. 6.28).

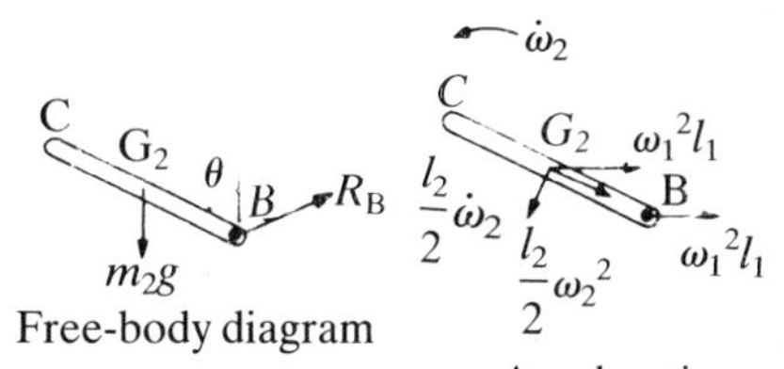

Figure 6.28

Since $\boldsymbol{a}_{G2} = \boldsymbol{a}_B + \boldsymbol{a}_{G2/B}$, the acceleration of G_2 has the three components shown. There are only two unknowns, R_B and $\dot{\omega}_2$, so we can find the latter by taking moments about B.

$$\Big[\sum M_B = I_{G2}\dot{\omega}_2 + (\text{moment of components of } ma_{G2} \text{ about B})\Big]$$

$$\left(\frac{l_2}{2}\sin\theta\right)(m_2 g) = \frac{m_2 l_2^2}{12}\dot{\omega}_2 + m_2\left[\frac{l_2}{2}\left(\frac{l_2}{2}\dot{\omega}_2\right) - \left(\frac{l_2}{2}\cos\theta\right)(\omega_1^2 l_1)\right]$$

Dividing by m_2 and substituting numerical values,

$$\frac{0.5}{2}\left(\frac{\sqrt{3}}{2}\right)(9.81) = \left(\frac{0.5}{12}\right)^2\dot{\omega}_2 + \frac{(0.5)^2}{2}\dot{\omega}_2 - \frac{0.5}{2}(\tfrac{1}{2})(25)^2(1)$$

$$\dot{\omega}_2 = 963.0 \text{ rad/s}^2$$

If we now combine the free-body diagrams for the two links the internal contact force at B will not appear and by taking moments about A for the whole system using equation (ii) we can find Q (see Fig. 6.29).

Free-body diagram

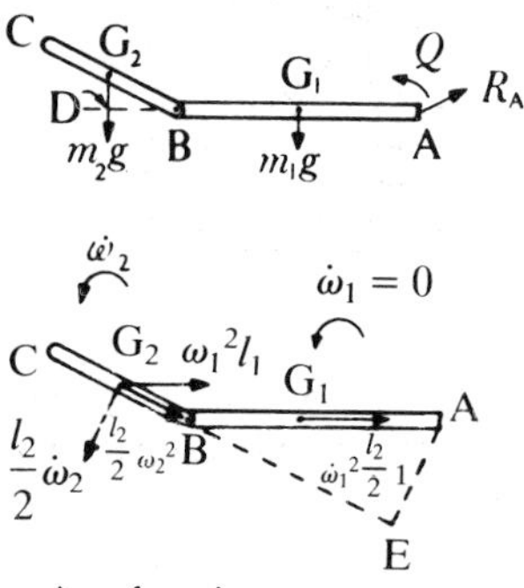

Accelerations

Figure 6.29

From equation (ii),

$$[\sum M_A = I_{G1}\dot{\omega}_1 + I_{G2}\dot{\omega}_2 + (\text{moment of } m_1 a_{G1} \text{ about A}) + (\text{moment of } m_2 a_{G2} \text{ about A})]$$

$$(AG_1)m_1 g + (AD)m_2 g + Q = 0 + I_{G2}\dot{\omega}_2 + 0 + m_2\left[(EG_2)\frac{l_2}{2}\dot{\omega}_2 + (AE)\frac{l_2}{2}\omega_2{}^2 - (DG_2)\omega_1{}^2 l_1\right]$$

$$0.5(2)(9.81) + \left(1 + 0.25\frac{\sqrt{3}}{2}\right)(1)(9.81) + Q = 0.25\frac{(0.5)^2}{12}(963) + 1\left[\left(\frac{\sqrt{3}}{2} + 0.25\right) \times \left(\frac{0.5}{2}\right)(963) + 0.5\left(\frac{0.5}{2}\right)(60)^2 - (0.125)(25)^2(1)\right]$$

Hence $Q = -1383$ N m

Problems

6.1 A thin uniform rod has a length l and a mass m. Show that the moments of inertia about axes through the mass centre and one end, perpendicular to the rod, are $ml^2/12$ and $ml^2/3$ respectively.

6.2 The uniform rectangular block shown in Fig. 6.30 has a mass m. Show that the moments of inertia for the given axes are

$$I_{11} = \frac{1}{12}m(l^2 + b^2),$$

$$I_{22} = \frac{1}{12}m(a^2 + b^2),$$

$$I_{33} = m\left(\frac{1}{3}l^2 + \frac{1}{12}a^2\right)$$

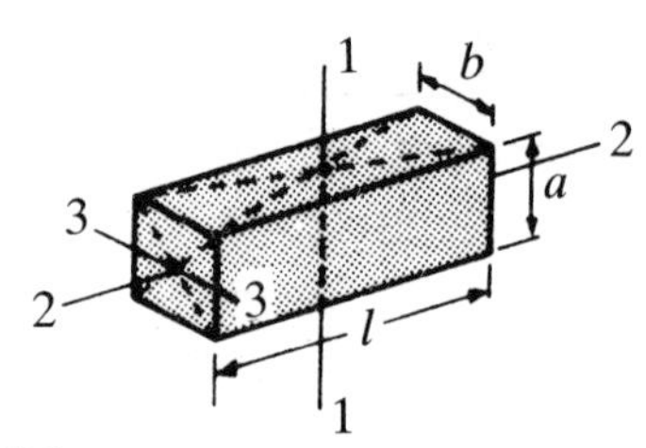

Figure 6.30

6.3 A uniform solid hemisphere has a radius r. Show that the mass centre is a distance $r_G = 3r/8$ from the flat surface.

6.4 Determine the location of the mass centre of the uniform thin plate shown in Fig. 6.31.

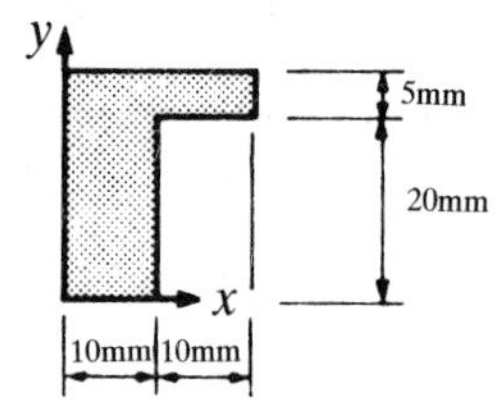

Figure 6.31

6.5 A couple $C = k\theta$ is applied to a flywheel of moment of inertia I whose angle of rotation is θ, and k is a constant. When the flywheel has rotated through one revolution, show that the angular velocity is $2\pi\sqrt{(k/I)}$ and the angular acceleration is $2\pi k/I$.

6.6 A flywheel consists of a uniform disc of radius R and mass m. Friction at the axle is negligible, but motion is restrained by a torsional spring of stiffness k so that the couple applied to the flywheel is $k\theta$ in the opposite sense to θ, the angle of rotation. If the system is set into motion, show that it oscillates with periodic time $2\pi R\sqrt{[m/(2k)]}$.

6.7 A light cord is wrapped round a pulley of radius R and axial inertia I and supports a body of mass m. If the system is released from rest, assuming that the cord does not slip on the pulley, show that the acceleration a of the body is given by

$$a = mgR^2/(I + mR^2)$$

if friction at the axle is negligible.

6.8 Repeat problem 6.7 assuming that there is a friction couple C_O at the axle which is insufficient to prevent motion and show that

$$a = (mgR^2 - C_O R)/(I + mR^2)$$

6.9 See Fig. 6.32. The coefficient of friction between body A and the horizontal surface is μ. The pulley has a radius R and axial moment of inertia I. Friction at the axis is such that the pulley will not rotate unless a couple of magnitude C_O is applied to it. If the rope does not slip on the pulley, show that the acceleration a is given by

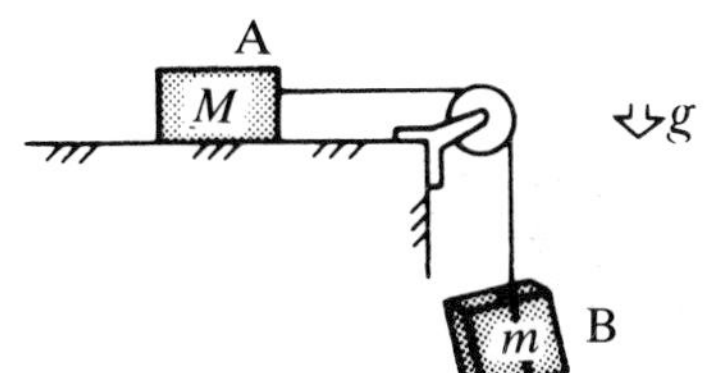

Figure 6.32

$$a = \frac{g(m - \mu M)R^2 - C_O R}{(m + M)R^2 + I}$$

provided that $m > \mu M + C_O/(gR)$.

6.10 Figure 6.33 shows a small service lift of 300 kg mass, connected via pulleys of negligible mass to two counterweights each of 100 kg. The cable drum is driven directly by an electric motor, the mass of all rotating parts being 40 kg and their combined radius of gyration being 0.5 m. The diameter of the drum is 0.8 m.

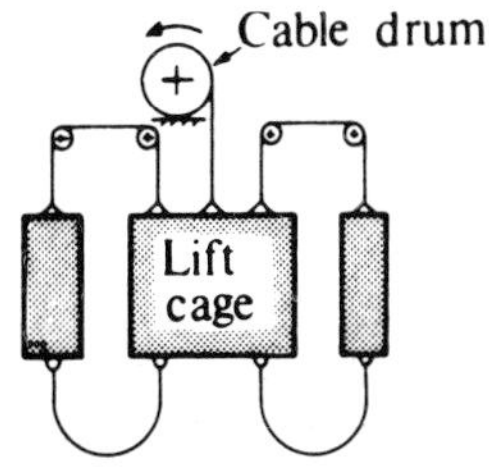

Figure 6.33

If the torque supplied by the motor is 50 N m, calculate the tensions in the cables.

6.11 The jet aircraft shown in Fig. 6.34 uses its engines E to increase speed from 5 m/s to 50 m/s in a distance of 500 m along the runway, with constant acceleration. The total mass of the aircraft is 120000 kg, with centre of mass at G.

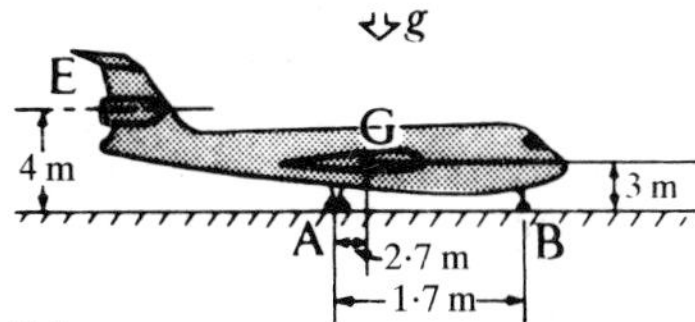

Figure 6.34

Find, neglecting aerodynamic forces and rolling resistance, (a) the thrust developed by the engines and (b) the normal reaction under the nose wheel at B during this acceleration.

6.12 See Fig. 6.35. The crank OB, whose radius is 100 mm, rotates clockwise with uniform angular speed $\dot{\theta} = +5$ rad/s. A pin on the crank at B engages with a smooth slot S in the member A, of mass 10 kg, which is thereby made to reciprocate on the smooth horizontal guides D. The effects of gravity may be neglected.

For the position $\theta = 45°$, sketch free-body diagrams for the crank OB and for the member A, and hence find

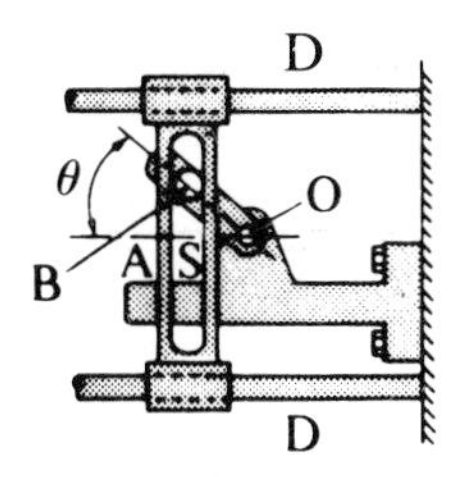

Figure 6.35

(a) the magnitude of the force exerted by the pin on the member A and (b) the driving couple which must be applied to the crank OB.

6.13 In problem 5.3, the spring S has a stiffness of 4 kN/m and is pre-compressed such that when the line OA is perpendicular to the motion of the follower F the compressive force in the spring is 150 N. The mass of the follower is 0.2 kg. The eccentricity $e = 10$ mm.

Neglecting friction and the mass of the spring, determine the maximum speed at which the cam C can run so that the follower maintains continuous contact.

6.14 The distance between the front and rear axles of a motor vehicle is 3 m and the centre of mass is 1.2 m behind the front axle and 1 m above ground level. The coefficient of friction between the wheels and the road is 0.4.

Assuming front-wheel drive, find the maximum acceleration which the vehicle can achieve on a level road.

During maximum acceleration, what are the vertical components of the forces acting on the road beneath the front and rear wheels if the mass of the vehicle is 1000 kg?

Neglect throughout the moments of inertia of all rotating parts.

6.15 The car shown in Fig. 6.36 has a wheelbase of 3.60 m and its centre of mass may be assumed to be midway between the wheels and 0.75 m above ground level. All wheels have the same diameter and the braking system is designed so that equal braking torques are applied to front and rear wheels. The coefficient of friction between the tyres and the road is 0.75 under the conditions prevailing.

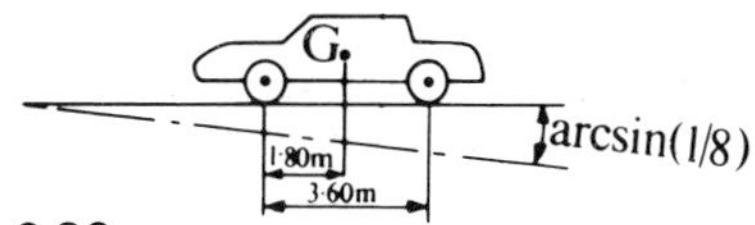

Figure 6.36

When the car is coasting down the gradient of 1 in 8 at 45 km/h, the brakes are applied as fully as possible without producing skidding at any of the wheels. Calculate the distance the car will travel before coming to rest.

6.16 The track of the wheels of a vehicle is 1.4 m and the centre of gravity G of the loaded vehicle is located

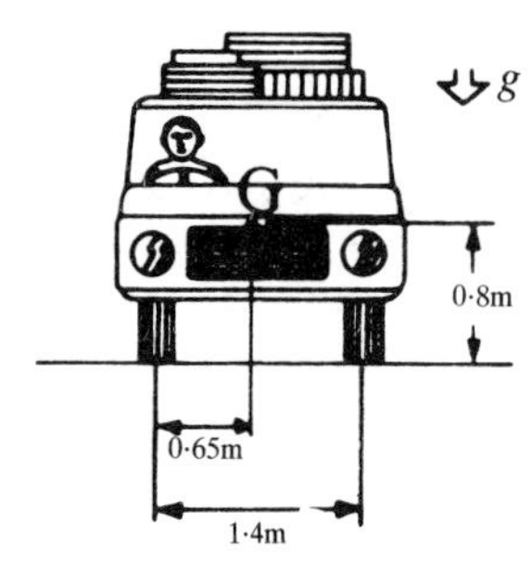

Figure 6.37

as shown in Fig. 6.37. The vehicle is travelling over a horizontal surface and negotiating a left-hand bend. The radius of the path traced out by G is 30 m and the steady speed of G is v. The coefficient of friction between tyres and road is 0.85.

As a first estimate, the effects of the suspension system can be neglected. Determine the maximum value of v such that the vehicle neither tips nor slips.

6.17 The vehicle shown in Fig. 6.38 travels along a level road and the friction coefficient between tyres and road is μ. When the brakes are applied, the braking ratio R is given by

$$R = \frac{\text{couple applied to front wheels by brakes}}{\text{couple applied to rear wheels by brakes}}$$

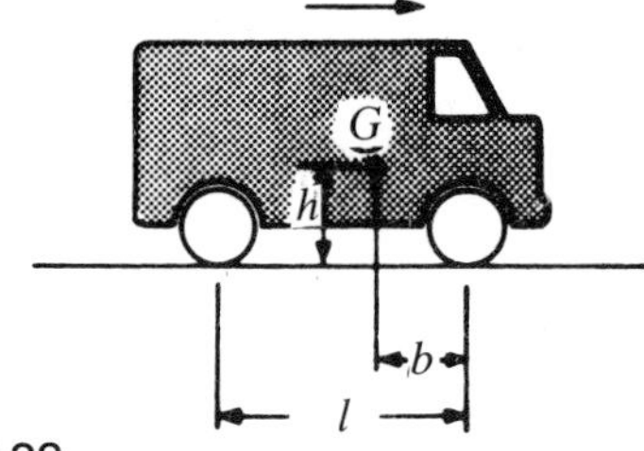

Figure 6.38

The inertia of the wheels and any braking effect of the engine are to be neglected.

a) Show that if only the rear wheels are locked the deceleration d is given by

$$d = \frac{\mu g b}{l/(1+R) + \mu h}$$

and if only the front wheels are locked then

$$d = \frac{\mu g (l-b)}{l/(1+1/R) - \mu h}$$

b) If R_m is the value of R for maximum deceleration, show that

$$R_m = l/(b - \mu h) - 1$$

and that this occurs when both front and rear wheels are locked.

c) If $l = 3.5$ m, $b = 1.6$ m and $h = 0.7$, plot d against R for values of R from 0 to 4, assuming that $\mu = 0.8$, and plot R_m against μ for values of μ from 0.2 to 1.1.

[It should be noted that in practice μ is not constant but varies with, amongst other things, relative slip speed. One of the consequences of this is that for a vehicle fitted with rubber tyres the maximum braking effect is normally obtained when the wheels are near to the point of slipping but do not actually slip. An idealisation of this effect is made in the next problem.]

6.18 Refer to problem 6.17 and assume that the brakes are applied to the rear wheels only ($R = 0$). The tyres are made from a material which, when in contact with the road surface, requires a tangential force to initiate slip of $\mu_s N$, where N is the normal force between tyre and road, but once slip has started the tangential force is $\mu_d N$ (μ_s and μ_d are known as the static and dynamic friction coefficients respectively). Assume that $\mu_s = 0.9$ and $\mu_d = 0.7 - 0.004\, v_s$, where v_s is the relative slip speed in (m/s), and that l, b and h have the same numerical values as in the previous problem.

If the vehicle is travelling at 30 m/s and the brakes are applied so that the rear wheels immediately lock, show that the stopping distance is about 177 metres. If, however, the brakes are applied so that slip does not quite occur, show that this distance is reduced by about 25.5 per cent.

[This problem not only shows the advantage of not allowing the wheels to slip but confirms the poor retardation available when only the rear-wheel brakes are operated.]

6.19 The motorcycle illustrated in Fig. 6.39 can be 'laid over' until $\theta = 40°$ before the footrest touches the ground.

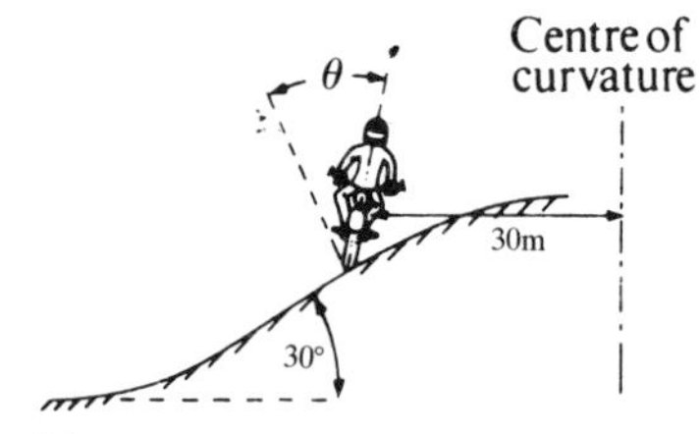

Figure 6.39

During a cross-country scramble the track runs at a constant height in a curved path around the side of a hill which slopes at 30° away from the centre of curvature of the path as shown. The radius of curvature to the centre of mass can be taken as 30 m. The coefficient of friction between the tyres and the ground is 0.65.

Find the theoretical maximum speed at which the curve can be negotiated. State whether at this speed the motorcycle would be on the point of slipping down the slope or of digging the footrest into the ground.

6.20 A hoist is driven by a motor and brake unit at E as shown in Fig. 6.40. The light cable passes over a drum

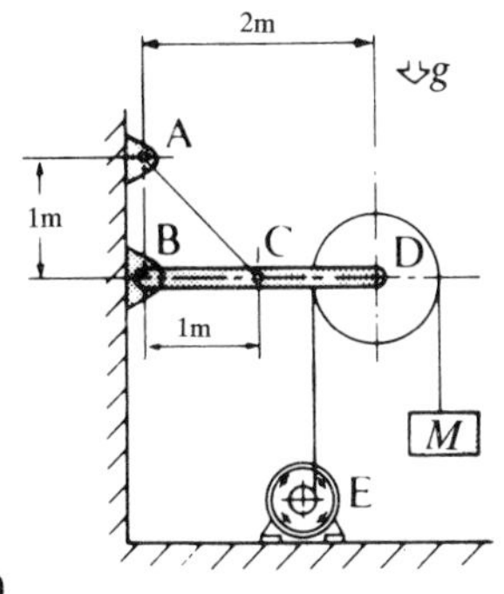

Figure 6.40

which is pivoted at D and which has a mass of 15 kg, a radius of 0.6 m and a radius of gyration about D of 0.5 m. A mass M of 20 kg is being lowered when the brake is applied such that the tension in the cable leaving the motor is 1.5 kN.

Calculate (a) the acceleration of the load and (b) the tension in the stay wire AC, neglecting the weight of the beam BD.

6.21 See Fig. 6.41. A lift cage with a mass of 2000 kg is supported by a cable wound around a 4 m diameter winding drum. Attached to the same shaft is another drum of diameter 2 m from which is suspended a counter-balance of mass 3000 kg. An electric motor drives the drum shaft through a 20 : 1 reduction gear.

The moments of inertia of the rotating parts about their respective axes of rotation are

rotor of the electric motor   60 kg m^2

winding drum   5000 kg m^2

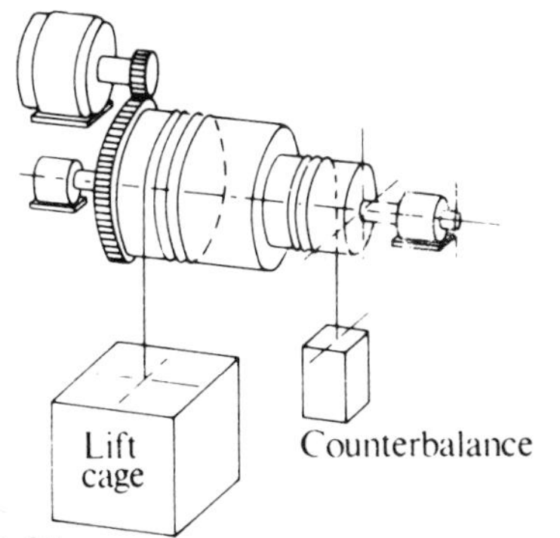

Figure 6.41

If the torque acting on the rotor is 900 N m, what is the tension in the lift cage cable during an ascent?

6.22 The winding cable for the crane illustrated in Fig. 6.42 passes over the light, frictionless pulleys in the trolley at A and B, under the 0.35 m diameter pulley at D, and is attached to the crane arm at C. The pulley D has a mass of 15 kg and a radius of gyration about the pivot axis of 0.1 m.

The load is being raised with an acceleration of

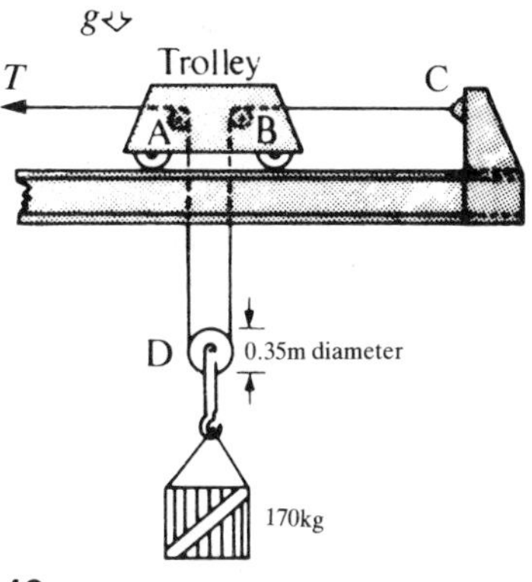

Figure 6.42

20 m/s^2. Calculate (a) the tension T in the cable leading to the winding gear; (b) the horizontal force, parallel to the crane arm, which must be applied to the trolley to prevent it from moving.

6.23 The dragster, complete with driver, illustrated in Fig. 6.43 has a total mass of 760 kg. Each rear wheel has a mass of 60 kg, a rolling radius of 0.4 m and a moment of inertia of 6 kg m^2. The moment of inertia of the front wheels may be neglected.

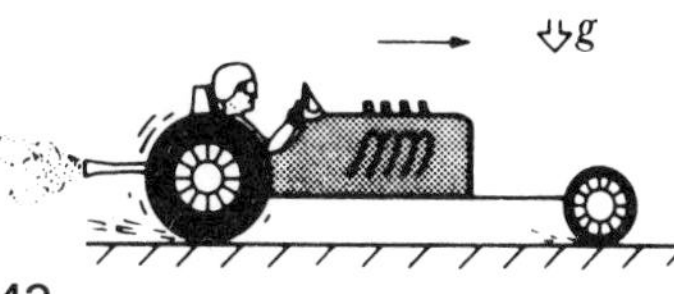

Figure 6.43

For the condition when the dragster is accelerating along a level road at 10.8 m/s^2, (a) draw free-body diagrams (i) for one rear wheel and (ii) for the dragster with rear wheels removed and (b) find the driving torque which is being applied to the hub of each rear wheel (assume that these torques are equal and that there is no slipping between tyres and road).

6.24 The excavator illustrated in Fig. 6.44 carries in its shovel a load of 400 kg with a centre of mass at G. The cab, arm and shovel assembly has a uniform angular acceleration from rest to $0.085\boldsymbol{k}$ rev/s during 90° of rotation. Simultaneously, the centre of mass G of the load is moved horizontally towards the axis of rotation at a steady rate of 0.2 m/s.

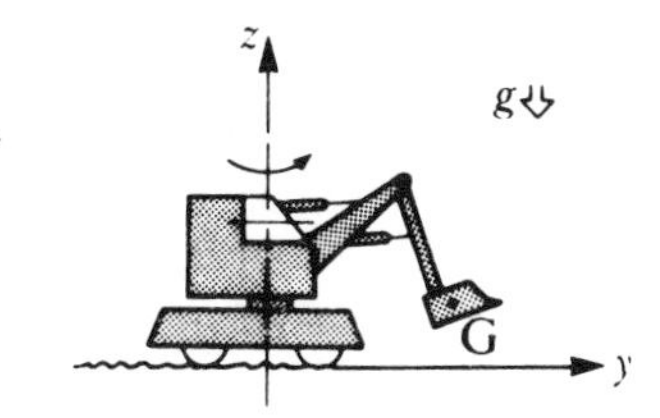

Figure 6.44

As the excavator passes through the 90° position, G is 3.5 m from the axis. Find the force exerted on the load at this instant.

6.25 Figure 6.45 shows an apparatus for performing an impact test on the specimen S. The rod AB of mass m swings from two light, parallel wires of length l, the inclination of the wires to the vertical being θ.

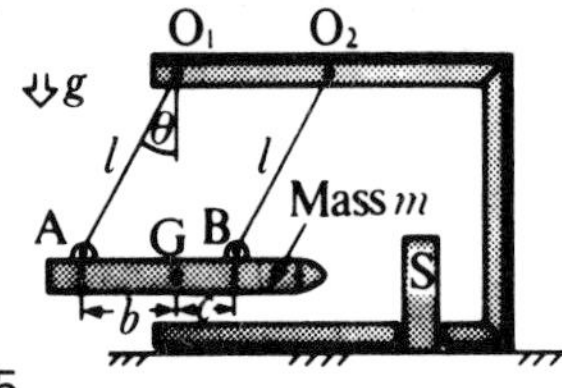

Figure 6.45

If the rod is released from rest at $\theta = 30°$ and strikes the specimen just after $\theta = 0$, find for $\theta = 0$ (a) the angular velocity and angular acceleration of the wires and (b) the tension in each wire.

6.26 Refer to problem 5.7. Figure 6.46 shows one of the cylinders C of a petrol engine. The crankshaft AB is rotating anticlockwise at a constant speed of 3000 rev/min about A, which is its mass centre. The connecting rod BD has a mass of 2.0 kg and its mass centre is at G. The moment of inertia of the connecting rod about G is 5×10^{-3} kg m^2. The mass of the piston E is 0.5 kg and the diameter of the cylinder C in which the piston slides is 90 mm.

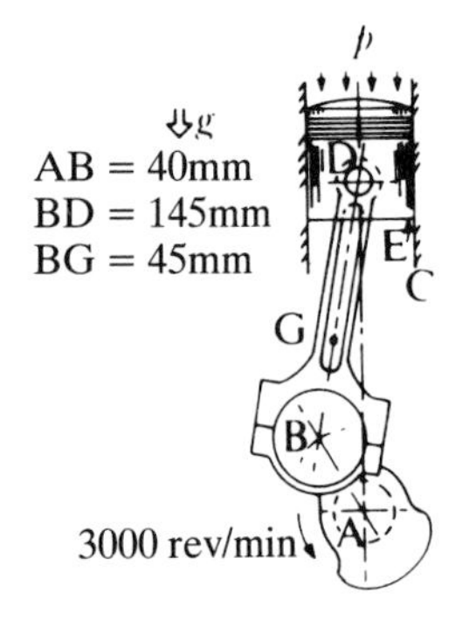

Figure 6.46

If the pressure p on top of the piston is 2.1 MPa when angle DAB = 30°, determine for this angle (a) the force in the gudgeon pin D and (b) the turning moment being applied to the crankshaft. Neglect friction.

6.27 The four-bar chain mechanism shown in Fig. 6.47 consists of a light crank AB of length 100 mm, a light rocker arm CD of length 300 mm, and a uniform connecting rod BC of length 400 mm, mass 4 kg and moment of inertia $I_{Gz} = 0.06$ kg m^2. AD = 400 mm. In the position shown, AB and BC are collinear, and angle ADC = 90°. The crank AB has a constant angular velocity given by $\omega_{AB} = -10\boldsymbol{k}$ rad/s. The effects of gravity and friction are to be neglected.

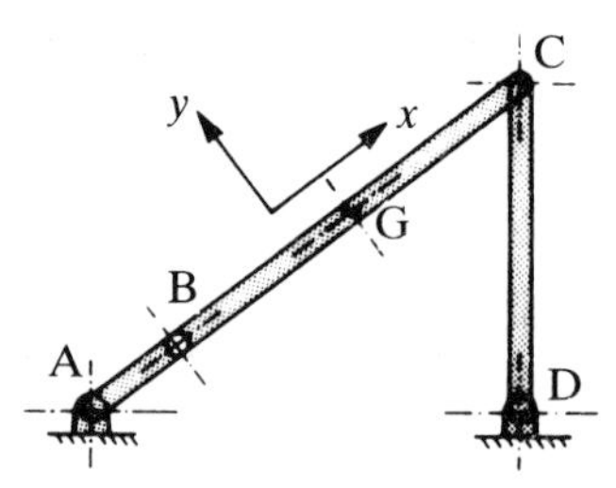

Figure 6.47

a) Draw the velocity and acceleration vector diagrams for the mechanism at this instant and hence determine the acceleration of the mass centre of the connecting rod BC and the angular acceleration of BC.
b) Write equations of motion for the connecting rod BC using the axes indicated and hence determine the forces acting on it at C and B.

6.28 A uniform slender rigid beam of mass 800 kg and length 3 m is pivoted at one end and rests on an elastic support at the other. In the position of static equilibrium the beam is horizontal. Details of the beam are shown in Fig. 6.48.

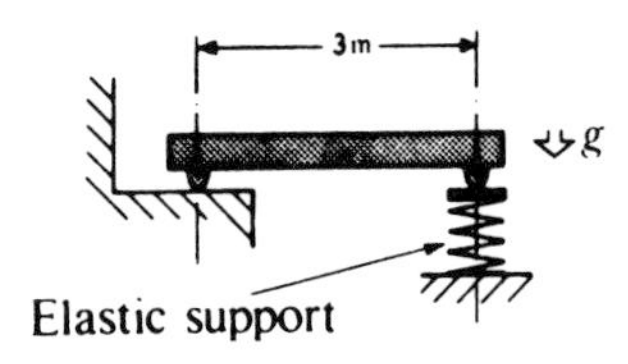

Figure 6.48

It is observed that, if disturbed, the beam performs small oscillations in the vertical plane with S.H.M. of frequency 5 Hz. What is the stiffness of the elastic support?

6.29 An impact testing machine has a pendulum which pivots about the z-axis, as shown in Fig. 6.49. It consists of a bar B, whose mass may be neglected, and a cylindrical bob C, of mass 50 kg.

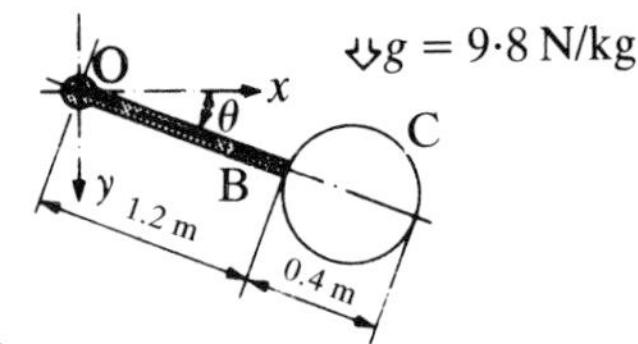

Figure 6.49

a) Calculate the moment of inertia, I_{Oz}, for the pendulum.
b) Write the moment equation for rotation of the pendulum about the fixed z-axis when it is swinging freely and is at an angle θ from the horizontal.
c) The pendulum is released from rest in the position $\theta = 0$. At the instant when $\theta = 60°$, determine the acceleration of the mass centre of the bob, and the angular velocity and angular acceleration of the pendulum.

6.30 A uniform rectangular trapdoor of mass m, hinged at one edge, is released from a horizontal position.

Show that the maximum value of the horizontal component of the force at the hinge occurs when the trapdoor has fallen through 45°. At this angle, calculate the magnitude of the total force in the hinge.

6.31 Two uniform rods, AB and BC, each of length 1 m and mass 1 kg, are pinned to each other at B and to supports at A and C as shown in Fig. 6.50. Determine the force acting on the support at A before and immediately after the pin at C is withdrawn. Take g to be 10 N/kg.

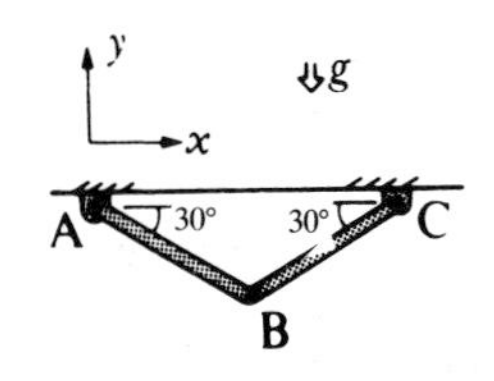

Figure 6.50

7 Energy

7.1 Introduction

Energy is one of the most important concepts encountered in a study of mechanics because it can appear in many guises in almost all disciplines of physics and chemistry. In mechanics we are mainly concerned with energy due to motion of material objects – that is kinetic energy – and energy associated with configuration, concentrating usually on gravitational potential energy and elastic strain energy; changes in other forms of energy such as thermal, electromagnetic and chemical are regarded as 'losses'. However, in other disciplines the description of useful energy and loss of energy may be different. The question of loss or gain of useful energy depends on the point of view, just as debit or credit in book-keeping depends on whose account we are considering, the same transaction appearing as a credit on one account yet as a debit on another.

Historically kinetic energy, called *vis viva* or 'living force' by Leibnitz, was a rival to momentum or 'quantity of motion' as favoured by Newton. The controversy was over which quantity was the true measure of the 'power' of a body to overcome resistance. It had been observed that if the speed of a body were doubled, the body could rise to quadruple the original height in a gravitational field; however, the time to reach that maximum height was only doubled. The fact that the difference between the two approaches was really only one of terminology was pointed out by d'Alembert, who showed that the 'living force' methods could be obtained from momentum considerations. Later Lagrange generalised the treatment of mechanical energy of systems, and his work forms the basis for some of the more advanced techniques in theoretical mechanics.

7.2 Work and energy for a system of particles

In Chapter 3 the equations of motion for a single particle were integrated with respect to displacement to give

$$\int_1^2 \boldsymbol{F} \cdot \mathrm{d}\boldsymbol{s} = \tfrac{1}{2}mv_2^2 + \tfrac{1}{2}mv_1^2 \qquad (7.1)$$

or, work done on the particle equals the change in kinetic energy. It must be emphasised that this result is just the outcome of a mathematical manipulation and does not introduce any new principle. We now generalise to a system of particles, where we use the notation

$\boldsymbol{F}_i \equiv$ force on ith particle acting directly from some external agency

$\boldsymbol{f}_{ij} \equiv$ force on ith particle due to the action of the jth particle.

Note that, from Newton's third law, $\boldsymbol{f}_{ij} = -\boldsymbol{f}_{ji}$ and these forces are collinear. Hence, for the ith particle (Fig. 7.1),

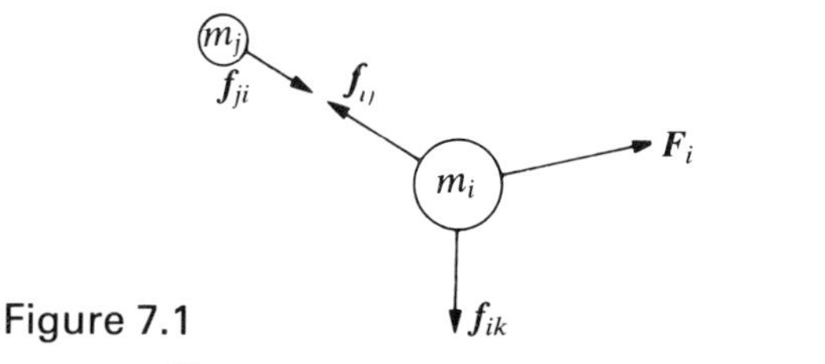

Figure 7.1

$$\boldsymbol{F}_i + \sum_j \boldsymbol{f}_{ij} = m_i\ddot{\boldsymbol{r}}_i \qquad (7.2)$$

The summation is over all particles in the system; however, it should be noted that $\boldsymbol{f}_{ii}$ has no meaning in this context.

If we now form the scalar product of both sides of equation 7.2 with the velocity $\dot{\boldsymbol{r}}_i$, we obtain

$$\boldsymbol{F}_i \cdot \dot{\boldsymbol{r}}_i + \dot{\boldsymbol{r}}_i \cdot \sum_j \boldsymbol{f}_{ij} = m_i\ddot{\boldsymbol{r}}_i \cdot \dot{\boldsymbol{r}}_i = \frac{\mathrm{d}}{\mathrm{d}t}\left[\frac{m_i}{2}\dot{\boldsymbol{r}}_i \cdot \dot{\boldsymbol{r}}_i\right] \qquad (7.3)$$

It is now required to sum for all particles in the system. In this summation we find that for every term of the form $\dot{\boldsymbol{r}}_i \cdot \boldsymbol{f}_{ij}$ there will occur a term $\dot{\boldsymbol{r}}_j \cdot \boldsymbol{f}_{ji}$

and adding these two terms gives

$$\dot{\boldsymbol{r}}_i \cdot \boldsymbol{f}_{ij} + \dot{\boldsymbol{r}}_j \cdot \boldsymbol{f}_{ji} = \boldsymbol{f}_{ij} \cdot (\dot{\boldsymbol{r}}_i - \dot{\boldsymbol{r}}_j)$$

The term $(\dot{\boldsymbol{r}}_i - \dot{\boldsymbol{r}}_j)$ is the relative velocity between the particles i and j, to which we will give the symbol $\dot{\boldsymbol{r}}_{ij}$.

Summing equation 7.3 for all particles gives

$$\sum_i \boldsymbol{F}_i \cdot \dot{\boldsymbol{r}}_i + \sum_{ij} \boldsymbol{f}_{ij} \cdot \dot{\boldsymbol{r}}_{ij} = \frac{\mathrm{d}}{\mathrm{d}t}\left[\sum_i \frac{m_i}{2} \dot{\boldsymbol{r}}_i \cdot \dot{\boldsymbol{r}}_i\right] \tag{7.4}$$

where $\sum_{ij}$ signifies that all combinations of terms, other than ii, are to be summed.

Integrating with respect to time gives

$$\sum_i \int_1^2 \boldsymbol{F}_{\mathrm{i}} \cdot \mathrm{d}\boldsymbol{r}_i + \sum_{ij} \int_1^2 \boldsymbol{f}_{ij} \cdot \mathrm{d}\boldsymbol{r}_{ij}$$
$$= \Delta\left[\sum_i \frac{m_i}{2} \dot{\boldsymbol{r}}_i \cdot \dot{\boldsymbol{r}}_i\right] \tag{7.5}$$

where Δ indicates a finite difference.

The first term on the left-hand side of the equation is the work done by external agencies, either by contact at the surface or by long-range body forces such as gravity and electromagnetic forces. The second term is the work done by internal forces and these, in general, are complex relationships. The right-hand side is, of course, just the change in the total kinetic energy.

Equation 7.5 is quite general, but to make use of this expression we must first consider some special cases, the first of which is the rigid body.

7.3 Kinetic energy of a rigid body

The kinetic energy of a particle has been defined as $\frac{1}{2}mv^2$ or $\frac{1}{2}m\dot{\boldsymbol{r}} \cdot \dot{\boldsymbol{r}}$, so for any collection of particles the kinetic energy is

$$\tfrac{1}{2} \sum_i m_i \dot{\boldsymbol{r}}_i \cdot \dot{\boldsymbol{r}}_i.$$

We have seen (equation 5.3) that for a rigid body in plane motion the particle velocity can be written in the form

$$\dot{\boldsymbol{r}}_i - \dot{\boldsymbol{r}}_{\mathrm{G}} = \boldsymbol{\omega} \times \boldsymbol{\rho}_i$$

so that

$$\tfrac{1}{2} \sum_i m_i \dot{\boldsymbol{r}}_i \cdot \dot{\boldsymbol{r}}_i = \tfrac{1}{2} \sum_i m_i (\dot{\boldsymbol{r}}_{\mathrm{G}} + \boldsymbol{\omega} \times \boldsymbol{\rho}_i) \cdot (\dot{\boldsymbol{r}}_{\mathrm{G}} + \boldsymbol{\omega} \times \boldsymbol{\rho}_i) \tag{7.6}$$
$$= \tfrac{1}{2}(\dot{r}_{\mathrm{G}})^2 \sum_i m_i + \dot{\boldsymbol{r}}_{\mathrm{G}} \cdot \sum_i m_i \boldsymbol{\omega} \times \boldsymbol{\rho}_i + \tfrac{1}{2} \sum_i m_i (\boldsymbol{\omega} \times \boldsymbol{\rho}_i) \cdot (\boldsymbol{\omega} \times \boldsymbol{\rho}_i)$$

The first term is simply $\frac{1}{2}M\dot{r}_{\mathrm{G}}{}^2$ (where $M = \sum m_i$), that is the kinetic energy if all the mass were at the centre of mass. The second term vanishes by reason of the definition of the centre of mass, viz. $\dot{\boldsymbol{r}}_{\mathrm{G}} \cdot \boldsymbol{\omega} \times \sum m_i \boldsymbol{\rho}_i = 0$, since $\sum m_i \boldsymbol{\rho}_i = 0$.

The last term may be simplified by writing $\boldsymbol{\rho}_i = \boldsymbol{a}_i + \boldsymbol{b}_i$, where $\boldsymbol{a}_i$ is parallel to $\boldsymbol{\omega}$ and $\boldsymbol{b}_i$ is perpendicular to $\boldsymbol{\omega}$ (Fig. 7.2). Hence

$$\boldsymbol{\omega} \times \boldsymbol{\rho}_i = \boldsymbol{\omega} \times (\boldsymbol{a}_i + \boldsymbol{b}_i)$$
$$= \omega b_i \boldsymbol{e}$$

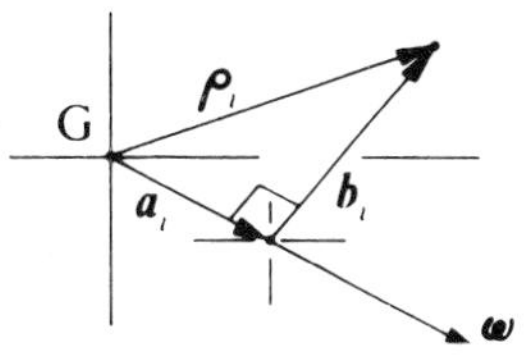

Figure 7.2

where $\boldsymbol{e}$ is a unit vector perpendicular to $\boldsymbol{a}_i$ and $\boldsymbol{b}_i$ so that

$$(\boldsymbol{\omega} \times \boldsymbol{\rho}_i) \cdot (\boldsymbol{\omega} \times \boldsymbol{\rho}_i) = \omega^2 b_i^2$$

The total kinetic energy now becomes

$$\tfrac{1}{2}M(\dot{r}_{\mathrm{G}})^2 + \tfrac{1}{2}\omega^2 \sum_i m_i b_i^2$$

$\sum_i \mathrm{m_i b_i}^2$ is defined as in Chapter 6, as the moment of inertia about an axis through the centre of mass and parallel to the axis of rotation.

Writing $I_{\mathrm{G}} \equiv \sum m_i b_i^2$, we obtain the kinetic energy:

$$\text{k.e.} = \tfrac{1}{2}M(\dot{r}_{\mathrm{G}})^2 + \tfrac{1}{2}I_{\mathrm{G}}\omega^2 \tag{7.7}$$

The reader should notice that once again the use of the centre of mass has enabled us to separate the effects of translation and rotation.

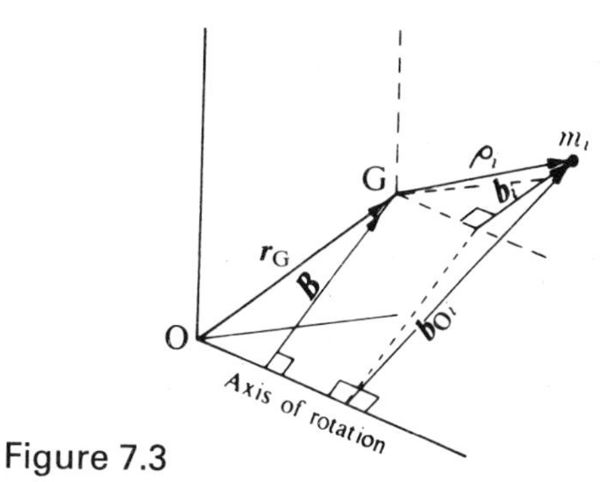

Figure 7.3

For the special case of rotation about a fixed axis (Fig. 7.3), equation 7.7 reduces to

$$\text{k.e.} = \tfrac{1}{2}M(\boldsymbol{\omega}\times\boldsymbol{B})\cdot(\boldsymbol{\omega}\times\boldsymbol{B}) + \tfrac{1}{2}I_G\omega^2$$
$$= \tfrac{1}{2}MB^2\omega^2 + \tfrac{1}{2}I_G\omega^2$$
$$= \tfrac{1}{2}\omega^2(I_G + MB^2)$$
$$= \tfrac{1}{2}I_O\omega^2 \qquad (7.8)$$

where $I_O = \sum m_i b_{Oi}{}^2 = \sum m_i b_i^2 + MB^2 \qquad (7.9)$

This is the parallel-axes theorem and is easily verified from equation 7.6 by putting $\dot{\boldsymbol{r}}_G = \boldsymbol{\omega}\times\boldsymbol{r}_G$.

7.4 Potential energy

For a rigid body there is no change in the separation between any two particles ($\mathrm{d}\boldsymbol{r}_{ij} = 0$), hence the work done by the internal forces is zero.

The left-hand side of equation 7.5 is now $\sum_i\int_1^2 \boldsymbol{F}_i\cdot\mathrm{d}\boldsymbol{r}_i$, which is simply the work done by the external forces. However, in cases where the forces are conservative, a simplification is possible.

A conservative force is defined as one for which the work done is independent of the path taken and depends solely on the limits, so that for conservative forces we may write

$$\sum_i \int_1^2 \boldsymbol{F}_i\cdot\mathrm{d}\boldsymbol{r}_i \equiv \mathcal{W}_2 - \mathcal{W}_1 \qquad (7.10)$$

where $\mathcal{W}_1$ is a function of $\boldsymbol{r}_1$ only and $\mathcal{W}_2$ is a function of $\boldsymbol{r}_2$ only.

Equation 7.5 is now

$$\mathcal{W}_2 - \mathcal{W}_1 = (\text{k.e.})_2 - (\text{k.e.})_1$$

($\mathcal{W}_2 - \mathcal{W}_1$) is the work done on the system by the conservative forces. It is convenient to regard this work as due to a reduction of some form of stored energy called potential energy (see Appendix 4) and given the symbol V. Therefore we may write

$$\mathcal{W}_2 - \mathcal{W}_1 = -(V_2 - V_1) \qquad (7.11)$$

So, for a system of particles acted on only by conservative forces, we have

$$\mathcal{W}_2 - \mathcal{W}_1 = -(V_2 - V_1) = (\text{k.e.})_2 - (\text{k.e.})_1$$

or

$$0 = (\text{k.e.} + V)_2 - (\text{k.e.} + V)_1 \qquad (7.12)$$

The two most common forms of potential energy encountered in engineering mechanics are gravitational potential energy and elastic strain energy.

Gravitational potential energy

If a force system is conservative, then the potential energy is defined by

$$V = -\int \boldsymbol{F}\cdot\mathrm{d}\boldsymbol{r} + \text{arbitrary constant}$$

In a gravitational field we have from Newton's law of gravitation that the force on m is given by

$$\boldsymbol{F} = -\frac{Gm_0m}{r^2}\boldsymbol{e}_r \qquad (7.13)$$

where m_0 and m are two masses (Fig. 7.4), the displacement of m from m_0 is $r\boldsymbol{e}_r$, and G is the universal gravitational constant. Hence we have

$$V = -\left[\int\left(-G\frac{m_0m}{r^2}\right)\boldsymbol{e}_r\cdot\mathrm{d}\boldsymbol{s}\right] + \text{constant}$$
$$= \int Gm_0m\frac{\mathrm{d}r}{r^2} + \text{constant}$$
$$= -\frac{Gm_0m}{r} + \text{constant}$$

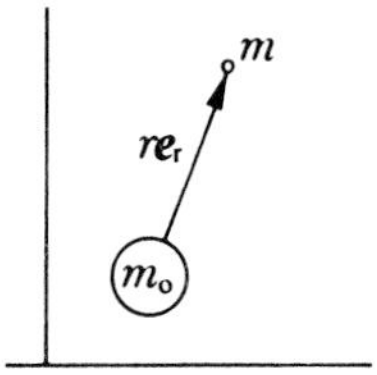

Figure 7.4

For orbital-motion problems it is convenient to consider the potential energy to be zero when r is infinity, in which case

$$V = -\frac{Gm_0m}{r} \qquad (7.14)$$

In problems where the variations in r are small, such as for motion close to the Earth's surface, then with $r = R + h$, where R is the radius of the Earth and m_0 is its mass, we have

$$V = -\frac{Gm_0m}{(R+h)} + \text{constant}$$

Expanding by means of the binomial theorem,

$$V \approx \frac{Gm_0m}{R}\left(-1 + \frac{h}{R}\right) + \text{constant}$$

or

$$V_2 - V_1 \approx \frac{Gm_0m}{R}\left(-1 + \frac{h_2}{R}\right) - \frac{Gm_0m}{R}\left(-1 + \frac{h_1}{R}\right)$$

$$= \left[\frac{Gm_0}{R^2}\right] m(h_2 - h_1) \qquad (7.15)$$

The quantity $[Gm_0/R^2]$ is the gravitational field constant g, which is loosely called the acceleration due to gravity. (It should be remembered that the values usually quoted are apparent values derived as a result of considering the surface of the Earth to be unaccelerated.)

Strain energy
Another force law of great importance is when the force between two particles is proportional to the change in separation; that is (Fig. 7.5)

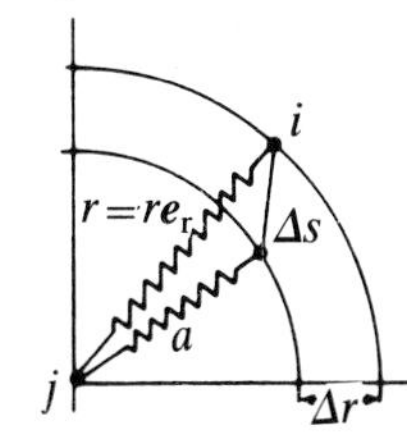

Figure 7.5

$$\boldsymbol{F}_i = -k\{|\boldsymbol{r}| + |\boldsymbol{a}|\}\boldsymbol{e}_r$$
$$= -k\Delta r\boldsymbol{e}_r$$

where k is a constant.
Hence

$$V = -\int -k\Delta r\boldsymbol{e}_r \cdot \mathrm{d}\boldsymbol{s} + \text{constant}$$
$$= \tfrac{1}{2}k(\Delta r)^2 + \text{constant} \quad \text{since } \boldsymbol{e}_r \cdot \mathrm{d}\boldsymbol{s} = \mathrm{d}(\Delta r)$$

It is usual to consider the energy to be zero when Δr is zero; this situation exists for deformations in a material obeying Hooke's law and is applicable to the deformation of linear springs. This form of potential energy is referred to as strain energy.
Therefore the strain energy of a uniform linear spring having a stiffness k is

$$V = \tfrac{1}{2}k(\Delta r)^2 \qquad (7.16)$$

Δr being measured from the free length of the spring.

7.5 Non-conservative systems

The most common non-conservative force in mechanics is that of friction. When friction is present in a system, processes are irreversible and the work done will probably depend on the path taken. A system which has non-conservative forces acting within its boundary is termed a non-conservative system, since the mechanical energy is not conserved but is changed into some other form which is not recoverable.

Let us consider the case of a block being dragged along a plane by a constant force. We can draw the free-body diagram (Fig. 7.6) and write down the equation of motion as follows.

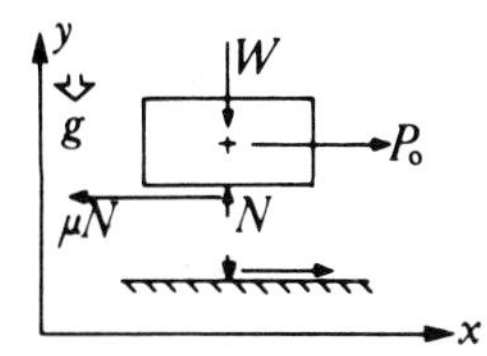

Figure 7.6

For the x-direction,

$$P_0 - \mu N = M\ddot{x}_G$$

and for the y-direction

$$N - W = 0$$

giving

$$P_0 - \mu W = Mv\frac{\mathrm{d}v}{\mathrm{d}x}$$

where $v = \dot{x}$.
Integrating with respect to x gives

$$\int_1^2 P_0\,\mathrm{d}x - \int_1^2 \mu W\,\mathrm{d}x = \int_1^2 Mv\,\mathrm{d}v$$
$$= \tfrac{1}{2}Mv_2^2 - \tfrac{1}{2}Mv_1^2 \qquad (7.17)$$

The first term is the work done by the external force P_0 and the right-hand side is the change in kinetic energy of the system. However, the second term on the left-hand side is not a work-done term as formally defined as we do not know the detailed movement of the particle on which the force is acting. As an exercise, consider the two extreme cases in Fig. 7.7.

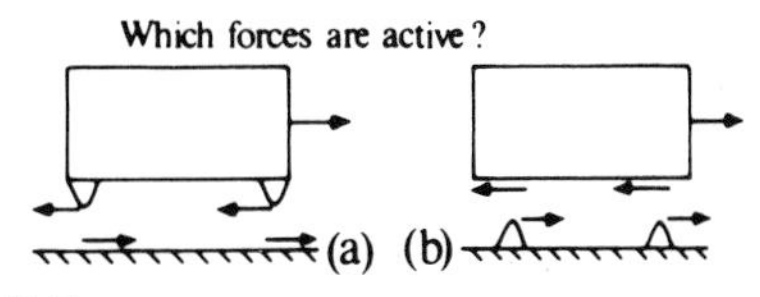

Figure 7.7

Equation 7.17 is, however, completely valid since it was derived by integrating the equations of motion, but it is not yet a new principle. Further consideration of the physics of the problem based on experience suggests that other measurable changes are taking place. In the first place, one would expect there to be a change in temperature and also one would expect some local vibration giving rise to the production of noise. It is also possible that changes of state of the material would take place – i.e. melting – electrostatic charges might be developed and other changes of a chemical nature might occur.

In a study of mechanics these latter changes represent a loss to the system, but in other disciplines such 'losses' might well be 'gains'.

It is now convenient to propose the general energy principle.

7.6 The general energy principle

Although integration of the equations of motion with respect to displacement leads to a form of the energy equation, the general energy principle may be considered as a fundamental law of mechanics. Terms which appear as losses in the general energy principle can sometimes have a numerical value ascribed to them by comparison with the integrated forms of the equations of motion.

In the context of engineering mechanics, the general energy principle may be stated as

> *the work done on a system is equal to the change in kinetic energy plus potential energy plus losses.*

The kinetic and potential energies are energies which are stored inside the system and are recoverable; all other energy forms are therefore losses. The above principle is stated with respect to a specific system, therefore in any problem we must carefully define the system boundaries and consider only forces which *do work across these boundaries*.

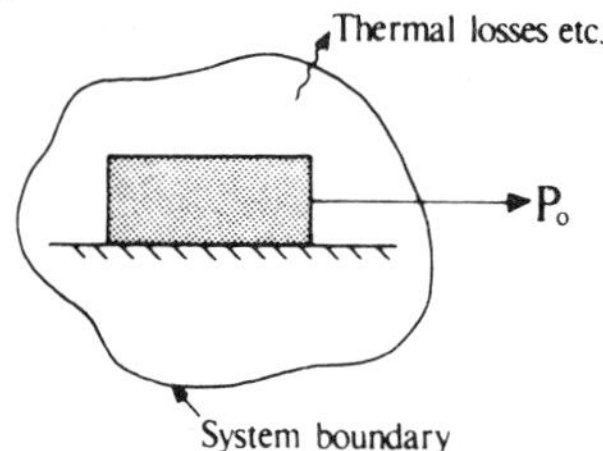

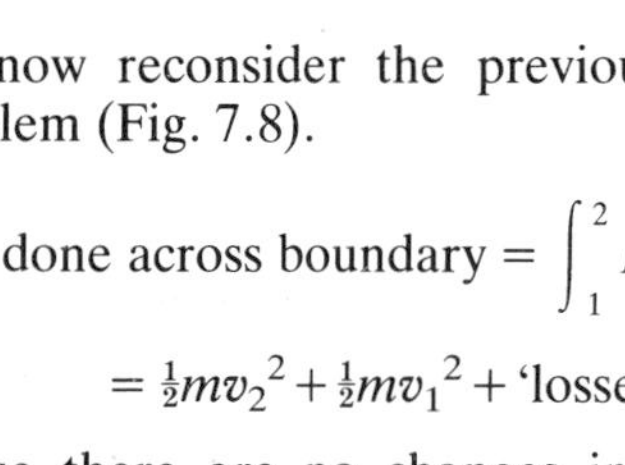

Figure 7.8

Let us now reconsider the previous sliding-block problem (Fig. 7.8).

$$\text{Work done across boundary} = \int_1^2 P_0 \, dx$$

$$= \tfrac{1}{2}mv_2^2 + \tfrac{1}{2}mv_1^2 + \text{'losses'} \quad (7.18)$$

In this case there are no changes in potential energy.

If we compare equations 7.17 and 7.18 we see that the 'losses' are equivalent to $\int \mu W \, dx$, i.e. the product of the frictional force and the slipped distance.

As a further example of the difference between the integrated form of the equations of motion and the general energy principle, consider the case of a smooth block being pulled along via a light spring (Fig. 7.9).

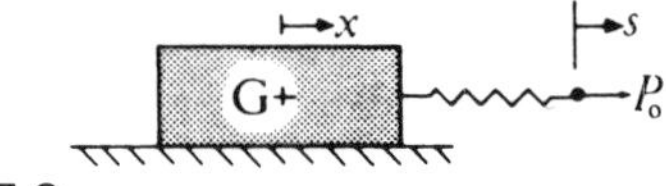

Figure 7.9

From the free-body diagram (Fig. 7.10) the equation of motion is

$$P_0 = m\ddot{x} = mv\frac{dv}{dx}$$

Integrating,

$$\int_1^2 P_0 \, dx = \tfrac{1}{2}mv_2^2 - \tfrac{1}{2}mv_1^2 \quad (7.19)$$

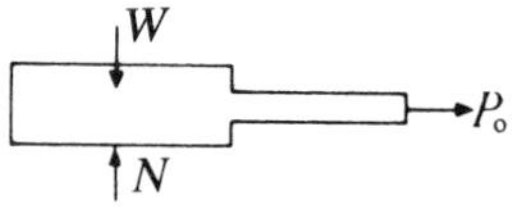

W
P₀
N

Figure 7.10

From the system diagram (Fig. 7.11) the energy principle gives

$$\int_1^2 P_0 \, ds = \left[\frac{mv_2^2}{2} + \frac{k}{2}(s_2 - x_2)^2\right] - \left[\frac{mv_1^2}{2} + \frac{k}{2}(s_1 - x_1)^2\right] \quad (7.20)$$

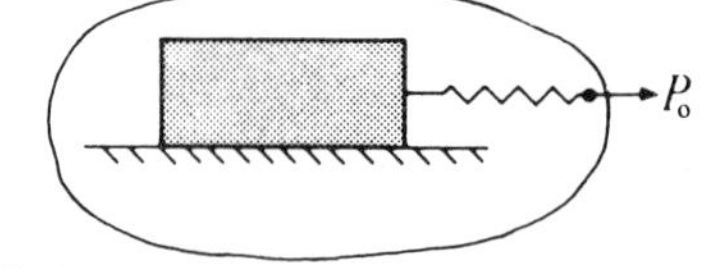

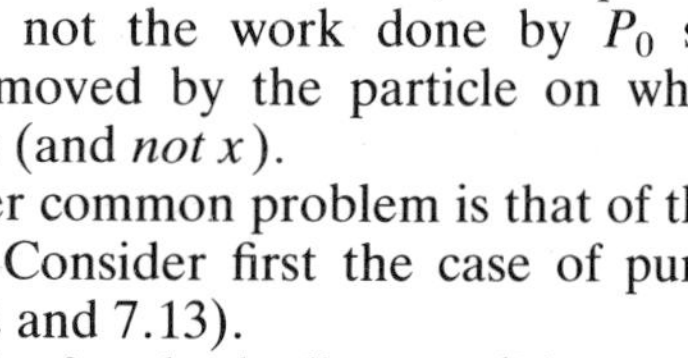

Figure 7.11

It should be noted that, in equation 7.19, $\int P_0 \, dx$ is not the work done by P_0 since the distance moved by the particle on which P_0 is acting is s (and *not* x).

Another common problem is that of the rolling cylinder. Consider first the case of pure rolling (Figs 7.12 and 7.13).

From the free-body diagram (Fig. 7.13),

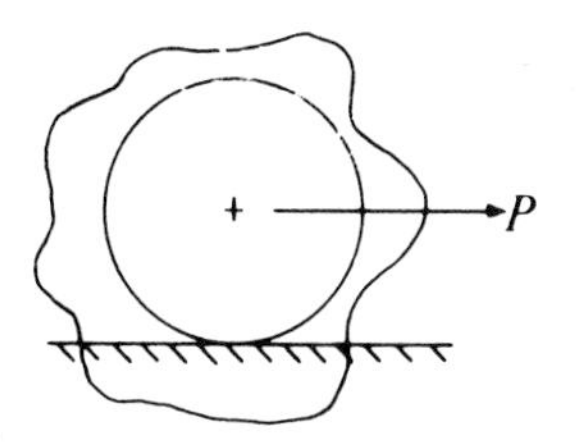

Figure 7.12

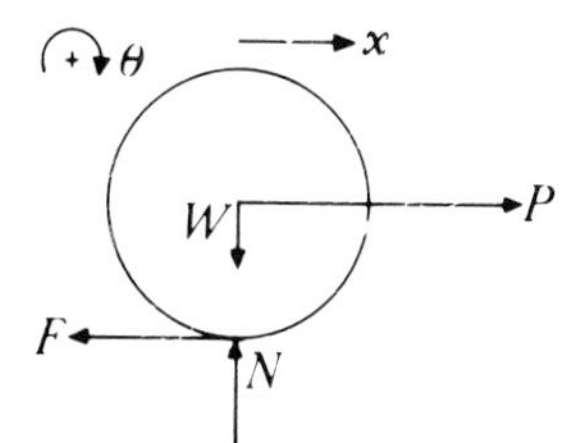

Figure 7.13

$$P - F = M\ddot{x} = Mv\frac{\mathrm{d}v}{\mathrm{d}x} \quad (7.21)$$

and $$Fr = I_G\ddot{\theta} = I_G\omega\frac{\mathrm{d}\omega}{\mathrm{d}\theta} \quad (7.22)$$

The kinematic constraint for no slipping is

$$\dot{x} = \dot{\theta}r$$

or $$\mathrm{d}x = r\,\mathrm{d}\theta \quad (7.23)$$

Eliminating F between equations 7.21 and 7.22 gives

$$P = Mv\frac{\mathrm{d}v}{\mathrm{d}x} + \frac{I_G}{r}\omega\frac{\mathrm{d}\omega}{\mathrm{d}\theta}$$

and, using the constraint, equation 7.23 leads to

$$P = Mv\frac{\mathrm{d}v}{\mathrm{d}x} + \frac{I_G}{r}\frac{v}{r}\frac{\mathrm{d}v}{\mathrm{d}x} = \left(M + \frac{I_G}{r^2}\right)v\frac{\mathrm{d}v}{\mathrm{d}x}$$

Integrating with respect to x gives

$$\int_1^2 P\,\mathrm{d}x = \tfrac{1}{2}\left(M + \frac{I_G}{r^2}\right)(v_2{}^2 - v_1{}^2) \quad (7.24)$$

The energy equation gives

$$\int_1^2 P\,\mathrm{d}x = \left[\tfrac{1}{2}Mv^2 + \frac{I_G\omega_2{}^2}{2}\right]_1^2$$

$$= \tfrac{1}{2}\left(M + \frac{I_G}{r^2}\right)(v_2{}^2 - v_1{}^2) \quad (7.25)$$

Here it should be noted that the friction force F progresses as the wheel rolls but the particle to which it is applied on the wheel moves at right angles to the force, thus F does no work.

If slipping is occurring then there is no kinematical restraint equation but the friction force is μN. Thus, as $W - N = 0$,

$$P - \mu N = Mv\frac{\mathrm{d}v}{\mathrm{d}x}$$

and $$\mu Nr = I_G\omega\frac{\mathrm{d}\omega}{\mathrm{d}\theta}$$

The energy equation now gives

$$\int_1^2 P\,\mathrm{d}x = [\tfrac{1}{2}Mv^2 + \tfrac{1}{2}I_G\omega^2]_1^2 + \text{'losses'}$$

In this case,

$$\text{'losses'} = \mu W|(x - r\theta)|$$

the modulus of $\mu W(x - r\theta)$ is necessary as the 'loss' must always be positive, irrespective of the direction of slip.

7.7 Summary of the energy method

The general energy principle, or first law of thermodynamics, which has as a corollary the conservation of energy, is a very powerful principle which has applications in all branches of physics. Since it has such wide interpretation it means that all forms of energy must be considered when forming the equation and care must be taken not to exclude changes such as thermal effects. The selection of a system boundary, which may not be a clear physical surface, requires experience and practice.

The main points are as follows.

Work The elemental work $\mathrm{d}\mathscr{W}$ is force times the elemental distance moved by the particle on which the force acts, in the direction of the force.

Kinetic energy of a rigid body

$$\text{k.e.} = \tfrac{1}{2}Mv_G{}^2 + \tfrac{1}{2}I_G\omega^2 \quad (7.26)$$

Gravitational potential energy

$$V_G = -\frac{Gm_0}{r} \quad \text{or} \quad Mgh \quad (7.27)$$

Linear elastic strain energy

$V_E = \tfrac{1}{2}k(\Delta r)^2$ for a spring having a stiffness k

$$(\text{Work done})_{\text{external}} = [\text{k.e.} + V_G + V_E]_2 \quad (7.28)$$
$$- [\text{k.e.} + V_G + V_E]_1$$
$$+ \text{'losses'} \quad (7.29)$$

7.8 The power equation

If the work–energy relationship is written for a small time interval Δt, then we have

$$\Delta \mathcal{W} = \Delta(\text{k.e.} + V) + \Delta(\text{losses})$$

Dividing by Δt and going to the limt $\Delta t \to 0$ leads to

$$\frac{d\mathcal{W}}{dt} = \frac{d}{dt}(\text{k.e.} + V) + \frac{d}{dt}(\text{losses}) \qquad (7.30)$$

or, power input equals the time rate of change of the internal energy plus power 'lost'.

Let us consider a simple case of a single particle acted upon by an external force $\boldsymbol{P}$ and also under the influence of gravity, then

$$\boldsymbol{P} \cdot \boldsymbol{v} = \frac{d}{dt}[\tfrac{1}{2}m\boldsymbol{v} \cdot \boldsymbol{v} + mgz]$$
$$= m\boldsymbol{v} \cdot \boldsymbol{a} + mg\dot{z}$$

If the motion is planar,

$$\boldsymbol{v} = \dot{x}\boldsymbol{i} + \dot{z}\boldsymbol{k}$$
$$\boldsymbol{a} = \ddot{x}\boldsymbol{i} + \ddot{z}\boldsymbol{k}$$

and $\quad \boldsymbol{P} = P_x\boldsymbol{i} + P_z\boldsymbol{k}$

so that

$$P_x\dot{x} + P_z\dot{z} = m(\dot{x}\ddot{x} + \dot{z}\ddot{z}) + mg\dot{z}$$

If $z = x\tan\alpha$ (Fig. 7.14) then dividing by x gives

$$P_x + P_z\tan\alpha = m\tan^2\alpha + 1)\ddot{z} + mg\tan\alpha$$

Hence $\ddot{z}$ may be found without considering the workless constraints.

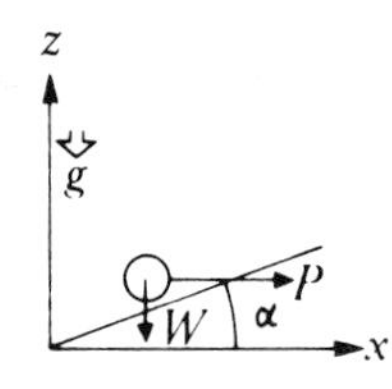

Figure 7.14

7.9 Virtual work

The concept of virtual work is one which saves a considerable amount of labour when dealing with complex structures, since there is no need to dismember the structure and draw free-body diagrams. Basically we shall be using the method as an alternative way of presenting the conditions for equilibrium and also to form a basis for the discussion of stability. In the early stages of understanding the principle the main advantages are not apparent, but the reader is asked to be patient as later examples will show some of its rewards.

A virtual displacement is defined as any small displacement which is possible subject to the constraints. The word virtual is used because the displacement can be *any* displacement and not necessarily an actual displacement which may occur during some specific time interval.

The notation used for a virtual displacement of some co-ordinate, say u, is δu. This form of delta is the same as is used in mathematics to signify a variation of u; indeed the concepts are closely related.

The work performed by the forces in the system over this displacement is the virtual work and is given the notation $\delta\mathcal{W}$.

Conditions for equilibrium

Let us first consider a single particle which is free to move in a vertical plane subject to the action of two springs as shown in Fig. 7.15.

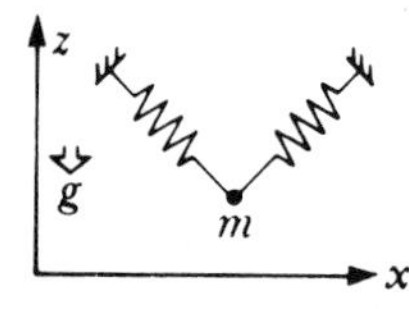

Figure 7.15

From the free-body diagram (Fig. 7.16) the condition for equilibrium is

$$\sum \boldsymbol{P} = \boldsymbol{P}_1 + \boldsymbol{P}_2 = \boldsymbol{W} = 0$$

or $\quad P_2\cos\alpha_2 - P_1\cos\alpha_1 = 0 \qquad (7.31)$

and $\quad P_1\sin\alpha_1 + P_2\sin\alpha_2 - mg = 0 \qquad (7.32)$

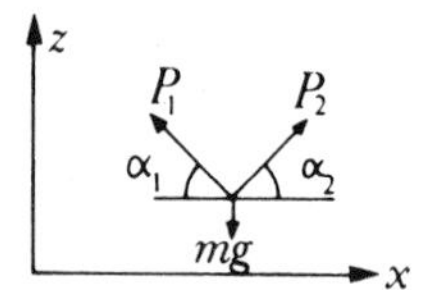

Figure 7.16

If equations 7.31 and 7.32 are multiplied by δx and δz respectively, then

$$\left.\begin{aligned} P_2\cos\alpha_2\,\delta x - P_1\cos\alpha_1\,\delta x = 0 \\ P_1\sin\alpha_1\,\delta z + P_2\sin\alpha_2\,\delta z - mg\,\delta z = 0 \end{aligned}\right\}$$

$$\text{or } \sum \boldsymbol{P} \cdot \delta\boldsymbol{s} = 0 \qquad (7.33)$$

These are the equations for the virtual work for the arbitrary displacements δx and δz – note, *arbitrary* displacements. In both cases we may state that the virtual work done by the forces over an arbitrary small displacement from the equilibrium position is zero.

If a system comprises many particles then the total virtual work done on all particles over any virtual displacement (or combination of displacements) is zero when the system is in equilibrium.

Principle of virtual work

We may now state the principle of virtual work as follows.

> If a system of particles is in equilibrium then the virtual work done over any arbitrary displacement, consistent with the constraints, is zero:
>
> $$\delta\mathscr{W} = 0 \qquad (7.34)$$

Application to a system with a single degree of freedom

Consider a rigid body freely pinned at A and held in equilibrium by a spring attached at B (Fig. 7.17). This body has one degree of freedom, that is the displacement of all points may be expressed in terms of one displacement such as θ, the angular rotation. If the spring is unstrained when AB is horizontal, then in a general position the active forces are the weight and the spring force; the forces at the pin do no work if friction is negligible.

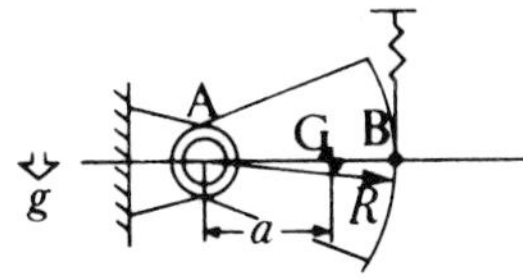

Figure 7.17

For a small displacement $\delta\theta$, the displacement of G is $a\,\delta\theta$ (Fig. 7.18) and the virtual work is

$$\delta\mathscr{W} = Mg(a\,\delta\theta\cos\theta) - kR\theta(R\,\mathrm{d}\theta)$$

For equilibrium,

$$\delta\mathscr{W} = 0 = (Mga\cos\theta - kR^2\theta)\,\delta\theta$$

and, as $\delta\theta$ is arbitrary,

$$Mga\cos\theta - kR^2\theta = 0 \qquad (7.35)$$

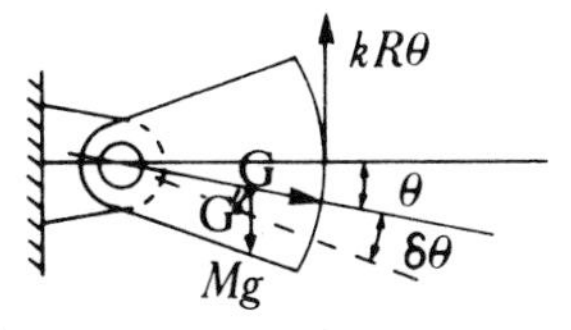

Figure 7.18

If, as in the previous example, the forces are conservative then equation 7.34 may be interpreted as

$$\delta\mathscr{W} = \delta\mathscr{W}_E + \delta\mathscr{W}_G + \delta\mathscr{W}' \qquad (7.36)$$

where $\delta\mathscr{W}_G = -\delta V_G = -$(variation of gravitational potential energy)

$\delta\mathscr{W}_E = -\delta V_E = -$variation of elastic strain energy)

and $\delta\mathscr{W}' =$ virtual work done by external forces

Therefore

$$\delta\mathscr{W} = \delta\mathscr{W}' - \delta V_E - \delta V_G = 0$$

or $\delta\mathscr{W}' = \delta(V_E + V_G) \qquad (7.37)$

Reworking the last problem,

$$0 = \delta[-Mga\sin\theta + \tfrac{1}{2}k(R\theta)^2]$$
$$0 = [-Mga\cos\theta + kR^2\theta]\,\delta\theta$$

Stability

Consideration of some simple situations shown in Fig. 7.19 will show that not all equilibrium configurations are stable. However, we cannot always rely on common sense to tell us which cases are stable. We have demonstrated that for equilibrium $\delta\mathscr{W} = 0$, but further consideration of

g

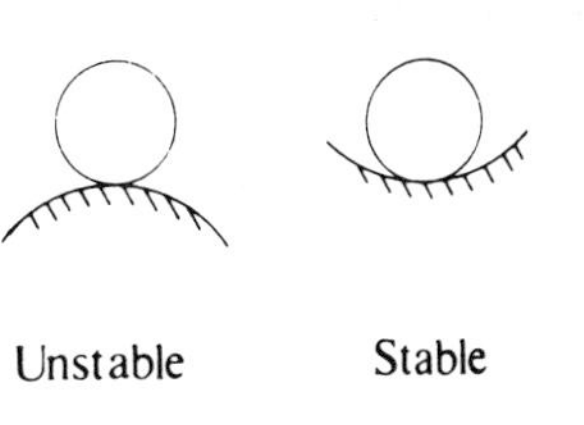

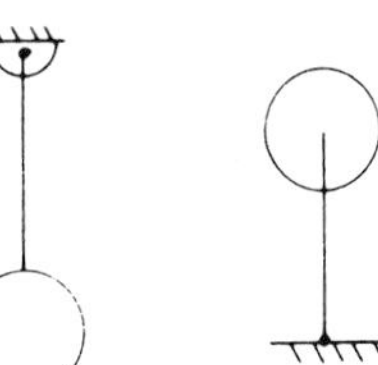

Stable

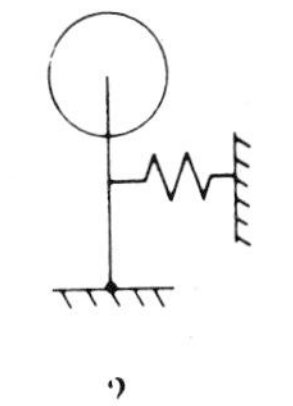

?

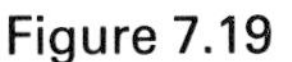

Figure 7.19

the value of δW as the virtual displacement becomes large will lead to the conclusion that if δW becomes negative then the force will be in the opposite direction to the displacement, showing that the forces are tending to return the system to the equilibrium configuration, which is therefore one of stable equilibrium. In mathematical notation, for stability

$$\delta(\delta W) < 0$$

or $$\delta^2 W < 0 \tag{7.38}$$

Looking at our previous case once again,

$$\delta W = (Mga\cos\theta - kR^2\theta)\,\delta\theta$$

therefore $$\delta(\delta W) = (-Mga\sin\theta - kR^2)(\delta\theta)^2$$

For $0 < \theta < \pi$ it is seen that any equilibrium state is stable.

For a conservative system, the equilibrium configuration is defined by

$$\delta V = 0$$

and stability is given by

$$\delta(\delta W) < 0$$

But, since $\delta W = -\delta V$,

$$\delta(\delta V) > 0$$

If V can be expressed as a continuous function of θ, then

$$\delta V = \frac{\partial V}{\partial \theta}\,\delta\theta$$

and

$$\delta^2 V = \frac{\partial}{\partial \theta}\left(\frac{\partial V}{\partial \theta}\,\delta\theta\right)\delta\theta = \frac{\partial^2 V}{\partial \theta^2}(\delta\theta)^2$$

Hence for equilibrium

$$\frac{\partial V}{\partial \theta} = 0 \tag{7.39}$$

and $$\frac{\partial^2 V}{\partial \theta^2} > 0 \quad \text{for stability} \tag{7.40}$$

Systems having two degrees of freedom

The configuration of a system having two degrees of freedom can be defined by any two independent co-ordinates q_1 and q_2. The virtual work for arbitrary virtual displacements δq_1 and δq_2 may be written in the form

$$\delta W = Q_1\,\delta q_1 + Q_2\,\delta q_2 \tag{7.41}$$

Since the virtual displacements are arbitrary, we may hold all at zero except for one and, as $\delta W = 0$ for equilibrium, we have

$$Q_1 = 0$$

and $$Q_2 = 0$$

The stability of a system having two degrees of freedom will be discussed for a conservative system.

It will be remembered that constant forces are conservative, therefore the majority of cases may be considered to be of this type.

If the independent co-ordinates – referred to as generalised co-ordinates – are q_1 and q_2, then the total potential energy (gravitational plus strain) is $V = V(q_1, q_2)$; hence

$$\delta V = \frac{\partial V}{\partial q_1}\,\delta q_1 + \frac{\partial V}{\partial q_2}\,\delta q_2$$

and, since $\delta V = 0$ for equilibrium, we have

$$\frac{\partial V}{\partial q_1} = \frac{\partial V}{\partial q_2} = 0 \tag{7.42}$$

For stable equilibrium we must have $\delta^2 V > 0$ for all possible values of δq_1 and δq_2. The second variation may be written

$$\delta^2 V = \left[\frac{\partial^2 V}{\partial q_1^2}\,\delta q_1 + \frac{\partial^2 V}{\partial q_1\,\partial q_2}\,\delta q_2\right]\delta q_1 + \left[\frac{\partial^2 V}{\partial q_2^2}\,\delta q_2 + \frac{\partial^2 V}{\partial q_1\,\partial q_2}\,\delta q_1\right]\delta q_2$$

or, since $\partial^2 V/\partial q_1\,\partial q_2 = \partial^2 V/\partial q_2\,\partial q_1$, then

$$\delta^2 V = \frac{\partial^2 V}{\partial q_1^2}(\partial q_1)^2 + \frac{\partial^2 V}{\partial q_2^2}(\delta q_2)^2 + 2\frac{\partial^2 V}{\partial q_1\,\partial q_2}(\delta q_1\,\delta q_2)$$

It is clear that, if $\delta q_2 = 0$, then

$$\frac{\partial^2 V}{\partial q_1^2} > 0$$

and, if $\delta q_1 = 0$, then

$$\frac{\partial^2 V}{\partial q_2^2} > 0$$

These are necessary conditions for stability, but not sufficient. To fully define stability, $\delta^2 V$ must be >0 for any linear combination of δq_1 and δq_2.

Let us assume that $\delta q_2 = a\,\delta q_1$, where a is an arbitrary real constant. This gives

$$\left[\frac{\partial^2 V}{\partial q_1{}^2} + \frac{\partial^2 V}{\partial q_2{}^2}\boldsymbol{a}^2 + 2\frac{\partial^2 V}{\partial q_1 \partial q_2}\boldsymbol{a}\right](\delta q_1)^2 > 0 \quad (7.43)$$

Using the notation $V_{,ij} = \partial^2 V/\partial q_i \partial q_j$, it is required that

$$V_{,11} + V_{,22}a^2 + V_{,12}2a > 0$$

since

$$V_{,11} > 0 \quad \text{and} \quad V_{,22} > 0$$

$$a^2 + 2a\left(\frac{V_{,12}}{V_{,22}}\right) + \left(\frac{V_{,11}}{V_{,22}}\right) > 0$$

$$\left[a + \frac{V_{,12}}{V_{,22}}\right]^2 - \left(\frac{V_{,12}}{V_{,22}}\right)^2 + \left(\frac{V_{,11}}{V_{,22}}\right) > 0$$

$$\frac{V_{,11}V_{,22}}{V_{,22}{}^2} - \frac{V_{,12}{}^2}{V_{,22}{}^2} > -\left[a + \frac{V_{,12}}{V_{,22}}\right]^2$$

Because a is real, the highest value the right-hand side can have is zero; therefore

$$V_{,11}V_{,22} - V_{,12}{}^2 > 0$$

The necessary and sufficient conditions for stable equilibrium are

$$\left.\begin{aligned} &\frac{\partial^2 V}{\partial q_1{}^2} > 0 \\ &\frac{\partial^2 V}{\partial q_2{}^2} > 0 \\ \text{and}\quad &\frac{\partial^2 V}{\partial q_1{}^2}\cdot\frac{\partial^2 V}{\partial q_2{}^2} - \left(\frac{\partial^2 V}{\partial q_1 \partial q_2}\right)^2 > 0 \end{aligned}\right\} \quad (7.44)$$

7.10 D'Alembert's principle

In essence, d'Alembert's principle extends the methods of virtual work to include dynamic problems. The principle may be interpreted as asserting that the virtual work done by external forces plus the fictitious inertia forces ($-m\ddot{\boldsymbol{r}}$) is zero. If we consider a system at a particular instant of time, then the inertia forces are regarded as constant external body forces. In this context we see why the virtual displacements are so called, since the inertia forces are 'frozen' as if time has stopped, while the virtual displacements take place.

For a single particle, the inertia force is $-m\ddot{\boldsymbol{r}}$; hence d'Alembert's principle gives

$$\delta\mathscr{W} = (\boldsymbol{F} - m\ddot{\boldsymbol{r}})\cdot\delta\boldsymbol{r} = 0 \quad (7.45)$$

which implies that, since $\delta\boldsymbol{r}$ is arbitrary,

$$\boldsymbol{F} - m\ddot{\boldsymbol{r}} = 0$$

or $\boldsymbol{F} = m\ddot{\boldsymbol{r}}$

This of course yields the expression which is given directly by Newton's laws. However, for more complex systems some advantages can be claimed.

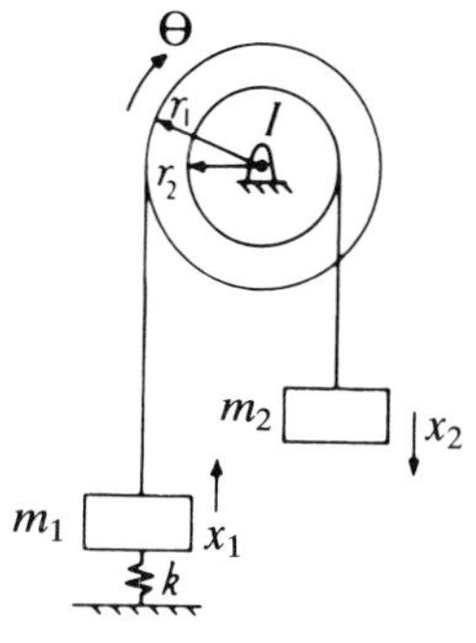

Figure 7.20

Consider the conservative system shown in Fig. 7.20; the active forces, real and fictitious, are shown in Fig. 7.21.

$$\begin{aligned}\delta\mathscr{W} &= \delta V - m_1\ddot{x}_1\,\delta x_1 - m_2\ddot{x}_2\,\delta x_2 - I\ddot{\theta}\,\delta\theta \\ &= \delta[\tfrac{1}{2}kx_1{}^2 + m_1 g x_1 - m_2 g x_2 + \text{const.}]\end{aligned}$$

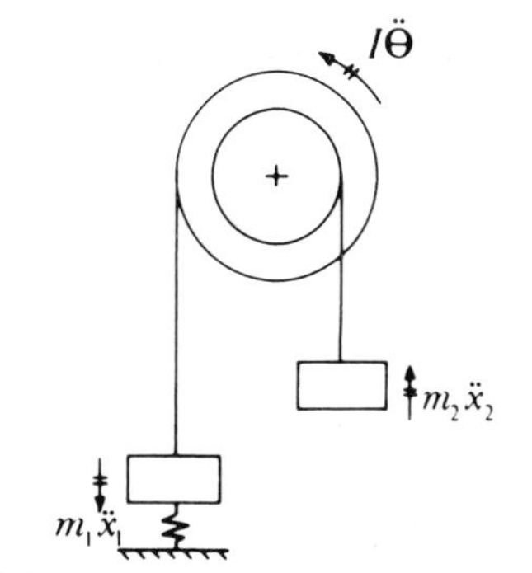

Figure 7.21

The kinematic equations are

$$\frac{x_1}{r_1} = \theta = \frac{x_2}{r_2}$$

hence $$\frac{\delta x_1}{r_1} = \delta\theta = \frac{\delta x_2}{r_2}$$

In terms of δx_2, the virtual-work equation becomes

$$\left[-m_1\left(\frac{r_1}{r_2}\right)^2 - m_2 - I\frac{1}{r_2{}^2}\right]\ddot{x}^2\,\delta x_2$$

$$= \left[k\left(\frac{r_1}{r_2}\right)x_2 + m_1 g\frac{r_1}{r_2} - m_2 g\right]\delta x_2$$

or
$$\left[m_2 + m_1\left(\frac{r_1}{r_2}\right)^2 + \frac{I}{r_2^{\,2}}\right]\ddot{x}_2 + k\left(\frac{r_1}{r_2}\right)^2 x_2$$
$$= m_2 g - m_1\left(\frac{r_1}{r_2}\right)g \qquad (7.46)$$

This approach does not involve the internal forces, such as the tension in the ropes or the workless constraints, but these may be brought in by dividing the system so that these forces appear as external forces.

Equation 7.46 could have been derived by the application of the power equation with a similar amount of labour, but for systems having more than one degree of freedom the power equation is not so useful.

Discussion examples

Example 7.1

A block of mass m can slide down an inclined plane, the coefficient of friction between block and plane being μ. The block is released from rest with the spring of stiffness k initially compressed an amount x_c (see Fig. 7.22). Find the speed when the block has travelled a distance equal to $1.2x_c$.

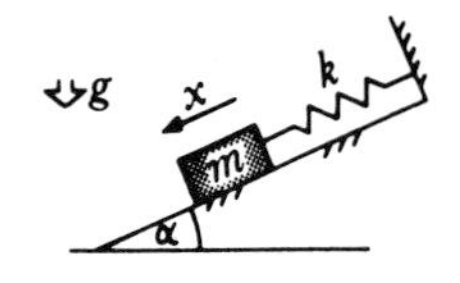

Figure 7.22

Solution If a free-body-diagram approach is used to solve this problem, the equation of motion will be in terms of an arbitrary displacement x (measured from, say, the initial position) and the acceleration $\ddot{x}$. Integration of this equation will be necessary to find the speed.

If an energy method is used, consideration of the initial and final energies will give the required speed. The two methods are compared below.

a) *Integration of equations of motion.* The free-body diagram (Fig. 7.23) enables us to write the following equations:

$[\sum F_y = m\ddot{y}_G]$

$$W\cos\alpha - N = 0 \quad \therefore N = W\cos\alpha \qquad \text{(i)}$$

$[\sum F_x = m\ddot{x}_G]$

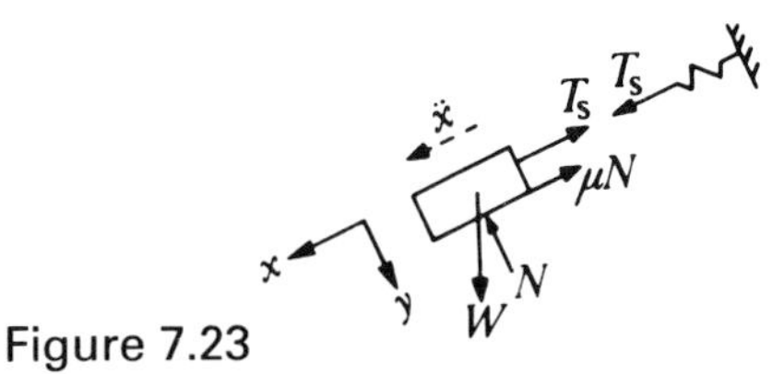

Figure 7.23

$$W\sin\alpha - \mu N - T_s = m\ddot{x} \qquad \text{(ii)}$$

If we measure x from the initial position, the spring tension T_s is given by $T_s = k(x - x_c)$ and we shall be integrating between the limits 0 and $1.2\,x_c$. We could, on the other hand, choose to measure x from the position at which there is no force in the spring, giving $T_s = kx$, and the limits of integration would be from $-x_c$ to $+0.2x_c$.

Using the former, (i) and (ii) combine to give

$$W\sin\alpha - \mu W\cos\alpha - k(x - x_c) = m\ddot{x} \qquad \text{(iii)}$$

Since we are involved with displacements, velocities and accelerations, the appropriate form for $\ddot{x}$ is $v\,\mathrm{d}v/\mathrm{d}x$: the direct form $\ddot{x} = \mathrm{d}v/\mathrm{d}t$ is clearly of no help here.

Hence equation (iii) becomes

$$\int_0^{1.2x_c} \{W(\sin\alpha - \mu\cos\alpha) - k(x - x_c)\}\,\mathrm{d}x = \int_0^v mv\,\mathrm{d}v$$

$$\left[\{W(\sin\alpha - \mu\cos\alpha) + kx_c\}x - \tfrac{1}{2}kx^2\right]_0^{1.2x_c} = m\left[\tfrac{1}{2}v^2\right]_0^v$$

$$\{mg(\sin\alpha - \mu\cos\alpha) + kx_c\}\,1.2x_c - \tfrac{1}{2}k(1.2x_c)^2 = \tfrac{1}{2}mv^2$$

and thus v can be found.

The reader should check this result by measuring x from some other position, for instance the position at which the spring is unstrained, as suggested previously.

b) *Energy method.* Since energy is lost due to the friction, we use equation 7.29 (see Fig. 7.24):

$$[\text{work done}]_{\text{external}} = [\text{k.e.} + V_G + V_E]_2 - [\text{k.e.} + V_G + V_G]_1 + \text{'losses'} \qquad \text{(iv)}$$

where the 'losses' will be $\mu N(1.2x_c)$ as explained in section 7.6. For the general case of both μ and N varying, this loss will be $\int_0^{1.2x_c}\mu N\,\mathrm{d}x$. None of the external forces does any work, according to our definitions, and thus the left-hand side of

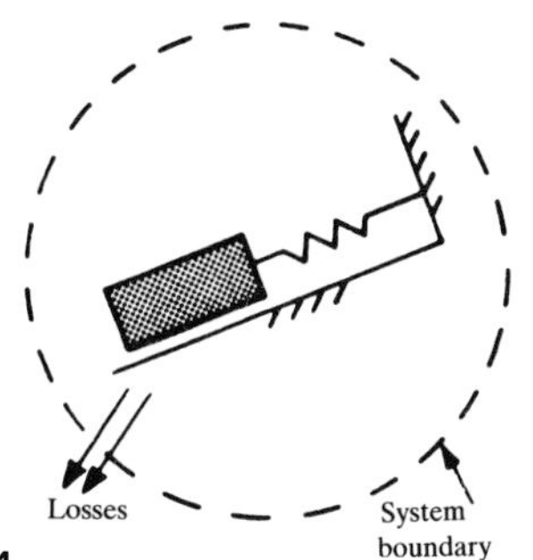

Figure 7.24

equation (iv) is zero.

It should be pointed out that the correct result *can* be obtained by treating the friction force as external to the system and saying that this force does negative work since it opposes the motion. The left-hand side of equation (iv) would then be $-\mu N(1.2x_c)$ and the 'losses' term on the right would be omitted. This is a common way of dealing with the friction force but is not considered to be a true energy method.

Kinetic energy. In the initial position ($x = 0$) the speed and thus the k.e. are zero. In the final position ($x = 1.2x_c$) the k.e. is $\frac{1}{2}mv^2$, from equation 7.26.

Gravitational energy, V_G. The datum for measuring gravitational energy is arbitrary and we may take as a convenient level that through the initial position; thus the initial g.e. is zero. Since the block then *falls* through a vertical distance of $1.2x_c \sin\alpha$, the final gravitational energy is, from equation 7.27, $-mg(1.2x_c \sin\alpha)$.

Strain energy, V_E. In the initial position, the spring is compressed an amount x_c and thus, from equation 7.28, the strain energy is $\frac{1}{2}kx_c^2$. In the final position the spring is extended by an amount $0.2x_c$ and so the final s.e. is $\frac{1}{2}k(0.2x_c)^2$.

Note that only the gravitational energy can have a negative value.

Equation (iv) becomes

$$0 = [\tfrac{1}{2}mv^2 - mg(1.2x_c \sin\alpha) + \tfrac{1}{2}k(0.2x_c^2)] - [0 + 0 + \tfrac{1}{2}kx_c^2] + \mu N(1.2x_c)$$

We still need a free-body diagram to determine that $N = mg\cos\alpha$, as in equation (i), and then v can be found directly.

For this particular problem there is little to choose between the free-body-diagram approach and the energy method. In the energy method we avoided the integration of the first method, which however presented no difficulty.

If, however, the block had been following a known curved path, the spring tension T_s could have been a complicated function of position giving rise to difficult integrals, possibly with no analytical solution. The energy method requires only the initial and final values of the spring energy and so the above complication would not arise. Variation in N could cause complications in both methods. In some cases the path between the initial and final positions may not be defined at all; here it would not be possible to define T_s as a general function of position. An energy method would give a solution directly for cases where friction is negligible (see, for example, problem 7.2).

Example 7.2

See Fig. 7.25(a). The slider B of mass m is constrained to move in vertical guides. A pin P fixed to the slider engages with the slot in link OA. The moment of inertia of the link about O is I_O and its mass is M, the mass centre being a distance a from O. The spring of stiffness k is attached to B and is unstrained when $\theta = 0$.

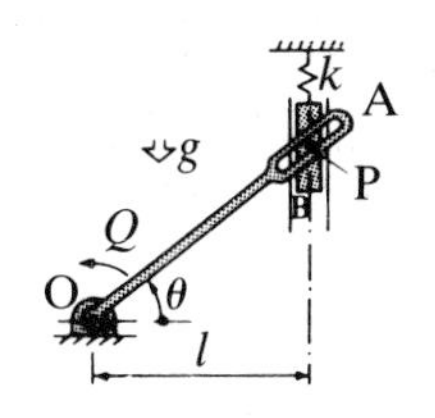

Figure 7.25

The system is released from rest at $\theta = 0$ under the action of the torque Q which is applied to link OA. The variation of Q with θ is shown in Fig. 7.25(b).

Determine the angular speed of OA when $\theta = 45°$, neglecting friction.

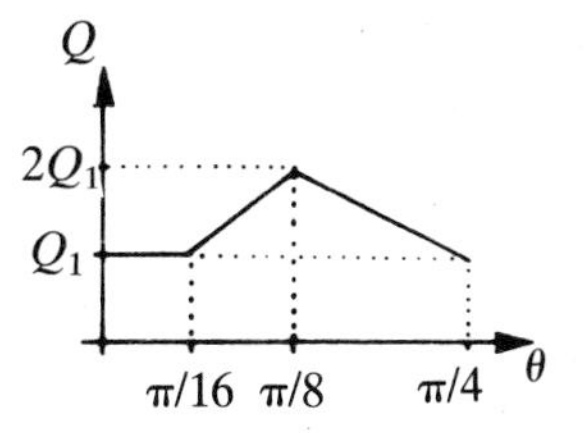

Solution This problem has been approached in example 6.5 by drawing two free-body diagrams and writing two equations of motion involving the contact force at the pin P. Since we are here concerned only with the angular velocity at a given position, and details of internal forces are not required, an energy method is indicated and will be seen to be easier than the method of Chapter 6.

Equation 7.29 becomes

$$(\text{work done})_{\text{external}} = [\text{k.e.} + V_G + V_E]_2 - [\text{k.e.} + V_G + V_E]_1 \quad \text{(i)}$$

since there are no losses. The left-hand side of this equation is the work done by the external forces or couples on the system during the motion and is thus $\int_0^{\pi/4} Q\,d\theta$. This is simply the area under the curve of Fig. 7.25(b), which is found to be $(11/32)\,\pi Q_1$.

The normal reaction N between the slider and the guides is perpendicular to its motion and the force R in the pin at O does not move its point of application: thus neither of these forces does work (see Fig. 7.26).

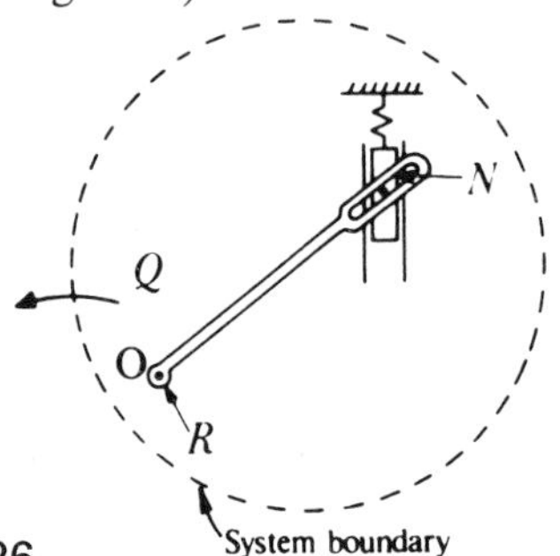

Figure 7.26

Kinetic energy. As the mechanism is initially at rest, the initial k.e. is zero. Since the motion of OA is rotation about a fixed axis, the final k.e. of OA, from equation 7.8, is $\frac{1}{2}I_O\,\omega_{OA}{}^2$. The slider B has no rotation and its final kinetic energy, from equation 7.7, is simply $\frac{1}{2}mv_B{}^2$.

Gravitational energy, V_G. We will take as datum levels the separate horizontal lines through the mass centres of link and slider when $\theta = 0$ and thus make the initial value of V_G zero. When $\theta = 45°$ the mass centre of the link has risen through a height $a/\sqrt{2}$ and that of the slider through a height l, and so the final value of V_G is $Mga/\sqrt{2} + mgl$.

Strain energy, V_E. Initially the strain energy is zero and in the final position the spring has been compressed an amount l; the final value of V_E is thus $\frac{1}{2}kl^2$.

Substituting in (i) gives

$$\frac{11}{32}\pi Q_1 = \left[\tfrac{1}{2}I_O\,\omega_{OA}{}^2 + \tfrac{1}{2}mv_B{}^2 + Mg\frac{a}{\sqrt{2}} + mgl + \tfrac{1}{2}kl^2\right] - \left[0\right] \quad \text{(ii)}$$

Before we can evaluate ω_{OA} we need to express v_B in terms of ω_{OA}. Since $y = l\tan\theta$,

$$v_B = dy/dt = l\sec^2\theta\,d\theta/dt = l\sec^2\theta\,\omega_{OA}$$

and at $\theta = \pi/2$, $v_B = 2l\omega_{OA}$.

Substitution in (ii) gives

$$\frac{11}{32}\pi Q_1 = \tfrac{1}{2}(I_O + 4ml^2)\,\omega_{OA}{}^2 + g\left(M\frac{a}{\sqrt{2}} + ml\right) + \tfrac{1}{2}kl^2$$

from which ω_{OA} can be found.

Comparison of this method with the free-body-diagram approach and the difficulty associated with integrating the equation of motion shows the superiority of the energy method for this problem.

What if the force S on the pin P has been required? This force does not appear in the energy method, but this does not mean that the energy method is of no help. Often an energy method can be used to assist in determining an unknown acceleration and then a free-body-diagram approach may be employed to complete the solution.

Example 7.3

A four-bar chain ABCD with frictionless joints is shown in Fig. 7.27.

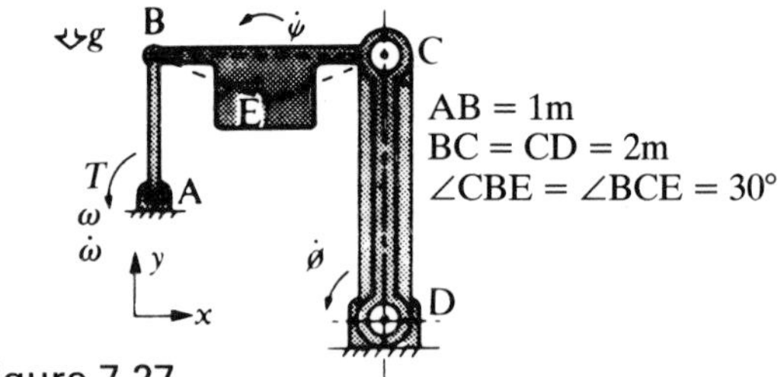

Figure 7.27

CD has a moment of inertia about D of 6 kg m^2 and its mass is 4.5 kg. BC has a moment of inertia about its mass centre E of 1.5 kg m^2 and its mass is 4 kg. At the instant when both AB and CD are vertical, the angular velocity of AB is 10 rad/s and its angular acceleration is 50 rad/s^2, both measured in an anticlockwise sense.

Neglecting the inertia of AB, determine the torque T which must be applied at A to produce the above motion.

Solution The velocity of B is $\boldsymbol{\omega} \times \overrightarrow{AB} = 10\boldsymbol{k} \times 1\boldsymbol{j} = -10\boldsymbol{i}$ m/s. The velocity diagram is shown in Fig. 7.28 and it can be seen that link BC is not rotating ($\dot{\psi} = 0$).

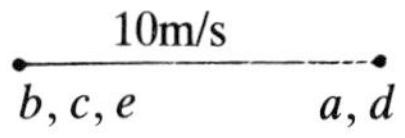

10m/s

b, c, e $\qquad$ a, d

Figure 7.28

The kinetic energy of link BG is thus $\frac{1}{2}M{v_E}^2$ and that of CD, for fixed-axis rotation about D, is $\frac{1}{2}I_D\dot{\phi}^2$. It would clearly *not* be correct to write the power equation (section 7.8) as

$$T \cdot \boldsymbol{\omega} = T\omega = \frac{\mathrm{d}}{\mathrm{d}t}(\tfrac{1}{2}M{v_E}^2 + \tfrac{1}{2}I{\phi_D}^2)$$

since $\frac{1}{2}M{v_e}^2$ is not a *general* expression for the k.e. of link BC (it is the particular value when AB is vertical). As $\boldsymbol{v}_e$ and $\boldsymbol{a}_e$ do not have the same direction, the correct power equation is

$$T \cdot \boldsymbol{\omega} = T\omega = \frac{\mathrm{d}}{\mathrm{d}t}(\tfrac{1}{2}M\boldsymbol{v}_E \cdot \boldsymbol{v}_E + \tfrac{1}{2}I_E\dot{\psi}^2 + \tfrac{1}{2}I_D\dot{\phi}^2)$$

$$= M\boldsymbol{v}_E \cdot \boldsymbol{a}_E + I_D\dot{\phi}\ddot{\phi} \qquad \text{(i)}$$

since $\dot{\psi} = 0$.

The acceleration of B is

$$\boldsymbol{a}_B = [-1(50)\boldsymbol{i} - 1(10)\boldsymbol{j}^2] \text{ m/s}^2$$

From the velocity diagram we find $\dot{\phi} = 10\boldsymbol{k}/2 = 5\boldsymbol{k}$ rad/s and thus

$$\boldsymbol{a}_C = [-2\ddot{\phi}\boldsymbol{i} - 2(5)^2\boldsymbol{j}] \text{ m/s}^2$$

The acceleration diagram is shown in Fig. 7.29 and $\ddot{\phi}$ is given by

$$\ddot{\phi} = \frac{cc'}{\mathrm{CD}} = \frac{50}{2} = 25 \text{ rad/s}^2$$

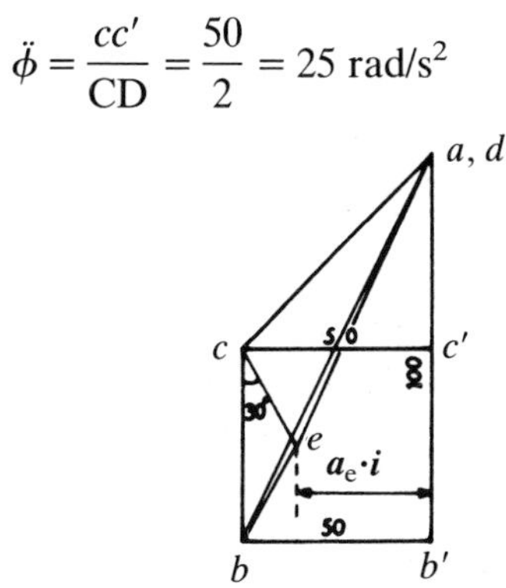

Figure 7.29

From the velocity diagram we see that $\boldsymbol{v}_E = -10\boldsymbol{i}$ m/s; the component of $\boldsymbol{a}_E$ in the same direction is $\boldsymbol{a}_E \cdot \boldsymbol{i} = -(50 - 25/\sqrt{3})$ m/s². Substituting into the power equation (i) gives

$$T10 = 4(10)(50 - 25/\sqrt{3}) + 6(5)(25)$$

and $\quad T = 217.2$ N m

Example 7.4

A slider-crank chain PQR is shown in Fig. 7.30 in its equilibrium position, equilibrium being maintained by a spring (not shown) at P of torsional stiffness k. Links PQ and QR are of mass m and $2m$ and their mass centres are at G_1 and G_2 respectively. The slider R has a mass M. The moment of inertia of PQ about P is I_P and that of QR about G_2 is I_{G2}; also, $PG_1 = G_1Q = QG_2 = G_2R = a$.

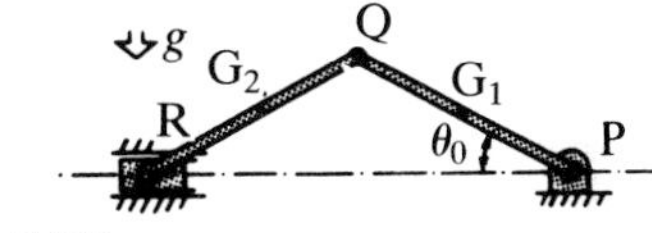

Figure 7.30

Find the frequency of small oscillations of the system about the equilibrium position, $\theta = \theta_0$, neglecting friction.

Solution Equations of motion for the links can of course be obtained from a free-body-diagram approach, but this would involve the forces in the pins and would be extremely cumbersome.

Use of the power equation leads directly to the required result. In this case we have power = d(energy)/dt = 0, since the energy is constant for the conservative system and clearly no power is fed into or taken out of the system.

Let the link PQ rotate clockwise from the equilibrium position through a *small* angle β as shown in Fig. 7.31. The new positions of the various points are shown by a prime.

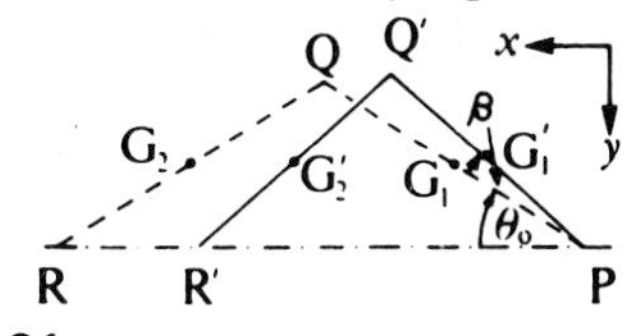

Figure 7.31

Kinetic energy. Link PQ has fixed-axis rotation about P and its k.e. is thus $\frac{1}{2}I_P\dot{\beta}^2$. By symmetry, the angular speed of QR is also $\dot{\beta}$. The k.e. of QR is given by $\frac{1}{2}I_{G2}\dot{\beta}^2 + \frac{1}{2}(2m){v_{G2}}^2$, where

$$v_{G2} = \frac{d}{dt}(\overrightarrow{PG_2'})$$

$$= \frac{d}{dt}[3a\cos(\theta_0+\beta)\boldsymbol{i} - a\sin(\theta_0+\beta)\boldsymbol{j}]$$

$$= -3a\sin(\theta_0+\beta)\dot{\beta}\boldsymbol{i} - a\cos(\theta_0+\dot{\beta})\boldsymbol{j}$$

The k.e. of slider R is $\frac{1}{2}Mv_R^2$ where

$$v_R = \frac{d}{dt}(\overrightarrow{PR'}) = \frac{d}{dt}[4a\cos(\theta_0+\beta)\boldsymbol{i}]$$

$$= -4a\sin(\theta_0+\beta)\dot{\beta}\boldsymbol{i}$$

Gravitational energy. A convenient datum level is the horizontal through PR. The mass centre of each of the links PQ and QR is at a height $a\sin(\theta_0+\beta)$ above the datum and their gravitational energy is thus, from equation 7.27,

$$V_G = mga\sin(\theta_0+\beta) + (2m)ga\sin(\theta_0+\beta)$$

The slider R moves the datum level and thus has no gravitational energy with respect to this level.

Strain energy. The couple applied by the spring to the link PQ in the equilibrium position is clockwise and equal to $k\gamma_0$, where γ_0 is the angle of twist. As link PQ rotates clockwise through an angle β, the angle of twist is *reduced* to $(\gamma_0-\beta)$ and thus the strain energy, from equation 7.28, is $\frac{1}{2}k(\gamma_0-\beta)^2$.

The total energy E is thus

$$E = \{\text{k.e.}\} + \{V_G\} + \{V_E\}$$

$$= \{\tfrac{1}{2}I_P\dot{\beta}^2 + \tfrac{1}{2}I_{G2}\dot{\beta}^2 + \tfrac{1}{2}(2m) \times [9a^2\sin^2(\theta_0+\beta) + a^2\cos^2(\theta_0+\beta)]\dot{\beta}^2 + \tfrac{1}{2}M.16a^2\sin^2(\theta_0+\beta)\dot{\beta}^2\} + \{3mga\sin(\theta_0+\beta)\} + \{\tfrac{1}{2}k(\gamma_0-\beta)^2\}$$

$= \text{constant}$ (since the system is conservative)

Since the above is a general expression for the energy, it can be differentiated to give the power equation. The term $\dot{\beta}^3$ arises which is negligible for small oscillations. We note that, since β is small, $\sin(\theta_0+\beta) \approx \sin\theta_0$ and $\cos(\theta_0+\beta) \approx \cos\theta_0$, but these approximations must not be made before differentiating. After dividing throughout by $\dot{\beta}$, we find, since $\dot{\beta}^2$ is small,

$$I\ddot{\beta} + k\beta = k\gamma_0 - 3mga\cos\theta_0 \qquad \text{(i)}$$

where $I = I_P + I_{G2} + 2m(9a^2\sin^2\theta_0 + a^2\cos^2\theta_0) + 16Ma^2\sin^2\theta_0$.

It can be shown by the method of virtual work, or otherwise that $k\gamma_0 = 3mga\cos\theta_0$ so that equation (i) reduces to $I\ddot{\beta} + k\beta = 0$. Thus, for small β, the motion about the equilibrium position is simple harmonic with a frequency of $(1/2\pi)\sqrt{(k/I)}$.

Example 7.5

The mechanism shown in Fig. 7.32 is in equilibrium. Link AB is light and the heavy link BC weighs 480 N, its mass centre G being midway between B and C. Friction at the pins A and C is negligible. The limiting friction couple Q_f in the hinge at B is 10 N m.

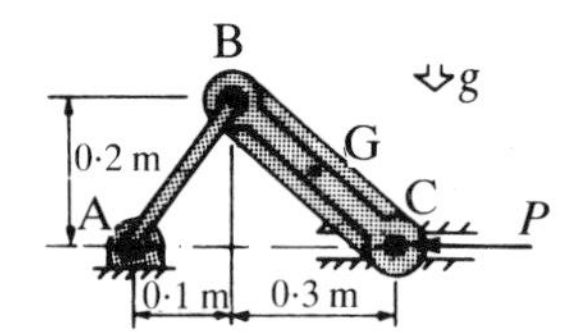

Figure 7.32

Pin C can slide horizontally, and the horizontal force P is just sufficient to prevent the collapse of the linkage. Find the value of P.

Solution This problem has been solved earlier in Chapter 4 (example 4.3). There a free-body diagram was drawn for each of links AB and BC and the unknown forces were eliminated from the moment equations. It will now be solved by the method of virtual work and the two methods will be compared.

If in the virtual-work method we treat forces due to gravity and springs and friction as being externally applied, the total virtual work done may then be equated to zero. In order to obtain the correct sign for the virtual work done by the internal friction couple Q_f in the present problem, we may use the following rule: *the virtual displacements must be chosen to be in the same direction as the actual or impending displacements and the virtual work done by friction is given a negative sign.*

Applying this rule, we let the virtual displacement of C be δx to the *right*, since this is the direction in which it would move if the mechanism were to collapse.

If a mechanism has a very small movement, the displacement vector of any point on the mechanism will be proportional to the velocity vector. Thus we can draw a small-displacement diagram which is identical in form with the corresponding velocity diagram. This results in

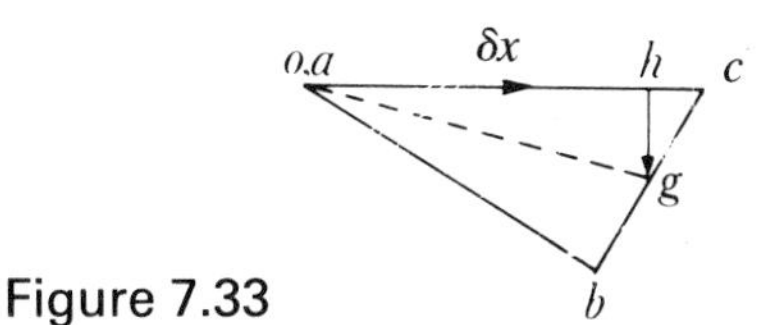

Figure 7.33

Fig. 7.33, where ab is drawn perpendicular to AB, bc is drawn perpendicular to BC and oc is of length δx.

Since the weight W acts vertically downwards and the vertical component (hg) of the displacement of G is also downwards, the virtual work done by W is positive and given by $+W(hg)$.

The virtual work done by P is $-P(oc)$, since the force P is opposite in sense to the assumed virtual displacement.

The virtual work done by the friction couple Q_f is $-Q_f|\delta\phi|$, where $|\delta\phi|$ is the magnitude of the change in the angle ABC. AB rotates clockwise through an angle ab/AB and BC rotates anticlockwise through an angle bc/BC. [If Fig. 7.33 had been a velocity diagram then of course the angular *speed* of AB would have been given by ab/AB, and so on.] The change in the angle ABC is thus

$$\delta\phi = \frac{ab}{\mathrm{AB}} + \frac{bc}{\mathrm{BC}}$$

Summing the virtual work to zero gives

$$W(hg) - P(oc) - Q_f\left(\frac{ab}{\mathrm{AB}} + \frac{bc}{\mathrm{BC}}\right) = 0$$

The virtual displacements ab, bc and hg are scaled directly from the diagram to give

$$480(0.1875\,\delta x) - P(\delta x)$$
$$-10\left(\frac{0.838\,\delta x}{0.2235} + \frac{0.451\,\delta x}{0.3605}\right) = 0$$

which, on dividing throughout by δx, gives $P = 40.0$ N.

Comparing the virtual-work solution of this problem with that of the normal statics/free-body-diagram approach of Chapter 4, it can be seen that here we are not concerned with the forces at A and B and the vertical component of the force at C. However, for this simple problem the more straightforward approach of Chapter 4 is to be preferred.

The virtual-work method comes into its own when many links are connected together. In such cases, drawing separate free-body diagrams for each link and writing the relevant equations is a very lengthy means of solution, whereas the virtual-work method disposes of the problem relatively quickly (see, for example, problem).

Example 7.6

A roller of weight W is constrained to roll on a circular path of radius R as shown in Fig. 7.34. The centre C of the roller is connected by a spring of stiffness k to a pivot at O. The position of the roller is defined by the angle θ and the spring is unstretched when $\theta = 90°$.

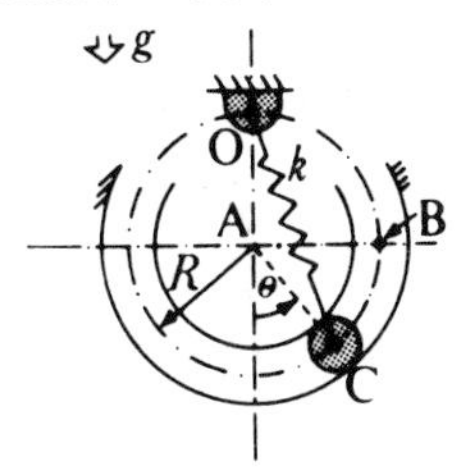

Figure 7.34

a) Show that the position $\theta = 0$ is one of stable equilibrium only if $W/(Rk) > 0.293$.
b) If $W/(Rk) = 0.1$, determine the positions of stable equilibrium.

Solution The strain energy V_E in the spring is zero when centre C is at B. We can also make the gravitational energy V_G zero for this position by taking AB as the datum level.

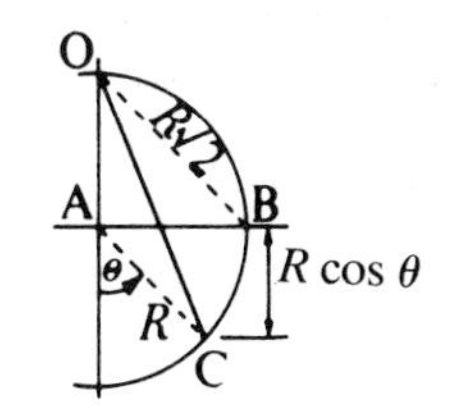

Figure 7.35

From Fig. 7.35, the stretch in the spring is $\mathrm{OC} - \mathrm{OB} = 2R\cos(\theta/2) - R\sqrt{2}$ and C is a vertical distance $R\cos\theta$ *below* AB. Thus, using equations 7.27 and 7.28, the total potential energy V is given by

$$V = V_G + V_E$$
$$= -WR\cos\theta + \tfrac{1}{2}k[2R\cos(\theta/2) - R\sqrt{2}]^2$$

The equilibrium positions are given, from equation 7.39, by

$$\mathrm{d}V/\mathrm{d}\theta = WR\sin\theta + k[2R\cos(\theta/2) - R\sqrt{2}] \times [-R\sin(\theta/2)]$$
$$= R^2k\{\sin\theta[W/(Rk) - 1] + \sqrt{2}\sin(\theta/2)\} = 0 \qquad \text{(i)}$$

a) One solution to equation (i) is clearly $\theta = 0$. Now,

$$\frac{d^2V}{d\theta^2} = R^2k\{\cos\theta[W/(Rk)-1] + (\sqrt{2}/2)\cos(\theta/2)\} \quad \text{(ii)}$$

and when $\theta = 0$,

$$\frac{d^2V}{d\theta^2} = R^2k[W/(Rk)-1+0.707]$$

$$= R^2k[W/(Rk)-0.293]$$

For stability, from equation 7.40,

$$d^2V/d\theta^2 > 0 \quad \text{i.e. } W/Rk > 0.293$$

b) Substituting $W/(Rk) = 0.1$ in equation (i) gives

$$dV/d\theta = R^2k[\sin\theta(-0.9) + \sqrt{2}\sin(\theta/2)] = 0$$

$$\therefore \sqrt{2}\sin(\theta/2) = 0.9\sin\theta = 1.8\sin(\theta/2)\cos(\theta/2)$$

We know that the solution $\theta = 0$ represents unstable equilibrium. The other solutions are given by

$$\cos(\theta/2) = \sqrt{2}/1.8 = 0.786$$

hence, $\theta = \pm 76.4°$.

The type of stability at these two positions can be confirmed by substituting for θ in equation (ii):

$$d^2V/d\theta^2$$
$$= R^2k[0.235(0.1-1)+0.707(0.786)]$$
$$= R^2k[-0.211+0.556] > 0$$

Thus, at $\theta = \pm 76.4°$, the system is in stable equilibrium.

Example 7.7

During the erection of a structure, three beams are connected as shown in Fig. 7.36. Beam ABC may be considered as fixed at A and to deform in torsion only. The vertical beams BD and CE may be considered as equal rigid uniform members. The torsional stiffnesses of AB and BC are k_1 and k_2 respectively and the weights of BD and of CE are each W. Determine the torsional stiffness requirements for AB and BC so that the structure is stable in the position shown.

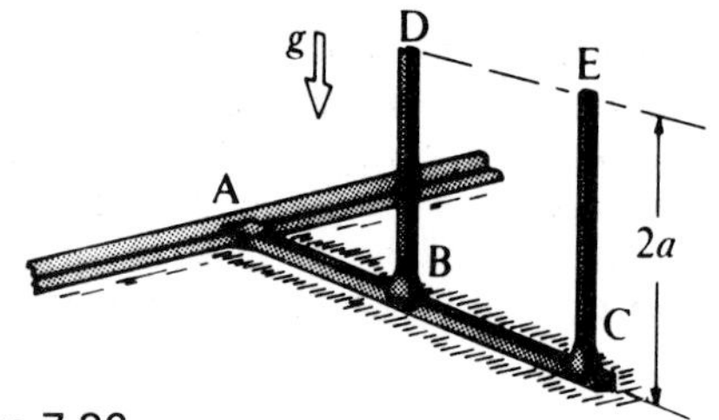

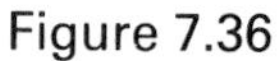
Figure 7.36

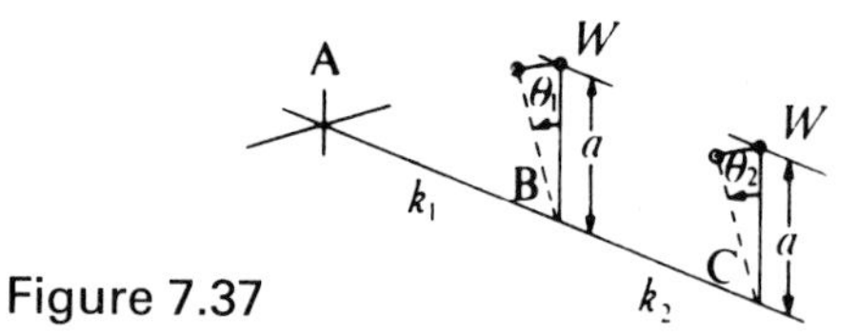

Figure 7.37

Solution See Fig. 7.37. The elastic strain energy for a torsional spring is

$$V_E = \int_0^\theta (\text{torque})\, d\theta$$

but torque $= k\theta$ and therefore

$$V_E = \int_0^\theta k\theta\, d\theta = k\theta^2/2$$

For this system we have

$$V_E = k_1\theta_1^2/2 + k_2(\theta_2-\theta_1)^2/2$$

The gravitational potential energy, taking AC as datum, is

$$V_G = Wa\cos\theta_1 + Wa\cos\theta_2$$

Hence $V = k_1\theta_1^2/2 + k_2(\theta_2-\theta_1)^2/2 + Wa\cos\theta_1 + Wa\cos\theta_2$

For equilibrium we have, from equation 7.42,

$$\partial V/\partial\theta_1 = 0 = k_1\theta_1 + k_2(\theta_2-\theta_1)(-1) - Wa\sin\theta$$

and $\quad \partial V/\partial\theta_2 = 0 = k_2(\theta_2-\theta_1) - Wa\sin\theta_2$

By inspection it is clear that $\theta_1 = \theta_2 = 0$ is one condition for equilibrium. To test for stability we use inequalities 7.44:

$$\frac{\partial^2V}{\partial\theta_1^2} = k_1 + k_2 - Wa\cos\theta_1 > 0$$

$$\frac{\partial^2V}{\partial\theta_2^2} = k_2 - Wa\cos\theta_2 > 0$$

$$\frac{\partial^2V}{\partial\theta_1\partial\theta_2} = -k_2$$

and $\quad \dfrac{\partial^2V}{\partial\theta_1^2}\cdot\dfrac{\partial^2V}{\partial\theta_2^2} - \left(\dfrac{\partial^2V}{\partial\theta_1\partial\theta_2}\right)^2 > 0$

Therefore, for stability when $\theta_1 = \theta_2 = 0$, we must have

$$k_1 + k_2 - Wa > 0 \quad \text{(i)}$$

$$k_2 - Wa > 0 \qquad \text{(ii)}$$

and $(k_1 + k_2 - Wa)(k_2 - Wa) - k_2^2 > 0$ (iii)

Expanding this last inequality and dividing by $(Wa)^2$ gives

$$\left(\frac{k_1}{Wa}\right)\left(\frac{k_2}{Wa}\right) - \left(\frac{k_1}{Wa} + \frac{2k_2}{Wa}\right) + 1 > 0$$

or $\left(\frac{k_1}{Wa}\right) > \frac{2(k_2/Wa) - 1}{(k_2/Wa) - 1}$ (iv)

If we now plot (k_1/Wa) against (k_2/Wa) we can see the range of values of stiffness (shown cross-hatched in Fig. 7.38) which will ensure a stable structure.

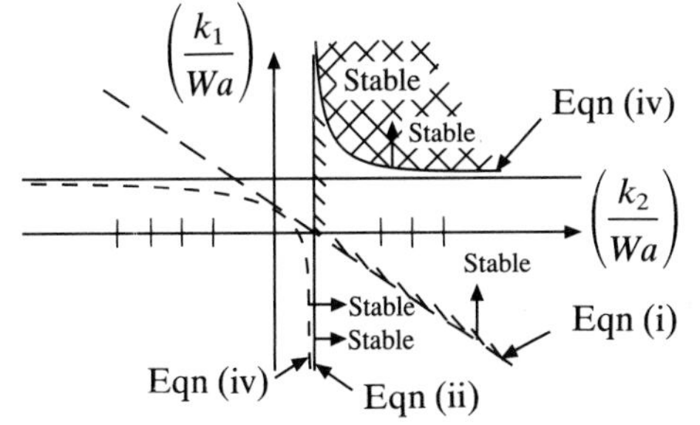

Figure 7.38

Example 7.8

A Simpson gearset as shown in Fig. 5.23(b) has forward gear ratios of 2.84, 1.60, 1.00 and a reverse ratio of 2.07. The maximum input torque is 200 Nm.

Assuming that the efficiency of the gearbox is 100% determine the output torque and the torque on the clutches and/or band brake, which are operative at a steady engine speed for each gear.

Solution We can apply the power equation to the gearbox and, since there is no change in kinetic energy and no losses, the net power into the system must be zero, that is

$$\text{power in} - \text{power out} = 0$$

Now

$$\text{power} = \text{torque} \times \text{angular speed} = Q \times \omega$$

so

$$Q_{in}\,\omega_{in} - Q_{out}\,\omega_{out} = 0$$

also ω_{in}/ω_{out} = the gear ratio G, therefore

$$Q_{in} = Q_{out} \times G.$$

If we consider the torques about the central axis applied to the gear train less the case, we have the input torque, the output torque and the holding torque between the fixed case and that part of the gear which has been fixed by the operation of the various clutches and band brake.

For first and second forward gears the forward clutch carries the input torque of 200 Nm, but for third gear the input torque is divided between the forward and direct clutches. In reverse the input torque is applied through the direct clutch.

For first gear the output torque $= -200 \times 2.84 = -568$ Nm and the sum of the torques is zero for steady speed running, thus

$$200 - 568 + Q_H = 0$$

or $Q_H = 368$ Nm.

This torque is transmitted either through the Sprag clutch or through the low/reverse clutch depending on whether lock down or normal drive has been selected. (See the description of the operation of the gearbox in section 5.11.)

For second gear the output torque $= -200 \times 1.60 = -320$ Nm so the moment equation gives

$$200 - 320 + Q_h = 0$$

hence $Q_H = 120$ Nm.

This torque is transmitted through the intermediate band.

For third gear the output torque is equal to the input torque so it follows that the holding torque is zero. It is left as an exercise for the reader to show that the proportion of the input torque carried by the forward clutch is 1/(1.60). (This is achieved by considering the equilibrium of the input planets.)

In reverse the output torque is $2.07 \times 200 = 414$ Nm. Now the moment equation gives

$$200 + 414 + Q_H = 0$$

or $Q_H = -614$ Nm

and this torque is carried by the low/reverse clutch.

An alternative method for finding the holding torques is to assume that the whole gear assembly is rotating at an arbitrary speed Ω, this means that the stationary parts of the gearbox are rotating at Ω and the input and output shafts have their speeds increased by Ω. The power equation now becomes

$$Q_{in}(\omega_{in} + \Omega) + Q_{out}(\omega_{out} + \Omega) + Q_H\Omega = 0$$

or $(Q_{in}\omega_{in} + Q_{out}\omega_{out})$
$$+\Omega(Q_{in} + Q_{out} + Q_H) = 0$$

and since Ω is arbitrary this equation must be true for all values of Ω, including zero, it follows that both bracketed terms must individually be equal to zero, thus repeating the previous results. It should be noted that this method assumes no internal friction whereas the moment equation is always true.

Problems

7.1 A slider B of mass 1 kg is released from rest and travels down an incline a distance of 2 m before striking a spring S of stiffness 100 N/m (see Fig. 7.39). The coefficient of friction between the slider and the plane is 0.1. Determine the maximum deflection of the spring.

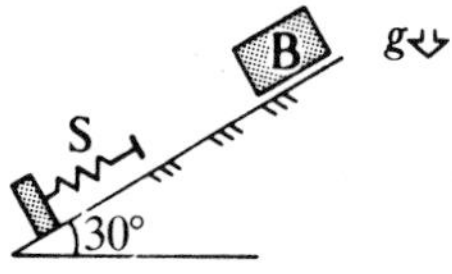

Figure 7.39

7.2 See Fig. 7.40. A light rope, passing over a light pulley P, connects the sliding collar C, mass 2 kg, to the spring of stiffness 50 N/m. The collar is released from rest in the position shown, the tension in the spring being 20 N in this position. Find the speed of the collar when it has travelled 40 mm down the inclined rod. Neglect friction.

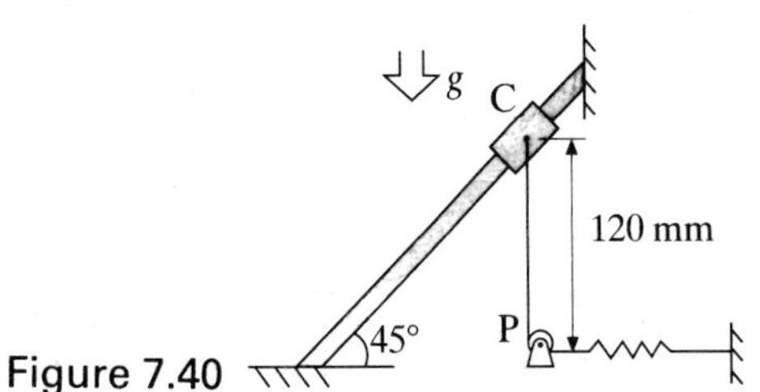

Figure 7.40

7.3 A small toy motor car A, mass 100 g, travels along a track T as shown in Fig. 7.41. The track consists of two circular arcs AB and BC which have centres O and O_1 respectively and which lie in the same vertical plane. The motor torque remains constant at 7×10^{-4} N m between A and C and the motor shaft rotates through one revolution while the car travels a distance of 1 cm.

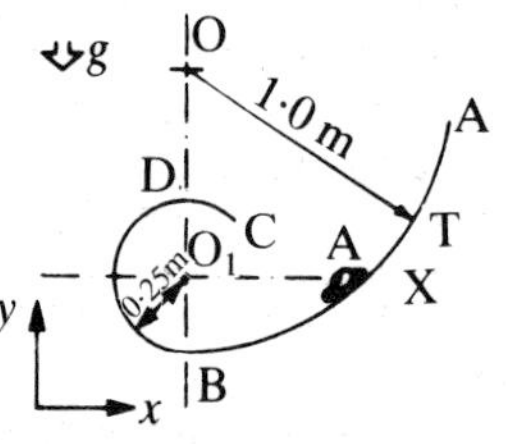

Figure 7.41

If the speed of the car at point X is 2 m/s, determine (a) the speed of the car and (b) the force on the track as the car passes the point D.

7.4 a) A satellite of mass m moves from a point P_1 at a height h_1 to a point P_2 at a height h_2 above the Earth's surface. The gravitational attraction between the Earth and the satellite obeys the inverse-square law, the distance concerned in this law being measured between the centres of Earth and satellite.

Prove that the work done by gravity on the satellite as it travels from P_1 to P_2 is $m[g_2(h_2+R) - g_1(h_1+R)]$, where g_1 and g_2 are the gravitational field strengths at the points P_1 and P_2 respectively, R is the effective radius of the Earth, and h is taken to be positive in a direction away from the centre of the Earth.

b) A satellite is in orbit around the Earth. At one point in its trajectory it is at a height of 8000 km above the Earth's surface and its speed is 4000 m/s. Determine its speed when it is at a point 1000 km above the Earth's surface.

Take the effective radius of the Earth to be 6370 km and g to be 10 N/kg at the surface of the Earth. Neglect air resistance.

7.5 A lunar module is jettisoned by the parent craft when its height above the lunar surface is 100 km and the speed is 600 km/h. Determine the speed of the module just prior to impact with the lunar surface (a) neglecting the variation of g with height and (b) taking into account the variation of g.

Take the value of g at the surface of the Moon to be 1.62 N/kg and the effective radius of the Moon to be 1.74×10^3 km.

7.6 A four-bar linkage consists of three similar uniform rods AB, BC and CD as shown in Fig.7.42. Each has a length of 0.5 m and a mass of 2.0 kg. A torsion spring (not shown) at A has a stiffness of 40 N m/rad, one end of the spring being fixed and the other end attached to AB.

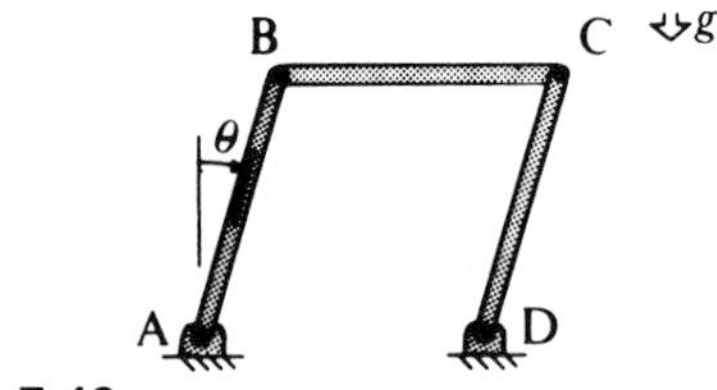

Figure 7.42

Initially the mechanism is held with AB vertical, and in this position the spring exerts a clockwise couple of 80 N m on AB. If the mechanism is then released, what is the angular speed of AB when $\theta = 30°$?

7.7 A roller of radius R has an axial moment of inertia I and a mass m. Initially the roller is at rest and then it is pulled along the ground by means of a horizontal rope attached to its axle C, the tension in the rope being a constant T_0. If the roller rolls without slipping show that, after it has travelled a distance l, its speed is $R[2T_0l/(I+mR^2)]^{1/2}$.

7.8 A motor drives a load through a reduction gearbox. the torque developed between the rotor and the stator of the motor is T_M. The total moment of inertia of the motor shaft is I_M and the damping torque is C_M times the motor shaft speed ω_M. The effective moment of inertia of the load shaft is I_L and the damping torque on this shaft is C_L times the load speed ω_L. The shaft drives a load torque T_L. If $\omega_M = n\omega_L$, show that

$$nT_M - T_L = (I_L + n^2I_M)\dot{\omega}_L + (C_L + n^2C_M)\omega_L$$

7.9 At a particular instant, the acceleration of a motor car is a and its speed is v. The engine power is P_E and the power used up in overcoming friction, rolling and wind resistance is P_L. The rear-axle ratio is n_a:1 and the gearbox ratio is n_g:1. The total mass of the car is M and the effective engine inertia I_e. The total wheel and axle inertia is I_w and other inertias can be neglected. The rolling radius of the tyres is R.

Show that

$$P_E - P_L = [I_e(n_g n_a/R)^2 + I_w(1/R^2) + M]va$$

7.10 In problem 5.18 a steady input torque of 25**k** N m is applied to the shaft attached to carrier C. Assuming that there is no loss in power, find (a) the output torque, (b) the power transmitted, (c) the fixing torque exerted by the casing on S.

7.11 An epicyclic gear consists of a fixed annulus A, a spider X which carries four planet wheels P and a sun wheel S as shown in Fig. 7.43. The power input is to the sun wheel, and the spider drives the output shaft. The numbers of teeth on the annulus, each planet and the sun wheel are 200, 50 and 100 respectively. The axial moment of inertia of the sun wheel and the associated shafting is 0.06 kg m^2 and that of the spider is 0.09 kg m^2. Each planet has a mass of 2.0 kg and an axial moment of inertia of 0.0025 kg m^2. The centres of the planet wheels are at a radius of 120 mm from the central axis of the gear.

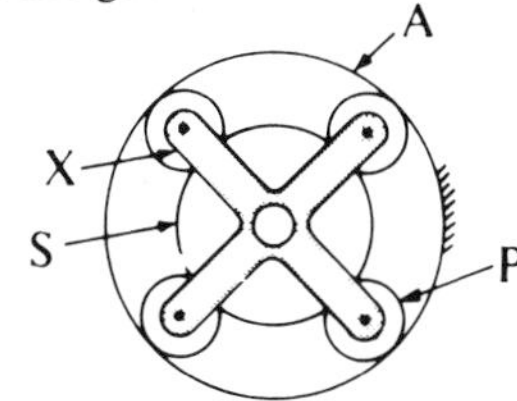

Figure 7.43

A torque of 30 N m is applied to the sun wheel, and the load torque in the output shaft is 60 N m. Determine the angular acceleration of the load.

7.12 The two mechanical systems shown in Fig. 7.44 are in their equilibrium positions. At (a) a uniform cylinder of mass m_1 and radius r rests at the bottom of a cylindrical surface of radius R, and at (b) a uniform rod of mass m_2 and length l rests at 30° to the vertical.

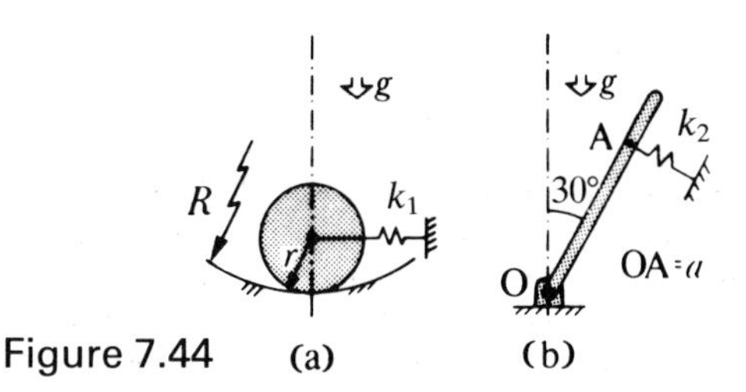

Figure 7.44 (a) (b)

Determine the natural frequency of each system for small oscillations.

7.13 An electric locomotive develops a constant power output of 4 MW while hauling a train up a gradient of slope arcsin (1/70). The mass of train and locomotive is 1×10^6 kg. The rotational kinetic energy is 10 per cent of the translational kinetic energy. The resistance to motion per unit mass of train is given by

$$R = (12.8 + 0.138v)10^{-3} \text{ N/kg}$$

where v is in m/s.

By use of the work energy principle, find (a) the acceleration of the train at the instant when its speed is 15 m/s; (b) the maximum speed at which the train is capable of ascending the incline.

7.14 The mechanism shown in Fig. 7.45 is used to transmit motion between shafts at A and D. The moment of inertia of AB about A is 0.45 kg m^2, that of BC about its mass centre G is 0.5 kg m^2 and that of CD about D is 0.12 kg m^2. The mass of BC is 20 kg.

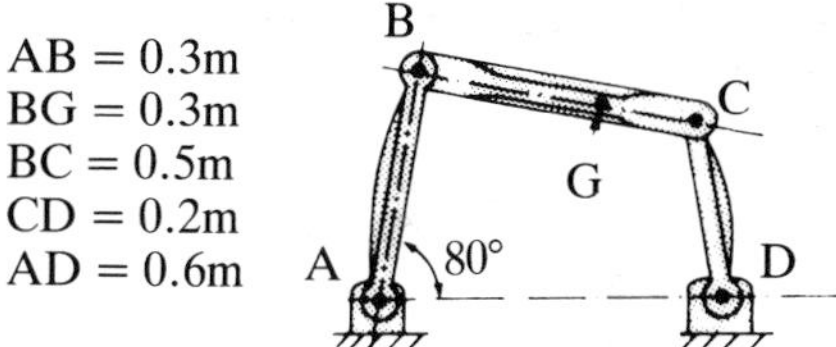

Figure 7.45

When the mechanism is in the configuration shown, the torque applied to shaft A is 40 N m, the angular velocity of AB is 10 rad/s and its angular acceleration is 30 rad/s^2, all measured in a clockwise sense.

Determine the torque applied to the shaft at D by the link CD. Neglect gravity.

7.15 A motor car with rear-wheel drive is fitted with a conventional bevel differential gear.

a) Neglecting inertial effects, show that at all times the torques applied to each rear wheel are equal and independent of the separate speeds of the wheels.

b) See Fig. 7.46. The car is being driven along a plane curve at a constant speed such that the path of each wheel centre is a circular arc. L and R are the left- and right-hand rear wheels and R_L and R_R are the corresponding radii of the paths. Denoting the angular speeds of L and R and the cage of the differential gear by ω_L, ω_R and ω_C respectively, find expressions for ω_L and ω_R in terms of ω_C. Hence show that the power

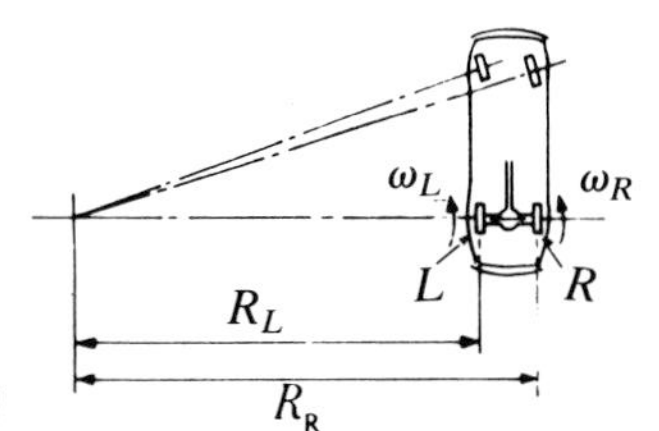

Figure 7.46

flows to wheels L and R are

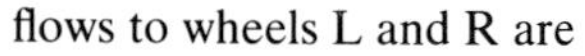

$$\frac{1}{1+G}P \quad \text{and} \quad \frac{G}{1+G}P$$

respectively, where P is the net power supplied by the engine and $G = R_R/R_L$. Neglect transmission losses and slip at the wheels.

7.16 Solve part (a) of problem 6.25 by an energy method.

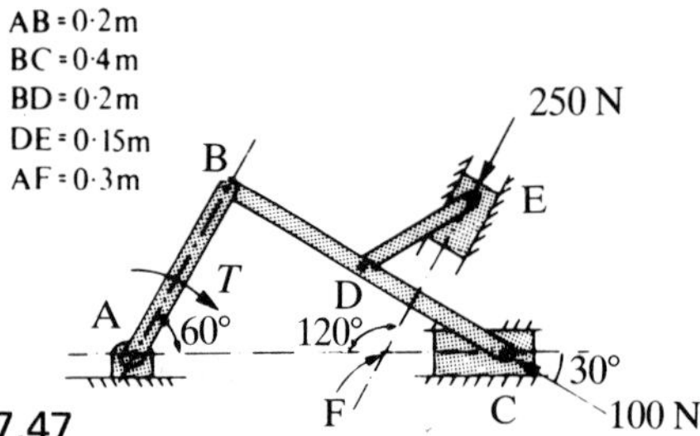

Figure 7.47

7.17 See Fig. 7.47. The 100 N and 250 N forces are applied to the mechanism as shown. Equilibrium is maintained by the application of the couple T. Determine the magnitude of T, neglecting the effects of friction and gravity.

7.18 Trace the car bonnet mechanism of example 4.2 and find the force in the springs by the method of virtual work.

7.19 A slider-crank chain ABC has attached to it at C a spring of stiffness 2600 N/m as shown in Fig. 7.48. The spring is unstrained when $\theta = 90°$. A constant couple of

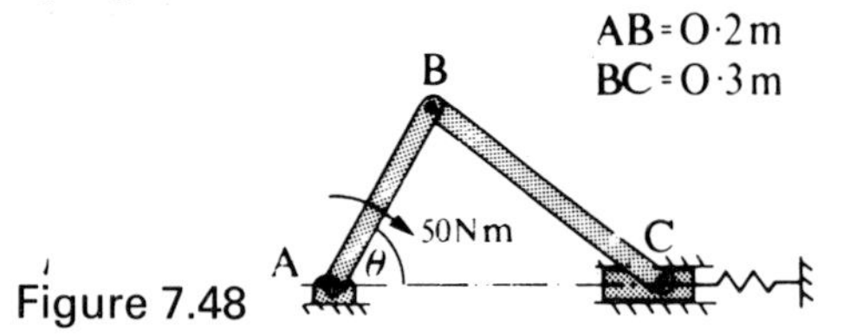

Figure 7.48

50 N m is applied to link AB and the system is in equilibrium. Determine the value of θ, neglecting gravity.

7.20 A beam of rectangular cross-section rests across a cylinder of radius R as shown in Fig. 7.49. Show that, for the position shown to be one of stable equilibrium, $R > h$.

The beam is then rolled without slipping around the cylinder to the unstable equilibrium position. If this

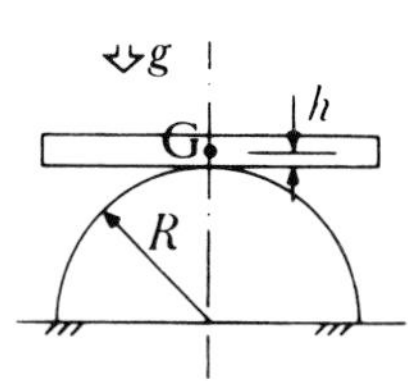

Figure 7.49

occurs when the beam is inclined at 22.5° to the horizontal, show that $h = 0.948R$. Also show that the least value of the coefficient of friction between the beam and the cylinder which prevents slip in the unstable equilibrium position is 0.414.

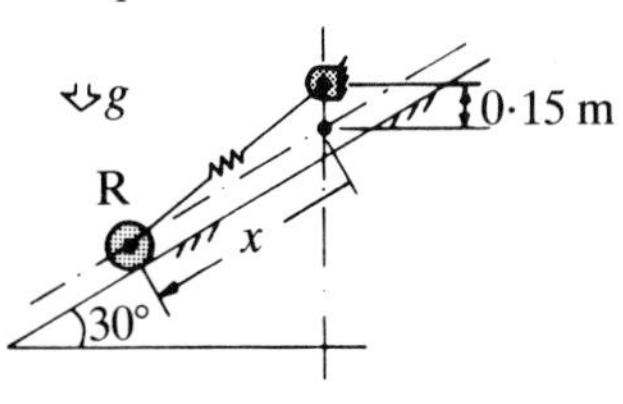

Figure 7.50

7.21 In the system shown in Fig. 7.50, the spring has a stiffness of 600 N/m and is unstrained when its length is 0.15 m. If the roller R has a mass of 3 kg, determine the value or values of x for an equilibrium configuration. State whether the equilibrium is stable or unstable.

7.22 A uniform rod of mass m and length l can pivot about a frictionless pin at O. The motion is controlled by a spring of torsional stiffness k.

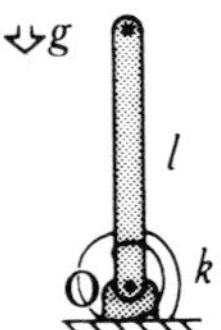

Figure 7.51

a) If the system is to be in stable equilibrium when the rod is vertical, as shown in Fig. 7.51, show that $k > mgl/2$.

b) If $k = mgl/4$, find the stable equilibrium positions.

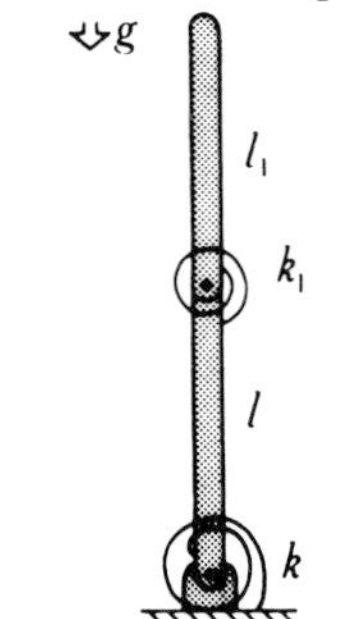

Figure 7.52

7.23 See problem 7.22. A second uniform rod of length l_1 and mass m_1 is pinned to the first as shown in Fig. 7.52 and relative motion between the two rods is restrained by a torsional spring of stiffness k_1. If the system is to be in stable equilibrium in the position shown, what are the conditions that ensure stability?

8
Momentum and impulse

8.1 Linear momentum

We have shown in Chapter 3 that, for any system of particles or rigid body,

$$\boldsymbol{F} = M\ddot{\boldsymbol{r}}_G \tag{3.15}$$

Integrating this equation with respect to time gives

$$\int_1^2 \boldsymbol{F}\,dt = \int_1^2 M\ddot{\boldsymbol{r}}_G\,dt = M\dot{\boldsymbol{r}}_{G2} - M\dot{\boldsymbol{r}}_{G1} \tag{8.1}$$

The integral $\int_1^2 \boldsymbol{F}\,dt$ is known as the impulse and is a vector quantity. Because

$$\sum m_i\dot{\boldsymbol{r}}_i = M\dot{\boldsymbol{r}}_G = \text{the linear momentum}$$

we can write

impulse = change in linear momentum

or, symbolically,

$$\boldsymbol{J} = \Delta\boldsymbol{G}$$

8.2 Moment of momentum

From Fig. 8.1 we see that the moment of momentum about the z-axis of a particle which is

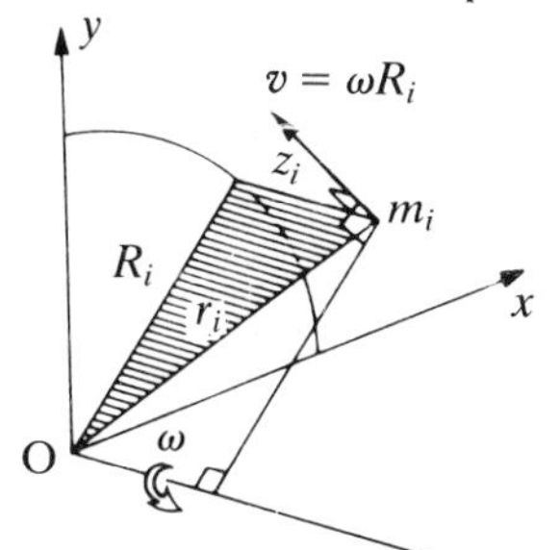

Figure 8.1

moving on a circular path, radius R_i, about the z-axis is

$$L_O = R_i(m_i\omega R_i)$$

For a rigid body rotating about the z-axis with angular velocity ω, the total moment of momentum is

$$L_O = \sum R_i(m_i\omega R_i)$$
$$= \omega\sum m_i R_i^2$$

Since $\sum m_i R_i^2$ is I_O, the moment of inertia of the body about the z-axis, the total moment of momentum for this case is

$$L_O = I_O\omega \tag{8.2}$$

In Chapter 6 we showed that, for a rigid body in general plane motion,

$$M_G = I_G\dot{\omega} \tag{6.11}$$

Integrating this equation with respect to time gives

$$\int_1^2 M_G\,dt = \int_1^2 I_G\dot{\omega}\,dt = I_G\omega_2 - I_G\omega_1 \tag{8.3}$$

that is,

moment of impulse = change in moment of momentum

or, symbolically,

$$K_G = \Delta L_G$$

If rotation is taking place about a fixed axis then equation 6.13 applies which, when integrated, leads to

$$\int_1^2 M_O\,dt = \int_1^2 I_O\dot{\omega}\,dt = I_O\omega_2 - I_O\omega_1 \tag{8.4}$$

or $K_O = \Delta L_O$

8.3 Conservation of momentum

If we now consider a collection of particles or rigid bodies interacting without any appreciable effects from bodies outside the system, then

$$\sum m_i\ddot{\boldsymbol{r}}_i = \boldsymbol{0}$$

so that $\sum m_i\dot{\boldsymbol{r}}_i = \text{constant}$ (8.5)

i.e. linear momentum is conserved.

Extending equation 6.12a for a system of bodies,

$$\sum I_G \dot{\omega} + \sum r_G M a_{G\theta} = 0 \quad (8.6)$$

Integrating with respect to time gives

$$\sum I_G \omega + \sum r_G M v_{G\theta} = \text{constant} \quad (8.7)$$

which is an expression of the conservation of moment of momentum. The term 'angular momentum' is often used in this context but is not used in this book since the term suggests that only the moments of momentum due to rotation are being considered whereas, for example, a particle moving along a straight line will have a moment of momentum about a point not on its path.

8.4 Impact of rigid bodies

We can make use of these conservation principles very effectively in problems involving impact. In many cases of collision between solid objects the time of contact is very small and hence only small changes in geometry take place during the contact period, although finite changes in velocity occur.

As an example, consider the impact of a small sphere with a rod as shown in Fig. 8.2. The rod is initially at rest prior to the impact, so that $u_1 = 0$ and $\omega_1 = 0$; u_2, v_2 and ω_2 are the velocities after impact.

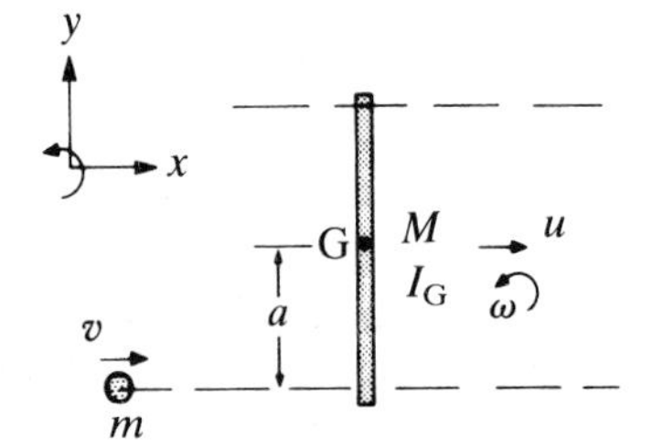

Figure 8.2

Conservation of linear momentum gives

$$mv_1 = Mu_2 + mv_2 \quad (8.8)$$

and conservation of moment of momentum about an axis through G gives

$$mv_1 a = I_G \omega_2 + mv_2 a \quad (8.9)$$

So far we have two equations, but there are three unknowns. To provide the third equation we shall make some alternative assumptions:

i) the rigid-body kinetic energies are conserved, or

ii) the two objects coalesce and continue as a single rigid body.

Case (i)

We shall call case (i) one of *ideal impact*. By conservation of the rigid-body kinetic energy,

$$\tfrac{1}{2}mv_1^2 = \tfrac{1}{2}Mu_2^2 + \tfrac{1}{2}I_G\omega_2^2 + \tfrac{1}{2}mv_2^2 \quad (8.10)$$

This case is often called elastic impact, but in many cases we have elastic deformation of the bodies during impact after which the objects are left in a state of vibration. This vibration energy may easily account for all the initial rigid-body kinetic energy.

An interesting consequence of equations 8.8 to 8.10 is the relationship which exists between the velocity of approach and the velocity of recession of the points of contact.

Velocity of approach $= v_1$

Velocity of recession $= (u_2 + \omega_2 a) - v_2$ (8.11)

Rewriting equations 8.8 to 8.10,

$$v_1 - v_2 = \frac{M}{m} u_2 \quad (8.12)$$

$$v_1 - v_2 = \frac{I_G}{ma^2}(\omega_2 a) \quad (8.13)$$

$$v_1^2 - v_2^2 = \frac{M}{m} u^2 + \frac{I_G}{ma^2}(\omega_2 a)^2 \quad (8.14)$$

Substituting for u_2 and $(\omega_2 a)$ in equation 8.14 gives

$$(v_1 - v_2)(v_1 + v_2) = \frac{m}{M}(v_1 - v_2)^2 + \frac{ma^2}{I_G}(v_1 - v_2)^2$$

or $$v_1 - v_2 = \left(\frac{m}{M} + \frac{ma^2}{I_G}\right)(v_1 - v_2)$$

But $$u_2 + \omega_2 a = (v_1 - v_2)\left(\frac{m}{M} + \frac{ma^2}{I_G}\right) = v_1 + v_2$$

therefore $$(u_2 + \omega_2 a - v_2) = v_1 \quad (8.15)$$

that is, the velocity of recession of the points of contact is equal to the velocity of approach.

This result, which is called Newton's relationship, is often quoted only for simple linear impact but it can be shown to be true for the general case of ideal impact of rigid bodies. For non-ideal impact a coefficient of restitution, e, is introduced, defined by

$$e = \frac{\text{velocity of recession}}{\text{velocity of approach}}$$

Great care must be exercised when using this coefficient since its value depends on the geometry of the colliding objects as well as on their material properties.

Case (ii)
This case is usually called *inelastic* or *plastic impact*. Here equation 8.14 is replaced by

$$v_2 = u_2 + \omega_2 a \tag{8.16}$$

which corresponds to $e = 0$.

We now have from equations 8.8, 8.9 and 8.16

$$v_2 = (u_2 + \omega_2 a) = v_1\left(\frac{m}{M} + \frac{ma^2}{I_G}\right) \times \left(1 + \frac{m}{M} + \frac{ma^2}{I_G}\right)^{-1} \tag{8.17}$$

Using equations 8.13, 8.17 and 7.26, it can be shown that

$$\frac{\text{final k.e.}}{\text{initial k.e.}} = \frac{1 + (a/k_G)^2}{M/m + 1 + (a/k_G)^2} < 1$$

where $k_G = (I_G/M)^{1/2}$.

8.5 Deflection of fluid streams

If we regard a fluid stream as a system of particles, then we can determine the forces required to deflect the stream when under steady flow conditions. It is possible to equate the force to the sum of the separate mass–acceleration terms for each particle, but it is much easier to use rate of change of momentum directly as follows.

If the fluid has a density ρ and is flowing at a constant rate then the velocity at any point in space will be constant in time. In Fig. 8.3 the boundary ABCD contains a specific number of particles which after a time Δt occupy a space A′B′C′D′. The momentum within the region

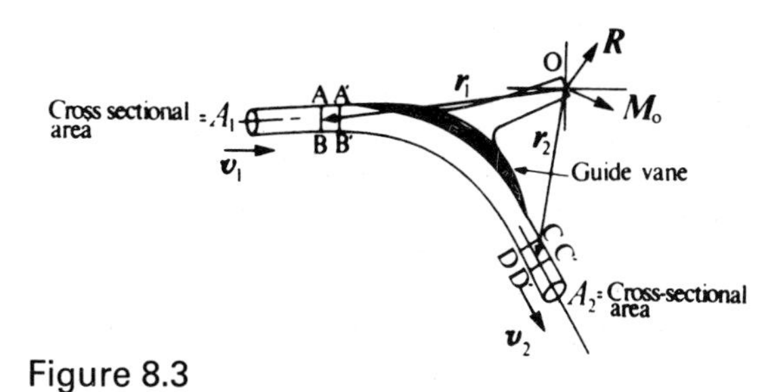

Figure 8.3

A′B′CD does not change, so the change in momentum is simply

momentum of DCC′D′ − momentum of ABB′A′

or $\rho A_2(\text{CC}')\boldsymbol{v}_2 - \rho A_1(\text{AA}')\boldsymbol{v}_1$

Since the mass within the boundary ABCD is the same as within A′B′C′D′,

$$\rho A_2(\text{CC}') = \rho A_1(\text{AA}')$$

But $\quad \text{AA}' = v_1 \Delta t$

and $\quad \text{CC}' = v_2 \Delta t$

therefore the change in momentum is

$$(\rho A_2 v_2 \boldsymbol{v}_2 - \rho A_1 v_1 \boldsymbol{v}_1)\Delta t \tag{8.18}$$

and this must equal the impulse $\boldsymbol{R}\Delta t$ supplied by the external forces supporting the vane.

The mass flow rate is

$$\rho A_2 v_2 = \rho A_1 v_1 = \dot{m} \tag{8.19}$$

hence $\quad \dot{m}(\boldsymbol{v}_2 - \boldsymbol{v}_1)\Delta t = \boldsymbol{R}\Delta t$

or

$$\boldsymbol{R} = (\text{mass flow rate})(\boldsymbol{v}_2 - \boldsymbol{v}_1) \tag{8.20}$$

In a similar manner the moment of the forces acting on the guide vane can be equated to the change in unit time of the moment of momentum, to give

$$\boldsymbol{M}_O = (\text{mass flow rate})(\boldsymbol{r}_2 + \boldsymbol{v}_2 - \boldsymbol{r}_1 \times \boldsymbol{v}_1) \tag{8.21}$$

8.6 The rocket in free space

The rocket is a device which depends for its operation on the ejection of mass, and again the mechanics is best understood by considering the rate of change of momentum.

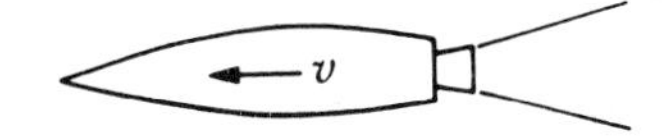

Figure 8.4

Referring to Fig. 8.4 and using the following notation:

m_0 = mass of rocket structure
m_f = mass of fuel
m_e = mass of exhaust
v = velocity of the rocket
v_j = velocity of the jet relative to the rocket
$\dot{m}$ = mass flow rate from rocket to exhaust

the momentum of the complete system of rocket and exhaust is

$$G = (m_0 + m_f)v + m_e v_e \quad (8.22)$$

where v_e is the average velocity of the exhaust gases.

After a time interval of Δt, the momentum becomes

$$G + \Delta G = (m_0 + m_f - \dot{m}\Delta t)(v + \Delta v) + m_e v_e + \dot{m}\Delta t(v - v_j)$$

thus the change in momentum is

$$\Delta G = (m_0 + m_f)\Delta v - \dot{m}\Delta t v_j - \dot{m}\Delta t \Delta v \quad (8.23)$$

giving $$\left(\frac{\Delta G}{\Delta t}\right)_{\Delta t \to 0} = (m_0 + m_f)\frac{dv}{dt} - \dot{m}v_j$$

Hence, the external force acting on the system

$$F = \frac{dG}{dt} = (m_0 + m_f)\frac{dv}{dt} - \dot{m}v_j \quad (8.24)$$

If the rocket is in free space then the external forces on the system will be zero; therefore

$$\frac{dv}{dt} = \frac{\dot{m}v_j}{m_0 + m_f} \quad (8.25)$$

Figure 8.5

If we draw the free-body diagram for the rocket and fuel only, as shown in Fig. 8.5, where T is the thrust of the rocket motor, then

$$T = (m_0 + m_f)\frac{dv}{dt} \quad (8.26)$$

because in this free-body diagram all particles have the same acceleration.

Comparing this with equation 8.24, we see that

$$T = \dot{m}v_j \quad (8.27)$$

8.7 Illustrative example

In problems such as that in the previous section, there is a temptation to write an expression for the momentum of the rocket plus fuel and then to differentiate this expression with respect to time in order to establish the external force required. The reason why this does not produce the correct result is simply that equations 8.24 and 8.26 apply to a specific group of particles whereas the mass of the rocket plus unburnt fuel is changing. A loss of momentum due solely to particles leaving a prescribed volume does not require that a force be applied to the boundary of that volume. The following example may help to illuminate this point.

Consider two trucks running, with negligible friction, on horizontal tracks one above the other as shown in Fig. 8.6. The upper truck is initially moving at a speed v_1 and then feeds sand at a constant rate via a hopper. The lower truck, empty mass M_2, is at rest when it starts to receive the sand. What is its subsequent motion?

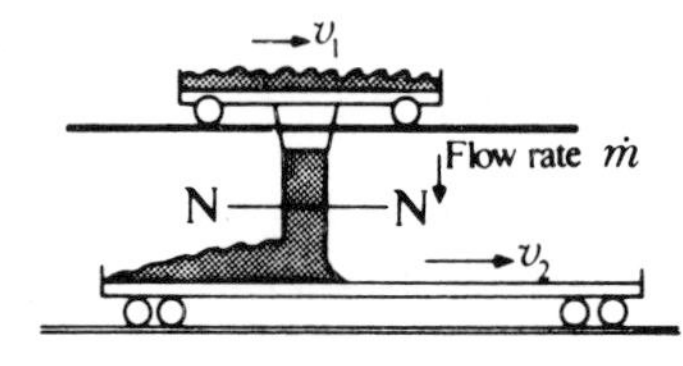

Figure 8.6

The line NN is moving at the same rate as the sand. Consider the mass of truck and sand above the line NN. The sand, once it has left the truck, continues with the same horizontal velocity throughout its free fall; therefore the horizontal component of momentum does not alter during fall and so no force is required to maintain the motion, even though the truck is itself losing momentum. The region below NN does exhibit a change in horizontal momentum G.

$$G = (\text{initial mass of sand in free fall})v_1 + (M_2 + \text{sand})v_2 \quad (8.28)$$

After time Δt,

$$G + \Delta G = (\text{initial mass of sand in free fall} - \dot{m}\Delta t)v_1 + (M_2 + \text{sand} + \dot{m}\Delta t)(v_2 + \Delta v_2)$$

Hence the change in momentum is

$$\Delta G = -\dot{m}\Delta t v_1 + (M_2 + \text{sand})\Delta v_2 + \dot{m}\Delta t \Delta v_2 + \dot{m}\Delta t v_2$$

so that $$\lim_{\Delta t \to 0}\left(\frac{\Delta G}{\Delta t}\right) = -\dot{m}(v_1 - v_2) + (M_2 + \text{sand})\frac{dv_2}{dt}$$

Since the external horizontal force is zero,

$$0 = -\dot{m}(v_1 - v_2) + (M_2 + \dot{m}t)\frac{dv_2}{dt} \quad (8.29)$$

Alternatively we can write

horizontal momentum of the system = constant

$$(M_1 - \dot{m}t)v_1 + (M_2 + \dot{m}t)v_2 = M_1 v_1 \qquad (8.30)$$

because we have already argued that v_1 is constant. Therefore

$$v_2 = \frac{\dot{m}tv_1}{M_2 + \dot{m}t} \qquad (8.31)$$

If we attempt to use $F = \sum ma$, we find difficulty in evaluating the acceleration of the sand particles as they hit the lower truck. However $F = \sum ma$ may be usefully employed when considering the upper truck since we know that the sand in free fall has no horizontal acceleration at any point prior to hitting the lower truck, therefore the horizontal force must be zero.

8.8 Equations of motion for a fixed region of space

A further way of forming the equations of motion is to consider a fixed region of space. Initially we use the impulse–momentum relationship for the known number of particles which originally occupied the prescribed region.

In Fig. 8.7, particles are leaving the region with a velocity v_a normal to the surface and entering the region with a velocity v_b. The density of the material is ρ and A is the area of the aperture. The unit vector $\boldsymbol{n}$ is in the direction of the outward normal to the surface of the region.

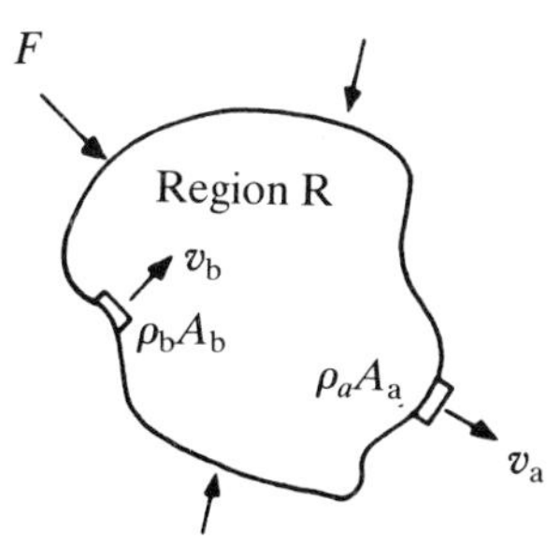

Figure 8.7

Now

$$\text{impulse} = \text{change of momentum}$$

so

$$\sum \boldsymbol{F}\Delta t = \Delta(\text{momentum within region R}) + (\rho_a A_a v_a \Delta t) v_a \boldsymbol{n}_a - (\rho A_b v_b \Delta t) v_b (-\boldsymbol{n}_b)$$

$$\sum \boldsymbol{F} = \frac{d}{dt}(\text{momentum within R}) + \rho_a A_a {v_a}^2 \boldsymbol{n}_a + \rho_b A_b {v_b}^2 \boldsymbol{n}_b$$

or $$\sum \boldsymbol{F} - \sum \rho A v^2 \boldsymbol{n} = \frac{d}{dt}(\text{momentum within R}) \qquad (8.32)$$

The terms $\rho A v^2$ could be regarded as fictitious forces acting on the surface of the region and directed towards the interior, for both inflow and outflow.

Because velocity is not absolute, it is permissible to choose a region of fixed shape but moving at a constant velocity relative to some other inertial frame of reference. In this case all velocities should be reckoned relative to the moving region.

The method just described may be applied to a jet engine moving at a constant speed v relative to the air. In Fig. 8.8 the inlet speed relative to the region R (often called the control volume) will be

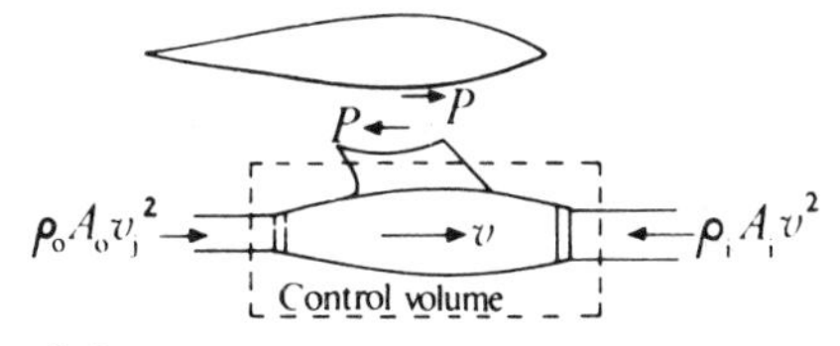

Figure 8.8

$v_i = v$, since the control volume is taken to have a velocity v. The exit speed is the speed of the jet relative to the nozzle, v_j. Thus

$$-P + \rho_O A_O {v_j}^2 - \rho_i A_i v^2 = 0 \qquad (8.33)$$

since in this case there is no momentum change within the region.

As we have steady flow conditions, the mass flow rate in must equal the mass flow rate out:

$$\rho_i A_i v = \rho_O A_O v_j = \dot{m} \qquad (8.34)$$

therefore equation 8.33 may be rewritten:

$$P = \dot{m}(v_j - v) \qquad (8.35)$$

Discussion examples

Example 8.1

Figure 8.9 shows two shafts AB and BC which can rotate freely in their respective bearings. Initially their angular velocities are ω_1 and ω_2 in the same sense. At time $t = 0$ the clutch B is operated and connects the shafts together. The clutch is spring-loaded and the maximum couple it can transmit is Q_O. The total axial inertia of shaft AB is I_1 and that of BC is I_2.

a) Find the final angular velocity and the time

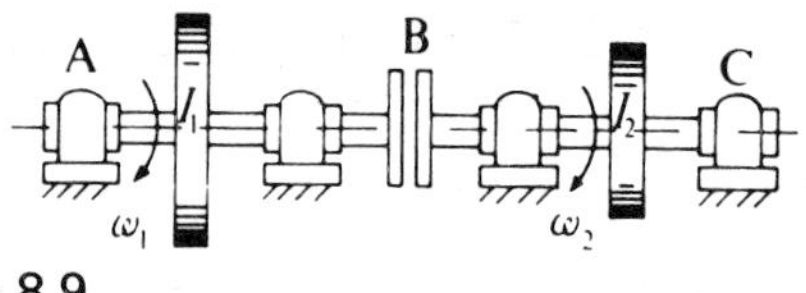

Figure 8.9

taken for slipping to cease.

b) Show that the energy lost is

$$\tfrac{1}{2}\left(\frac{I_1 I_2}{I_1+I_2}\right)(\omega_1-\omega_2)^2$$

Solution The horizontal and vertical forces acting on the system are shown in the free-body diagrams (Fig. 8.10) but are not relevant to this problem since they do not appear in the axial moment equations.

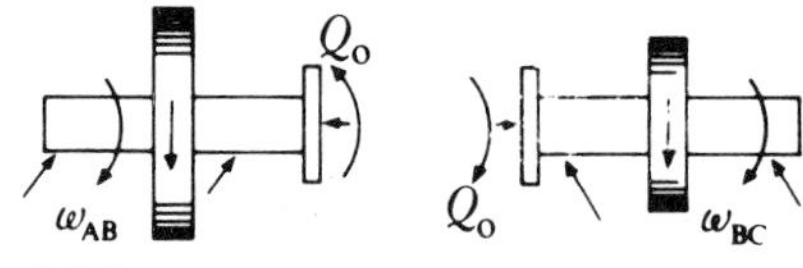

Figure 8.10

While the clutch is slipping, the couple it transmits is Q_O and when slipping ceases the shafts will have a common angular velocity, say ω_C. The directions for Q_O are marked on the free-body diagrams on the assumption that $\omega_1 > \omega_2$.

Shaft AB:

$$[M_G = I_G\dot{\omega}]$$

$$-Q_O = I_1\,\mathrm{d}\omega_{AB}/\mathrm{d}t$$

$$-\int_0^t Q_O\,\mathrm{d}t = \int_{\omega_1}^{\omega_C} I_1\,\mathrm{d}\omega_{AB} = I_1(\omega_C-\omega_1) \qquad \text{(i)}$$

Shaft BC:

$$[M_G = I_G\dot{\omega}]$$

$$+Q_O = I_2\frac{\mathrm{d}\omega_{BC}}{\mathrm{d}t}$$

$$\int_0^t Q_O\,\mathrm{d}t = \int_{\omega_2}^{\omega_C} I_2\,\mathrm{d}\omega_{BC} = I_2(\omega_C-\omega_2) \qquad \text{(ii)}$$

Adding equations (i) and (ii) we obtain

$$(I_1+I_2)\omega_C-(I_1\omega_1+I_2\omega_2)=0$$

and we note that, for this case, there is no change in moment of momentum. The final angular velocity is

$$\omega_C = (I_1\omega_1+I_2\omega_2)/(I_1+I_2) \qquad \text{(iii)}$$

Until time t, when slipping ceases, the transmitted couple Q_O is constant, so that

$$\int_0^t Q_O\,\mathrm{d}t = Q_O t \qquad \text{(iv)}$$

Combining equations (iii) and (iv) with either (i) or (ii), we find

$$t=\frac{I_1 I_2(\omega_1-\omega_2)}{Q_O(I_1+I_2)}$$

and note that, since we have already assumed $\omega_1 > \omega_2$, the time taken is positive!

The energy change (final energy minus initial energy) is

$$\tfrac{1}{2}(I_1+I_2)\omega_C^2-(\tfrac{1}{2}I_1\omega_1^2+\tfrac{1}{2}I_2\omega_2^2)$$

which after substitution of ω_C from equation (iii) and some manipulation is equal to

$$-\tfrac{1}{2}\left(\frac{I_1 I_2}{I_1+I_2}\right)(\omega_1-\omega_2)^2$$

Example 8.2

Figure 8.11 shows a box of mass m on a roller conveyor which is inclined at angle α to the horizontal. The conveyor consists of a set of rollers 1, 2, 3, . . ., each of radius r and axial moment of inertia I and spaced a distance l apart. The box is slightly longer than $3l$.

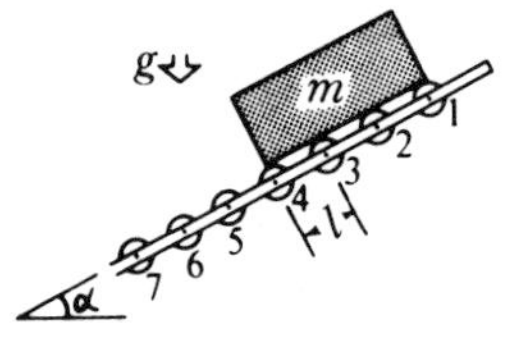

Figure 8.11

If $\alpha = 30°$, $r = 50$ mm, $I = 0.025$ kg m^2, $l = 0.3$ m and $m = 30$ kg, and if the box is released from rest with the leading edge just in contact with roller 4, (a) determine the velocity of the box just after the first impulsive reaction with roller 6 takes place and (b) show that, if the conveyor is sufficiently long, the box will eventually acquire a mean velocity of 2.187 m/s.

Assume that the box makes proper contact with each roller it passes over and that the time taken for the slip caused by each impact is extremely short. Also assume that friction at the axles is negligible.

Solution Let us consider a general case (Fig. 8.12) just after the front of the box has made

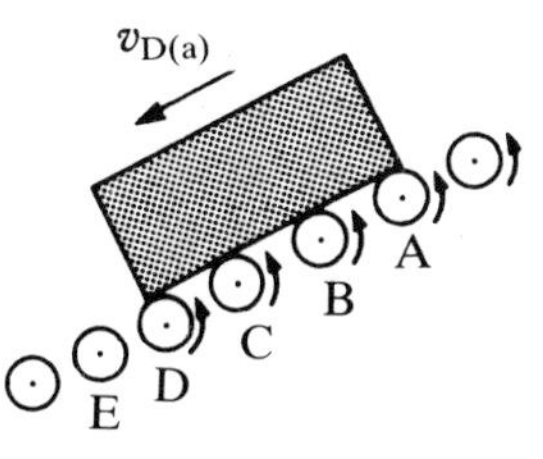

Figure 8.12

contact with a roller D and slip has ceased. The velocity of the box is $v_{D(a)}$ where the subscript '(a)' denotes 'after impact'. The angular velocity ω of rollers A, B, C and D will be $\omega = v_{D(a)}/r$. The box then immediately loses contact with roller A which, in the absence of friction, continues to rotate at the same angular velocity. Until the next impact, energy will be conserved. The box accelerates under the action of gravity to a velocity $v_{E(b)}$ just before it makes contact with roller E, where the subscript '(b)' denotes 'before impact'.

The kinetic-energy increase is

$$\left[\frac{1}{2}mv_{E(b)}{}^2 + \frac{3}{2}I\left(\frac{v_{E(b)}}{r}\right)^2\right] - \left[\frac{1}{2}mv_{D(a)}{}^2 + \frac{3}{2}I\left(\frac{v_{D(a)}}{r}\right)^2\right]$$

and the gravitational energy decrease is $mgl\sin\alpha$. Hence

$$v_{E(b)}{}^2 = v_{D(a)}{}^2 + \frac{2mgl\sin\alpha}{m + 3I/r^2}$$

$$= v_{D(a)}{}^2 + \frac{2(30)(9.81)(0.3)(\sin 30°)}{30 + 3(0.025)/(0.05)^2}$$

$$v_{E(b)} = [v_{D(a)}{}^2 + 1.4715]^{1/2} \qquad \text{(i)}$$

The box then contacts roller E (Fig. 8.13) which receives an impulsive tangential force P_E.

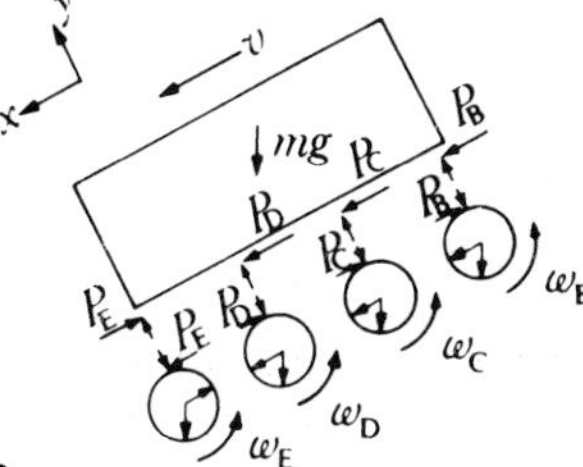

Figure 8.13

The equal and opposite force acting on the box rapidly changes its speed and at the same time impulsive reactions P_B, P_C, and P_D occur with rollers B, C and D. The velocity of the box just after the impact is denoted by $v_{E(a)}$.

For the box $[\sum F_x = m\ddot{x}_G]$,

$$mg\sin\alpha - P_E + P_D + P_C + P_B = m\frac{dv}{dt}$$

Integrating to obtain the impulse–momentum equation,

$$\int_0^{\Delta t} mg\sin\alpha\, dt - \int_0^{\Delta t} P_E\, dt + \int_0^{\Delta t} P_D\, dt + \int_0^{\Delta t} P_C\, dt + \int_0^{\Delta t} P_B\, dt = m(v_{E(a)} - v_{E(b)}) \qquad \text{(ii)}$$

Since the impact forces are large, we assume that the first integral is negligible.

For the rollers $[\sum M_G = I_G\dot{\omega}]$,

$$P_E r = I\frac{d\omega_E}{dt}, \quad -P_D r = I\frac{d\omega_D}{dt},$$

$$-P_C r = I\frac{d\omega_C}{dt} \quad \text{and} \quad -P_B r = I\frac{d\omega_B}{dt}$$

$$r\int_0^{\Delta t} P_E\, dt = I\left(\frac{v_{E(a)}}{r} - 0\right) - r\int_0^{\Delta t} P_D\, dt$$

$$= I\left(\frac{v_{E(a)}}{r} - \frac{v_{E(b)}}{r}\right)$$

$$= -r\int_0^{\Delta t} P_C\, dt = -r\int_0^{\Delta t} P_B\, dt$$

Substituting into equation (ii) gives

$$v_{E(a)} = v_{E(b)}\left(\frac{m + 3I/r^2}{m + 4I/r^2}\right)$$

$$= v_{E(b)}\frac{30 + 3(0.025)/(0.05)^2}{30 + 4(0.025)/(0.05)^2}$$

$$v_{E(a)} = 6v_{E(b)}/7 \qquad \text{(iii)}$$

Just before first contact with roller 5 [equation (i)],

$$v_{5(b)} = [0 + 1.4715]^{1/2} = 1.2131 \text{ m/s}$$

and, just after [equation (iii)],

$$v_{5(a)} = 6(1.2131)/7 = 1.0398 \text{ m/s}$$

Similarly,

$$v_{6(b)} = [(1.0398)^2 + 1.4715]^{1/2} = 1.5977 \text{ m/s}$$

and $v_{6(a)} = 6(1.5977)/7 = 1.3694$ m/s

If the box eventually acquires a steady mean velocity, then the velocity lost at each impact will be exactly regained at the end of the following impact-free motion. After a few trials we find that, if the velocity just before impact is 2.3551 m/s, the velocity just after the impact [equation (iii)] is

$$v = 6(2.3551)/7 = 2.0187 \text{ m/s}$$

and the velocity just before the next impact [equation (i)] is

$$v = [(2.0187)^2 + 1.4715]^{1/2} = 2.3551 \text{ m/s}$$

which is the same as just before the previous impact.

Since the acceleration between impacts is constant, the mean velocity v_m is

$$v_m = \tfrac{1}{2}(2.0187 + 2.3551) = 2.1869 \text{ m/s}$$

Example 8.3

A building block ABCD (Fig. 8.14) falls vertically and strikes the ground with corner A as shown. At the instant before impact the mass centre G has a downward velocity v_0 of 4 m/s and the angular velocity ω_0 of the block is 5 rad/s anticlockwise. The mass of the block is 36 kg and $I_G = 1.3 \text{ kg m}^2$.

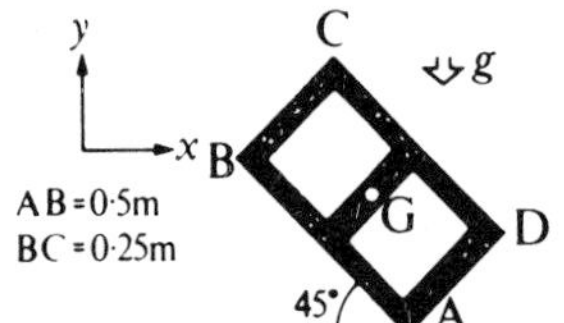

Figure 8.14

Assuming that the impact force at A is of very short duration and that after impact the block rotates about A, find (a) the angular velocity ω_1 just after impact, (b) the energy lost in the impact and (c) the angular velocity ω_2 just before corner B strikes the ground.

Solution The free-body diagram (Fig. 8.15) discloses the two forces acting on the block during the impact: the weight mg and the large impact force P.

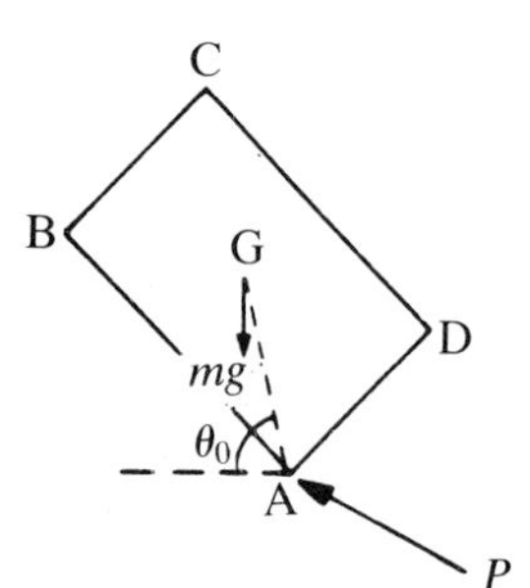

Figure 8.15

Now the moment of impulse about G for an impact time Δt is

$$\int_0^{\Delta t} M_G \, dt = I_G(\omega_1 - \omega_0)$$

which is the change in moment of momentum. This does not help, since we cannot determine M_G as all we know about P is its point of application.

The moment of impulse about point A is

$$\int_0^{\Delta t} M_A \, dt$$

$$= [\text{moment of momentum at } t = \Delta t] - [\text{moment of momentum at } t = 0]$$

$M_A = mg(\text{AG})\sin\theta_0$ and, since mg is a force of 'normal' magnitude, $\int_0^{\Delta t} M_A \, dt$ is negligible as Δt is very small. Thus there is no change in moment of momentum about A during the impact time Δt.

(In general we note that, if a body receives a *single* blow of very short duration, during the blow the moment of momentum about a point on the line of action of the blow does not change.)

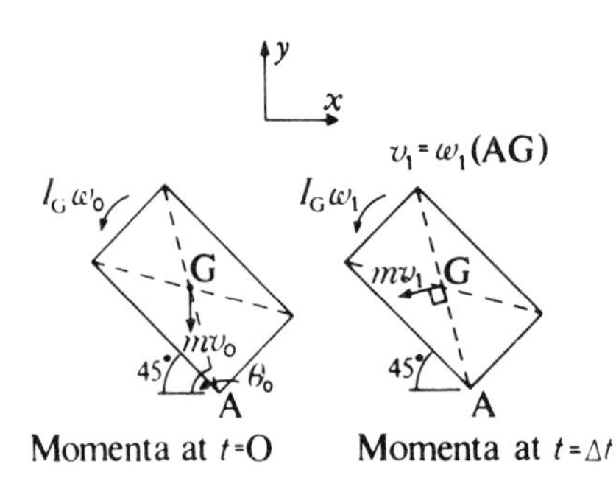

Figure 8.16

From Fig. 8.16 we can write for the moments of momentum

$$[\boldsymbol{L}_A]_{t=0} = I_G\omega_0\boldsymbol{k} + \boldsymbol{r}_G \times m\boldsymbol{v}_{G0}$$

and $[L_A]_{t=0} = I_G\omega_0 + (\text{AG})\cos\theta_0(mv_0)$

$$[\boldsymbol{L}_A]_{t=\Delta t} = I_G\omega_1\boldsymbol{k} + \boldsymbol{r}_G \times m\boldsymbol{v}_{G1}$$

and $[L_A]_{t=\Delta t} = I_G\omega_1 + (\text{AG})mv_1$

Equating the moments of momentum about A, we have

$$\omega_1 = \frac{I_G\omega_0 + (\text{AG})\cos\theta_0(mv_0)}{I_G + m(\text{AG})^2}$$

$$= \frac{1.3(5)+(\sqrt{5/8})\cos[45°+\arctan(\frac{1}{2})]36(4)}{1.3+36(\sqrt{5/8})^2}$$

$$= 4.675 \text{ rad/s}$$

At time $t = 0$, the energy is

$$\tfrac{1}{2}mv_0{}^2+\tfrac{1}{2}I_G\,\omega^2 = \tfrac{1}{2}(36)(4)^2+\tfrac{1}{2}(1.3)(5)^2 = 304.3 \text{ J}$$

At time $t = \Delta t$, the energy is

$$\tfrac{1}{2}mv_1{}^2+\tfrac{1}{2}I_G\,\omega_1{}^2 = [\tfrac{1}{2}(36)(\sqrt{5/8})^2+\tfrac{1}{2}(1.3)] \times(4.675)^2 = 45.0 \text{ J}$$

and the energy lost is $304.3-45.0 = 259.3$ J.

There is no energy lost from time $t = \Delta t$ to the instant just before corner B strikes the ground, when the angular velocity is ω_2. In this interval, the centre of mass G falls through a vertical distance

$$h_0 = (AG)(\sin\theta_0-\sin 45°) = (\sqrt{5/8})[\sin\{45°+\arctan(\tfrac{1}{2})\}-\sin 45°] = 0.06752 \text{ m}$$

The gravitational energy lost is

$$mgh_0 = 36(9.81)(0.06752) = 23.85 \text{ J}$$

The kinetic energy when the angular velocity is ω_2 is

$$\tfrac{1}{2}[I_G+m(AG)^2]\,\omega_2{}^2 = 2.056\omega_2{}^2$$

Equating the total energies at the beginning and end of the interval,

$$2.056\omega_2{}^2 = 45.0+23.85$$

$$\omega_2 = 5.7842 \text{ rad/s}$$

Example 8.4

A fluid jet of density ρ and cross-sectional area A is ejected from a nozzle N with a velocity v and strikes the flat blade B as shown in Fig. 8.17. Determine the force exerted on the blade by the fluid stream when (a) the blade is stationary and (b) the blade has a velocity u in the same direction as v ($u<v$). Assume that after impact the fluid flows along the surface of the blade.

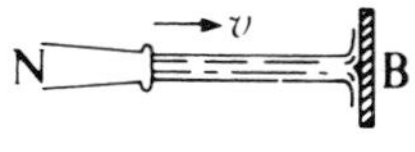

Figure 8.17

Solution

a) The free-body diagram on the left of Fig. 8.18 is for the fluid which is outside the nozzle. The force $P_1 = 0$ since the pressure in the fluid just outside the nozzle is assumed to be zero. The mass flow rate at the blade is ρAv, the velocity change of the fluid stream is $(0-v)$ and from equation 8.20 the force acting on the fluid stream at the blade is $P_2 = \rho Av(-v) = -\rho Av^2$ to the right, that is a force of ρAv^2 to the left. The force acting on the blade to the right is thus ρAv^2. The force R which holds the blade in equilibrium is also equal to ρAv^2.

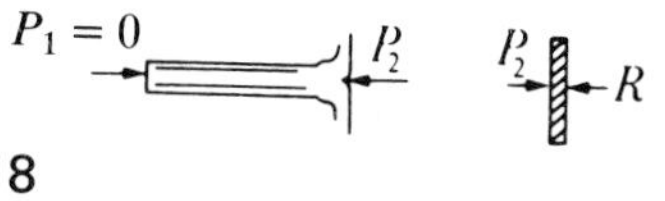

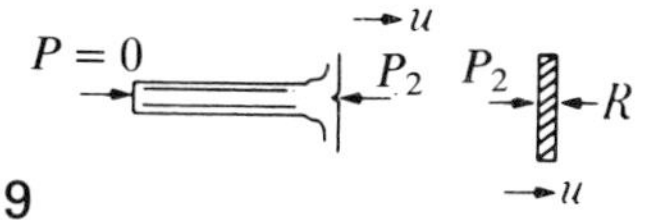

Figure 8.18

Figure 8.19

b) The free-body diagram on the left of Fig. 8.19 is for a fixed quantity of fluid. If we now change the frame of reference to one moving at a constant velocity u with the plate, then the left-hand boundary will have a constant velocity $(v-u)$. Thus the change in momentum is

$$-\rho A\,(v-u)(v-u) = -\rho A\,(v-u)^2$$

and

$$P_1-P_2 = -\rho A\,(v-u)^2$$

But $P_1 = 0$ and therefore

$$P_2 = \rho A\,(v-u)^2 = R$$

Alternatively, a control volume moving with the plate could be used, in which case the actual and 'fictitious' forces are as shown in Fig. 8.20.

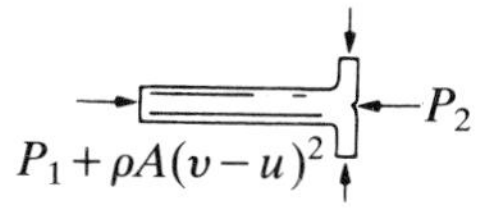

Figure 8.20

Hence,

$$P_1+\rho A\,(v-u)^2-P_2 = 0$$

since the change in momentum within the control volume is zero.

Example 8.5

An open-linked chain is piled over a hole in a horizontal surface and a length l_1 of chain hangs below the hole as shown in Fig. 8.21. Motion is prevented by a restraining device just below the hole which is just capable of preventing motion if the length of chain below it is l_0, the mass/unit

Figure 8.21

length of the chain being ρ. An object of mass m_0 is hooked on to the lower end of the chain and is then released.

If $l_1 = 1$ m, $l_0 = 3$ m, $m_0 = 5$ kg, $\rho = 1$ kg/m and $g = 10$ N/kg, show that the velocity v of the object after it has fallen a distance x is given by

$$v = \left[\frac{20(18 + 4\tfrac{1}{2}x + \tfrac{1}{3}x^2)}{(6+x)^2}\right]^{1/2}$$

Neglect frictional effects apart from those in the restraining device and ignore any horizontal or vertical motion or clashing of the links above the hole.

Solution If we consider the forces acting on the complete chain and attached object (Fig. 8.22), which is a system of constant mass, then we can write

$$\sum \text{forces} = \frac{\mathrm{d}}{\mathrm{d}t}(\text{momentum}) \qquad \text{(i)}$$

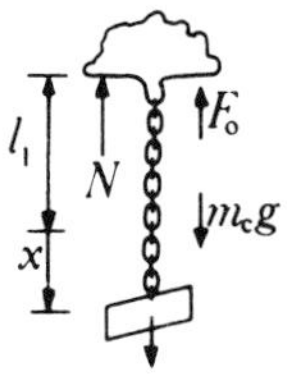

Figure 8.22

The free-body diagram for the chain and attached object shows the weights $m_0 g$ and $m_C g$ acting downwards, m_C being the mass of the complete chain. The restraining force F_0, which we assume to be constant throughout the motion, and N, the resultant contact force with the surface, both act upwards. $F_0 = \rho l_0 g$ and it is reasonable to assume that N is equal and opposite to the weight of the chain above the hole: $N = [m_C - \rho(l_1 + x)]g$. We note that the motion takes place since, when $x = 0$, numerically,

$$(m_0 + m_C)g > (N + F_0)$$

The mass which is in motion is $m_0 + \rho(l_1 + x)$ and its downward velocity is v. Thus, for equation (i) we can write

$$(m_0 + m_C)g - N - F_0 = \frac{\mathrm{d}}{\mathrm{d}t}[\{m_0 + \rho(l_1 + x)\}v]$$

$$[m_0 + \rho(l_1 + l_0 + x)]g = [m_0 + \rho(l_1 + x)]\frac{\mathrm{d}v}{\mathrm{d}t} + \rho v^2 \qquad \text{(ii)}$$

Substituting numerical values and replacing $\mathrm{d}v/\mathrm{d}t$ by $v\,\mathrm{d}v/\mathrm{d}x$ we have

$$10(3 + x) = v\left[(6 + x)\frac{\mathrm{d}v}{\mathrm{d}x} + v\right]$$

It is not necessary to use numerical methods with an equation of this type as it can readily be solved by making a substitution of the form $z = (6 + x)v$ [note that $\mathrm{d}z/\mathrm{d}x = (6 + x)\mathrm{d}v/\mathrm{d}x + v$] and multiplying both sides of the equation by $(6 + x)$. This leads to

$$10(3 + x)(6 + x) = z\frac{\mathrm{d}z}{\mathrm{d}x}$$

which when integrated gives the desired result.

Example 8.6

A rocket-propelled vehicle is to be fired vertically from a point on the surface of the Moon where the gravitational field strength is 1.61 N/kg. The total mass m_R of the rocket and fuel is 40 000 kg. Ignition occurs at time $t = 0$ and the exhaust gases are ejected backwards with a constant velocity $v_j = 3000$ m/s relative to the rocket. The rate $\dot{m}$ of fuel burnt varies with time and is given by $\dot{m} = 600(1 - \mathrm{e}^{-0.05t})$ kg/s. Determine when lift-off occurs and also find the velocity of the rocket after a further 6 s.

Solution From equation 8.27, the effective upthrust T on the rocket is

$$T = \dot{m}v_j = 600(1 - \mathrm{e}^{-0.05t})(3000) \qquad \text{(i)}$$

The weight W of the rocket plus fuel at time t is

$$W = (m_R - \dot{m}t)g = [40\,000 - 600(1 - \mathrm{e}^{-0.05t})t](1.61) \qquad \text{(ii)}$$

Motion begins at the instant T acquires the value of W. Using a graphical method or a trial-and-error numerical solution we find that lift-off begins at time $t = 0.728$ s. Thereafter the equation of motion is (see the free-body diagram, Fig. 8.23)

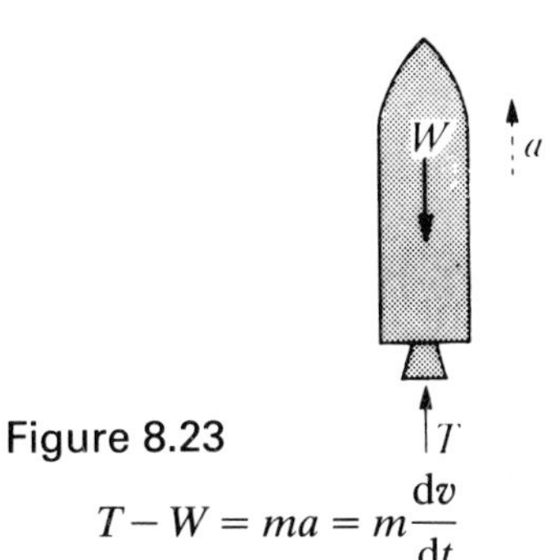

Figure 8.23

$$T - W = ma = m\frac{dv}{dt}$$

Substituting for T and W, re-arranging and integrating we find that

$$v = \int_{t_0}^{t} f(t)\,dt$$

where

$$f(t) = \left[\frac{(1 - e^{-0.05t})(18 \times 10^5)}{40\,000 - 600(1 - e^{-0.05t})t} - 1.61\right]$$

and $t_0 = 0.728$ s.

We shall evaluate the integral numerically using Simpson's rule (Appendix 3) and calculate the values of $f(t)$ at 1 s intervals from $t = 0.728$ s to $t = 6.728$ s.

t/s	0.728	1.728	2.728	3.728	4.728	5.728	6.728
$f(t)$/ ms^{-2}	0	2.124	4.159	6.117	8.008	9.842	11.63

The velocity at $t = 6.728$ s is given by

$$v = \tfrac{1}{3}[0 + 11.63 + 4(2.124 + 6.117 + 9.842) + 2(4.159 + 8.008)]$$
$$= 36.1 \text{ m/s}$$

Example 8.7

A sphere of mass m_1 is moving at a speed u_1 in a direction which makes an angle θ with the x axis. The sphere then collides with a stationary sphere mass m_2 such that at the instant of impact the line joining the centres lies along the x axis.

Derive expressions for the velocities of the two spheres after the impact. Assume ideal impact.

For the special case when $m_1 = m_2$ show that after impact the two spheres travel along paths which are 90° to each other, irrespective of the angle θ.

Solution This is a case of oblique impact but this does not call for any change in approach providing that we neglect any frictional effects during contact. Referring to Fig. 8.24 we shall

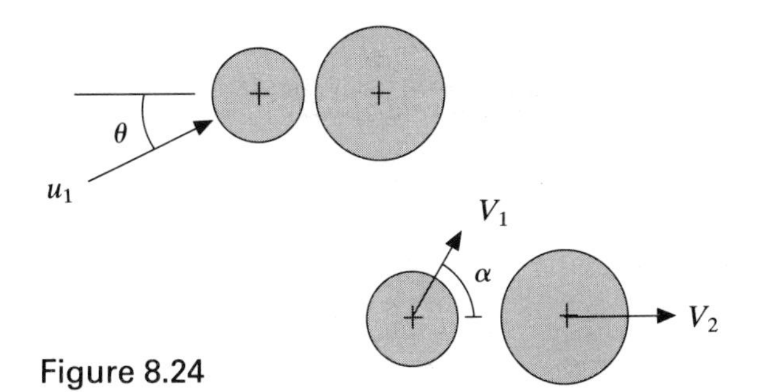

Figure 8.24

apply conservation of linear momentum in the x and y directions and for the third equation we shall assume that, for ideal impact, the velocity of approach will equal the velocity of recession.

Conservation of momentum in the x direction gives

$$m_1 u_1 \cos\theta = m_1 v_1 \cos\alpha + m_2 v_2 \qquad \text{(i)}$$

and in the y direction

$$m_1 u_1 \sin\theta = m_1 v_1 \sin\alpha \qquad \text{(ii)}$$

Equating approach and recession velocities gives

$$u_1 \cos\theta = v_2 - v_1 \cos\alpha \qquad \text{(iii)}$$

Note that the velocities are resolved along the line of impact.

Substituting equation (iii) into (i)

$$m_1 u_1 \cos\theta = m_1[v_2 - u_1\cos\theta] + m_2 v_2$$

thus

$$v_2 = \frac{2m_1 u_1 \cos\theta}{(m_1 + m_2)}$$

From (iii)

$$v_1 \cos\alpha = v_2 - u_1\cos\theta = u_1\cos\theta\frac{(m_1 - m_2)}{(m_1 + m_2)}$$

and from (ii)

$$v_1 \sin\alpha = u_1 \cos\theta$$

therefore as

$$v_1{}^2 = (v_1\sin\alpha)^2 + (v_1\cos\alpha)^2$$

$$v_1 = u_1\left[\sin^2\theta + \cos^2\theta\left(\frac{m_1 - m_2}{m_1 + m_2}\right)^2\right]^{1/2}$$

and

$$\tan\alpha = \tan\theta\frac{(m_1 + m_2)}{(m_1 - m_2)}.$$

From this equation it follows that if $m_1 = m_2$ $\alpha = 90°$ for all values of θ except when $\theta = 0$; which is of course the case for collinear impact.

Example 8.8

A solid cylindrical puck has a mass of 0.6 kg and a diameter of 50 mm. The puck is sliding on a frictionless horizontal surface at a speed of 10 m/s and strikes a rough vertical surface, the direction of motion makes an angle of 30° to the normal to the surface.

Given that the coefficient of limiting friction between the side of the puck and the vertical surface is 0.2, determine the subsequent motion after impact. Assume that negligible energy is lost during the impact.

Solution It is not immediately obvious how to use the idea of equating velocity of approach with that of recession, so we shall in this case use energy conservation directly.

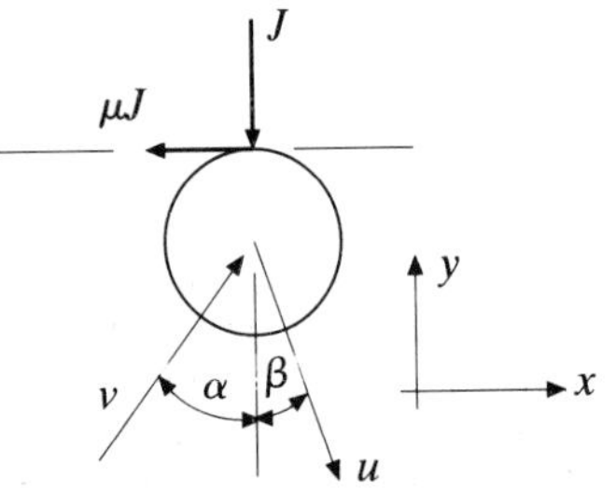

Figure 8.25

Referring to Fig. 8.25, which is a plan view, and resolving in the x and y directions

$$-\mu J = mu\sin\beta - mv\sin\alpha \qquad \text{(i)}$$

$$-J = -mu\cos\beta - (mv\cos\alpha) \qquad \text{(ii)}$$

and considering moments about the centre of mass

$$\mu Jr = I\omega = mk_G{}^2\omega. \qquad \text{(iii)}$$

Equating energy before impact to that after gives

$$\frac{m}{2}v^2 = \frac{m}{2}u^2 + \frac{mk_G{}^2}{2}\omega^2 \qquad \text{(iv)}$$

From (i) and (ii) with $\bar{J} = J/m$

$$u^2 = (v\sin\alpha - \mu\bar{J})^2 + (\bar{J} - v\cos\alpha)^2$$
$$= v^2 + \bar{J}^2(1+\mu^2) - 2\mu\bar{J}(\cos\alpha + \mu\sin\alpha)$$

and from (iii)

$$\omega = \frac{\mu\bar{J}r}{k_G{}^2}$$

substituting into (iv) gives

$$v^2 = v^2 + \bar{J}^2(1+\mu^2) - 2v\bar{J}(\cos\alpha + \mu\sin\alpha)$$
$$+ \mu^2\bar{J}^2\left(\frac{r}{k_G}\right)^2$$

or $\quad 0 = \bar{J}[\bar{J}[1+\mu^2(1+(r/k_G)^2)] - 2v\omega(\cos\alpha + \mu\sin\alpha)]$

For a non-trivial solution

$$\bar{J} = \frac{2v(\cos\alpha + \mu\sin\alpha)}{1+\mu^2(1+(r/k_G)^2)}.$$

Inserting the numerical values (noting that $k_G = r/\sqrt{2}$)

$$\bar{J} = \frac{2\times 10\times(\cos 30° + 0.2\times\sin 30°)}{1+0.2^2(1+2)}$$
$$= 17.25 \text{ Ns/kg}$$

From (i)

$$u\sin\beta = v\sin\alpha - \mu\bar{J}$$
$$= 10\times\sin 30° - 0.2\times 17.25$$
$$= 1.55$$

and from (ii)

$$u\cos\beta = \bar{J} - v\cos\alpha$$
$$= 17.25 - 10\times\cos 30°$$
$$= 8.59$$

therefore

$$u = (1.55^2 + 8.59^2)^{1/2} = 8.73 \text{ m/s}$$

and

$$\beta = \arctan(1.55/8.59) = 10.23°$$

Also

$$\omega r = \mu\bar{J}(r/k_G)^2 = 0.2\times 17.25\times 2$$
$$= 6.9 \text{ m/s}$$

or

$$\omega = 6.9/0.025 = 276 \text{ rad/s}$$

We stated previously that the speed of recession equals the speed of approach for the contacting particles. In this case the direction is not obvious but we may suspect that velocities resolved along the line of the resultant impulse is the most likely. The angle of friction γ is the direction of the resultant contact force so $\gamma = \arctan(\mu) = \arctan(0.2) = 11.3°$. The angle of incidence being 30° lies outside the friction angle so we expect the full limiting friction to be developed. We therefore resolve the incident and reflected velocities along this line.

Component of approach velocity
$= 10 \times \cos(30° - 11.3°) = 9.473$ m/s

Component of recession velocity

$$= u\cos(\beta + \gamma) + \omega r \cos(90° - \gamma)$$
$$= 8.73\cos(10.23° + 11.3°) + 6.9\cos(90° - 11.3°)$$
$$= 9.473 \text{ m/s which justifies the assumption.}$$

If the angle of incidence is less than γ then the impulse will be in a direction parallel to the incident velocity.

Problems

8.1 A rubber ball is dropped from a height of 2 m on to a concrete horizontal floor and rebounds to a height of 1.5 m. If the ball is dropped from a height of 3 m, estimate the rebound height.

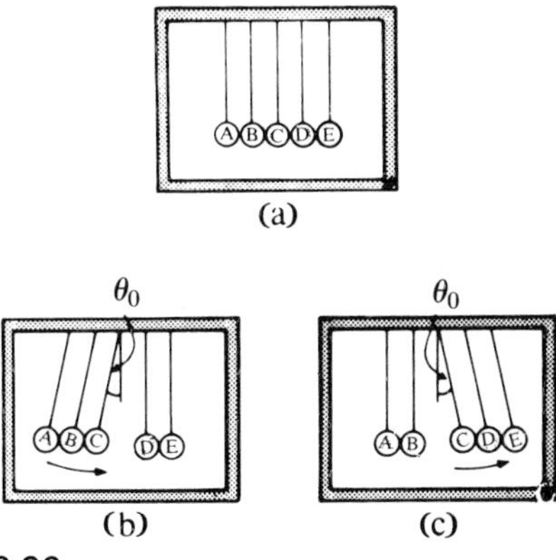

Figure 8.26

8.2 Figure 8.26 shows a toy known as a Newton's cradle. The balls A, B, C, D and E are all identical and hang from light strings of equal length as shown at (a). Balls A, B and C are lifted together so that their strings make an angle θ_0 with the vertical, as shown at (b), and they are then released. Show that, if energy losses are negligible and after impact a number of balls rise together, then after the first impact balls A and B will remain at rest and balls C, D and E will rise together until their strings make an angle θ_0 to the vertical as shown at (c).

8.3 Figure 8.27 shows two parallel shafts AB and CD which can rotate freely in their bearings. The total axial moment of inertia of shaft AB is I_1 and that of shaft CD is I_2. A disc B with slightly conical edges is keyed to shaft AB and a similar disc C on shaft CD can slide axially on splines. The effective radii of the discs are r_B and r_C respectively.

Initially the angular velocities are $\omega_1 \boldsymbol{k}$ and $\omega_2 \boldsymbol{k}$. A device (not shown) then pushes disc C into contact with disc B and the device itself imparts a negligible couple

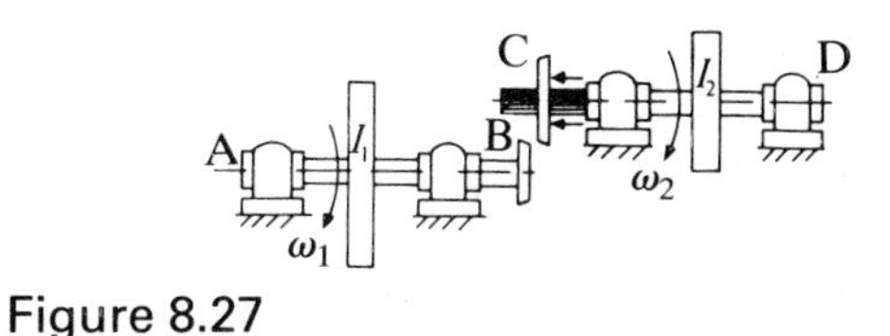

Figure 8.27

to shaft CD.

Show that, after slipping ceases, the angular velocity of shaft CD is

$$\boldsymbol{\omega}_{CD} = \frac{I_2\omega_2 - (r_C/r_B)I_1\omega_1}{I_2 + I_1(r_C/r_B)^2}\boldsymbol{k}$$

Why is the moment of momentum not conserved?

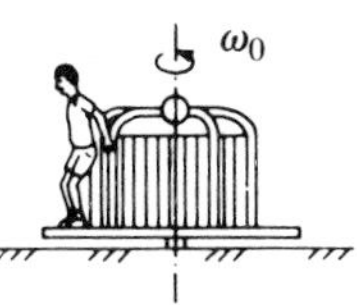

Figure 8.28

8.4 See Fig. 8.28. A roundabout can rotate freely about its vertical axis. A child of mass m is standing on the roundabout at a radius R from the axis. The axial moment of inertia of the roundabout is I_A. When the angular velocity is ω_0 the child leaps off and lands on the ground with no horizontal component of velocity. What is the angular velocity of the roundabout just after the child jumps?

8.5 Figure 8.29 shows the plan and elevation of a puck resting on ice. The puck receives an offset horizontal blow P as shown. The blow is of short duration and the horizontal component of the contact force with the ice is negligible compared with P. Immediately after the impact, the magnitude of $\boldsymbol{v}_G$, the velocity of the centre of mass G, is v_0.

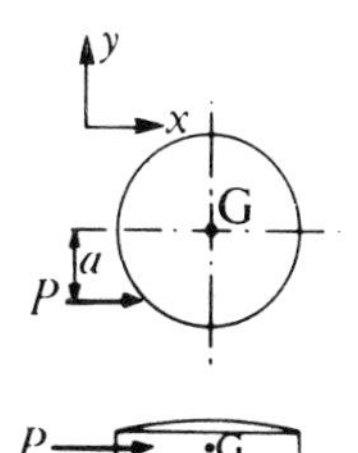

Figure 8.29

If the mass of the puck is m and the moment of inertia about the vertical axis through G is I, determine the angular velocity after the impact.

8.6 A uniform pole AB of length l and with end A resting on the ground rotates in a vertical plane about A and strikes a fixed object at point P, where AP = b. Assuming that there is no bounce, show that the minimum length b such that the blow halts the pole with no further rotation is $b = 2l/3$.

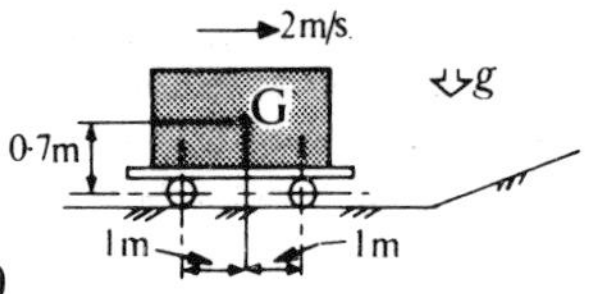

Figure 8.30

8.7 A truck is travelling on a horizontal track towards an inclined section (see Fig. 8.30). The velocity of the truck just before it strikes the incline is 2 m/s. The wheelbase is 2 m and the centre of gravity G is located as shown. The mass of the truck is 1000 kg and its moment of inertia about G is 650 kg m^2. The mass of the wheels may be neglected.

a) If the angle of the incline is 10° above the horizontal, determine the velocity of the axle of the leading wheels immediately after impact, assuming that the wheels do not lift from the track. Also determine the loss in energy due to the impact.

b) If the angle of the incline is 30° above the horizontal and the leading wheels remain in contact with the track, show that the impact causes the rear wheels to lift.

8.8 A jet of water issuing from a nozzle held by a fireman has a velocity of 20 m/s which is inclined at 70° above the horizontal. The diameter of the jet is 28 mm. Determine the horizontal and vertical components of the force that the fireman must apply to the nozzle to hold it in position. Also determine the maximum height reached by the water, neglecting air resistance.

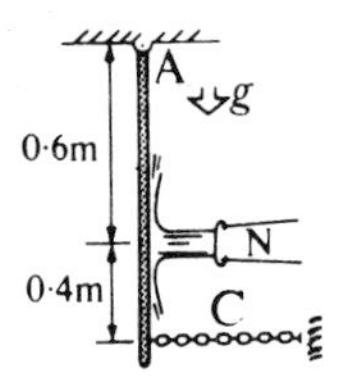

Figure 8.31

8.9 See Fig. 8.31. A jet of fluid of density 950 kg/m^3 emerges from the nozzle N with a velocity of 10 m/s and diameter 63 mm. The jet impinges on a vertical gate of mass 3.0 kg hanging from a horizontal hinge at A. The gate is held in place by the light chain C. Neglecting any horizontal velocity of the fluid after impact, determine the magnitude of the force in the hinge at A.

8.10 A jet of fluid is diverted by a fixed curved blade as shown in Fig. 8.32. The jet leaves the nozzle N at 25 m/s at a mass flow rate of 20 kg/s. Neglecting the effects of friction between the fluid and the blade, determine the direct force, the shear force and the bending moment in the support at section AA.

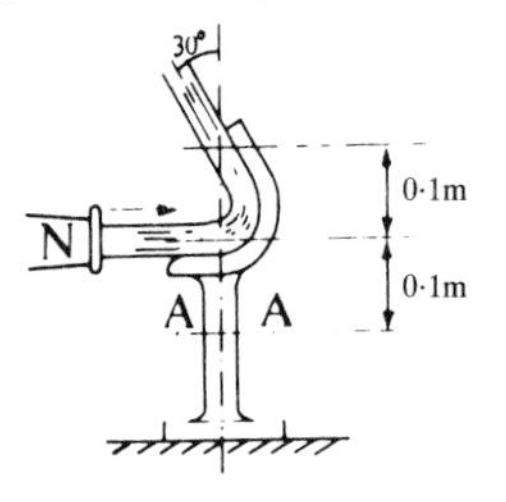

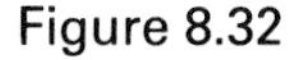
Figure 8.32

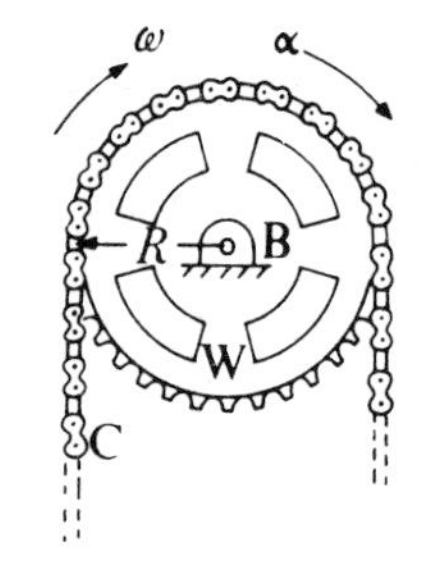

Figure 8.33

8.11 Figure 8.33 shows part of a transmission system. The chain C of mass per unit length ρ passes over the chainwheel W, the effective radius of the chain being R. The angular velocity and angular acceleration of the chainwheel are ω and α respectively, both in the clockwise sense.

a) Obtain an expression for the horizontal momentum of the chain.

b) Determine the horizontal component of the force the chainwheel exerts on its bearings B.

8.12 A container (Fig. 8.34) consists of a hopper H and a receiver R. Initially the hopper contains a quantity of grain and the receiver is empty, flow into the receiver being prevented by the closed valve V. The valve is then opened and grain flows through the valve at a constant mass rate r_O.

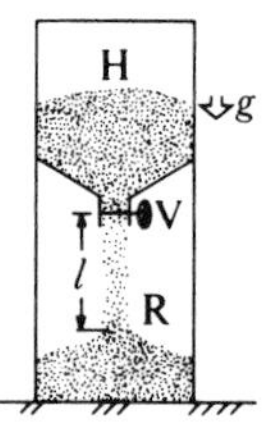

Figure 8.34

At a certain instant the column of freely falling grain has a length l. The remaining grain may be assumed to have negligible velocity and the rate of change of l with time is small compared with the impact velocity. Show that (a) the freely falling grain has a mass $r_O(2l/g)^{1/2}$ and (b) the force exerted by the container on the ground is the same as before the valve was opened.

8.13 An open-linked chain has a mass per unit length of 0.6 kg/m. A length of 2 m of the chain lies in a straight line on the floor and the rest is piled as shown in Fig. 8.35. The coefficient of friction between the chain and the floor is 0.5.

If a constant horizontal force of 12 N is applied to

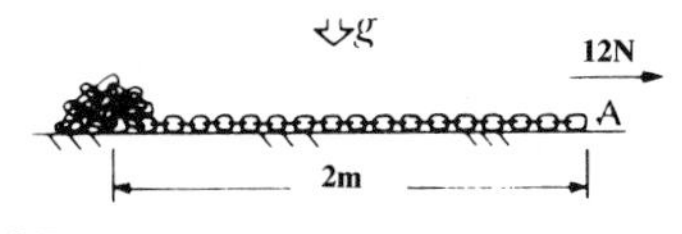

Figure 8.35

end A of the chain in the direction indicated, show, neglecting the effects of motion inside the pile, that motion will cease when A has travelled a distance of 3.464 m. (Take g to be 10 N/kg.)

8.14 Figure 8.36 shows a U-tube containing a liquid. The liquid is displaced from its equilibrium position and then oscillates. By considering the moment of

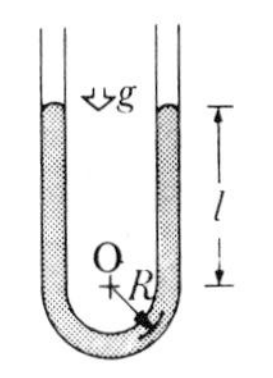

Figure 8.36

momentum about point O, show that the frequency f of the oscillation is given, neglecting viscous effects, by

$$f = \frac{1}{2\pi}\sqrt{\frac{2g}{2l + \pi R}}.$$

Also solve the problem by an energy method.

8.15 A length of chain hangs over a chainwheel as shown in Fig. 8.37 and its mass per unit length is 1 kg/m. The chainwheel is free to rotate about its axle and has an axial moment of inertia of 0.04 kg m^2.

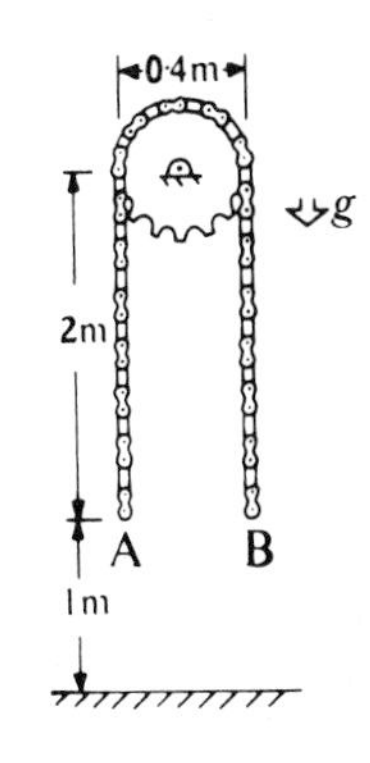

Figure 8.37

When the system is released from this unstable equilibrium position, end B descends. If the upward displacement in metres of end A is x, show that the downward force F on the ground for $1 < x < 2$ is given by

$$F = 9.81\left[(x-1) + \frac{4 - 3x - x^2}{6.628 + x}\right]\text{N}$$

8.16 a) A rocket burns fuel at a constant mass rate r and the exhaust gases are ejected backwards at a constant velocity u relative to the rocket. At time $t = 0$ the motor is ignited and the rocket is fired vertically and subsequently has a velocity v.

If air resistance and the variation in the value of g can both be neglected, and lift-off occurs at time $t = 0$, show that

$$v = u\ln\left(\frac{M}{M - rt}\right) - gt$$

Where M is the initial mass of the rocket plus fuel.
b) A space vehicle is fired vertically from the surface of the Moon ($g = 1.61$ N/kg). The vehicle is loaded with 30000 kg of fuel which after ignition burns at a steady rate of 500 kg/s. The initial acceleration is 36 m/s^2.

Find the mass of the rocket without fuel and the velocity after 5 s.

8.17 An experimental vehicle travels along a horizontal track and is powered by a rocket motor. Initially the vehicle is at rest and its mass, including 260 kg of fuel, is 2000 kg. At time $t = 0$ the motor is ignited and the fuel is burned at 20 kg/s, the exhaust gases being ejected backwards at 2020 m/s relative to the vehicle. The combined effects of rolling and wind resistance are equivalent to a force opposing motion of $(400 + 1.0v^2)$ N, where v is the velocity in m/s. At the instant when all the fuel is burnt, brakes are applied causing an additional constant force opposing motion of 6000 N.

Determine the maximum speed of the vehicle, the distance travelled to reach this speed and the total distance travelled.

9
Vibration

SECTION A
One-degree-of-freedom systems

9.1 Introduction

Mechanical vibration is said to occur when parts of a system execute periodic motion about a static-equilibrium configuration. In the simplest cases only one co-ordinate is required to describe the position of the system, which is thus defined to have one degree of freedom. If friction is very small, this co-ordinate performs oscillations with simple harmonic motion – at least for small oscillations.

Any real system can deform in many ways and therefore requires many co-ordinates to describe its position – it is said to possess many degrees of freedom. However, the result of analysis which follows in Section B indicates that, for linear systems, one many-degrees-of-freedom system is equivalent to many one-degree-of-freedom systems, thus a detailed study of one-degree-of-freedom systems is a necessary prerequisite.

9.2 Free vibration of undamped systems

One of the simplest systems is the combination of a rigid body and a light linear spring as shown in Fig. 9.1. The mass m is supported by a light spring which has a constant stiffness k. If x is the

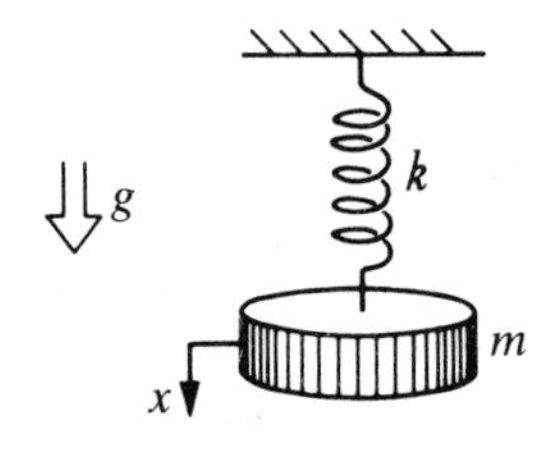

Figure 9.1

extension of the spring, the free-body diagram is as shown in Fig. 9.2. Thus the equation of motion is

$$mg - kx = m\ddot{x}$$

or $\quad m\ddot{x} + kx = mg \qquad (9.1)$

By letting $y = x - mg/k$, i.e. y is the deflection from the static equilibrium position, we obtain

$$m\ddot{y} + k(y + mg/k) = mg$$

or $\quad m\ddot{y} + ky = 0 \qquad (9.2)$

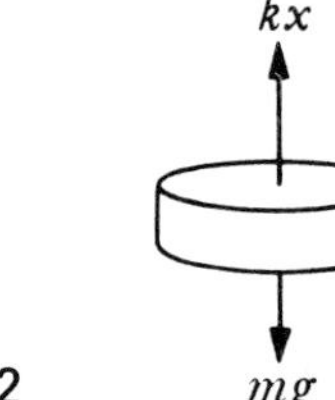

Figure 9.2

The solution of equation 9.2 is (see Chapter 2)

$$y = A\cos\omega_n t + B\sin\omega_n t \qquad (9.3)$$

where $\quad \omega_n = \surd(k/m) \qquad (9.4)$

Differentiation with respect to time gives

$$\dot{y} = \omega_n(-A\sin\omega_n t + B\cos\omega_n t) \qquad (9.5)$$

and $\quad \ddot{y} = -\omega_n^2(A\cos\omega_n t + B\sin\omega_n t) \qquad (9.6)$

$$= -\omega_n^2 y$$

thereby justifying the solution given in equation 9.3.

The constants A and B depend on the initial conditions so that if, when $t = 0$, $\dot{y} = V$ and $y = Y$, substitution of these values into equations 9.3 and 9.5 gives

$$Y = A$$

and $\quad V = \omega_n B$

thus $\quad y = Y\cos\omega_n t + (V/\omega_n)\sin\omega_n t \qquad (9.7)$

An alternative form for equation 9.7 is

$$y = C\cos(\omega_n t - \phi) \qquad (9.8)$$

where $\quad C = \surd(Y^2 + V^2/\omega_n^2)$

and $\quad \tan\phi = V/(\omega_n Y)$

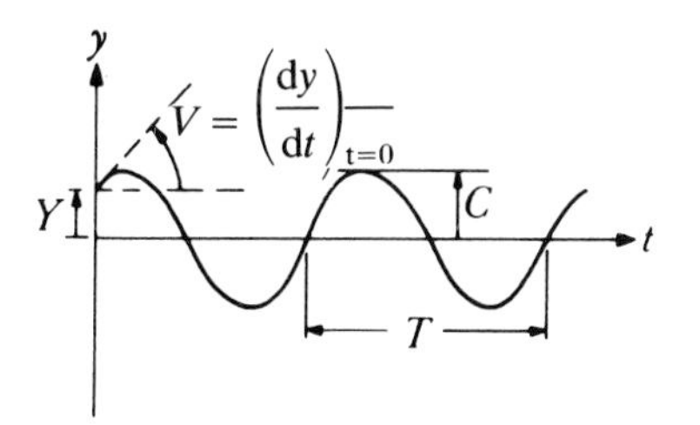

Figure 9.3

A plot of y against time is shown in Fig. 9.3, in which the following terms are defined.

i) Amplitude, C – the maximum displacement reckoned from the mean position. Twice C is referred to as the peak-to-peak amplitude.
ii) Periodic time, T – the minimum time interval after which the motion is repeated.
iii) Frequency, ν – the number of cycles performed in unit time; hence $\nu = 1/T$.

One cycle of this periodic motion may be represented by the projection of a line OA, rotating about O at an angular speed ω_n, on to a diameter of the circle as shown in Fig. 9.4.

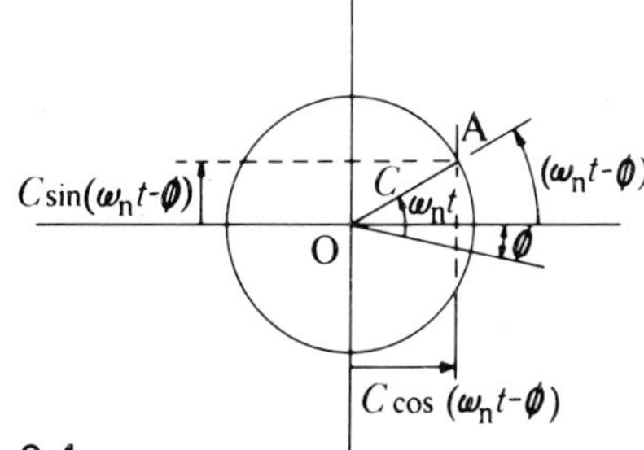

Figure 9.4

From this figure it is clear that the time for a complete cycle is given by

$$\omega_n T = 2\pi$$

thus $$T = 2\pi/\omega_n \qquad (9.9)$$

and $$v = \omega_n/(2\pi) \qquad (9.10)$$

From equation 9.8 we have

$$v = dy/dt = -\omega_n C\sin(\omega_n t - \phi) \qquad (9.11)$$

and from Fig. 9.4 we see that the projection on the vertical axis is

$$C\sin(\omega_n t - \phi) = -v/\omega_n$$

9.3 Vibration energy

The kinetic energy of the system discussed in section 9.2 is

$$\tfrac{1}{2}mv^2 = \tfrac{1}{2}mC^2\omega_n{}^2\sin^2(\omega_n t - \phi)$$

and, since $\omega_n{}^2 = k/m$, we have

$$\tfrac{1}{2}mv^2 = \tfrac{1}{2}kC^2\sin^2(\omega_n t - \phi)$$

the strain energy in the spring is

$$\begin{aligned}\tfrac{1}{2}kx^2 &= \tfrac{1}{2}k[C\cos(\omega_n t - \phi) + mg/k]^2 \\ &= \tfrac{1}{2}kC^2\cos^2(\omega_n t - \phi) \\ &\quad + kC\cos(\omega_n t - \phi)mg/k + \tfrac{1}{2}k(mg/k^2)\end{aligned}$$

and the gravitational potential energy is

$$-mgy = -mgC\cos(\omega_n t - \phi)$$

The total energy is

$$\begin{aligned}E &= \tfrac{1}{2}mv^2 + \tfrac{1}{2}kx^2 + (-mgy) \\ &= \tfrac{1}{2}kC^2 + \tfrac{1}{2}k(mg/k)^2 \qquad (9.12)\end{aligned}$$

The second term in equation 9.12 is a constant and is the strain energy in the spring when in the static-equilibrium position; hence the energy associated with the vibration is

$$E_v = \tfrac{1}{2}kC^2 = \tfrac{1}{2}m\omega_n{}^2C^2 \qquad (9.13)$$

or, maximum strain energy reckoned from the static-equilibrium position = maximum kinetic energy = E_v.

We see that constant forces, such as gravity, merely change the static-equilibrium position and do not affect the vibration, so it is customary to consider only motion and energies relative to the static-equilibrium configuration.

Reworking our example, we may write

$$-ky = m\ddot{y}$$

knowing that the weight is opposed by an equal but opposite spring force.

Alternatively, using energy, we write

$$E_v = \tfrac{1}{2}m\dot{y}^2 + \tfrac{1}{2}ky^2 = \text{constant}$$

$$dE_v/dt = m\dot{y}\ddot{y} + ky\dot{y} = 0$$

thus $$m\ddot{y} + ky = 0$$

9.4 Pendulums

A case in which gravity may not be neglected, of course, is in the study of pendulums. Here, however, the effect of gravity does not produce a constant effect – as we shall show.

For the simple pendulum shown in Fig. 9.5 we have, by considering moments about O,

$$-mgl\sin\theta = I_O\ddot{\theta}$$

For a simple pendulum (light rod with

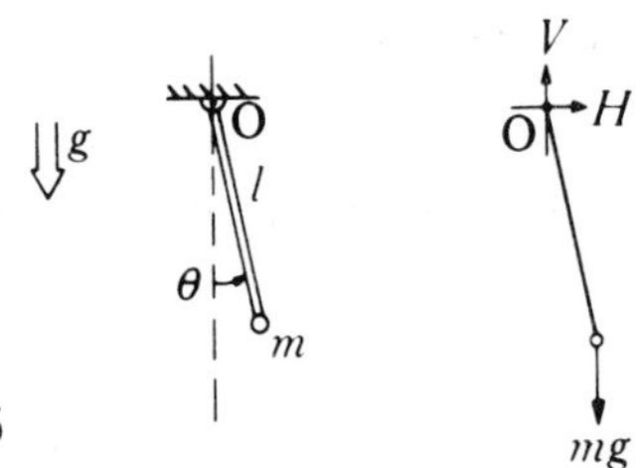

Figure 9.5

concentrated mass m), $I_O = ml^2$, so

$$-mgl\sin\theta = ml^2\ddot{\theta} \quad (9.14)$$

This equation is non-linear, but, as is true for most systems, if we consider only small oscillations about the equilibrium position the equation becomes linear. For small θ, $\sin\theta \to \theta$; hence equation 9.14 becomes

$$-mgl\theta = ml^2\ddot{\theta}$$

or $\ddot{\theta} + (g/l)\theta = 0$ (9.15)

thus $\omega_n = \sqrt{\dfrac{g}{l}}$ and $\nu_n = \dfrac{1}{2\pi}\sqrt{\dfrac{g}{l}}$

If the mass is not concentrated, then

$$-mgl\theta = I_O\ddot{\theta}$$

or $\ddot{\theta} + (mgl/I_O)\theta = 0$ (9.16)

where l is the distance of the centre of mass from O.

Using the parallel-axes theorem

$$I_O = I_G + ml^2 = mk_G{}^2 + ml^2$$

leads to

$$\omega_n{}^2 = \frac{mgl}{I_O} = \frac{mgl}{mk_G{}^2 + ml^2} = \frac{g}{l(1 + k_G{}^2/l^2)}$$

thus $\nu_n = \dfrac{1}{2\pi}\sqrt{\dfrac{g}{l(1 + k_G{}^2/l^2)}}$ (9.17)

9.5 Levels of vibration

At this point it is helpful to consider the orders of magnitude of vibration in terms of human comfort and machine tolerance. It is convenient to consider a plot of log(velocity amplitude) against log(frequency).

The velocity amplitude is given by

$$\hat{v} = 2\pi\nu(\text{displacement amplitude}) = 2\pi\nu\hat{x}$$

and the acceleration amplitude by

$$\hat{a} = 2\pi\nu\hat{v} = (2\pi\nu)^2\hat{x}$$

therefore $\log\hat{a} = \log\nu + \log\hat{v} + \log 2\pi$

and $\log\hat{x} = -\log\nu + \log\hat{v} - \log 2\pi$

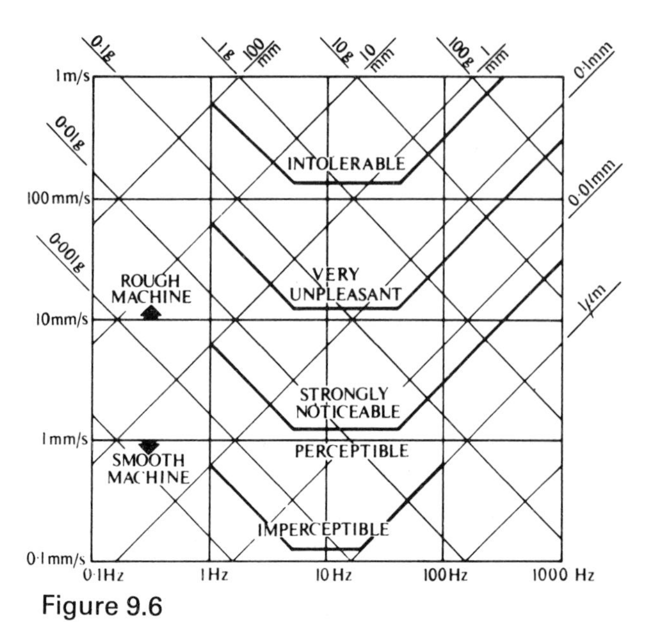

Figure 9.6

so that, on a graph of $\log\hat{v}$ against $\log\nu$, lines of constant $\hat{a}$ appear as straight lines of slope -1 and lines of constant $\hat{x}$ appear as lines of slope $+1$, as shown in Fig. 9.6.

9.6 Damping

In all mechanical systems there is some means by which the vibrational energy is reduced, so without external stimulus any system will eventually come to rest. The most common means are some form of internal friction which converts the vibrational energy into thermal energy or the dissipation of energy into the surroundings by the generation of sound and vibration in any supporting structure or surrounding fluid.

A system in which energy is dissipated is said to be damped. If the damping is large then periodic motion will not occur and the system, once disturbed, will return toward an equilibrium position without the velocity reversing. Such motion is called aperiodic.

One means of providing extra damping is to make use of the viscous properties of fluids. Figure 9.7(a) shows a damper of the dashpot type, in which oil is forced through holes in the piston by a force proportional to the velocity of the piston relative to the cylinder. The usual symbol for a viscous damper is shown in Fig. 9.7(b).

Another form of damping is eddy-current damping, in which a conductor is moved relative to a magnetic field. This also requires a force

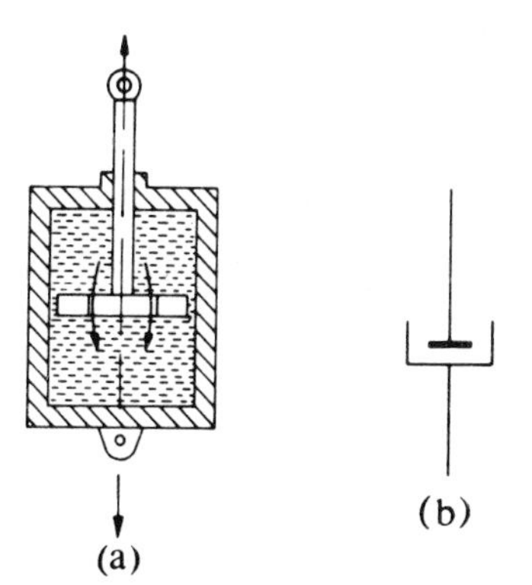

Figure 9.7

proportional to the relative velocity of the conductor and field.

9.7 Free vibration of a damped system

The system shown in Fig. 9.8 consists of a rigid body of mass m, a spring of stiffness k and a damper having a damping coefficient c such that the force exerted on the damper is $c\dot{x}$. The

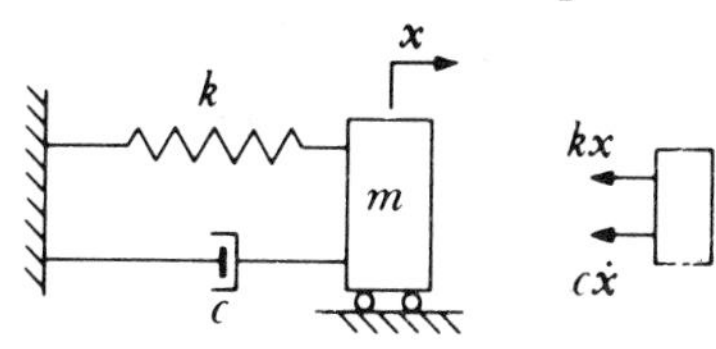

Figure 9.8

equation of motion for the mass is

$$-kx - c\dot{x} = m\ddot{x}$$

which is usually written as

$$m\ddot{x} + c\dot{x} + kx = 0 \qquad (9.18)$$

If $c = 0$ the motion is simple harmonic and, remembering that

$$e^{j\theta} = \cos\theta + j\sin\theta$$

where $j = \surd(-1)$, we may write the solution of equation 9.18 as

$$x = \tfrac{1}{2}(A - jB)e^{j\omega_n t} + \tfrac{1}{2}(A + jB)e^{-j\omega_n t}$$
$$= A\cos\omega_n t + B\sin\omega_n t$$

where $\omega_n = \surd(k/m)$.

For very large damping the inertia effects will be small, so that the motion is described by

$$c\dot{x} + kx = 0$$

or $\quad \mathrm{d}x/\mathrm{d}t = -(k/c)x$

Thus $\quad \int_{x_0}^{x} (1/x)\,\mathrm{d}x = -\int_0^t (k/c)\,\mathrm{d}t$

giving $\quad \ln(x/x_0) = -(k/c)t$

or $\quad x = x_0 e^{-(k/c)t}$

So for small and large damping the solution for x is of the form

$$x = A e^{\lambda t}$$

therefore we shall try this form as a general solution to equation 9.18. Thus if

$$x = A e^{\lambda t}$$

then $\quad \dot{x} = \lambda A e^{\lambda t}$

and $\quad \ddot{x} = \lambda^2 A e^{\lambda t}$

Substituting these terms into equation 9.18 gives

$$(m\lambda^2 + c\lambda + k)A e^{\lambda t} = 0$$

which, for a non-trivial solution, means that

$$m\lambda^2 + c\lambda + k = 0 \qquad (9.19)$$

Solving for λ gives

$$\lambda = -(c/2m) \pm \surd[(c/2m)^2 - k/m]$$

When $(c/2m)^2 > k/m$, both values of λ are real and negative so that the form of the solution is

$$x = A e^{-\lambda_1 t} + B e^{-\lambda_2 t}$$

and we see that the motion is aperiodic.

When $(c/2m)^2 < k/m$,

$$\lambda = -(c/2m) \pm j\surd[k/m - (c/2m)^2]$$

in which case

$$x = A'\exp\{-(c/2m)t + j[k/m - (c/2m)^2]^{1/2}t\}$$
$$+ B'\exp\{-(c/2m)t - j[k/m - (c/2m)^2]^{1/2}t\}$$

or $\quad x = \exp[-(c/2m)t]\{A\cos[k/m - (c/2m)^2]^{1/2}t$
$$+ B\sin(k/m - (c/2m)^2]^{1/2}t\} \qquad (9.20)$$

It is convenient to introduce some characteristic parameters so that equation 9.20 is readily applicable to other physically similar situations. We have noticed that when $c/2m > \surd(k/m)$ the motion is aperiodic and when $c/2m < \surd(k/m)$ the motion is periodic; hence *critical damping* is defined by

$$\frac{c_{\text{crit.}}}{2m} = \sqrt{\frac{k}{m}}$$

or $\quad c_{\text{crit.}} = 2\surd(km) \qquad (9.21)$

The *damping ratio*, ζ, is defined by

$$\zeta = \frac{c}{c_{crit.}} = \frac{c}{2\sqrt{(km)}}$$

Since $\sqrt{(k/m)} = \omega_n$, the undamped natural frequency, we have

$$c_{crit.} = 2m\sqrt{(k/m)} = 2m\omega_n$$

and $\quad c = \zeta c_{crit.} = m2\zeta\omega_n$

The equation of motion, equation 9.18, can now be written

$$m\ddot{x} + m2\zeta\omega_n x + m(k/m)x = 0$$

giving finally

$$\ddot{x} + 2\zeta\omega_n\dot{x} + \omega_n^2 x = 0 \qquad (9.22)$$

Noting that $c/2m = \zeta\omega_n$, equation 9.20 may be written

$$x = e^{-\zeta\omega_n t}[A\cos\omega_d t + B\sin\omega_d t] \qquad (9.23)$$

where $\omega_d = [k/m - (c/2m)^2]^{1/2}$
$= (\omega_n^2 - \zeta^2\omega_n^2)^{1/2}$
$= \omega_n\sqrt{(1-\zeta^2)} \qquad (9.24)$

Differentiating equation 9.23 with respect to time gives

$$\dot{x} = e^{-\zeta\omega_n t}[(B\omega_d - A\zeta\omega_n)\cos\omega_d t - (A\omega_d + B\zeta\omega_n)\sin\omega_d t] \qquad (9.25)$$

The constants A and B depend on the initial conditions. For example if, when $t = 0$, $x = x_0$ and $\dot{x} = 0$, then

$$x_0 = A$$

and $\quad 0 = B\omega_d - A\zeta\omega_n$
$B = \zeta\omega_n A/\omega_d = x_0\zeta/\sqrt{(1-\zeta^2)}$

Hence $\quad x = x_0 e^{-\zeta\omega_n t}[\cos\omega_d t + \{\zeta/\sqrt{(1-\zeta^2)}\}\sin\omega_d t] \qquad (9.26)$

A plot of x against t is shown in Fig. 9.9. The periodic time T is given by

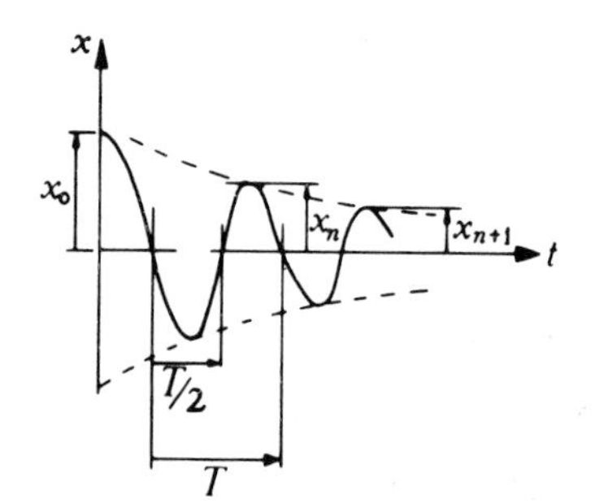

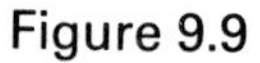
Figure 9.9

$$\omega_d T = 2\pi$$

so $\quad T = \frac{2\pi}{\omega_d} = \frac{2\pi}{\omega_n\sqrt{(1-\zeta^2)}}$

and the damped natural frequency

$$\nu_d = T^{-1} = \frac{\omega_n}{2\pi}\sqrt{(1-\zeta^2)}$$

The variation of ω_d/ω_n with ζ is shown in Fig. 9.10.

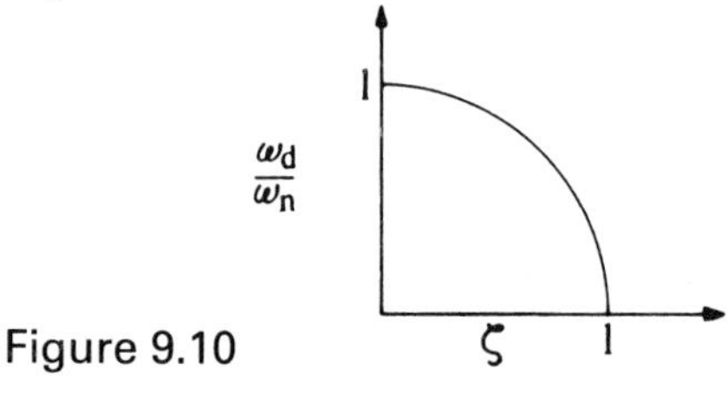

Figure 9.10

Logarithmic decrement

A convenient way of indicating the amount of damping is to quote the logarithmic decrement ('log.dec.') which, referring to Fig. 9.9, is defined as

$$\delta = \ln(x_n/x_{n+1})$$

But $\quad \frac{x_n}{x_{n+1}} = \frac{e^{-\zeta\omega_n t}f(\omega t)}{e^{-\zeta\omega_n(t+T)}f(\omega t + T)} = e^{\zeta\omega_n T}$

thus

$$\delta = \zeta\omega_n T = \zeta\omega_n 2\pi/[\omega_n\sqrt{(1-\zeta^2)}] = 2\pi\zeta/\sqrt{(1-\zeta^2)} \qquad (9.27)$$

and for small damping

$$\delta \approx 2\pi\zeta \qquad (9.28)$$

Specific loss

A further way of indicating the amount of damping in lightly damped systems is to evaluate the energy lost per cycle as a fraction of the energy at the start of the cycle.

Specific loss $= (\tfrac{1}{2}kx_n^2 - \tfrac{1}{2}kx_{n+1}^2)/(\tfrac{1}{2}kx_n^2)$
$= 1 - (x_{n+1}/x_n)^2$
$= 1 - \exp[-2\zeta\omega_n T]$

so, for small damping,

specific loss $= 1 - [1 - 2\zeta\omega_n T + \ldots]$
$\approx 2\zeta\omega_n T \approx 4\pi\zeta \approx 2\delta \qquad (9.29)$

Coulomb damping

When the damping force has a constant

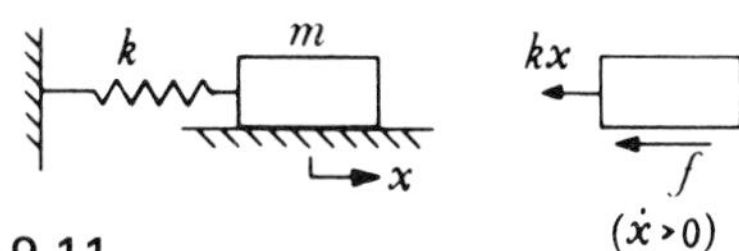

Figure 9.11

magnitude and always opposes the motion, it is known as Coulomb damping.

In Fig. 9.11 the coefficient of sliding friction is taken to be constant, so the equation of motion is

$$-kx - f\langle\text{sign of } \dot{x}\rangle = m\ddot{x}$$

$$m\ddot{x} + kx = \mp f \qquad (9.30)$$

9.8 Phase-plane method

A plot of velocity against displacement is known as a phase-plane diagram. The phase-plane method is readily adapted to give a graphical means of solving any single-degree-of-freedom vibration problem. In this book we shall be using it only for linear systems, or systems where the motion can be described by a number of linear differential equations (sometimes known as piecewise linear systems).

The phase-plane method is based on the fact that, for a constant external force, a graph of x against $\dot{x}/\omega_n$ is a circle. In Fig. 9.4 we saw that the projection of a rotating radius gave x and $-\dot{x}/\omega_n$ on the horizontal and vertical axes respectively. In order to plot $\dot{x}/\omega_n$ in the positive sense, it is simply necessary to reverse the sense of rotation. In general, the equation

$$a\ddot{x} + bx = A = \text{constant}$$

transforms to a circle with centre at $x = A/b$, $\dot{x}/\omega_n = 0$. If the initial conditions are given, then one point on a circle of known centre completely defines the circle, as shown in Fig. 9.12. If after a given interval of time or at a specific value of x or $\dot{x}$ the constant changes, this just alters the position of the centre of the circle, the radius, of course, changing so that the trajectory on the phase plane is continuous.

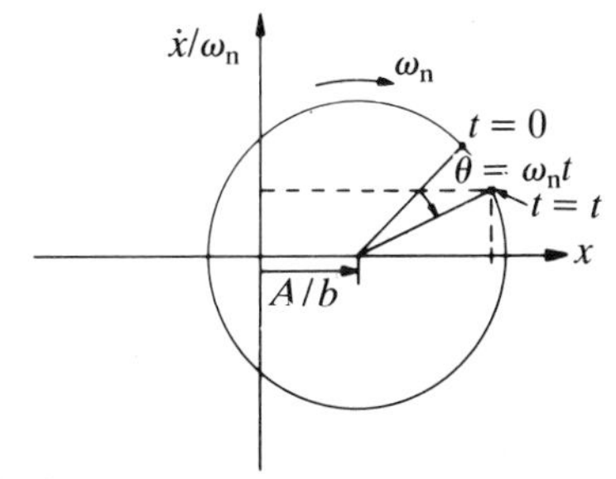

Figure 9.12

Application to Coulomb damping

From equation 9.30,

$$m\ddot{x} + kx = -f \quad (\dot{x}>0)$$

$$m\ddot{x} + kx = +f \quad (\dot{x}<0)$$

Assuming initial conditions $t = 0, x = x_0, \dot{x} = 0$, we may draw part of a circle on the phase plane, see Fig. 9.13, for $\dot{x}<0$. At point A the velocity changes sign and the centre of the circle moves

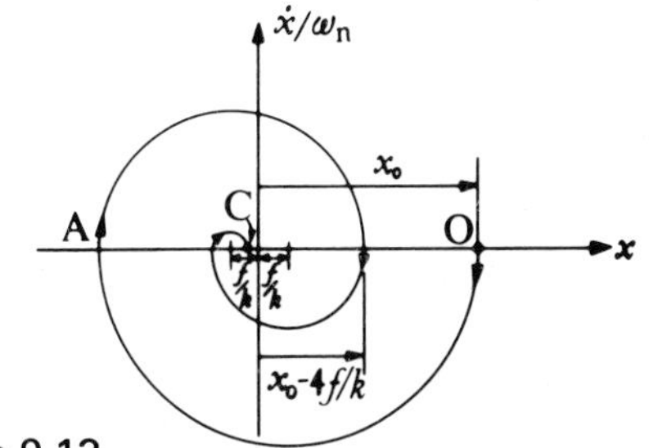

Figure 9.13

from $x = f/k$ to $x = -f/k$. This process continues until a point C is reached such that

$$-f/k < x < f/k$$

at which instant the motion ceases.

The amplitude of vibration drops by $2f/k$ each half cycle, thus the decay rate is linear with time and not exponential; also, it is seen that the periodic time is *not* affected by the damping.

9.9 Response to simple input forces

Consider the two systems shown in Fig. 9.14 in which $P = P(t)$ and $x_0 = x_0(t)$. In both cases x is

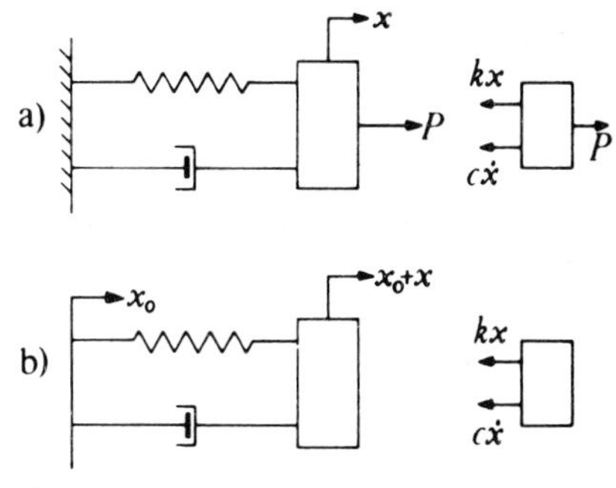

Figure 9.14

the extension of the spring. The equation of motion for (a) is

$$-kx - c\dot{x} + P = m\ddot{x}$$

or $$\ddot{x} + 2\zeta\omega_n\dot{x} + \omega_n{}^2x = P/m \qquad (9.31)$$

and for (b),

$$-kx - c\dot{x} = m(\ddot{x}_0 + \ddot{x})$$

or $$\ddot{x} + 2\zeta\omega_n\dot{x} + \omega_n{}^2 x = -\ddot{x}_0 \qquad (9.32)$$

We note that equations 9.31 and 9.32 are of the same form.

Response to a step input

Assume $P = P_0$ for $t>0$ and $P = 0$ for $t<0$.

The solution of equation 9.31 is

$$x = e^{-\zeta\omega_n t}[A\cos\omega_d t + B\sin\omega_d t] + P_0/m\omega_n{}^2$$
$$= \text{complementary function} + \text{particular integral}$$

It is seen that the particular integral $P_0/m\omega_n{}^2 = P_0/k$ is simply the final steady-state solution after the complementary function has become zero. There are many formal mathematical methods for determining the particular integral, but for these simple cases the result can be achieved by inspection.

If, when $t = 0$, $x = 0$ and $\dot{x} = 0$

then $\quad 0 = A + P_0/k$

and $\quad 0 = B\omega_d - A\zeta\omega_n$

hence $$x = -e^{-\zeta\omega_n t}(P_0/k) \times [\cos\omega_d t + \zeta/\surd(1-\zeta^2)\sin\omega_d t] + P_0/k$$

or $$x/(P_0/k) = (1 - e^{-\zeta\omega_n t}\cos\omega_d t) - \frac{e^{-\zeta\omega_n t}\zeta}{\surd(1-\zeta^2)}\sin\omega_d t \qquad (9.33)$$

and $$\frac{\dot{x}}{\omega_n(P_0/k)} = \frac{e^{-\zeta\omega_n t}}{\surd(1-\zeta^2)}\sin\omega_d t \qquad (9.34)$$

A graph of x against time is shown in Fig. 9.15.

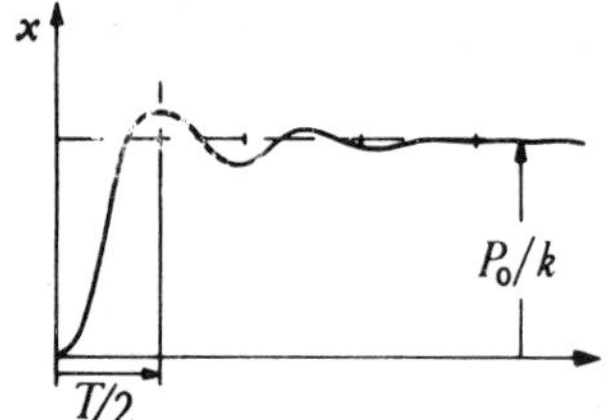

Figure 9.15

The overshoot $(x_{max} - P_0/k)$ occurs when the velocity is first zero, i.e. when $\omega_d t = \pi$, so

$$(x_{max} - P_0/k) = (P_0/k)\exp[-\zeta\pi/\surd(1-\zeta^2)]$$

or $$\frac{x_{max} - P_0/k}{P_0/k} \approx e^{-\zeta\pi} \text{ for small damping.}$$

Response to impulsive input

If at $t = 0$ a short-duration impulse $P\tau$ is applied, then at $t = 0$, $x = 0$ and, since impulse equals change in momentum,

$$\dot{x} = P\tau/m$$

giving

$$x = \frac{P\tau}{m\omega_n}\frac{e^{-\zeta\omega_n t}}{\surd(1-\zeta^2)}\sin\omega_d t \qquad (9.35)$$

a graph of which is shown in Fig. 9.16.

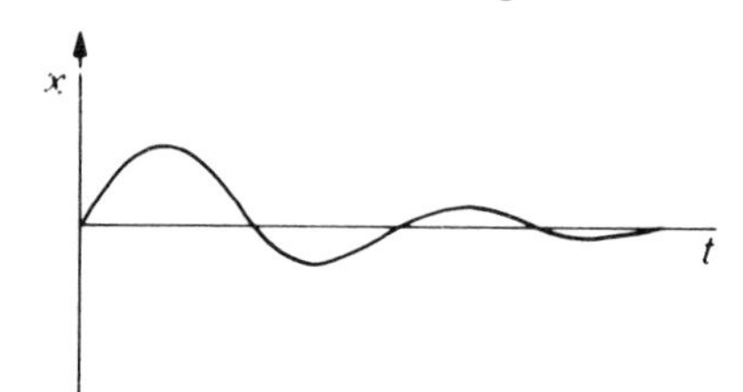

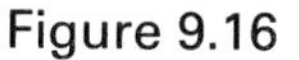

Figure 9.16

Response to a ramp input

A ramp input is of the form

$$P = 0 \quad \text{for} \quad t<0$$
$$= at \quad \text{for} \quad t\geqslant 0$$

thus the equation of motion is

$$m\ddot{x} + c\dot{x} + kx = at$$

or $$\ddot{x} + 2\zeta\omega_n\dot{x} + \omega_n{}^2 x = at/m$$

In this case the steady-state solution is a constant velocity V, or $x = Vt + b$, so for large t, when $\ddot{x} = 0$,

$$2\zeta\omega_n V + \omega_n{}^2(Vt + b) = at/m$$

Thus, equating the coefficients of powers of t,

$$2\zeta\omega_n V + \omega_n{}^2 b = 0$$

and $$\omega_n{}^2 V = a/m$$

therefore $$V = a/m\omega_n{}^2 = a/k$$

and $$-b = 2\zeta V/\omega_n = 2\zeta a/\omega_n k$$

Hence the general solution is

$$x = e^{-\zeta\omega_n t}[A\cos\omega_d t + B\sin\omega_d t] + at/k - 2\zeta a/\omega_n k$$

For initial conditions $x = 0$, $\dot{x} = 0$,

$$0 = A - 2\zeta a/\omega_n k$$

and $\quad 0 = (B\omega_d - A\zeta\omega_n) + a/k$

hence $\quad A = 2\zeta a/\omega_n k$

and $$B = \frac{2\zeta^2 a}{k\omega_n\sqrt{(1-\zeta^2)}} - \frac{a}{k\omega_n\sqrt{(1-\zeta^2)}}$$

$$= \frac{-a(1-2\zeta^2)}{k\omega_n\sqrt{(1-\zeta^2)}}$$

giving

$$x = \left\{ e^{-\zeta\omega_n t}\left[2\zeta\cos\omega_d t - \frac{(1-2\zeta^2)}{\sqrt{(1-\zeta^2)}}\sin\omega_d t\right] + \omega_n t - 2\zeta \right\} \frac{a}{\omega_n k} \qquad (9.36)$$

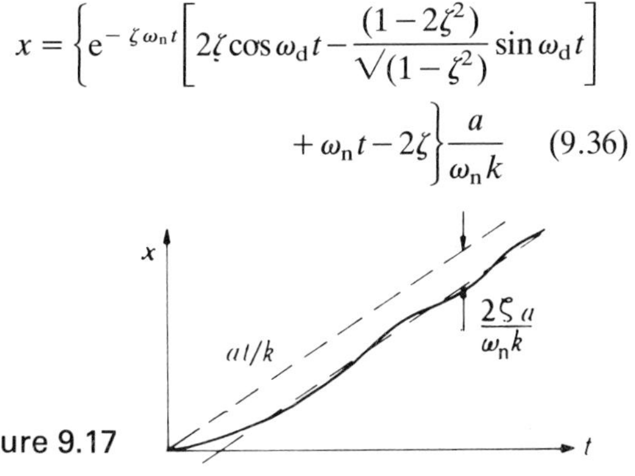

Figure 9.17

Figure 9.17 shows the form of the response.

9.10 Periodic excitation

By use of the Fourier theorem, any periodic function representing a physical disturbance may be replaced by a series of sinusoidal disturbances.

The total response to this excitation applied to a linear system is the sum of the individual responses to each of the sinusoidal disturbances – this is known as the *principle of superposition*.

Fourier series

Consider a function of time that repeats after a periodic time T,

i.e. $f(t) = f(t+T)$

Assume that

$$f(t) = a_0 + a_1\cos\omega_0 t + \ldots + a_n\cos n\omega_0 t + \ldots + b_1\sin\omega_0 t + \ldots + b_n\sin n\omega_0 t + \ldots \text{(where } \omega_0 = 2\pi/T\text{)}$$

This is known as a Fourier series.

Multiplying both sides by $\cos n\omega_0 t$ and integrating over one cycle gives

$$\int_0^T f(t)\cos n\omega_0 t\,dt = a_n\int_0^T \cos^2 n\omega_0 t\,dt$$

All other terms are zero since

$$\int_0^T (\cos n\omega_0 t\sin m\omega_0 t)\,dt = 0$$

$$\int_0^T (\cos n\omega_0 t\cos m\omega_0 t)\,dt = 0 \quad \text{for} \quad n \neq m$$

$$\int_0^T (\sin n\omega_0 t\sin m\omega_0 t)\,dt = 0 \quad \text{for} \quad n \neq m$$

However, when $m = n$,

$$\int_0^T \cos^2 n\omega_0 t\,dt = T/2$$

hence $$\int_0^T f(t)\cos n\omega_0 t\,dt = a_n T/2$$

and similarly

$$\int_0^T f(t)\sin n\omega_0 t\,dt = b_n T/2$$

To summarise, we have

$$a_0 = \frac{1}{T}\int_0^T f(t)\,dt \qquad (9.37)$$

$$a_n = \frac{2}{T}\int_0^T f(t)\cos n\omega_0 t\,dt \qquad (9.38)$$

$$b_n = \frac{2}{T}\int_0^T f(t)\sin n\omega_0 t\,dt \qquad (9.39)$$

where $\omega_0 = 2\pi/T$ (9.40)

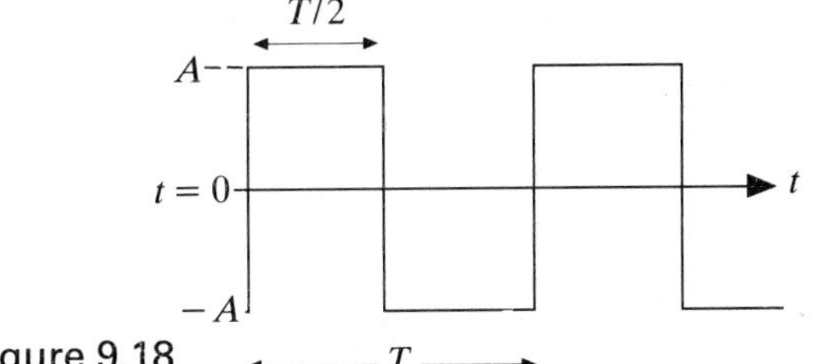

Figure 9.18

As an example consider the square wave shown in Fig. 9.18. By the Fourier theorem,

$$a_0 = 0$$

and $a_n = 0$

as the wave is asymmetrical about the $t = 0$ axis,

$$b_n = 2 \times \frac{2}{T}\int_0^{T/2} A\sin n\omega_0 t\,dt$$

$$= \frac{4A}{T}\left[-\frac{\cos n\omega_0 t}{n\omega_0}\right]_0^{T/2}$$

$$= \frac{2A}{\pi}\left[\frac{1-\cos n\pi}{n}\right]$$

Substituting integer values for n in the above expression gives

n	1	2	3	4	5
$b_n \frac{\pi}{2A}$	2	0	$\frac{2}{3}$	0	$\frac{2}{5}$

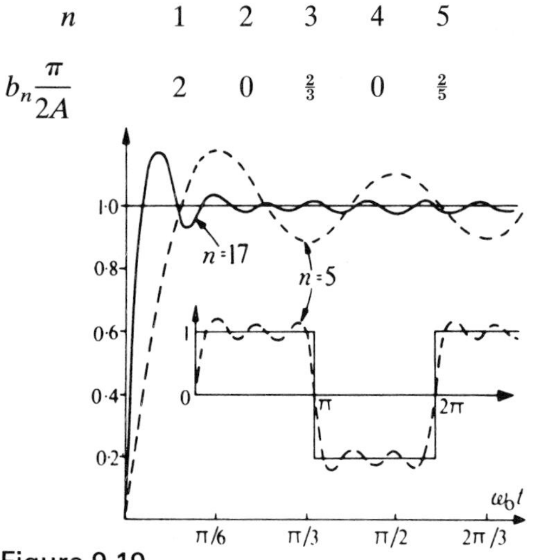

Figure 9.19

Figure 9.19 shows the result of taking the first three non-zero terms, i.e. $n = 5$, and also a plot using the first nine non-zero terms, i.e. $n = 17$.

9.11 Work done by a sinusoidal force

If a sinusoidal force $F = F_0 \cos(\omega t + \phi)$ acts on a particle moving with simple harmonic motion such that $x = X \cos \omega_0 t$, then the work done is

$$\begin{aligned}\int F \mathrm{d}x &= -\int F_0 \cos(\omega_0 t + \phi)\, \omega_0 X \sin \omega_0 t \,\mathrm{d}t \\ &= \omega_0 F_0 X[-\cos\phi \int(\cos \omega t \sin \omega_0 t)\,\mathrm{d}t \\ &\quad + \sin\phi \int(\sin \omega t \sin \omega_0 t)\,\mathrm{d}t]\end{aligned}$$

If the integration is taken over a long period of time then the integral will tend to zero unless $\omega = \omega_0$. (This statement may be proved by methods similar to those used in the development of the Fourier theorem.)

With $\omega = \omega_0$ the work done per cycle is

$$\int_{t=0}^{t=T} F \mathrm{d}x = \omega F_0 X[-\cos\phi(0) + \sin\phi(T/2)]$$

$$= \pi F_0 X \sin\phi \qquad (9.41)$$

Hence we see that the maximum work done per cycle occurs when $\phi = 90°$, i.e. when the force is in phase with the velocity.

The phasor diagram shown in Fig. 9.20 shows the displacement x as the projection on to a horizontal diameter of the rotating line ON. Similarly the velocity

$$v = -\omega X \sin \omega t = -V \sin \omega t$$

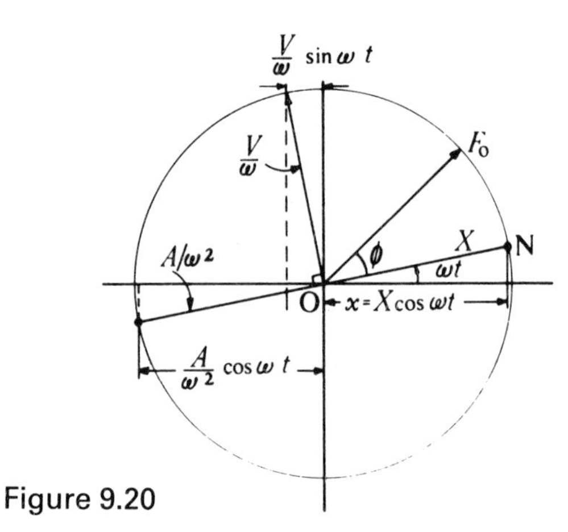

Figure 9.20

and the acceleration

$$a = -\omega^2 X \cos \omega t = -A \cos \omega t$$

can be depicted by rotating lines on the diagram. V and A are the velocity and acceleration amplitudes respectively.

9.12 Response to a sinusoidal force

The linear damped single-degree-of-freedom system shown in Fig. 9.14(a) has a sinusoidally varying force applied to the mass. Measuring x from the position of no strain in the spring, the equation of motion is

$$F_0 \cos \omega t - kx - c\dot{x} = m\ddot{x}$$

or $$m\ddot{x} + c\dot{x} + kx = F_0 \cos \omega t \qquad (9.42)$$

The solution of equation 9.42 is in two parts: the complementary function, which is a solution when $F_0 = 0$, plus the particular integral.

The complementary function has been discussed in section 9.7 and is seen to be a transient term leading to no motion as time increases. The particular integral is a steady-state solution which exists when the transient has died away.

There are many ways of finding the steady-state solution, but we will base our solution on physical reasoning.

We assume that the steady-state solution is of the form

$$x = X \cos(\omega t - \phi) \qquad (9.43)$$

where ω is the forcing frequency. Energy must be transferred to the system, since the damper is dissipating energy, hence the steady-state-response frequency is the same as the forcing frequency for reasons given in the preceding section. The amplitude X and the phase angle ϕ

are constants to be determined.

Substituting equation 9.43 into equation 9.42 gives

$$-m\omega^2 X\cos(\omega t-\phi)-\omega cX\sin(\omega t-\phi) + kX\cos(\omega t-\phi)=F_0\cos\omega t \quad (9.44)$$

This equation is represented on the phasor diagram shown in Fig. 9.21. From the diagram,

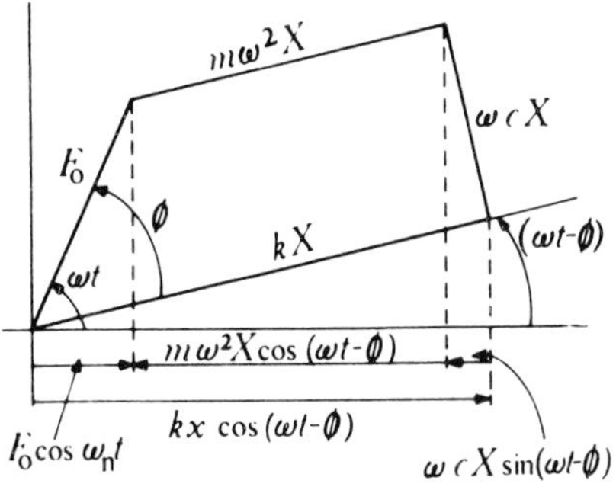

Figure 9.21

$$F_0^2=(kX-m\omega^2X)^2+(\omega cX)^2$$

thus $$X=F_0/\sqrt{[(k-m\omega^2)^2+\omega^2c^2]} \quad (9.45)$$

If ω is small then the maximum value of X is F_0/k. Dividing both numerator and denominator in equation 9.45 by k leads to

$$X=\frac{F_0/k}{\sqrt{\{[1-(\omega/\omega_n)^2]^2+(2\zeta\omega/\omega_n)^2\}}} \quad (9.46)$$

where $\omega_n=\sqrt{(k/m)}$

and $\zeta=c/c_{crit.}=c/(2\sqrt{km})$

From Fig. 9.21,

$$\tan\phi=\omega cX/(kX-m\omega^2X)=\omega c/(k-m\omega^2)$$
$$=2\zeta(\omega/\omega_n)/[1-(\omega/\omega_n)^2] \quad (9.47)$$

An alternative mathematical treatment using complex numbers will now be given. It is known that

$$e^{\pm j\theta}=\cos\theta\pm j\sin\theta$$

where $j=\sqrt{(-1)}$, so the real part of $e^{j\omega t}\equiv \mathrm{Re}(e^{j\omega t})=\cos\omega t$. Also

$$\mathrm{Re}[e^{j(\omega t-\phi)}]=\cos(\omega t-\phi)$$

Equation 9.42 may be written

$$\ddot{x}+2\zeta\omega_n\dot{x}+\omega_n^2x=(F_0/m)\mathrm{Re}(e^{j\omega t})$$

and its steady-state solution as

$$x=X\mathrm{Re}(e^{j\omega t}e^{-j\phi})$$

If the real part of one side of an equation is equal to the real part of the other side then so must the imaginary parts be equal; we may therefore drop the reference to real part and write

$$\ddot{x}+2\zeta\omega_n\dot{x}+\omega_n^2x=(F_0/m)e^{j\omega t} \quad (9.48)$$

$$x=Xe^{j\omega t}e^{-j\phi} \quad (9.49)$$

Substituting equation 9.49 into equation 9.48 gives

$$(-\omega^2+2\zeta\omega_n j\omega+\omega_n^2)Xe^{j\omega t}e^{-j\phi}=(F_0/m)e^{j\omega t}$$

thus $$-\omega^2X+2\zeta\omega_n j\omega X+\omega_n^2X=(F_0/m)e^{j\phi}$$

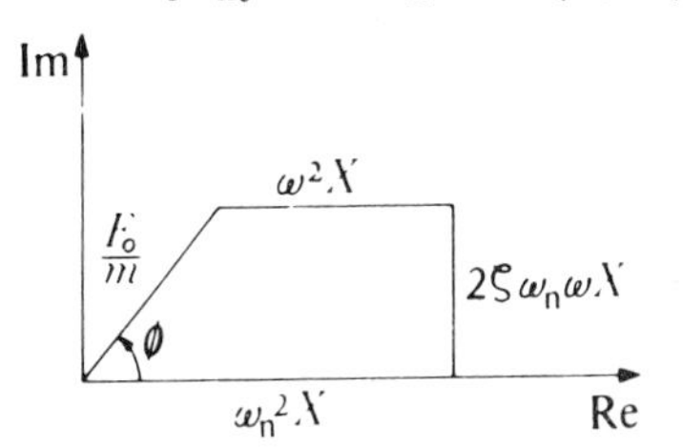

Figure 9.22

This may be represented on an Argand diagram as shown in Fig. 9.22. This figure is seen to have the same form as Fig. 9.21 and obviously equations 9.46 and 9.47 are obtained. Equation 9.45 may be written in non-dimensional form as

$$\frac{X}{(F_0/k)}=\frac{1}{\sqrt{\{[1-(\omega/\omega_n)^2]^2+(2\zeta\omega/\omega_n)^2\}}}=\mu \quad (9.50)$$

where μ is known as the dynamic magnifier. A plot of μ against ω/ω_n for various values of ζ is given in Fig. 9.23 and a plot of phase angle ϕ against ω/ω_n is shown in Fig. 9.24.

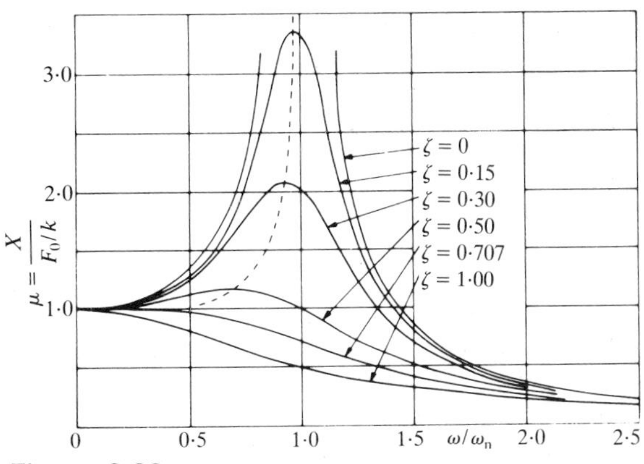

Figure 9.23

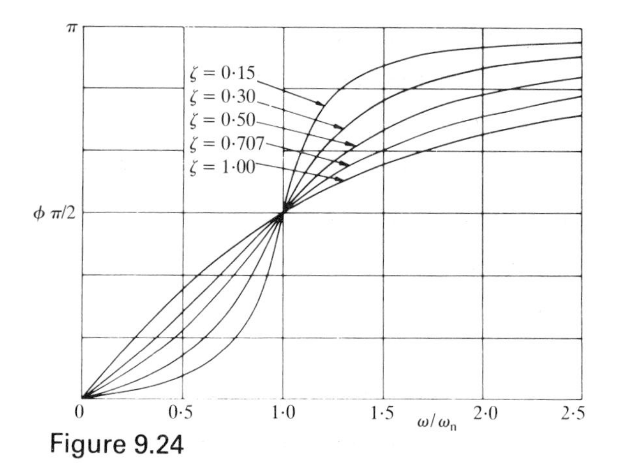

Figure 9.24

9.13 Moving foundation

In Fig. 9.25 the system is disturbed owing to the vibration of the foundation and it is assumed that the base movement x_0 is not affected by the subsequent motion of the mass. This type of vibration occurs in many situations ranging from the vibration of an instrument on a moving vehicle to the motion of a building during an earthquake.

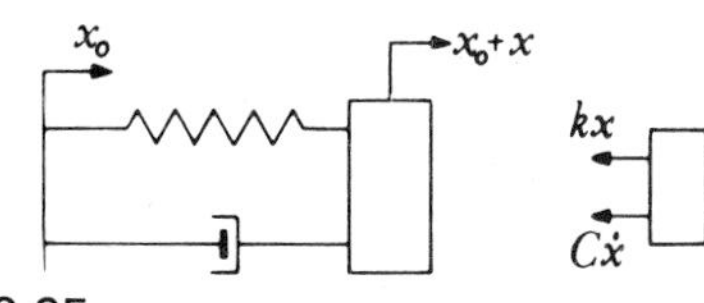

Figure 9.25

We have a choice of co-ordinate to specify the motion of the system. We can measure the motion of the mass relative either to the base or to an inertial frame of reference. Both are useful but we shall use the former as it is the strain in the elastic member which is usually of greater interest; hence x here will be the movement of the mass relative to the base, giving the absolute motion of the mass as $x+x_0$.

The equation of motion for the mass is

$$-kx - c\dot{x} = m(\ddot{x}_0 + \ddot{x})$$

or $m\ddot{x} + c\dot{x} + kx = -m\ddot{x}_0$

which may be written

$$\ddot{x} + 2\zeta\omega_n\dot{x} + \omega_n{}^2 x = -\ddot{x}_0 \qquad (9.51)$$

If the base movement is $x_0 = X_0 \cos \omega t$

then $\ddot{x}_0 = -\omega^2 X \cos \omega t$

hence $\ddot{x} + 2\zeta\omega_n\dot{x} + \omega_n{}^2 x = \omega^2 X_0 e^{j\,\omega t}$

it being understood that only the real part is finally required. Making the usual assumption for steady-state vibration that $x = Xe^{j\,\omega t}e^{-j\,\phi}$, we obtain

$$-\omega^2 X + 2\zeta\omega_n\omega X + \omega_n{}^2 X = \omega^2 X_0 e^{j\phi} \qquad (9.52)$$

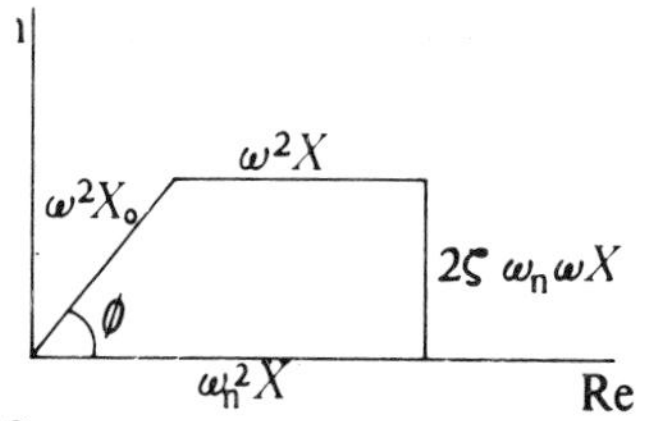

Figure 9.26

The Argand diagram representation of equation 9.52 is shown in Fig. 9.26. From the diagram we obtain

$$X = \frac{\omega^2 X_0}{\sqrt{[(\omega_n{}^2 - \omega^2)^2 + 4\zeta^2\omega_n{}^2\omega^2]}}$$

$$= \frac{(\omega/\omega_n)^2 X_0}{\sqrt{\{[1-(\omega/\omega_n)^2]^2 + 4\zeta^2(\omega/\omega_n)^2\}}} \qquad (9.53)$$

and $$\tan\phi = \frac{2\zeta\omega_n\omega X}{(\omega_n{}^2 - \omega^2)X} = \frac{2\zeta(\omega/\omega_n)}{[1-(\omega/\omega_n)^2]} \qquad (9.54)$$

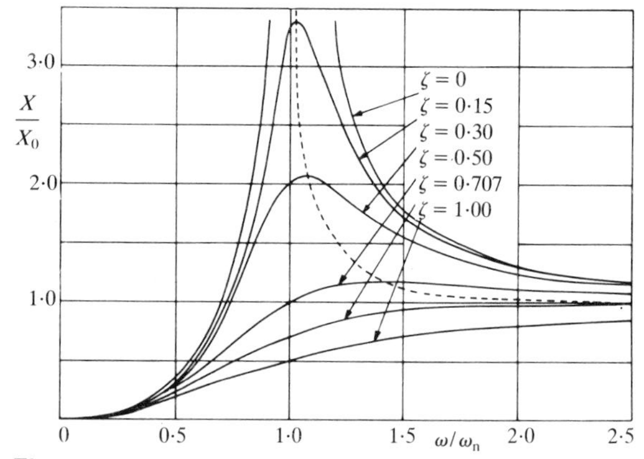

Figure 9.27

A plot of X/X_0 is shown in Fig. 9.27.

9.14 Rotating out-of-balance masses

The problem of vibration generated by rotating machinery is very common because it is impossible to manufacture a machine which is perfectly balanced. We will idealise the problem to that of a rigid frame carrying a rotating wheel with its centre of mass eccentric. In Fig. 9.28 the total mass is M and the mass of the rotating part is

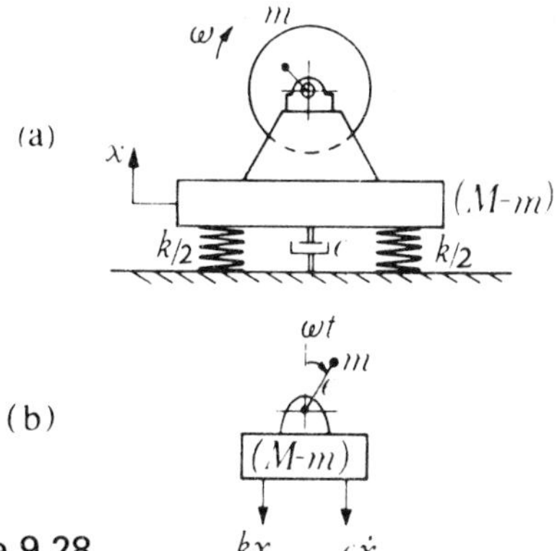

Figure 9.28

m, with its centre of mass eccentric by an amount e from the axis of rotation.

From the free-body diagram, the equation of motion is

$$-kx - c\dot{x} = (M-m)\ddot{x} + m\,\mathrm{d}^2(e\cos\omega t + x + \text{const.})/\mathrm{d}t^2$$

$$= (M-m)\ddot{x} + m(-\omega^2 e\cos\omega t + \ddot{x})$$

$$= M\ddot{x} - me\omega^2\cos\omega t$$

or $\quad M\ddot{x} + c\dot{x} + kx = me\omega^2\cos\omega t \qquad (9.55)$

or $\quad \ddot{x} + 2\zeta\omega_n\dot{x} + \omega_n^2 x = (me\omega^2/M)\cos\omega t \qquad (9.56)$

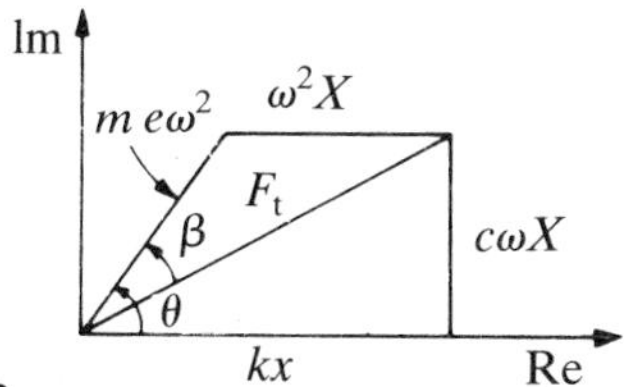

Figure 9.29

This equation has the same form as equation 9.51 therefore a detailed solution need not be given, but the resulting Argand diagram is shown in Fig. 9.29.

9.15 Transmissibility

It is often required to determine the magnitude of the force which is transmitted to the foundation. For the above case it will be the sum of the spring force and the damping force; thus

transmitted force

$$= kx + c\dot{x}$$

$$= kX\cos(\omega t - \phi) - cX\omega\sin(\omega t - \phi)$$

$$= F_t\cos(\omega t - \beta) \qquad (9.57)$$

where F_t is the modulus of the transmitted force and β is a phase angle.

Using complex notation,

$$kX\mathrm{e}^{\mathrm{j}(\omega t - \phi)} + \mathrm{j}cX\omega\mathrm{e}^{\mathrm{j}(\omega t - \phi)} = F_t\mathrm{e}^{\mathrm{j}(\omega t - \beta)}$$

which leads to

$$kX + \mathrm{j}cX = F_t\mathrm{e}^{\mathrm{j}(\phi - \beta)} \qquad (9.58)$$

or, dividing through by M,

$$\omega_n^2 X + \mathrm{j}2\zeta\omega_n\omega X = (F_t/M)\,\mathrm{e}^{\mathrm{j}(\phi - \beta)}$$

From Fig. 9.29,

$$F_t = X(k^2 + c^2\omega^2)^{1/2}$$

$$F_t/M = X[\omega_n^4 + (2\zeta\omega_n\omega)^2]^{1/2}$$

but $\quad X = (e\omega^2 m/M\omega_n^2)\{[1-(\omega/\omega_n)^2]^2 + (2\zeta\omega/\omega_n)^2\}^{-1/2}$

thus $\quad \dfrac{F_t}{m\omega^2 e} = \sqrt{\dfrac{1+(2\zeta\omega/\omega_n)^2}{[1-(\omega/\omega_n)^2]^2 + (2\zeta\omega/\omega_n)^2}} \qquad (9.59)$

The ratio of the transmitted force to the 'out-of-balance' force $F_t/m\omega^2 e$ is known as the *transmissibility* of the mounting. Curves of transmissibility against ω/ω_n are plotted in Fig. 9.30.

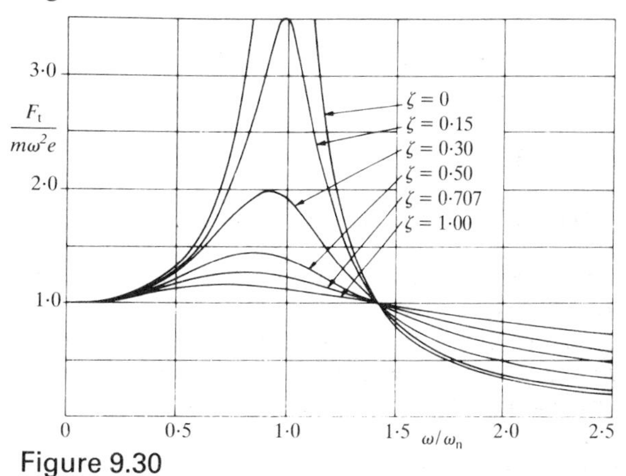

Figure 9.30

It is seen from Fig. 9.30 that the amplitude of the force transmitted is adversely affected by the presence of damping at the higher frequencies, although for $\omega/\omega_n > \sqrt{2}$ the force transmitted is still less than the 'out-of-balance' force exerted on the bearings of a rigidly mounted frame.

A similar problem exists when a body is to be isolated from a moving foundation. In this case the absolute amplitude of the mass compared to the amplitude of motion of the foundation is also known as the transmissibility – see example 9.5.

9.16 Resonance

Resonance is the condition of a system, subject to a sinusoidal excitation, exhibiting a maximum response to a given input. If the damping of the system is very small then resonance occurs when $\omega = \omega_n$, at which frequency all parameters – displacement, velocity, acceleration, spring deflections, etc. – tend to large values. However, when the damping is larger, $0.1 < \zeta < 0.4$ say, the various parameters reach their maxima at different frequencies.

In the case of a constant-amplitude applied force, the amplitude resonance occurs at a frequency lower than ω_n given by

$$\omega_{res.} = \omega_n \surd(1 - 2\zeta^2) \qquad (9.60)$$

For the moving-foundation case, the maximum amplitude of relative motion occurs at

$$\omega_{res.} = \omega_n / \surd(1 - 2\zeta^2) \qquad (9.61)$$

In both cases we see that resonance does not occur when $\zeta > 1/\surd 2$.

For small damping ($\zeta < 0.3$), the response at resonance is close to that when $\omega = \omega_n$ so for the first case

$$X/(F_0/k)_{max.} \approx 1/2\zeta \qquad (9.62)$$

and for the second case

$$(X/X_0)_{max.} \approx 1/2\zeta \qquad (9.63)$$

9.17 Estimation of damping from the width of the peak

Figure 9.31 is a sketch of a resonance peak for small damping. It is obvious that the lower the damping the sharper the peak. Equation 9.50 gives

$$X/(F_0/k) = 1/\surd\{[1 - (\omega/\omega_n)^2]^2 + 4\zeta^2(\omega/\omega_n)^2\}$$

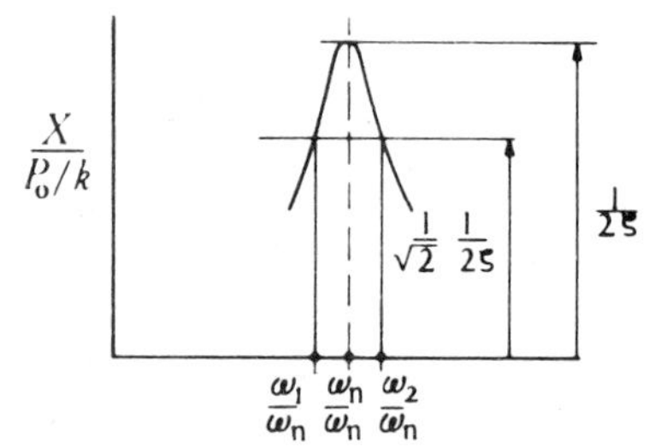

Figure 9.31

The values of ω when $X/(F_0/k) = (1/2\zeta)/\surd 2$ are found from

$$\frac{1}{8\zeta^2} = \frac{1}{[1 - (\omega/\omega_n)^2]^2 + 4\zeta^2(\omega/\omega_n)^2}$$

or $(\omega/\omega_n)^4 - (2 - 4\zeta^2)(\omega/\omega_n)^2 + (1 - 8\zeta^2) = 0$

The two roots of the quadratic are

$$(\omega/\omega_n)^2_{1,2} = (1 - 2\zeta^2) \pm \surd[(1 - 2\zeta^2)^2 - (1 - 8\zeta^2)]$$
$$= 1 - 2\zeta^2 \pm \surd(4\zeta^4 + 4\zeta^2)$$

hence
$$\left(\frac{\omega_2}{\omega_n}\right)^2 - \left(\frac{\omega_1}{\omega_n}\right)^2 = 2\surd(4\zeta^4 + 4\zeta^2) = 4\zeta\surd(1 + \zeta^2)$$

so that
$$\left(\frac{\omega_2 - \omega_1}{\omega_n}\right)\left(\frac{\omega_1 + \omega_2}{\omega_n}\right) = 4\zeta\surd(1 + \zeta^2)$$

If the damping is small, $\zeta^2 \ll 1$, and $(\omega_1 + \omega_2)/2 \approx \omega_n$, we obtain

$$\frac{\omega_2 - \omega_1}{\omega_n} = 2\zeta \qquad (9.64)$$

In other words, the width of the resonance peak at 0.707 of the height of the peak equals $2\zeta\omega_n$.

SECTION B Two-degree-of-freedom systems

9.18 Free vibration

If a system requires two independent co-ordinates to define its configuration then it is said to have two degrees of freedom. The simplest example is the two-mass two-spring system shown in Fig. 9.32. The choice of co-ordinates is arbitrary, the most obvious ones being x_1 and x_2 which are the displacements of the individual masses. However another convenient set would be the extensions of the springs which are x_1 and $(x_2 - x_1)$ respectively. We shall use both of these sets in turn.

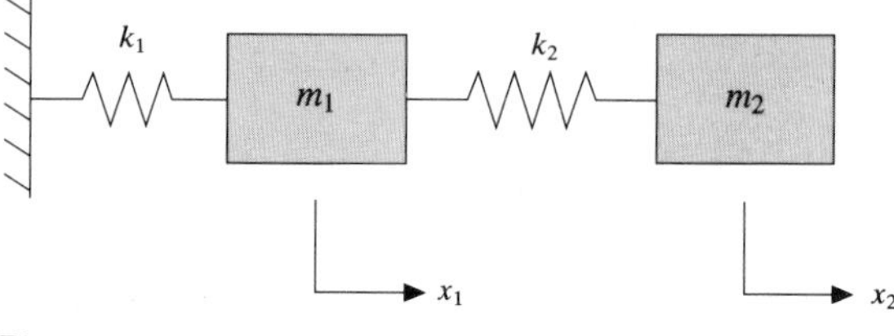

Figure 9.32

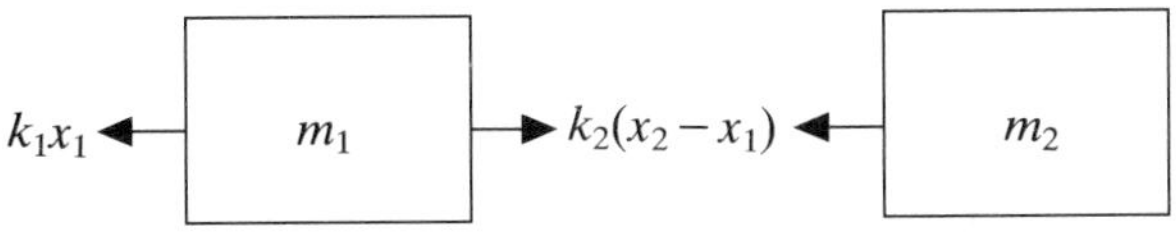

k_1x_1 ← m_1 → $k_2(x_2-x_1)$ ← m_2

Figure 9.33

Using the first arrangement we see from Fig. 9.33 that the equations of motion are

$$-k_1x_1+k_2(x_2-x_1)=m_1\ddot{x}_1 \tag{9.65}$$

and $$-k_2(x_2-x_1)=m_2\ddot{x}_2 \tag{9.66}$$

Re-arranging gives

$$m_1\ddot{x}_1+(k_1+k_2)x_1-k_2x_2=0 \tag{9.67}$$

and $$m_2\ddot{x}_2-k_2x_1+k_2x_2=0 \tag{9.68}$$

These equations may also be written in matrix form as

$$\begin{bmatrix} m_1 & 0 \\ 0 & m_2 \end{bmatrix}\begin{bmatrix} \ddot{x}_1 \\ \ddot{x}_2 \end{bmatrix} + \begin{bmatrix} (k_1+k_2) & -k_2 \\ -k_2 & k_2 \end{bmatrix}\begin{bmatrix} x_1 \\ x_2 \end{bmatrix} = \begin{bmatrix} 0 \\ 0 \end{bmatrix} \tag{9.69}$$

In compact form the above equation may be written

$$[\boldsymbol{M}](\ddot{x})+[\boldsymbol{K}](x)=(0)$$

where the square matrix $[\boldsymbol{M}]$ is known as the mass matrix and the square matrix $[\boldsymbol{K}]$ as the stiffness matrix.

We shall now assume that both masses oscillate with simple harmonic motion at the same frequency. Our intention is to see what conditions must prevail if the assumption is valid.

Thus we assume that

$$x_1=A\mathrm{e}^{\lambda t}$$

and $$x_2=B\mathrm{e}^{\lambda t}.$$

We have again used the complex form knowing that with λ imaginary the motion is sinusoidal. $x=X\sin(\omega t)$ would be a suitable form of assumption in this case but the complex form is preferred as it simplifies the algebra for more advanced problems.

Substituting into equations 9.67 and 9.68 and dividing through by the common factor $\mathrm{e}^{\lambda t}$ we obtain

$$[m_1\lambda^2+(k_1+k_2)]A-k_2B=0$$
$$-k_2A+[m_2\lambda^2+k_2]B=0.$$

These are a pair of simultaneous homogeneous equations and so cannot be solved for A and B directly, however two expressions for the ratio B/A can be found, i.e.

$$\frac{B}{A}=\frac{m_1\lambda^2+k_1+k_2}{k_2}=\frac{k_2}{m_2\lambda^2+k_2} \tag{9.70}$$

From the second equality we find

$$m_1m_2\lambda^4+[m_1k_2+m_2k_1+m_2k_2]\lambda^2+k_2k_1=0 \tag{9.71}$$

This is a quadratic in λ^2 with positive real coefficients. From the theory of linear equations both roots will be real and negative; let these roots be $-\omega_1{}^2$ and $-\omega_2{}^2$.

Thus the four roots are $\pm j\omega_1$, $\pm j\omega_2$ where $j=\sqrt{-1}$, and the general solution is

$$x_1=A_1\mathrm{e}^{j\omega_1 t}+A_2\mathrm{e}^{-j\omega_1 t}+A_3\mathrm{e}^{j\omega_2 t}+A_4\mathrm{e}^{-j\omega_2 t}$$
$$x_2=B_1\mathrm{e}^{j\omega_1 t}+B_2\mathrm{e}^{-j\omega_1 t}+B_3\mathrm{e}^{j\omega_2 t}+B_4\mathrm{e}^{-j\omega_2 t}$$

Since $\mathrm{e}^{j\theta}=\cos\theta+j\sin\theta$, the above two equations can be written in the form

$$x_1=[E\cos(\omega_1 t)+F\sin(\omega_1 t)]+[G\cos(\omega_2 t)+H\sin(\omega_2 t)]$$
$$x_2=\mu_1[E\cos(\omega_1 t)+F\sin(\omega_1 t)]+\mu_2[G\cos(\omega_2 t)+H\sin(\omega_2 t)] \tag{9.72}$$

The constants E, F, G and H are functions of the A and B coefficients, their values depending on the initial displacements and velocities of the masses; μ_1 and μ_2 are defined in the next paragraph.

The ratio B/A has two values one when $\omega=\omega_1$ and the other when $\omega=\omega_2$. Using the first equality in equation 9.72 the amplitude ratios μ_1 and μ_2 are defined (the second equality would lead to the same result).

$$\frac{-m_1\omega_1{}^2+k_1+k_2}{k_2}=\frac{B_1}{A_1}=\frac{B_2}{A_2}=\mu_1$$

and $$\frac{-m_1\omega_2{}^2+k_1+k_2}{k_2}=\frac{B_3}{A_3}=\frac{B_4}{A_4}=\mu_2$$

From the above argument it is seen that it is possible to choose the starting conditions such that $G=H=0$, in which case both masses vibrate with a frequency $\omega_1/(2\pi)$ and the ratio of their amplitudes is μ_1. Similarly, if initial conditions are chosen such that E and F are zero then the system may oscillate at a frequency ω_2

and with an amplitude ratio of μ_2.

It now follows that for any arbitrary starting conditions the motion may be considered as the sum of the two special cases just mentioned, the proportion of each case depending on the actual values of the starting conditions.

This fact is the cornerstone of the analysis of multi-degree-of-freedom vibrating systems. We see that for the two-degree-of-freedom system there are just two distinct frequencies at which vibration can occur, these are known as the *principal natural frequencies* and associated with each is a unique amplitude ratio known as a *principal mode shape*.

Example

We shall now consider a simple problem in detail in order to fix our ideas of principal modes.

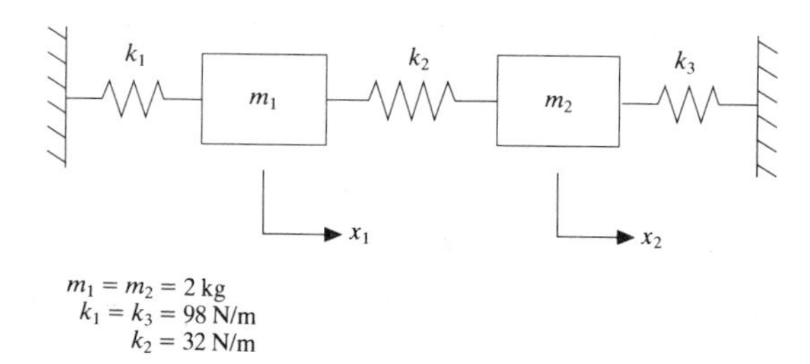

Figure 9.34

Figure 9.34 shows a symmetrical two mass three spring system. The equations of motion using SI units are

$$-98x_1 + 32(x_2 - x_1) = 2\ddot{x}_1$$

and $$-98x_2 + 32(x_1 - x_2) = 2\ddot{x}_2$$

or $$\begin{bmatrix} 2 & 0 \\ 0 & 2 \end{bmatrix}\begin{bmatrix} \ddot{x}_1 \\ \ddot{x}_2 \end{bmatrix} + \begin{bmatrix} 130 & -32 \\ -32 & 130 \end{bmatrix}\begin{bmatrix} x_1 \\ x_2 \end{bmatrix} = \begin{bmatrix} 0 \\ 0 \end{bmatrix}$$

Substituting $x_1 = A\mathrm{e}^{\lambda t}$ and $x_2 = B\mathrm{e}^{\lambda t}$ we obtain

$$32B = (2\lambda^2 + 130)A$$
$$32A = (2\lambda^2 + 130)B$$

so $B/A = (2\lambda^2 + 130)/32 = 32/(2\lambda^2 + 130)$

yielding $(2\lambda^2 + 130)^2 = 32^2$

therefore $2\lambda^2 + 130 = \pm 32$

so $\lambda^2 = -49$ or -81

and $\lambda = \pm j7$ or $\pm j9$

hence $B/A = +1$ or -1.

The general solution is

$$x_1 = [E\cos(7t) + F\sin(7t)] + (+1)[G\cos(9t) + H\sin(9t)]$$
$$x_2 = [E\cos(7t) + F\sin(7t)] + (-1)[G\cos(9t) + H\sin(9t)]$$

Let us assume that the system starts from rest with initial displacement of $x_1 = 0$ and $x_2 = 0.01$ m to the right. Differentiating the expressions for displacement we obtain the velocities

$$\dot{x}_1 = 7[E\sin(7t) + F\cos(7t)] + (+9)[G\sin(9t) + H\cos(9t)]$$
$$\dot{x}_2 = 7[E\sin(7t) + F\cos(7t)] + (-9)[G\sin(9t) + H\cos(9t)]$$

Substituting for time $t = 0$

$$x_1 = 0 = E + G \quad \text{(a)}$$
$$\ddot{x}_1 = 0 = 7F + 9H \quad \text{(b)}$$
$$x_2 = 0.01 = E - G \quad \text{(c)}$$
$$\dot{x}_2 = 0 = 7F - 9H \quad \text{(d)}$$

From (b) and (d) $F = H = 0$ and from (a) and (b) we obtain

$$E = 0.005 \text{ and } G = -0.005$$

so finally

$$x_1 = 0.005\cos(7t) + 0.005\cos(9t)$$
$$x_2 = 0.005\cos(7t) - 0.005\cos(9t)$$

These two equations may be written in matrix form as

$$\begin{bmatrix} x_1 \\ x_2 \end{bmatrix} = \begin{bmatrix} 0.005 \\ 0.005 \end{bmatrix}\cos(7t) + \begin{bmatrix} +0.005 \\ -0.005 \end{bmatrix}\cos(9t)$$

from which we can see that the motion is the sum of an in-phase motion of equal amplitudes at a frequency of $7/(2\pi)$ Hz plus an out-of-phase motion with equal amplitudes at a frequency of $9/(2\pi)$ Hz.

The principal frequencies and mode shapes are shown in Fig. 9.35.

9.19 Coupling of co-ordinates

Let us return to the first system shown in Fig. 9.32. Instead of using the displacement of the masses as co-ordinates we are now going to use the deflection of the springs denoting the extension of the left spring by x and the extension of the right spring by y. The motions of the masses will be x and $(x + y)$ respectively.

The equations of motion are now

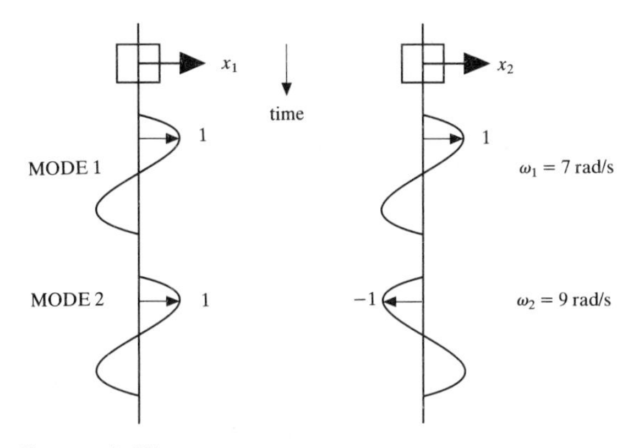

Figure 9.35

$$-k_1 x + k_2 y = m_1(\ddot{x}) \tag{9.73}$$

and
$$-k_2 y = m_2(\ddot{x} + \ddot{y}) \tag{9.74}$$

In comparison with the previous co-ordinates

$$x = x_1 \quad \text{and} \quad y = (x_2 - x_1).$$

If we look at the matrix equation 9.69 we see that both the square matrices are symmetrical but if we form the matrix equation from 9.73 and 9.74 this will not be so. It can be shown that by re-arranging the equations the mass and the stiffness matrices can both become symmetrical. In this case we can achieve this end by adding equation 9.73 to equation 9.74 to give equation 9.75 and forming the matrix expression from equations 9.75 and 9.73.

Thus
$$-k_1 x = (m_1 + m_2)\ddot{x} + m_2 \ddot{y} \tag{9.75}$$

and
$$\begin{bmatrix} m_1 + m_2 & m_2 \\ m_2 & m_2 \end{bmatrix} \begin{bmatrix} \ddot{x} \\ \ddot{y} \end{bmatrix} + \begin{bmatrix} k_1 & 0 \\ 0 & k_2 \end{bmatrix} \begin{bmatrix} x \\ y \end{bmatrix} = \begin{bmatrix} 0 \\ 0 \end{bmatrix} \tag{9.76}$$

It should be noted that the numerical method used for the previous example does not require the matrices to be in symmetrical form.

If there are no off-diagonal terms in the stiffness matrix then the co-ordinates are said to be statically uncoupled. Similarly, as in equation 9.69, if there are no off-diagonal terms in the mass matrix then the co-ordinates are said to be dynamically uncoupled. It is important to note that it is the *co-ordinates* which are coupled or uncoupled and coupling therefore depends only on the choice of co-ordinates and is not a function of the system.

Clearly it would be advantageous if co-ordinates could be chosen such that uncoupling occurs in both mass and stiffness matrices, in which case we would be left with two single-degree-of-freedom equations and the solution would be simple.

9.20 Normal modes

Since any motion of our system can be represented by the addition of the principal modes it is possible to write

$$\begin{bmatrix} x_1 \\ x_2 \end{bmatrix} = \begin{bmatrix} 1 & 1 \\ \mu_1 & \mu_2 \end{bmatrix} \begin{bmatrix} \eta_1 \\ \eta_2 \end{bmatrix} \tag{9.77}$$

or
$$x_1 = \eta_1(t) + \eta_2(t)$$
$$x_2 = \mu_1 \eta_1(t) + \mu_2 \eta_2(t)$$

The square matrix is known as the modal matrix since each column is a mode shape. The new time dependent variables η_1, η_2 are known as the principal co-ordinates. We will write equation 9.77 in a compact form as

$$(\boldsymbol{x}) = [\boldsymbol{A}](\boldsymbol{\eta})$$

and the equation of motion as

$$[\boldsymbol{m}](\ddot{\boldsymbol{x}}) + [\boldsymbol{k}](\boldsymbol{x}) = (\boldsymbol{0})$$

Substituting equation 9.77 into the equation of motion gives

$$[\boldsymbol{m}][\boldsymbol{A}](\ddot{\boldsymbol{\eta}}) + [\boldsymbol{k}][\boldsymbol{A}](\boldsymbol{\eta}) = (\boldsymbol{0})$$

The modified matrices will not be symmetrical so we now pre-multiply by $[A]^{\mathrm{T}}$, the transpose of the modal matrix, giving

$$[\boldsymbol{A}]^{\mathrm{T}}[\boldsymbol{m}][\boldsymbol{A}][\ddot{\boldsymbol{\eta}}][\boldsymbol{A}]^{\mathrm{T}} + [\boldsymbol{k}][\boldsymbol{A}][\boldsymbol{\eta}] = [0]$$

We next prove that the modified mass and stiffness matrices are diagonal.

9.21 Principle of orthogonality

Each of the principal modes is a solution to the equation of motion so it follows that

$$\lambda_1^2[\boldsymbol{m}](\boldsymbol{A}_1) + [\boldsymbol{k}](\boldsymbol{A}_1) = (\boldsymbol{0}) \tag{9.78}$$
$$\lambda_2^2[\boldsymbol{m}](\boldsymbol{A}_2) + [\boldsymbol{k}](\boldsymbol{A}_2) = (\boldsymbol{0}) \tag{9.79}$$

If we pre-multiply equation 9.78 by $(A_2)^{\mathrm{T}}$ and equation 9.79 by $(\boldsymbol{A}_1)^{\mathrm{T}}$ we get

$$\lambda_1^2(\boldsymbol{A}_2)^{\mathrm{T}}[\boldsymbol{m}](\boldsymbol{A}_1) + (\boldsymbol{A}_2)^{\mathrm{T}}[\boldsymbol{k}](\boldsymbol{A}_1) = 0 \tag{9.80}$$
$$\lambda_2^2(\boldsymbol{A}_1)^{\mathrm{T}}[\boldsymbol{m}](\boldsymbol{A}_2) + (\boldsymbol{A}_1)^{\mathrm{T}}[\boldsymbol{k}](\boldsymbol{A}_2) = 0 \tag{9.81}$$

Each of the terms in the above two equations is a scalar so must equal its own transpose. The transpose of a product is a product of transposed matrices taken in reverse order. Therefore the second terms in the equations are identical, providing that the mass and stiffness matrices are symmetrical. Similarly the first terms are the same apart from the different λ.

Subtracting equation 9.80 from 9.81 gives

$$(\lambda_2^2 - \lambda_1^2)(A_1)[m](A_2) = 0.$$

Since, in general, λ_2 does not equal λ_1, it follows that

$$(A_1)[m](A_2) = 0$$

and $(A_1)[k](A_2) = 0$

If we apply this principle to the numerical example

where $[k] = \begin{bmatrix} 36 & -32 \\ -32 & 36 \end{bmatrix}$ and $[m] = \begin{bmatrix} 2 & 0 \\ 0 & 2 \end{bmatrix}$

the modal matrix $[A] = [(A_1)(A_2)] = \begin{bmatrix} 1 & 1 \\ 1 & -1 \end{bmatrix}$

so $[A]^T[k][A] = \begin{bmatrix} 196 & 0 \\ 0 & 324 \end{bmatrix}$

and $[A]^T[m][A] = \begin{bmatrix} 4 & 0 \\ 0 & 4 \end{bmatrix}$

The corresponding scalar equations are

$$4\ddot{\eta}_1 + 196\eta_2 = 0$$

and $4\ddot{\eta}_2 + 324\eta_2 = 0.$

The two natural frequencies are $\sqrt{(196/4)} = 7$ rad/s and $\sqrt{(324/4)} = 9$ rad/s.

As we have already calculated the natural frequencies there is no need to diagonalise both mass and stiffness matrices: either one will do. Further, because of the orthogonality principle, we know that all off-diagonal terms are zero so only the diagonal terms require to be calculated. Thus, in general, the ith mode modal mass is $(A_i)^T[m](A_i)$. For the problem just considered

$$M_1 = (1 \quad 1)\begin{bmatrix} 2 & 0 \\ 0 & 2 \end{bmatrix}\begin{bmatrix} 1 \\ 1 \end{bmatrix} = 4$$

$$M_2 = (1 \quad -1)\begin{bmatrix} 2 & 0 \\ 0 & 2 \end{bmatrix}\begin{bmatrix} 1 \\ -1 \end{bmatrix} = 4$$

We have shown the application of principal or normal mode analysis to a simple free system but the method is also applicable to forced systems as shown in the next section.

9.22 Forced vibration

Consider again the system shown in Fig. 9.32 but this time with a forcing term $f_1(t)$ applied to mass 1. The matrix form of the equations of motion are as before except for the non-zero term on the right.

$$\begin{bmatrix} m_1 & 0 \\ 0 & m_2 \end{bmatrix}\begin{bmatrix} \ddot{x}_1 \\ \ddot{x}_2 \end{bmatrix} + \begin{bmatrix} k_1+k_2 & -k_2 \\ -k_2 & k_2 \end{bmatrix}\begin{bmatrix} x_1 \\ x_2 \end{bmatrix} = \begin{bmatrix} f_1 \\ 0 \end{bmatrix}$$

For the general case where f_1 is an arbitrary function of time it is easiest to use the normal mode analysis. So the first step is to solve for the free vibration case, to obtain the natural frequencies and the modal matrix, and use this to uncouple the equations. Hence

$$[A]^T[m][A](\ddot{\eta}) + [A]^T[k][A](\eta) = [A]^T\begin{bmatrix} f_1 \\ 0 \end{bmatrix}$$

leading to

$$M_1\ddot{\eta}_1 + \omega_1^2 M_1\eta_1 = (A_1)^T(f_1 \ 0)^T$$
$$M_2\ddot{\eta}_2 + \omega_2^2 M_2\eta_2 = (A_2)^T(f_1 \ 0)^T$$

These equations can be solved for η_1 and η_2 and then the values of x_1 and x_2 can be found by use of the modal matrix.

If the forcing term is sinusoidal then the steady-state solution can be found directly from the equations of motion by assuming that the response is also sinusoidal and at the same frequency as the forcing function. If $f_1 = F_1\cos(\omega t)$ then we assume that

$$\begin{bmatrix} x_1 \\ x_2 \end{bmatrix} = \begin{bmatrix} X_1 \\ X_2 \end{bmatrix}\cos(\omega t)$$

where X_1 and X_2 are the unknown amplitudes.

Substitution into the equation of motion and dividing through by the time function we obtain

$$-\omega^2\begin{bmatrix} m_1 & 0 \\ 0 & m_2 \end{bmatrix}\begin{bmatrix} X_1 \\ X_2 \end{bmatrix} + \begin{bmatrix} k_1+k_2 & -k_2 \\ -k_2 & k_2 \end{bmatrix}\begin{bmatrix} X_1 \\ X_2 \end{bmatrix} = \begin{bmatrix} F_1 \\ 0 \end{bmatrix}$$

$$\begin{bmatrix} k_1+k_2-\omega^2 m_1 & -k_2 \\ -k_2 & k_2-\omega^2 m_2 \end{bmatrix}\begin{bmatrix} X_1 \\ X_2 \end{bmatrix} = \begin{bmatrix} F_1 \\ 0 \end{bmatrix}$$

This is a pair of simultaneous equations and can be solved for X_1 and X_2. In full we have

$$(k_1+k_2-\omega^2 m_1)X_1 + (-k_2)X_2 = F_1$$
$$(-k_2)X_1 + (k_2-\omega^2 m_2)X_2 = 0$$

Hence

$$X_1 = \frac{F_1(k_2 - \omega^2 m_2)}{\omega^4 m_1 m_2 - \omega^2(m_2 k_1 + m_2 k_2 + m_1 k_2) + k_1 k_2}$$

and

$$X_2 = \frac{F_1 k_2}{\omega^4 m_1 m_2 - \omega^2(m_2 k_1 + m_2 k_2 + m_1 k_2) + k_1 k_2}$$

Notice that the denominators are the same in both expressions and are identical to the left-hand side of the frequency equation 9.71 with λ^2 replaced by $-\omega^2$. It follows that when the forcing frequency equals either of the natural frequencies ω_1 or ω_2 the amplitudes become infinite. A plot of the amplitude versus frequency is shown in Fig. 9.36 from which it should be noted that a negative amplitude indicates that the displacement is out of phase with the force.

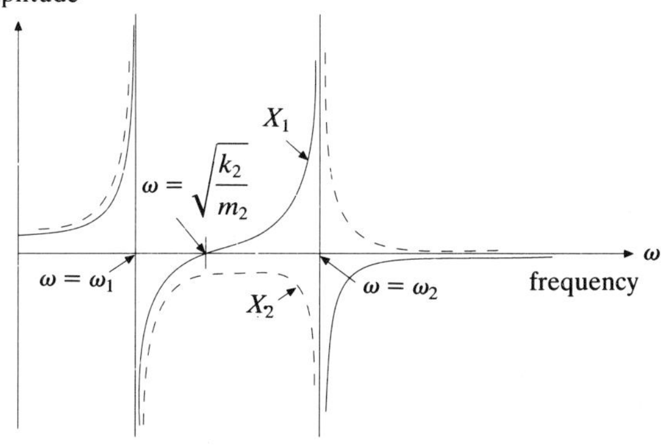

Figure 9.36

An interesting condition exists when ω equals $\sqrt{k_2/m_2}$. At this frequency the amplitude of the driven mass becomes zero, the second mass is acting as a *vibration absorber*.

In this section we have seen that a two-degree-of-freedom system has two natural frequencies and it follows that a system with n degrees of freedom will have n natural frequencies. Many of the analytical techniques introduced above can readily be applied to systems with three or more degrees of freedom – these are known as lumped parameter systems.

Continuous systems, such as vibrating beams or shafts, have an infinite number of degrees of freedom and can be shown to have an infinite number of discrete frequencies and mode shapes; they require different analytical techniques which are outside the scope of this book.

Discussion examples

Example 9.1

Find the natural frequency of small oscillations of the inverted simple pendulum shown in Fig. 9.37(a).

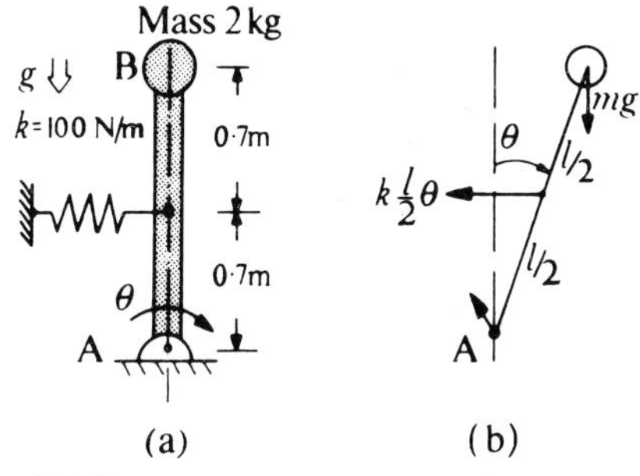

Figure 9.37

Solution As is usually the case, the motion is considered to consist of small oscillations about the static-equilibrium position. This implies that for small angles $\cos\theta \rightarrow 1$ and $\sin\theta \rightarrow \theta$, thus the spring force acting on the rod will be sensibly horizontal.

From the free-body diagram shown in Fig. 9.37(b) we have, by considering moments about A,

$$mgl\theta - k(l/2)\,\theta(l/2) = ml^2\ddot{\theta}$$

or $\quad \ddot{\theta} + (k/4m - g/l)\,\theta = 0 \qquad$ (i)

thus the natural frequency is given by

$$\nu_n = \frac{1}{2\pi}\sqrt{\left(\frac{k}{4m} - \frac{g}{l}\right)}$$

$$= \frac{1}{2\pi}\sqrt{\left(\frac{100}{4\times 2} - \frac{9.81}{1.4}\right)} = 0.37\text{ Hz} \qquad \text{(ii)}$$

Alternatively, an energy approach may be used:

gravitational potential energy $= mgl\cos\theta$
strain energy $= \frac{1}{2}kx^2$
kinetic energy $= \frac{1}{2}ml^2(\dot{\theta})^2$

$\therefore$ total energy
$E = mgl\cos\theta$
$+ \frac{1}{2}kx^2 + \frac{1}{2}ml^2(\dot{\theta})^2$

so $\quad dE/dt = -mgl\sin\theta\dot{\theta} + kx\dot{x} + ml^2\dot{\theta}\ddot{\theta} = 0$

For small angles,

$$x = \tfrac{1}{2}l\theta \quad \text{and} \quad \dot{x} = \tfrac{1}{2}l\dot{\theta}$$

thus $\quad -mgl\theta + k(l/2)^2\theta + ml^2\ddot{\theta} = 0$

or $\quad \ddot{\theta} + (k/4m - g/l)\theta = 0 \qquad \text{(iii)}$

Notice that in this method we cannot use the small-angle approximations until after the differentiation has been completed, otherwise the potential-energy term would have been lost.

Another interesting point is that if $k/4m < g/l$ the frequency becomes imaginary, in which case the solution of equation (iii) is

$$\theta = A\mathrm{e}^{\lambda t} + B\mathrm{e}^{-\lambda t}$$

where $\lambda = (g/l - k/4m)^{1/2}$. For example, if $\theta = \delta\theta$, $\dot{\theta} = 0$ when $t = 0$, then

$$\delta\theta = A + B$$
$$\dot{\theta} = 0 = \lambda A - \lambda B$$
$$\therefore \quad A = B = \tfrac{1}{2}\delta\theta$$

thus $\quad \theta = \tfrac{1}{2}\delta\theta(\mathrm{e}^{\lambda t} + \mathrm{e}^{-\lambda t}) = \delta\theta \cosh \lambda t$

From this result we see that for a small disturbance from the equilibrium position, θ increases with time showing that the system is unstable.

Example 9.2

A simple weighing machine is constructed as shown in Fig. 9.38(a). The beam and scale-pan together form a rigid body which has a moment of inertia about the pivot of 0.084 kg m^2. The spring stiffness is 7000 N/m.

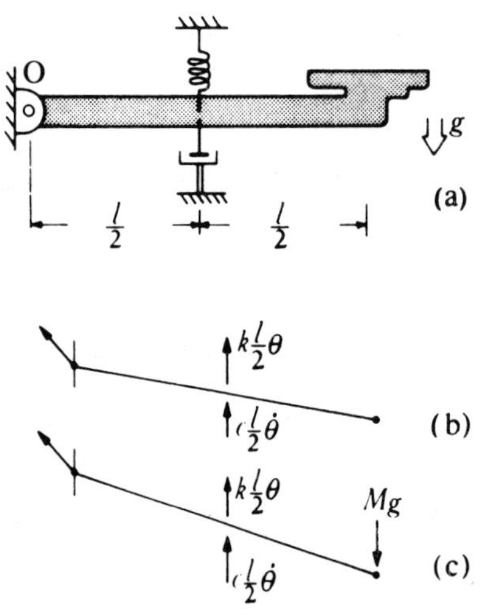

Figure 9.38

a) Find the force per unit velocity for a viscous damper placed in the position indicated such that the motion is just aperiodic when there is no load on the scale-pan.

b) With the system as specified in part (a), a mass of 4.5 kg is placed in contact with the scale-pan and is then released. Find the maximum deflection of the beam. Assume that the system performs small oscillations.

Solution In this example the effect of gravity, for small displacements, is merely to change the datum position. Hence we have the choice of measuring the deflection either from the position of zero strain in the spring or from the static-equilibrium configuration. Using the latter choice, the free-body diagram (Fig. 9.38(b)) shows only the forces additional to the self-balancing static set.

Taking moments about O gives

$$-k\tfrac{1}{2}l\theta\tfrac{1}{2}l - c\tfrac{1}{2}l\dot{\theta}\tfrac{1}{2}l = I_O\ddot{\theta}$$

or $\quad \ddot{\theta} + [cl^2/(4I_0)]\dot{\theta} + [kl^2/(4I_0)]\theta = 0 \qquad \text{(i)}$

From equation 9.21 we see that the critical value of $(cl^2/4I_O)$ is $2\surd(kl^2/4I_O)$

or $\quad c_{\text{crit.}} = c = 8\dfrac{I_O}{l^2}\sqrt{\left(\dfrac{kl^2}{4I_O}\right)}$

$$= 4\sqrt{\left(\frac{kI_O}{l^2}\right)} = 4\sqrt{\left(\frac{7000 \times 0.084}{0.300^2}\right)}$$

$$= 323 \text{ N s/m}$$

When a mass is placed on the scale-pan it has two effects: firstly the moment of inertia of the beam is increased and secondly the equilibrium position is altered, so once again we have a choice of datum position. Let us keep the same datum as used in part (a) of the example so that now $\theta = 0$ is not an equilibrium position. The free-body diagram is as shown in Fig. 9.38(c).

The equation of motion is

$$Mgl - (kl^2/4)\theta - (cl^2/4)\dot{\theta} = (I_O + Ml^2)\ddot{\theta}$$

where the parallel-axes theorem has been used to determine the new moment of inertia about O. Hence

$$\ddot{\theta} + [cl^2/4(I_O + Ml^2)]\dot{\theta} + [kl^2/4(I_O + Ml^2)]\theta = Mgl/(I_O + Ml^2) \qquad \text{(ii)}$$

but $\quad \dfrac{c(l/2)^2}{I_O + Ml^2} = \dfrac{323 \times 0.150^2}{0.084 + 4.5 \times 0.3^2} = 14.86 = 2\zeta\omega_n$

and $\dfrac{k(l/2)^2}{I_O + Ml^2} = \dfrac{7000 \times 0.15^2}{0.084 + 4.5 \times 0.3^2} = 322.1 \equiv \omega_n^{\,2}$

thus $\nu_n = \dfrac{1}{2\pi}\sqrt{322.1} = 2.86$ Hz

The damping ratio $\zeta = 14.86/(2\sqrt{322.1}) = 0.414$.

The particular integral of equation (ii) is the steady-state solution, which is just the final equilibrium position, and is, by inspection,

$$\theta_{ss} = \frac{Mgl}{k(l/2)^2} = \frac{4.5 \times 9.81 \times 0.3}{7000 \times 0.150^2}$$
$$= 0.084 \text{ rad} = 4.82°$$

The general solution is

$$\theta = e^{-\zeta\omega_n t}(A\cos\omega_d t + B\sin\omega_d t) + \theta_{ss}$$

The initial conditions are $t = 0$, $\theta = -\theta_{ss}$, $\dot{\theta} = 0$; hence

$$0 = A + \theta_{ss}$$

Now $\dot{\theta} = e^{-\omega\zeta_n t}\{(-\zeta\omega_n A + B\omega_d)\cos\omega_d t + (-\zeta\omega_n B - A\omega_d)\sin\omega_d t\}$

thus $0 = -\zeta\omega_n A + B\omega_d$

so that $A = -\theta_{ss}$ and $B = -\zeta\omega_n\theta_{ss}/\omega_d$

The maximum value of θ will occur when $\dot{\theta} = 0$,

i.e. $\dot{\theta} = e^{-\zeta\omega_n t}\theta_{ss}(\zeta^2\omega_n^{\,2}/\omega_d + \omega_d)\sin\omega_d t = 0$

$\dot{\theta}$ will be zero when $\sin\omega_d t = 0$, and so the first maximum will be when $\omega_d t = \pi$.

Hence $\theta_{max.} = e^{-\zeta\omega_n\pi/\omega_d}(-\theta_{ss}\cos\pi) + \theta_{ss}$
$= \theta_{ss}[e^{-\zeta\omega_n\pi/\omega_d} + 1]$

so

$\theta_m = 4.82°\{\exp[-\pi 0.414/\sqrt{(1 - 0.414^2)}] + 1\}$
$= 1.15° + 4.82° = 5.97°$

Example 9.3

The basic construction of a vibration velocity transducer is shown in Fig. 9.39(a). The mass of 0.042 kg is suspended by a spring having a stiffness of 52.9 N/m and the viscous damping force is 0.707 of the critical value. If the transducer is vibrating with s.h.m., derive an expression for the ratio

$$\frac{\text{velocity amplitude of the mass relative to the case}}{\text{velocity amplitude of the base}}$$

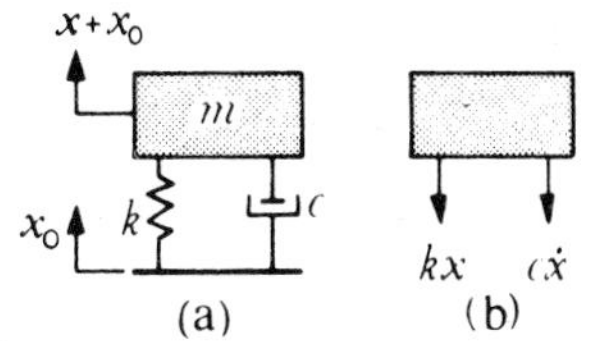

Figure 9.39

Over which frequency range may the instrument be used if the above ratio is to be in the range $1.0 \pm 10\%$?

Solution This type of transducer is designed to operate at frequencies higher than the natural frequency of the internal spring–mass system such that, in the operating range, the mass is sensibly stationary. The output signal, which is proportional to the velocity of the mass relative to the case, is closely proportional to the velocity of the case.

In the free-body diagram, Fig. 9.39(b), x is the displacement of the mass relative to the case and x_0 is the displacement of both the case and the object to which the transducer is attached.

The equation of motion is

$$-kx - c\dot{x} = m(\ddot{x}_0 + \ddot{x})$$

or $\quad m\ddot{x} + c\dot{x} + kx = -m\ddot{x}_0 \qquad$ (i)

Given that

$$x_0 = X_0\cos\omega t = \text{Re}(X_0 e^{j\omega t})$$

the steady-state solution is assumed to be

$$x = X\cos(\omega t - \phi) = \text{Re}(Xe^{j(\omega t - \phi)})$$

Substitution into equation (i) gives

$$(-m\omega^2 X + cj\omega X + kx)e^{j(\omega t - \phi)} = \omega^2 m X_0 e^{j\omega t}$$

thus $\quad kX + jc\omega X - m\omega^2 X = mX_0\omega^2 e^{j\phi}$

which is shown on an Agrand diagram in Fig. 9.40.

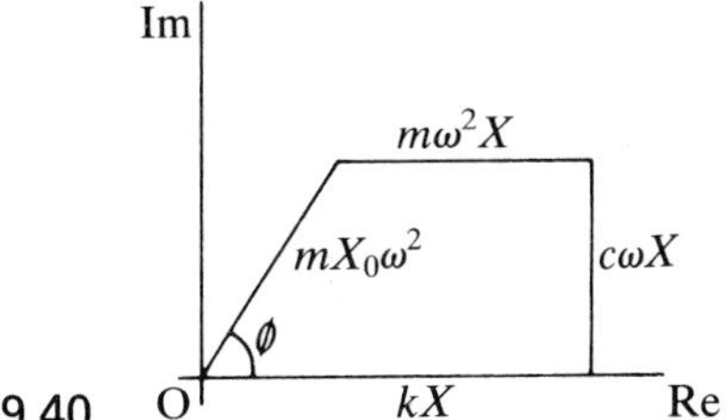

Figure 9.40

From the diagram we see that

$$m\omega^2 X_0 = X\sqrt{[(k - m\omega^2)^2 + c^2\omega^2]}$$

The ratio required is

$$\frac{\omega X}{\omega X_0} = \frac{X}{X_0} = \frac{m\omega^2}{\sqrt{[(k - m\omega^2)^2 + c^2\omega^2]}}$$

but $k/m = \omega_n^2 = 52.5/0.042 = 1250$

and $c/m = 2\zeta\sqrt{(km)}/m = 2\zeta\omega_n$

$$\text{thus } \frac{X}{X_0} = \frac{\omega^2}{\sqrt{[(\omega_n^2 - \omega^2)^2 + 4\zeta^2\omega_n^2\omega^2]}}$$

$$= \frac{(\omega/\omega_n)^2}{\sqrt{\{[1 - (\omega/\omega_n)^2]^2 + 4\zeta^2(\omega/\omega_n)^2\}}}$$

With $\zeta = 1/\sqrt{2}$,

$$X/X_0 = (\omega/\omega_n)^2/\sqrt{[1 + (\omega/\omega_n)^4]}$$

which, by inspection, is less than unity, therefore putting $X/X_0 = 0.9$ gives

$$0.9^2[1 + (\omega/\omega_n)^4] = (\omega/\omega_n)^4$$

thus $\omega = 50.867$

and $\nu = 50.867/2\pi = 8.1$ Hz

Thus, above 8.1 Hz the response is such that $0.9 < \omega X/\omega X_0 < 1.0$.

The same mechanical arrangement is suitable for an accelerometer, but the transducer is now used below the natural frequency so that the mass and case have small relative movement. The output is made proportional to the strain in the spring which, in an ideal situation, is proportional to the acceleration of the instrument. The sensitivity of the accelerometer is proportional to X/ω^2X_0 where

$$\frac{X}{X_0}\left(\frac{\omega_n}{\omega}\right)^2 = \frac{1}{\sqrt{\{[1 - (\omega/\omega_n)^2]^2 + 4\zeta^2(\omega/\omega_n)^2\}}}$$

see Fig. 9.23.

It is interesting to note that in this transducer the choice of $\zeta = 1/\sqrt{2}$ has the added advantage of delaying the output relative to the input by a constant time lag, due to the fact that the phase angle is nearly proportional to the frequency (see Fig. 9.24).

If the output is $x = X\cos(\omega t - \phi)$, then if $\phi = a\omega$, where a is a constant, we have

$$x = X\cos\omega(t - a)$$

hence a is a time lag, which is independent of frequency. This implies that if the waveform is not sinusoidal no distortion of the waveform will take place since all the harmonics, within the working range, will be delayed by the same time lag. Thus the output waveform will be that of the input.

Example 9.4

A rack of electronic instruments (Fig. 9.41(a)) is isolated from a vibrating floor by four antivibration mountings, each with the same stiffness and damping coefficient.

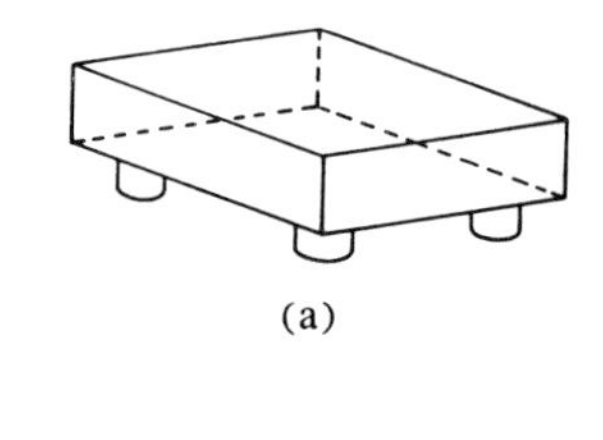

(a)

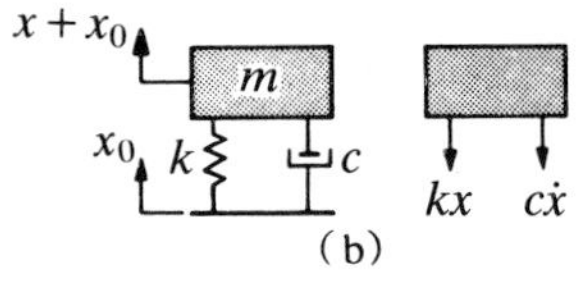

(b)

Figure 9.41

A free-vibration test is carried out and it is recorded that the amplitude of vertical vibrations drops by 90% in 4 cycles or 1.5 seconds. If the floor is vibrating vertically what will be the amplitude of vibration of the rack (a) when the frequency is 3 Hz and the amplitude is 1 mm, (b) when the frequency is 30 Hz and the amplitude is 0.1 mm?

Solution The characteristics of the system are determined by the natural frequency and the damping ratio, both of which may be found from the free-vibration trace.

From the definition of log. dec.,

$$\delta = \ln(X_1/X_2) = \ln(X_2/X_3) \quad \text{etc.}$$

but $\ln(X_1/X_4) = \ln(X_1X_2)(X_2/X_3)(X_3/X_4) = 3\delta$

thus $\delta = \frac{1}{3}\ln(10/1) = 0.767$

From equation 9.27,

$$\delta = 2\pi\zeta/\sqrt{(1 - \zeta^2)}$$

Since ζ is small,

$$\zeta = \delta/2\pi = 0.767/2\pi = 0.122$$

The damped natural frequency is $4/1.5 = 2.67$

Hz and, since ζ is small, $\omega_n \approx \omega_d = 2.67 \times 2\pi = 8.38\ \text{s}^{-1}$.

The equation of motion for the rack (see Fig. 9.41(b)) is

$$-kx - c\dot{x} = m(\ddot{x}_0 + \ddot{x})$$

or $\quad \ddot{x} + 2\zeta\omega_n\dot{x} + \omega_n^2 x = -\ddot{x}_0 \qquad$ (i)

where x is the movement of the rack relative to the base.

If the movement of the base is $x_0 = X_0 e^{j\omega t}$, then we assume that $x = Xe^{j(\omega t - \phi)}$, which, when substituted into (i), gives

$$-\omega^2 X + 2\zeta\omega_n j\omega X + \omega_n^2 X = X_0 e^{j\phi} \qquad \text{(ii)}$$

The Argand diagram is shown in Fig. 9.42.

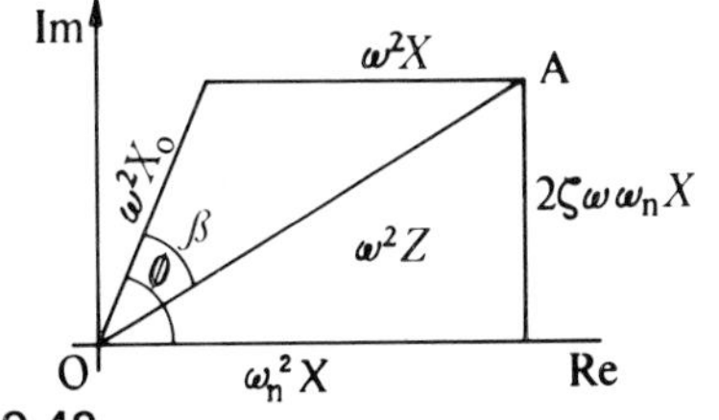

Figure 9.42

We require the absolute motion of the rack, $z = x + x_0$. Letting

$$z = Ze^{j(\omega t - \beta)}$$

then $\quad ze^{j(\omega t - \beta)} = Xe^{j(\omega t - \phi)} + Xe^{j\omega t}$

or $\quad Ze^{-j\beta} = Xe^{-j\phi} + X_0$

thus $\quad \omega^2 Ze^{j(\phi - \beta)} = \omega^2 X + \omega^2 X_0^{j\phi}$

so on the diagram $\omega^2 Z$ is the vector $\overrightarrow{OA}$.

The diagram yields

$$\omega^2 Z = X\sqrt{[\omega_n^4 + (2\zeta\omega_n\omega)^2]}$$

but $\quad \omega^2 X_0 = X\sqrt{[(\omega_n^2 - \omega^2)^2 + (2\zeta\omega_n\omega)^2]}$

thus
$$\frac{Z}{X_0} = \sqrt{\left(\frac{\omega_n^4 + (2\zeta\omega_n\omega)^2}{(\omega_n^2 - \omega^2)^2 + (2\zeta\omega_n\omega)^2}\right)}$$
$$= \sqrt{\left(\frac{1 + 4\zeta^2(\omega/\omega_n)^2}{[1 - (\omega/\omega_n)^2]^2 + 4\zeta^2(\omega/\omega_n)^2}\right)}$$

This equation is identical to equation 9.59 and therefore gives the transmissibility.

For part (a) of the example,

$$\omega/\omega_n = \nu/\nu_n = 3/2.67 = 1.125 \text{ and } \zeta = 0.122$$

so
$$\frac{Z}{X_0} = \sqrt{\left[\frac{1 + 4 \times 0.122^2 \times 1.125^2}{(1 - 1.125^2)^2 + 4 \times 0.122^2 \times 1.125^2}\right]}$$
$$= 2.71$$

thus the amplitude of motion of the rack is $2.71 \times 1 = 2.71$ mm.

For part (b),

$$\frac{Z}{X_0} = \sqrt{\left[\frac{1 + 4 \times 0.122^2 \times 11.25^2}{(1 - 11.25^2)^2 + 4 \times 0.122^2 \times 11.25^2}\right]}$$
$$= 0.023$$

thus $\quad Z = 0.023 \times 0.1 = 0.0023$ mm

Example 9.6

Figure 9.43(a) shows a light elastic structure in the form of a quadrant which is supporting an object which may be modelled as a concentrated mass of 16 kg. Static tests on the structure at A were carried out to determine the elastic characteristics. The results of the tests are as follows.

A steady force of 100 N in the y direction produced a deflection of 19.6 mm in the y direction and 12.5 mm in the x direction. A steady force of 100 N in the x direction produced a deflection of 8.9 mm in the x direction and 12.5 mm in the y direction.

Determine the natural frequencies of free vibration in the xy plane and the associated mode shapes.

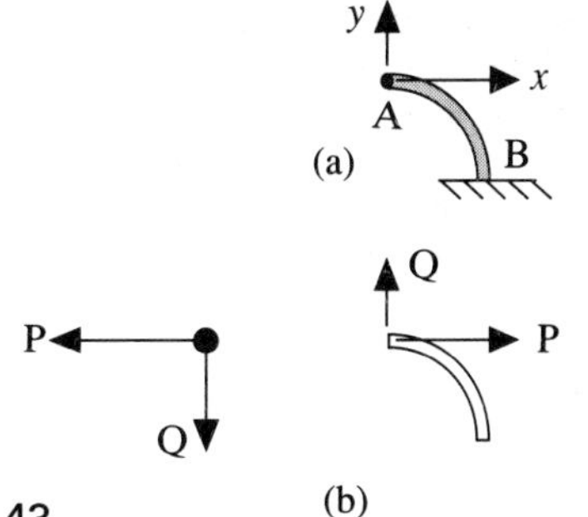

Figure 9.43

Solution In the free-body diagrams, Fig. 9.43(b), P and Q are the forces (in Newtons) acting on the structure in the x and y directions respectively, so from the test data the deflections in the x and y directions will be

$$x = (0.089P + 0.125Q)10^{-3}\ \text{m} \qquad \text{(i)}$$
$$y = (0.125P + 0.196Q)10^{-3}\ \text{m} \qquad \text{(ii)}$$

For the mass

$$-P = 16\ddot{x}$$

and $\quad -Q = 16\ddot{y}$

Substituting these values of P and Q into

equations (i) and (ii) gives

$$x = -16\times 10^{-6}(89\ddot{x}+125\ddot{y})$$
$$y = -16\times 10^{-6}(125\ddot{x}+196\ddot{y})$$

We now assume that $x = Xe^{\lambda t}$ and $y = Ye^{\lambda t}$ from which $\ddot{x} = X\lambda^2 e^{\lambda t}$ and $\ddot{y} = X\lambda^2 e^{\lambda t}$. Substitution gives

$$X(1+89\lambda^2 16\times 10^{-6}) + Y125\lambda^2 16\times 10^{-6} = 0 \quad \text{(iii)}$$
$$X125\lambda^2 16\times 10^{-6} + Y(1+196\lambda^2 16\times 10^{-6}) = 0 \quad \text{(iv)}$$

For ease of calculation let $\Lambda^2 = \lambda^2 16\times 10^{-6}$ so that from equation (iii) we have

$$X/Y = -125\Lambda^2/(1+89\Lambda^2) \quad \text{(v)}$$

and from (iv)

$$X/Y = -(1+196\Lambda^2)/125\Lambda^2 \quad \text{(vi)}$$

Equating the two expressions for X/Y given by (v) and (vi) we have

$$(125\Lambda^2)^2 = (1+89\Lambda^2)(1+196\Lambda^2)$$

and this leads to a quadratic equation in Λ^2

$$(196\times 89 - 125^2)\Lambda^4 + (196+89)\Lambda^2 + 1 = 0$$

or $\quad 1819\Lambda^4 + 285\Lambda^2 + 1 = 0$

The roots of this equation are $\Lambda^2 = -0.00359$ and -0.153 so $\lambda^2 = -3590/16$ and $-153000/16$ therefore $\lambda = \pm j14.98$ and $\pm j97.79$.

This means that the circular frequencies ($\omega = j\lambda$) are 14.98 and 97.79 ($\omega^2 = 224$ and 9563) from which the frequencies ($\omega/(2\pi)$) are

2.38 Hz and 15.56 Hz.

From equation (vi) (or we could have used (v)) we obtain two values for X/Y, one for $\Lambda^2 = -0.00359$ and one for $\Lambda^2 = -0.153$

so $\quad X/Y = -(1-196\times 0.00359)/(125\times 0.00359)$
$= -0.660$
and $\quad = -(1-196\times 0.153)/(125\times 0.153)$
$= +1.516$

To visualise the meaning of mode shape in this example the direction of motion for the two modes is plotted in Fig. 9.44. From this figure it is clear that the directions are at right angles to each other: that is they are orthogonal. This is also demonstrated by the condition for two lines to be normal to each other that is, the product of their gradients shall be -1. In this case $(-0.660)\times(1.516) = -1.00$.

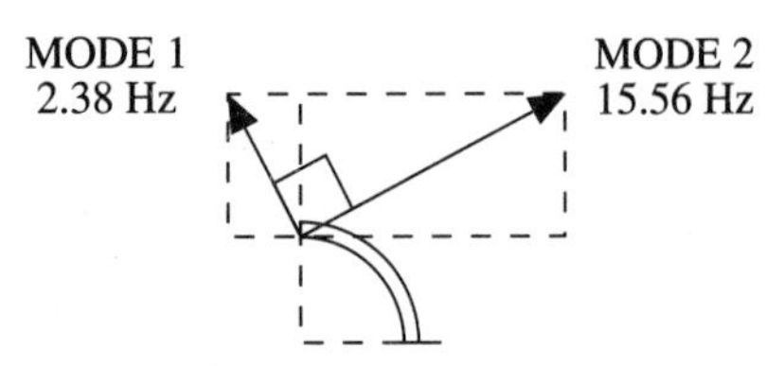

Figure 9.44

It is helpful to re-order the data in standard matrix format. The structural data may be written as

$$\begin{bmatrix} x \\ y \end{bmatrix} = 10^{-6}\begin{bmatrix} 89 & 125 \\ 125 & 196 \end{bmatrix}\begin{bmatrix} P \\ Q \end{bmatrix}$$

inverting this equation gives

$$\begin{bmatrix} P \\ Q \end{bmatrix} = \begin{bmatrix} 107751 & -68719 \\ -68719 & 48928 \end{bmatrix}\begin{bmatrix} x \\ y \end{bmatrix}$$

From the free-body diagram

$$\begin{bmatrix} -P \\ -Q \end{bmatrix} = \begin{bmatrix} 16 & 0 \\ 0 & 16 \end{bmatrix}\begin{bmatrix} \ddot{x} \\ \ddot{y} \end{bmatrix}$$

therefore by eliminating P and Q

$$\begin{bmatrix} 16 & 0 \\ 0 & 16 \end{bmatrix}\begin{bmatrix} \ddot{x} \\ \ddot{y} \end{bmatrix} + \begin{bmatrix} 107751 & -68719 \\ -68719 & 48928 \end{bmatrix}\begin{bmatrix} x \\ y \end{bmatrix} = \begin{bmatrix} 0 \\ 0 \end{bmatrix}$$

The matrix equation is solved by assuming that

$$\begin{bmatrix} x \\ y \end{bmatrix} = e^{\lambda t}\begin{bmatrix} X \\ Y \end{bmatrix}$$

and substituting into the previous equation gives

$$\begin{bmatrix} 16\lambda^2 + 107751 & -68719 \\ -68719 & 16\lambda^2 + 48928 \end{bmatrix}\begin{bmatrix} X \\ Y \end{bmatrix} = \begin{bmatrix} 0 \\ 0 \end{bmatrix}$$

Putting the determinant of the square matrix equal to zero yields the same quadratic in λ^2 as before and hence the same natural frequencies. The mode shapes are found from one of the two scalar equations which form the matrix equation.

The modal matrix with the x deflection taken as unity is used to transform to the principal co-ordinates η_1 and η_2

$$\begin{bmatrix} x \\ y \end{bmatrix} = \begin{bmatrix} 1 & 1 \\ -0.660 & 1.516 \end{bmatrix}\begin{bmatrix} \eta_1 \\ \eta_2 \end{bmatrix}$$

Transforming the mass matrix

$$\begin{bmatrix} 1 & -0.660 \\ 1 & 1.516 \end{bmatrix}\begin{bmatrix} 16 & 0 \\ 0 & 16 \end{bmatrix}\begin{bmatrix} 1 & 1 \\ -0.660 & 1.516 \end{bmatrix}$$

$$= \begin{bmatrix} 23.0 & 0.0 \\ 0.0 & 52.3 \end{bmatrix}$$

As a check we will transform the stiffness matrix

$$\begin{bmatrix} 1 & -0.660 \\ 1 & 1.516 \end{bmatrix} \begin{bmatrix} 107751 & -68719 \\ -68719 & 48928 \end{bmatrix} \begin{bmatrix} 1 & 1 \\ -0.660 & 1.516 \end{bmatrix}$$

$$= \begin{bmatrix} 220 & 0 \\ 0 & 11.8 \end{bmatrix} 10^3$$

The two uncoupled scalar equations are

$$23.0\ddot{\eta}_1 + 220 \times 10^3 \eta_1 = 0$$

and $52.3\ddot{\eta}_2 + 11.8 \times 10^3 \eta_2 = 0$

from which we find that $\omega_1{}^2 = 220 \times 10^3/23.0 = 9565$ (rad/s)^2 and $\omega_2{}^2 = 11.8 \times 10^3/52.3 = 226$ (rad/s)^2. These two values compare well with the values obtained previously, the small differences being due entirely to rounding errors.

Example 9.7

The system shown in Fig. 9.45 consists of two wheels connected by a flexible shaft. The whole assembly is free to rotate in two journal bearings. This arrangement is a model of an electric motor driving a load. The load, A, has a moment of inertia $I_A = 0.03$ kg m^2 and the moment of inertia of the rotor, B, is $I_B = 0.05$ kg m^2. The torsional stiffness of the shaft is 5000 Nm/rad.

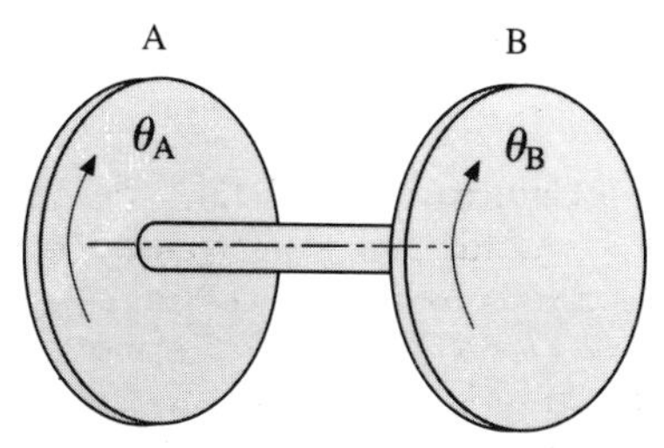

Figure 9.45

The system is initially at rest when a constant torque of 16 Nm is suddenly applied to the rotor, B. Derive an expression for the torque in the shaft.

Solution This example has a feature not specifically mentioned in the text, which is that one of the natural modes is that of rigid body rotation. This does not interfere with setting up the equations of motion and finding the natural frequencies, but the forcing term is best treated by the use of the normal mode method. Therefore we will start by considering the free vibration problem. However we shall set up the equations with the forcing term included for future convenience.

From the free-body diagrams shown in Fig. 9.46 we obtain, by moments about the axis of rotation, for wheel A

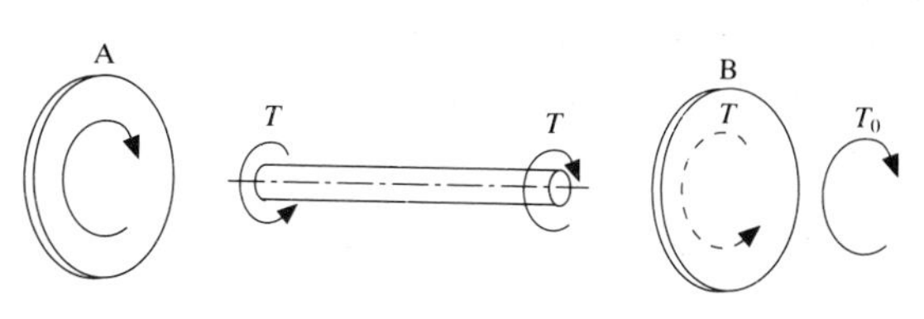

Figure 9.46

$$T = I_A \ddot{\theta}_A$$

and for wheel B

$$T_0 - T = I_B \ddot{\theta}_B$$

For the shaft we have $T = k(\theta_B - \theta_A)$ which when substituted into the two equations of motion lead to

$$0 = I_A \ddot{\theta}_A + k\theta_A - k\theta_B \qquad \text{(i)}$$

$$T_0 = -k\theta_A + k\theta_B + I_B \ddot{\theta}_B \qquad \text{(ii)}$$

or in matrix form

$$\begin{bmatrix} I_A & 0 \\ 0 & I_B \end{bmatrix} \begin{bmatrix} \ddot{\theta}_A \\ \ddot{\theta}_B \end{bmatrix} + \begin{bmatrix} k & -k \\ -k & k \end{bmatrix} \begin{bmatrix} \theta_A \\ \theta_B \end{bmatrix} = \begin{bmatrix} 0 \\ T_0 \end{bmatrix}$$

Assume $\begin{bmatrix} \theta_A \\ \theta_B \end{bmatrix} = \begin{bmatrix} \mathrm{A} \\ \mathrm{B} \end{bmatrix} e^{\lambda t}$ for the case when $T_0 = 0$

Substitution into the previous equation and dividing through by $e^{\lambda t}$ gives

$$\begin{bmatrix} I_A\lambda^2 + k & -k \\ -k & I_B\lambda^2 + k \end{bmatrix} \begin{bmatrix} \mathrm{A} \\ \mathrm{B} \end{bmatrix} = \begin{bmatrix} 0 \\ 0 \end{bmatrix}$$

The characteristic equation is given by putting the determinant of the square matrix equal to zero

$$(I_A\lambda^2 + k)(I_B\lambda^2 + k) - k^2 = 0$$

which leads to

$$I_A I_B \lambda^4 + (I_A + I_B)k\lambda^2 = 0$$

therefore

$$\lambda^2 = 0 \quad \text{or} \quad \lambda^2 = -\frac{(I_A + I_B)}{I_A I_B} k$$

So we see that one natural frequency is zero corresponding to the rigid body mode and the other frequency is

$$\frac{1}{2\pi}\sqrt{\frac{(I_A + I_B)}{I_A I_B} k}$$

The corresponding mode shapes are found from either of the two equations involving A and B, i.e.

$$\frac{B}{A} = \frac{I_A \lambda^2 + k}{k} = 1 \quad \text{or} \quad -I_A/I_B$$

The first value of unity shows that for the rigid body mode both wheels move in unison as expected. The second vibratory mode shows that the motions are out of phase.

Putting numbers to the expressions we have

First mode frequency = 0
mode shape B : A = 1 : 1

Second mode frequency

$$= \frac{1}{2\pi}\sqrt{\frac{0.03 + 0.05}{0.03 \times 0.05} 5000} = 82.19 \text{ Hz}$$

mode shape B : A = −0.003 : 0.005 = −3 : 5

In order to solve the forced vibration problem we shall in this case use modal analysis. We know that using the modal matrix to change to principal co-ordinates will enable us to reduce the coupled equations to two single degree of freedom equations.

The transformation relationship is

$$(\boldsymbol{\theta}) = [\boldsymbol{A}](\boldsymbol{\eta})$$

where the modal matrix $[A] = \begin{bmatrix} 1 & 1 \\ 1 & -3/5 \end{bmatrix}$

and we now use this to diagonalise the equations of motion

$$[\boldsymbol{M}](\ddot{\boldsymbol{\theta}}) + [\boldsymbol{K}](\boldsymbol{\theta}) = (\boldsymbol{F})$$
$$[\boldsymbol{A}]^T[\boldsymbol{M}][\boldsymbol{A}](\ddot{\boldsymbol{\eta}}) + [\boldsymbol{A}]^T[\boldsymbol{K}][\boldsymbol{A}](\boldsymbol{\eta}) = [\boldsymbol{A}]^T(\boldsymbol{F})$$

For the mass matrix

$$\begin{bmatrix} 1 & 1 \\ 1 & -3/5 \end{bmatrix}\begin{bmatrix} 0.03 & 0 \\ 0 & 0.05 \end{bmatrix}\begin{bmatrix} 1 & 1 \\ 1 & -3/5 \end{bmatrix}$$

$$= \begin{bmatrix} 0.08 & 0 \\ 0 & 0.048 \end{bmatrix}$$

Similarly the transformed stiffness matrix is

$$\begin{bmatrix} 0 & 0 \\ 0 & 12800 \end{bmatrix}$$

The right-hand side of the equation becomes

$$\begin{bmatrix} 1 & 1 \\ 1 & 3/5 \end{bmatrix}\begin{bmatrix} 0 \\ 16 \end{bmatrix} = \begin{bmatrix} 16 \\ 9.6 \end{bmatrix}$$

So the two uncoupled equations are now

$$0.08\ddot{\eta}_1 + 0 = 16$$
$$0.048\ddot{\eta}_2 + 12800\eta_2 = 9.6$$

or $\quad \ddot{\eta}_1 = 200$

and $\quad \ddot{\eta}_2 + 516.4^2\eta_2 = 200$

Now $\theta_A = \eta_1 + \eta_2$ and $\theta_B = \eta_1 - 0.6\eta_2$

so the torque in the shaft $T = k(\theta_B - \theta_A) = k(-1.6\eta_2)$ therefore we only need to solve for η_2.

The initial condition is a state of rest hence at $t = 0$, $\eta_2 = 0$ and $\dot{\eta}_2 = 0$. The complementary function plus the particular solution of the equation is

$$\eta_2 = G\cos(516.4t) + H\sin(516.4t) + 200/516.4^2$$

at $t = 0 \quad \eta_2 = 0$ therefore $G = -200/516.4^2$.

$$\dot{\eta}_2 = 516.4(-G\sin(516.4t) + H\cos(516.4t))$$

at $t = 0 \quad \dot{\eta}_2 = 0$ therefore $H = 0$

hence $\quad \eta_2 = 200/516.4^2(1 - \cos(516.4t))$

and finally the torque

$$T = -5000 \times 1.6 \times (200/516.4^2)(1 - \cos(516.4t))$$
$$\underline{T = -6.0(1 - \cos(516.4t)) \text{ Nm.}}$$

This method of solution is general and may be followed for more complex problems, however for this particular case, and with the benefit of hindsight the co-ordinates can be changed by inspection to remove the rigid body mode.

If we add equations (i) and (ii) we have

$$I_A\ddot{\theta}_A + I_B\ddot{\theta}_B = T_0$$

and if we subtract $I_B \times$ (i) from $I_A \times$ (ii)

$$I_A I_B(\ddot{\theta}_B - \ddot{\theta}_A) + k(I_A + I_B)(\theta_B - \theta_A) = I_A T_0$$

Let $\theta_B - \theta_A = \phi$, the twist in the shaft

thus $\ddot{\phi} + \dfrac{k(I_A + I_B)}{I_A I_B}\phi = \dfrac{T_0}{I_B}$

This equation can now be solved for ϕ in the same way as in the previous method, the torque in the shaft simply being $k\phi$.

Example 9.8

The trailer shown in Fig. 9.47(a) consists of a body supported by springs on an axle having two wheels. The mass of the body is 360 kg and the wheels plus axle assembly has a mass of 90 kg. The stiffness of the suspension is 72 kN/m and the stiffness of the tyres is 180 kN/m (both stiffnesses are total).

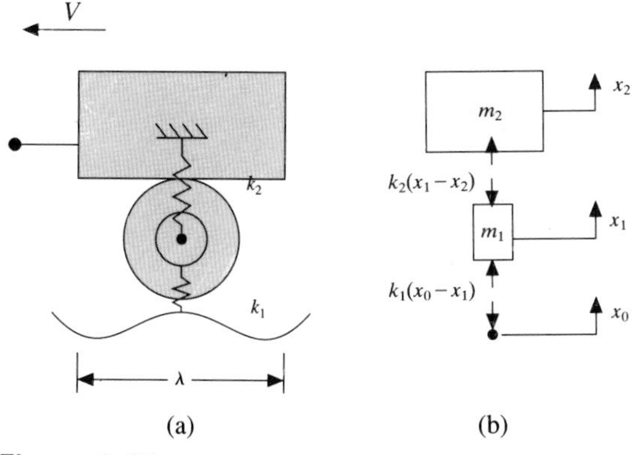

Figure 9.47

The trailer is drawn along a road which has a sinusoidal surface with a wavelength of 3 m and an amplitude of 10 mm. Determine the critical speeds of the trailer and the speed at which the axle will have no vertical motion. Calculate also the response at a steady 50 km/hr.

Solution Figure 9.47(b) shows the free-body diagrams for the system considering vertical motion and forces. The displacement at the bottom of the tyre is $x_0 = X_0 \cos(\omega t)$. The frequency ν depends on the speed on the trailer V and the wavelength λ. These quantities are related by $V = \lambda\nu$.

The equations of motion are

$$k_1(x_0 - x_1) - k_2(x_1 - x_2) = m_1\ddot{x}_1$$
$$k_2(x_1 - x_2) = m_2\ddot{x}_2$$

Now in complex form $x_0 = X_0 e^{j\omega t}$, therefore in matrix form the equations of motion are

$$\begin{bmatrix} m_1 & 0 \\ 0 & m_2 \end{bmatrix}\begin{bmatrix} \ddot{x}_1 \\ \ddot{x}_2 \end{bmatrix} + \begin{bmatrix} k_1 + k_2 & -k_2 \\ -k_2 & k_2 \end{bmatrix}\begin{bmatrix} x_1 \\ x_2 \end{bmatrix} = \begin{bmatrix} k_1 X_0 \\ 0 \end{bmatrix} e^{j\omega t} \quad \text{(i)}$$

The solution of this equation is in two parts, the complementary function which is the free vibration response and the particular integral or steady state response. Any real system will have sufficient inherent damping to ensure that the free vibration will die away leaving only the steady state motion.

For the steady state solution we assume

$$\begin{bmatrix} x_1 \\ x_2 \end{bmatrix} = \begin{bmatrix} X_1 \\ X_2 \end{bmatrix} e^{j\omega t}$$

Substituting into (i) and dividing through by the common factor $e^{j\omega t}$ we obtain the algebraic equations

$$\begin{bmatrix} -m_1\omega^2 + k_1 + k_2 & -k_2 \\ -k_2 & -m_2\omega^2 + k_2 \end{bmatrix}\begin{bmatrix} X_1 \\ X_2 \end{bmatrix} = \begin{bmatrix} k_1 X_0 \\ 0 \end{bmatrix}$$

Inserting the numerical values

$$\begin{bmatrix} -90\omega^2 + 252000 & -72000 \\ -72000 & -360\omega^2 + 72000 \end{bmatrix}\begin{bmatrix} X_1 \\ X_2 \end{bmatrix} = \begin{bmatrix} 1800 \\ 0 \end{bmatrix}$$

Evaluation of X_1 and X_2 may be achieved by pre-multiplying both sides of the equation by the inverse of the square matrix.

$$\begin{bmatrix} X_1 \\ X_2 \end{bmatrix} = \frac{1}{\Delta}\begin{bmatrix} 72000 - 360\omega^2 & 72000 \\ 72000 & 252000 - 90\omega^2 \end{bmatrix}\begin{bmatrix} 1800 \\ 0 \end{bmatrix} \quad \text{(ii)}$$

where $\Delta = (72000 - 360\omega^2)(252000 - 90\omega^2) - (72000)^2$

is the determinant of the square matrix.

It should be noticed that putting $\Delta = 0$ leads to a quadratic in ω^2 which is identical to the characteristic equation with $\lambda^2 = -\omega^2$. As might be expected the two frequencies for which $\Delta = 0$ are the natural frequencies of free vibration with $x_0 = 0$. At these frequencies the amplitudes tend to infinity and the system is said to resonate. Note that resonance simply means that the amplitude of vibration is a maximum.

Expanding the equation $\Delta = 0$ gives a quadratic in ω^2

$$360 \times 90\omega^4 - (72000 \times 90 + 252000 \times 360)\omega^2 + (72000 \times 252000 - 72000^2) = 0$$

$$32.4 \times 10^3 \omega^4 - 97.2 \times 10^6 \omega^2 + 12.96 \times 10^9 = 0$$

the roots of which are $\omega^2 = 140$ or 2860

The corresponding frequencies (ν) are

$$\sqrt{140}/(2\pi) = 1.88 \text{ Hz}$$

and $\sqrt{2860}/(2\pi) = 8.51$ Hz.

The corresponding road speeds are ($V = \lambda\nu$)

$$3 \times 1.88 = 5.64 \text{ m/s (20.3 km/hr)}$$

and $3 \times 8.51 = 25.53$ m/s (91.9 km/hr).

From equation (ii) we see that

$$X_1 = (72000 - 360\omega^2)1800/\Delta$$

and $X_2 = 72000 \times 1800/\Delta$

thus $X_1 = 0$ when

$$\omega^2 = 72000/360 = 200$$

or $\omega = 14.14$

so that frequency $= 14.14/(2\pi) = 2.25$ Hz the road speed being $3 \times 2.25 = 6.75$ m/s (24.3 km/hr).

At 50 km/hr $V = 50/3.6 = 13.9$ m/s and $\omega = (13.9/3)2\pi = 29.11$

so $\Delta = 32.4 \times 10^3(29.11)^4 - 97.2 \times 10^6(29.11)^2 + 12.96 \times 10^9 = -46 \times 10^9$

therefore

$$X_1 = (72000 - 360 \times 29.11^2)1800/\Delta = 9.12 \text{ mm}$$

and $X_2 = 72000 \times 1800/\Delta = -2.82$ mm

The amplitude of compression of the tyre $= X_0 - X_1 = 10 - 9.12 = 0.88$ mm and the compression of the spring $= X_1 - X_2 = 9.12 - (-2.82) = 11.94$ mm.

Problems

9.1 Determine the frequency of oscillation of the systems shown in Figs 9.48(a) and (b).

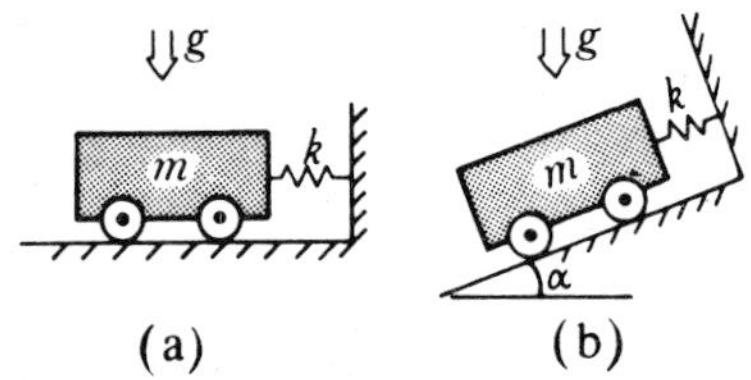

Figure 9.48

9.2 A body of mass m hangs from a hinged support at O. The centre of mass G is at a distance a from O. When the body is given a slight disturbance from the equilibrium position, it oscillates with a frequency ν. Determine the moment of inertia about O.

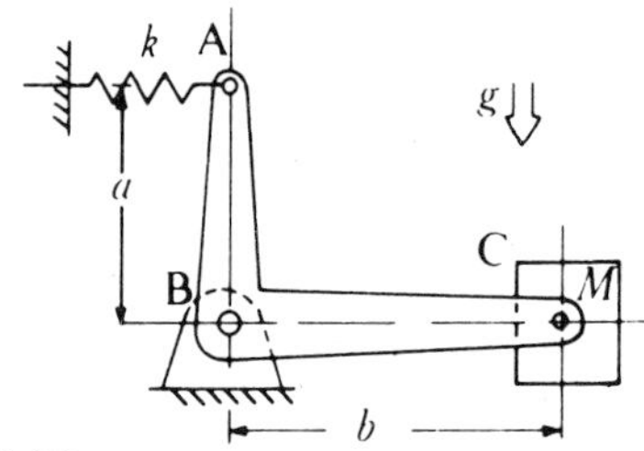

Figure 9.49

9.3 Determine the natural frequency of small oscillations of the bell-crank lever shown in Fig. 9.49. Neglect the mass of the arms. Take BC as horizontal when in the static equilibrium position.

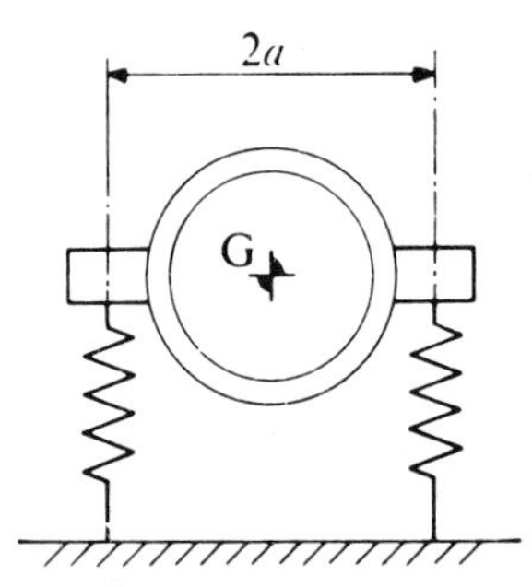

Figure 9.50

9.4 An electric motor of mass m is supported by four springs, each having a stiffness k, as shown in Fig. 9.50. If the polar moment of inertia is J_O, find the natural frequency of small oscillations (a) for vertical motion and (b) for torsional motion about G.

9.5 A flywheel with a polar moment of inertia of 0.65 kg m^2 is supported in frictionless bearings and is under the control of a torsion spring having a torsional stiffness of 4 N m/rad and a viscous damper. When the flywheel is displaced from its equilibrium position and released, the ratio of successive amplitudes of oscillation of the flywheel in the same sense is 10:1.

a) Determine the frequency of the damped oscillations of the flywheel.

b) The flywheel is displaced through 1 radian from its position of static equilibrium and held at rest; it is then released. Determine its angular velocity after a time interval equal to one quarter of the damped periodic time.

9.6 A homogeneous solid sphere, of radius a, rolls in a hemispherical cup of radius R. Determine the natural frequency of small oscillations about the equilibrium position.

9.7 A balanced wheel, with its axle contacting two parallel rails bent into circular arcs, oscillates in a vertical plane as shown in Fig. 9.51. The radius of the axle is r, that of the arcs is R, and the radius of gyration for the wheel about its axis is k.

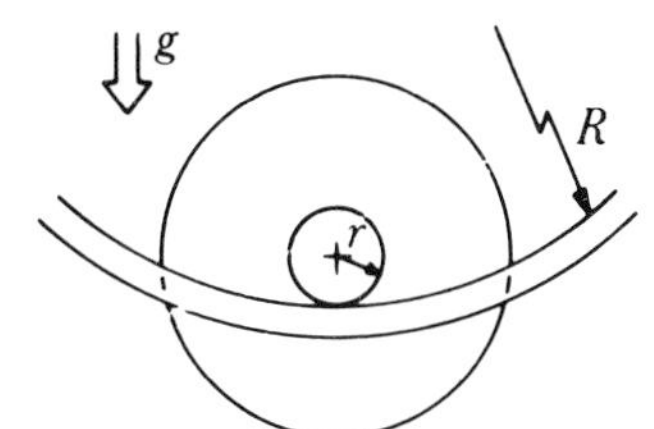

Figure 9.51

Assuming that the amplitude of the oscillation is kept small, and that no slip occurs, derive from first principles a differential equation representing the motion of the wheel and hence show that the frequency of the oscillation is the same as that of a simple pendulum of length

$$L=(R-r)[1+(k^2/r^2)]$$

9.8 A mass of 0.2 kg is supported by a spring of stiffness 500 N/m. A mass of 0.6 kg is dropped through a height of 20 mm on to the smaller mass and does not rebound. Find the frequency and amplitude of vibration.

9.9 When a package of mass 1.2 kg is placed on the platform of a weighing machine, the platform comes to rest 60 mm below its unloaded position when all oscillations have ceased. In order to reduce oscillation and expedite the weighing procedure, a dashpot is added to the mechanism, providing viscous damping of magnitude 14.4 N at a platform velocity of 1 m/s.

Identical packages are delivered to the platform for checking by means of a chute, and it may be assumed that a package has a vertical velocity of 35 mm/s when it makes contact with the weighing platform.

Determine the greatest height to which the platform will rise above the static-loaded position on the first upward swing after acceptance of a package.

9.10 A mirror galvanometer is constructed as shown diagrammatically in Fig. 9.52, damping being provided by a fluid in the casing. The instrument has an undamped natural frequency of 100 Hz, damping is 64% of the critical damping, and the d.c. sensitivity is 16 rad/mA. Determine (a) the damped natural

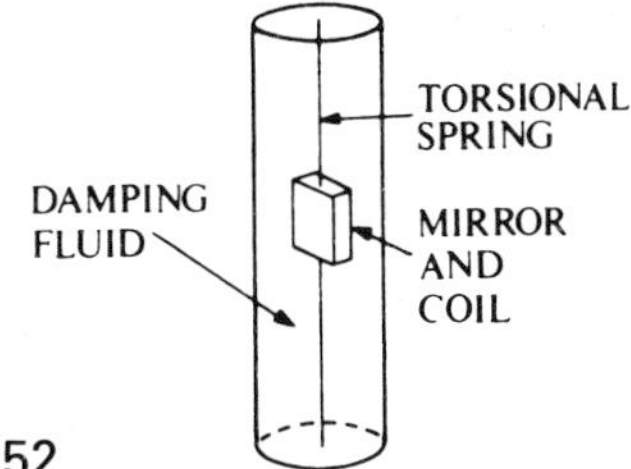

Figure 9.52

frequency, (b) the maximum deflection if a current of 50 μA is suddenly passed through the instrument.

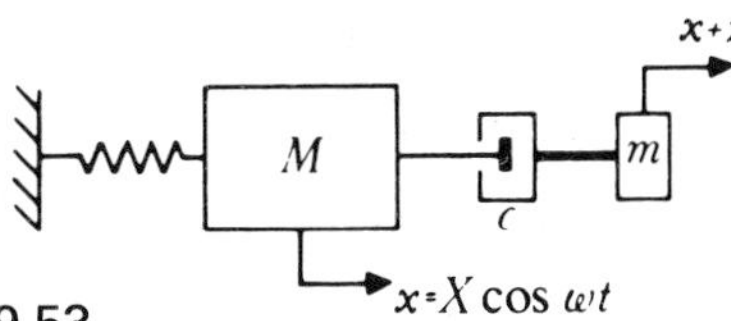

Figure 9.53

9.11 A large structure has a mass M and a natural frequency ω. As shown in Fig. 9.53 damping is provided by a damper connected to a rigid body of mass m. Assuming that the motion of the structure is still approximately simple harmonic, determine the amount of energy absorbed per cycle by the damper. What value of c will give maximum energy absorption per cycle?

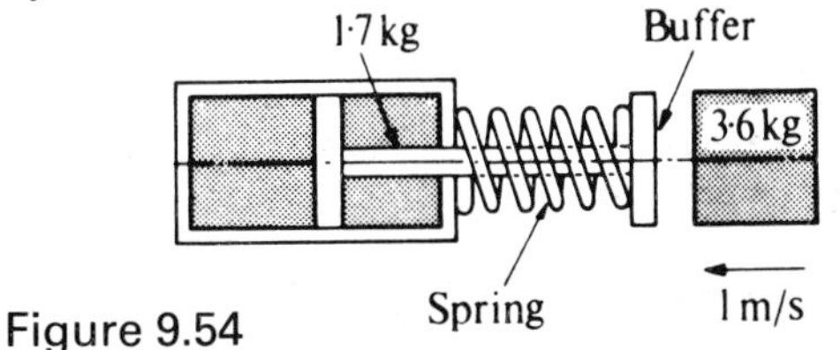

Figure 9.54

9.12 A shock-absorber consists of a solid buffer, a spring, and a piston rod and piston as indicated in Fig. 9.54. The piston moves in an oil-filled cylinder and this arrangement may be assumed to give a force, resisting motion, that is proportional to velocity. The shock-absorber is to be tested by placing it horizontally and projecting a body of mass 3.6 kg against it. This body is to strike the buffer at a speed of 1 m/s and may be assumed to remain in contact with the buffer until brought to a halt. Assuming the damping to be adjusted to give the critical value for this arrangement, and using the data given below, estimate the distance the buffer should travel. The spring has no initial compression and its stiffness is 525 N/m. The mass of the moving parts of the shock absorber is 1.7 kg. The mass of the spring is negligible.

9.13 A support platform carrying a machine can be represented by a 5200 kg mass mounted on two relatively long light pillars as shown in Fig. 9.55. The platform is constrained to move in the plane containing the centre-lines of the pillars. Measurements show that,

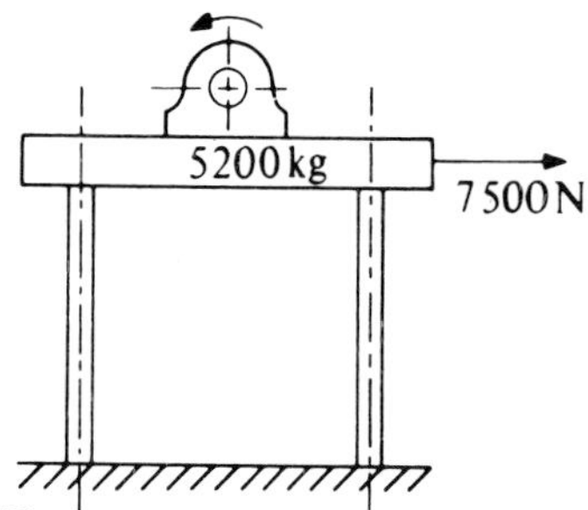

Figure 9.55

when a horizontal force of 7500 N is applied to the mass, the resulting horizontal displacement of the mass is 1.5 mm.

a) Calculate the operating speed of the machine which should be avoided to reduce the risk of appreciable vibration of the platform.

b) During tests with the machine switched off, the platform is made to vibrate at its natural frequency and a velocity pick-up indicates a peak horizontal velocity of 0.09 m/s. Find (i) the amplitude of horizontal vibrations and (ii) the peak value of the horizontal acceleration of the platform.

9.14 A machine which is subjected to a vertical sinusoidal force of constant amplitude is to be mounted on a rigid foundation. Discuss the problem of limiting the force transmitted to the foundation for each of the following cases (a) the forcing frequency has a fixed value which is rapidly attained after starting; (b) the forcing frequency is a variable, having a known upper limit.

Sketch vector diagrams and response curves to illustrate your answer.

9.15 A light shaft, which has a stiffness in torsion of 108 N m/rad, is fixed at one end and carries a wheel with a moment of inertia of 0.84 kg m^2 midway along its length. If the free end of the shaft is given a torsional oscillation with an amplitude of 1° and a frequency of 3 Hz, find the amplitude of steady-state oscillations of the flywheel.

9.16 A recording instrument is used in a location where the floor is subject to vertical simple harmonic vibrations of frequency 6.37 Hz and amplitude 1 mm.

The mass of the instrument is 4.5 kg and it is separated from the vibrating floor by spring mountings of stiffness 1800 N/m and some internal viscous damping. The amplitude of steady-state vertical vibrations of the instrument is observed to be 0.412 mm.

Write down the equation of motion of the instrument and find the degree of damping in the mountings, expressed as a percentage of the critical value.

9.17 A punching machine of mass 510 kg is supported by a mounting which has a stiffness of 240 kN/m and exerts a viscous damping force of 9600 N at 1 m/s. Part of the machine moves vertically with simple harmonic motion through a stroke of 40 mm and has a mass of 8 kg.

Calculate the amplitude of forced vibrations when the machine is running at 380 strokes/min.

9.18 A motor is mounted on flexible supports which permit it to oscillate through a small angle about the axis of the motor frame. The torsional stiffness of the supports acting on the motor frame is 3 kN m/degree and the moment of inertia of the motor frame about the axis of rotation is 14.4 kg m^2. When the motor is running, the frame is subjected to a reaction torque of $1500 \sin 3\omega t$ N m, where ω is the rotor speed and t is the time.

Determine the amplitude of steady-state torsional oscillation of the motor frame when the rotor speed is 400 rev/min. If, under resonant conditions, this amplitude must not exceed 5 degrees, what percentage of critical damping must be applied?

9.19 A velocity pick-up is constructed as shown in Fig. 9.56. The output voltage is 0.2 V for a relative velocity between the seismic mass and the case of 1 m/s. The details of the instrument are seismic mass 0.01 kg, spring stiffness 160 N/m, and viscous damping co-efficient 1.75 N s/m. The case is rigidly attached to a surface which is vibrating sinusoidally at 25 Hz.

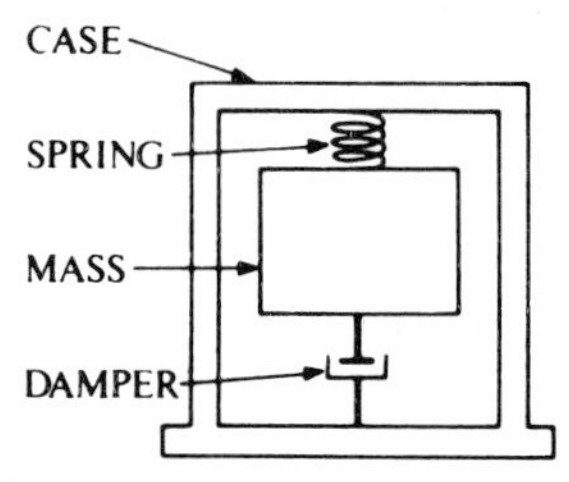

Figure 9.56

What is the peak output voltage when the peak velocity of the surface is 1 m/s?

9.20 Figure 9.57 shows part of a recording instrument. The light spring AB of torsional stiffness S is connected to the pointer DE via the coupling C. Backlash in the coupling permits a relative angular rotation β_0 between shafts BC and CD. The pointer

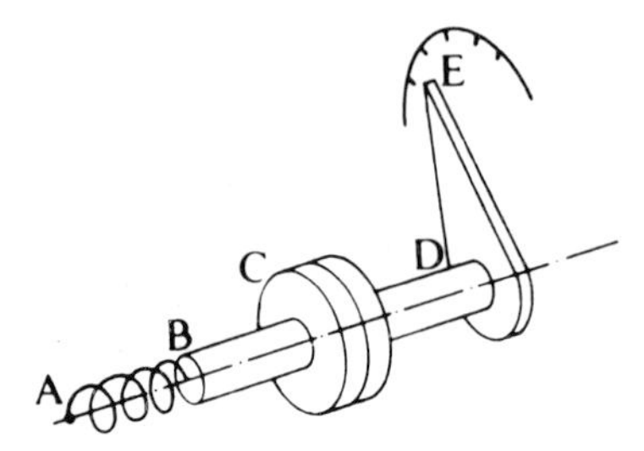

Figure 9.57

DE has an axial moment of inertia I, and all other inertias are to be neglected. End A is held fixed, and the pointer is oscillating with an angular amplitude α_0. Determine the periodic time T of the oscillation and sketch the graph of T against α_0.

9.21 A spring of torsional stiffness 0.200 N m/rad of twist is fixed at one end and attached to a flywheel of moment of inertia 5×10^{-4} kg m^2 at the other end. The motion of the flywheel is opposed by dry friction such that the torque required to initiate motion is 0.022 N m but once motion has started the resisting torque is 0.015 N m. Viscous damping is negligible.

The flywheel is held in a position where the torque in the spring is 0.200 N m and is then released. Find (a) the time taken for all motion to cease and (b) the residual torque in the spring.

9.22 A simple model of a motor vehicle suitable for investigating the relationship between pitch and bounce is shown in Fig. 9.58. The mass of the vehicle is 1300 kg and its moment of inertia about a transverse axis through the centre of mass is 1500 kg m^2. The combined spring stiffness at the front axle is 28 kN/m and that at the rear axle is 24 kN/m.

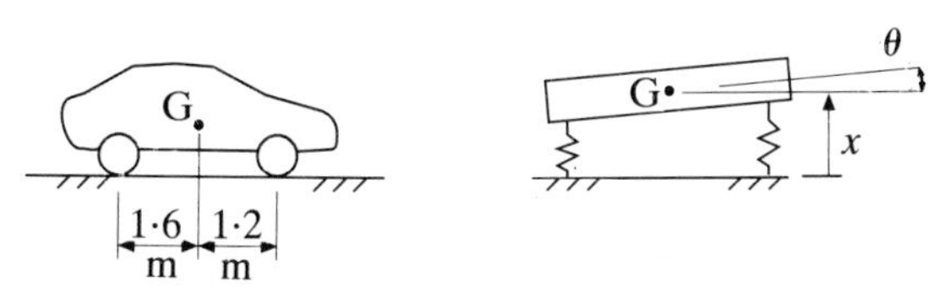

Figure 9.58

Show that, if the effects of damping are neglected, the natural frequencies for motion in the fore and aft vertical plane are 1.0 Hz and 1.31 Hz. Show also that the corresponding mode shapes expressed in terms of angle of pitch to amplitude of bounce, θ/x, are +0.114 rad/m and −7.654 rad/m.

9.23 Determine the natural frequencies of free torsional vibration of the rotor system shown in Fig. 9.59. The moment of inertia of the rotor at A is sufficiently large for the end at A to be considered as fixed.

The moments of inertia of rotors B and C are 0.5 kgm^2 and 1.5 kgm^2 respectively. The shaft is hollow with an outside diameter D_0 of 50 mm and an internal diameter D_i of 45 mm. The length L of shaft between A and B is 1.0 m and between B and C it is 0.3 m. The shear modulus G for the material of the shaft is 80 GN/m^2.

Note that from Chapter 12 the torsional stiffness of a shaft is $G\pi(D_0^4 - D_i^4)/(32L)$.

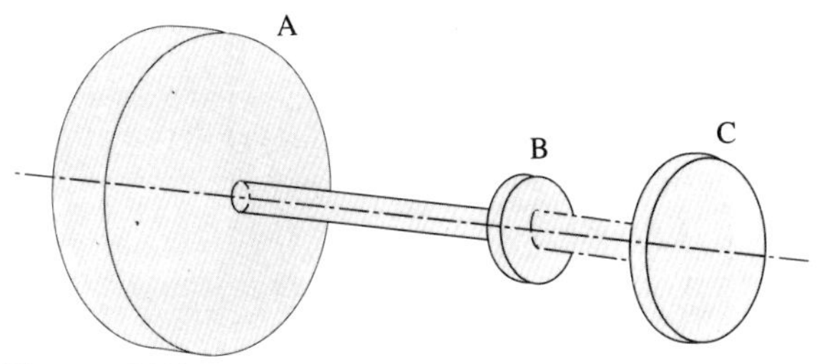

Figure 9.59

9.24 An overhead gantry crane has a trolley, of mass 225 kg, which is driven along a horizontal track by means of a cable and is shown in Fig. 9.60. When the drive mechanism to the cable is held stationary the free vibrations of the trolley along the track have a natural frequency of 1.65 Hz.

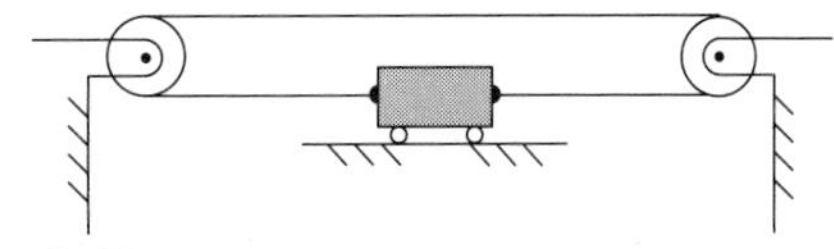

Figure 9.60

Determine the possible natural frequencies of vibration of the trolley along the track when it is supporting a load of mass 225 kg by means of a light inextensible cable having a free length of 0.6 m, the drive mechanism to the cable again being stationary.

9.25 An instrument is to be mounted on a foundation which is vibrating at 50 Hz. A spring is inserted between the foundation and the instrument, and a vibration absorber having the same mass as the instrument is connected to the instrument by a spring identical to that used between the instrument and the foundation.

If the mass of the instrument is 0.6 kg, determine the stiffness of the springs so that the amplitude of the vibration of the instrument is zero.

What will then be the natural frequencies of the complete system?

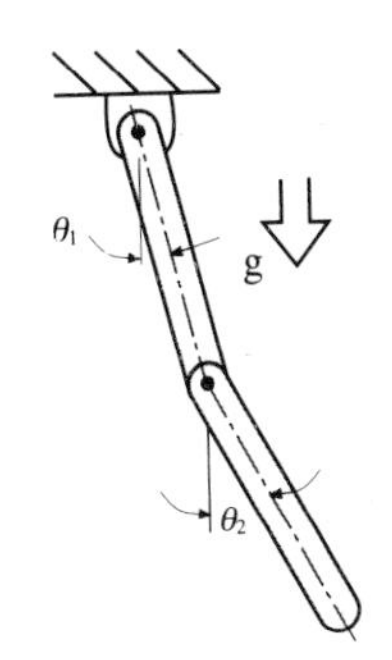

Figure 9.61

9.26 Figure 9.61 shows a double pendulum consisting of two equal uniform slender bars each of mass m and length L. Friction at the pivots is to be neglected.

Determine the natural frequencies for *small* oscillations about the equilibrium position.

(*Hint*: The small angle approximation implies that all non-linear terms can be excluded. Such terms are the centripetal accelerations and any product of co-ordinates.)

9.27 A structure carries two heavy machines, one at point A and one at point B. The static deflections when the machine at A is installed are 5 mm at A and 2 mm at B. When the machine at B is added the deflection at B is increased to 6 mm and that at A becomes 8 mm. If a 7 Hz sinusoidal force of amplitude equal to 1% of the weight of machine A is applied to the machine at A what is the amplitude of motion of the structure at A and B? Find also the natural frequencies of the system.

10
Introduction to automatic control

10.1 Introduction

This chapter is devoted to an examination of elementary mechanical control systems. The discussion will be limited to the class of systems whose motion can be described by linear differential equations with constant coefficients. In practice many control systems have non-linear elements, but the overall motion can very often be closely approximated to that of a purely linear system. The main features of all control systems can be introduced by discussing specific examples.

Let us consider the position control of a machine tool which has only straight-line motion. Let the actual position of the tool be defined by x_o and the desired position by x_i. The variables x_o and x_i are referred to as the *system output* and the *system input* respectively.

The system *error*, x_e, is formally defined by

$$x_e = x_i - x_o \tag{10.1}$$

and it is the object of the control system to take corrective action and reduce this error to zero.

Assume that the tool is initially at rest and that the system has zero error. If a new position is required, the appropriate input is applied, giving rise to an error in position. The controller then acts, attempting to reduce the error to zero, and, for a linear system, the motion of the tool will be described by a linear differential equation.

A human operator often forms part of a control system. As an example of this consider the case of a man driving a car at a speed which he wishes to remain constant at 100 km/h. This constitutes a speed-control system where the desired speed or input, v_i, is 100 km/h. The output, v_o, is the actual speed of the car, and the error, v_e, is the difference between input and output.

If, for example, the car meets a headwind, the drop in speed (the error) will be noticed by the driver who, among other things, is acting as an error-sensing device, and he will take corrective action by adjusting the position of the accelerator pedal in an attempt to reduce the error to zero. If the head wind is such as to cause a rapid increase of error, the corrective action will not be the same as that for a slow change. Thus we observe that the driver's control action takes account not only of the magnitude of the error, v_e, but also the rate of change of error, dv_e/dt.

Later we shall see that in some control systems a measure of the integral $\int v_e \, dt$ is useful. When a human operator is part of the control process, his reaction time introduces a finite delay into the system, making it non-linear. Such systems are not discussed further here.

10.2 Position-control system

We can now examine in some detail a particular elementary position-control system and use it to introduce the block-diagram notation by which control systems are often represented.

A rotatable radar aerial has an effective moment of inertia I. The aerial is driven directly by a d.c. motor which produces a torque T_m equal to k_1 times the motor voltage V; thus

$$T_m = k_1 V \tag{10.2}$$

The motor voltage V is effectively the difference between two voltages V_a and V_b which are applied to the two motor terminals so that

$$V = V_a - V_b \tag{10.3}$$

and, of course, if V_a and V_b were identical the motor would have zero output torque. A potentiometer-and-amplifier system produces the voltage V_a proportional to the desired angular position θ_i of the aerial, the constant of proportionality being k_2. Thus

$$V_a = k_2 \theta_i \tag{10.4}$$

A position transducer, attached to the aerial whose angular position is θ_o (the system output) produces the voltage V_b such that

$$V_b = k_3 \theta_o \tag{10.5}$$

If θ_i and θ_o are equal, then the position error defined by

$$\theta_e = \theta_i - \theta_o \tag{10.6}$$

is zero and for this condition it is required that the voltage V and hence the torque T_m be zero. The voltage V_a represents the desired position or input, and the voltage V_b represents the actual position or output. The voltage V thus represents the error, and we conclude that k_2 must equal k_3 and equations 10.3 to 10.6 can be combined to give

$$V = k_2\theta_e \tag{10.7}$$

Equation 10.2 can be written as

$$T_m = k_1 k_2 \theta_e \tag{10.8}$$

and we see that the motor torque is proportional to the error.

Equation 10.8 represents the *control action* of the system. In order to determine the motion of the system for a particular input θ_i, we need to incorporate the dynamics of the aerial itself. (In mechanical control systems, the object whose position or speed is being controlled is usually referred to as the *load*.)

If the aerial has negligible damping, the only torque applied to it is that from the motor; thus

$$T_m = I\,\mathrm{d}^2\theta_o/\mathrm{d}t^2 \tag{10.9a}$$

or

$$T_m = I\mathrm{D}^2\theta_o \tag{10.9b}$$

where D is the operator d/dt.

Eliminating T_m from equations 10.8 and 10.9(b),

$$k_1 k_2 \theta_e = I\mathrm{D}^2\theta_o \tag{10.10}$$

For any control system, the relationship between input and output is of major importance. From equations 10.6 and 10.10,

$$k_1 k_2(\theta_i - \theta_o) = I\mathrm{D}^2\theta_o$$

$$(I\mathrm{D}^2 + k_1 k_2)\theta_o = k_1 k_2 \theta_i \tag{10.11a}$$

or

$$I\ddot{\theta}_o + k_1 k_2 \theta_o = k_1 k_2 \theta_i \tag{10.11b}$$

By solving equations 10.11, we can find the output θ_o as a function of time for a given function θ_i.

Note that a purely mechanical analogue of this system could consist of a flywheel of moment of inertia I connected to a shaft of torsional stiffness $K = k_1 k_2$, as shown in Fig. 10.1.

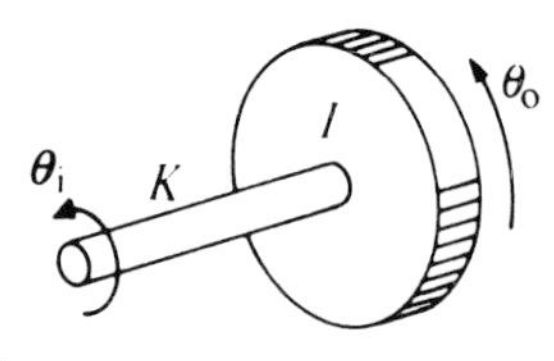

Figure 10.1

10.3 Block-diagram notation

It is common practice to represent control systems in block-diagram form. There are three basic elements: an adder/subtracter, a multiplier, and a pick-off point as shown in Figs 10.2, 10.3, and 10.4.

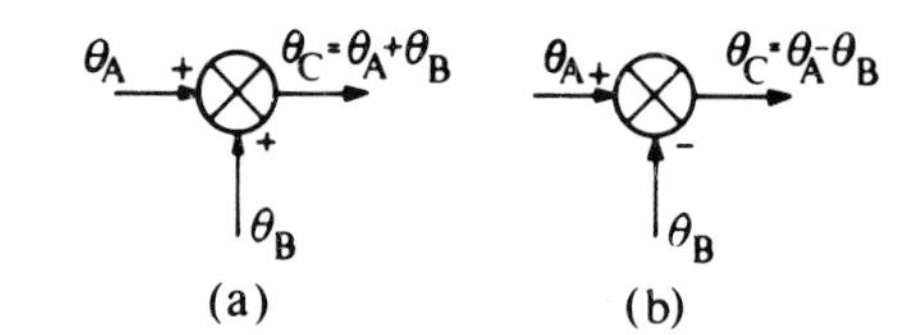

Figure 10.2

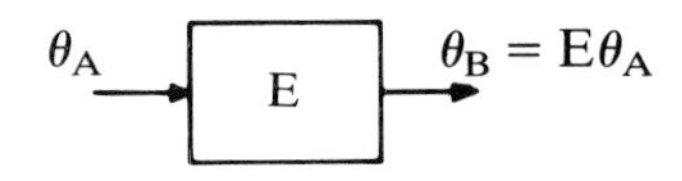

Figure 10.3

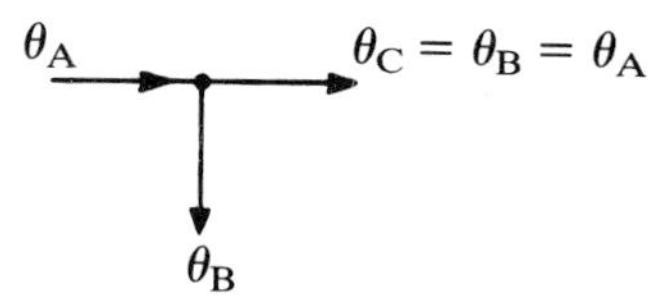

Figure 10.4

In Fig. 10.3, the simplest form of the multiplier E will be a constant, and the most complicated form can always be reduced to a ratio of two polynomials in operator D. We can write

$$\frac{\theta_A}{\theta_B} = \mathrm{E}$$

and E is called the *transfer operator* between θ_A and θ_B. Note that if $\theta_2 = \mathrm{E}\theta_1$ and $\theta_3 = \mathrm{F}\theta_2$, as shown in Figs 10.5(a) and (b), then $\theta_3 = \mathrm{EF}\theta_1$, as shown in Fig. 10.5(c).

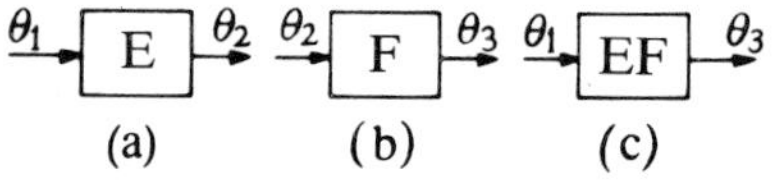

θ_1 E θ_2 (a) θ_2 F θ_3 (b) θ_1 EF θ_3 (c)

Figure 10.5

Equations 10.2, 10.6, 10.7, and 10.9(b) are

represented by the block-diagram elements shown in Fig. 10.6. Note that there is an implication of cause and effect: the output of a block-diagram element is the result of applying the input(s). In equation 10.9(b), the angular rotation θ_o is the result of applying the torque T_m. The equation is thus rewritten as $T_m(1D^2) = \theta_o$, so that Fig. 10.6(d) can be drawn with T_m as input.

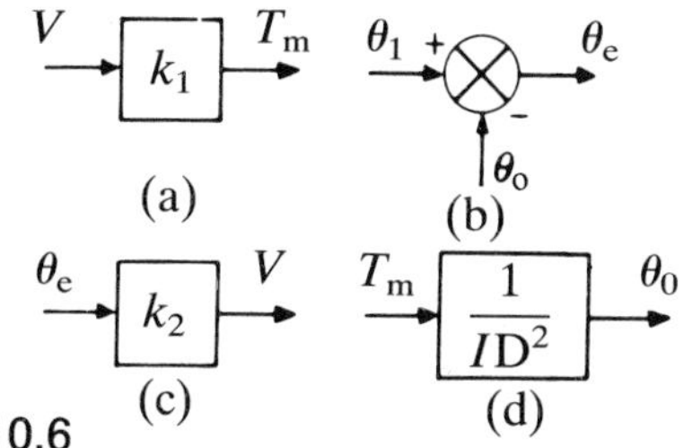

Figure 10.6

Noting that the output of Fig. 10.6(d) is one of the inputs to Fig. 10.6(b), and connecting the four elements in the appropriate order, we obtain the system block diagram shown in Fig. 10.7. From this figure we note that a control system is a *closed-loop* system. One of the variables (θ_o) is subtracted from a variable (θ_i) which precedes it; this is known as *negative feedback*.

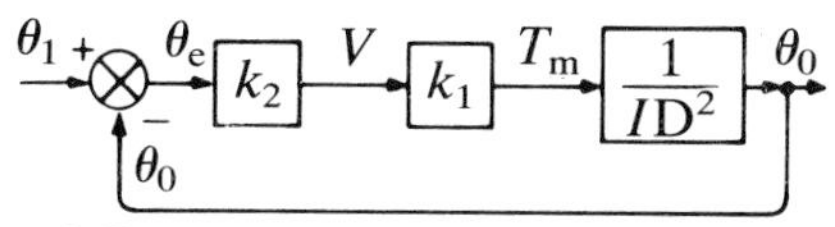

Figure 10.7

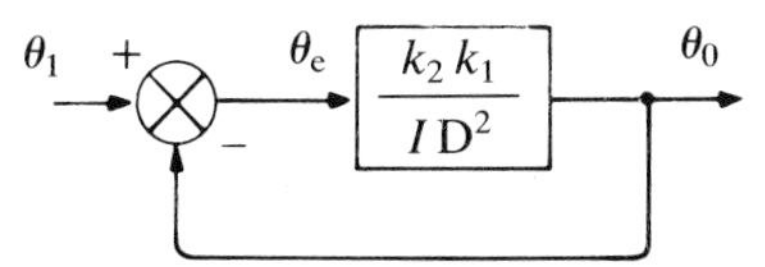

Figure 10.8

Using the techniques of Fig. 10.5, Fig. 10.7 can be reduced to Fig. 10.8.

10.4 System response

Returning to equations 10.11, we can determine the response of the system to particular inputs θ_i. Suppose we want the load suddenly to rotate through an angle α_0 at time $t = 0$. This corresponds to the step input $\theta_i = 0, t < 0$; $\theta_i = \alpha_0$, $t \geq 0$ shown in Fig. 10.9. It is left as an exercise for the reader to show that the response to this input is given by

$$\theta_o = \alpha_0(1 - \cos \omega_n t) \quad (10.12)$$

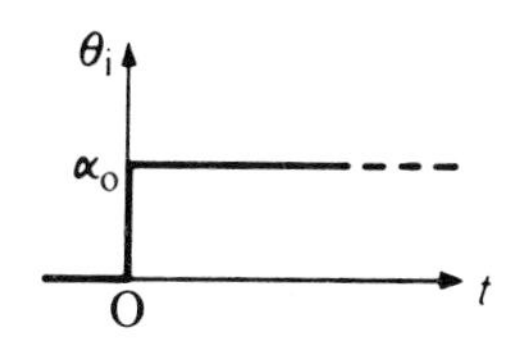

Figure 10.9

where $\omega_n = (k_1 k_2 / I)^{1/2}$, as shown in Fig. 10.10.

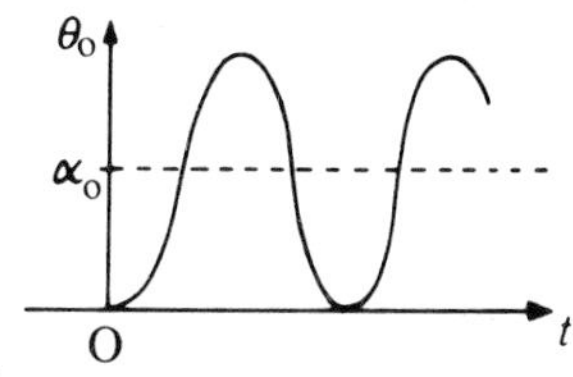

Figure 10.10

Rather than taking up the required position $\theta_o = \alpha_0$, the load oscillates about this position with circular frequency ω_n. This performance is clearly unsatisfactory and it is evident that some form of damping must be introduced. The response would then take the form of either Fig. 10.11(a) or (b), depending on the amount of damping.

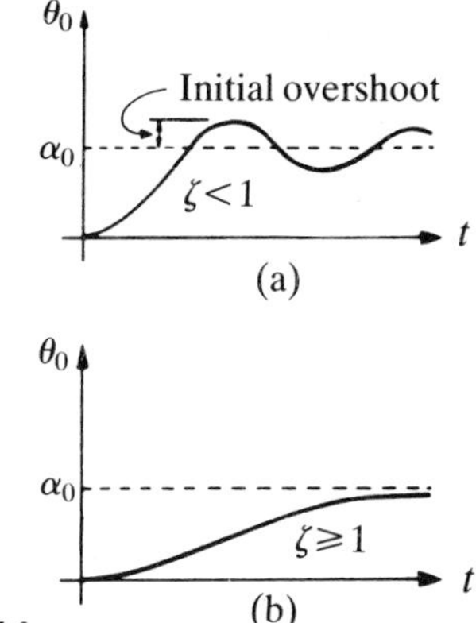

Figure 10.11

One way of providing damping is to attach a damper to the load. If the damper provides a torque T_d which opposes the motion of the load and is proportional to the velocity (viscous damping), the constant of proportionality being C, then equation 10.9(a) is replaced by

$$T_m - T_d = I\mathrm{d}^2\theta_o \mathrm{d}t^2 \quad (10.13)$$

$$T_m - CD\theta_o = ID^2\theta_o \quad (10.14)$$

$$T = (ID^2 + CD)\theta_o$$

$$T_m/(ID^2 + CD) = \theta_o$$

The block diagram for the damped load is shown in Fig. 10.12. We note that the effect of

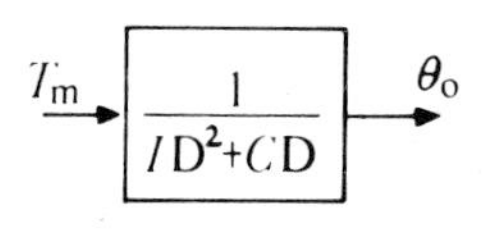

Figure 10.12

adding the damper is to replace ID^2 in the undamped system by $ID^2 + CD$. Hence, for the damped system of Fig. 10.13 (cf. equation 10.11(a)),

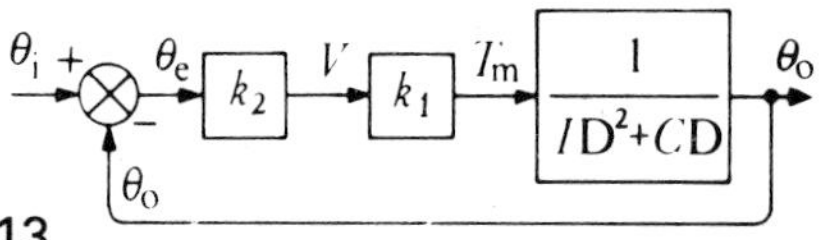

Figure 10.13

$$(ID^2 + CD + K)\theta_o = K\theta_i \qquad (10.15)$$

where $K = k_1 k_2$. Dividing by I to obtain the standard form (see equation 9.22) we have

$$(D^2 + 2\zeta\omega_n D + \omega_n{}^2)\theta_o = \omega_n{}^2\theta_i \qquad (10.16)$$

where $\omega_n{}^2 = K/I$ and $\zeta = C/2\sqrt{(KI)}$.

For the same input, Fig. 10.9, the solutions to equation 10.16 are (see also equation 9.33)

$$\left.\begin{aligned} \theta_o &= \alpha_0\{1 - e^{-\zeta\omega_n t}[\cos\omega_d t \\ &\quad + (\zeta/\sqrt{1-\zeta^2})\sin\omega_d t]\} & \zeta<1 \\ &= \alpha_0\{1 - e^{-\zeta\omega_n t}[1 + \omega_n t]\} & \zeta=1 \\ &= \alpha_0\{1 - e^{-\zeta\omega_n t}[\cosh\omega_e t \\ &\quad + (\zeta/\sqrt{\zeta^2-1})\sinh\omega_e t]\} & \zeta>1 \end{aligned}\right\} \quad (10.17)$$

where $\omega_d = \omega_n\sqrt{1-\zeta^2}$ and $\omega_e = \omega_n\sqrt{\zeta^2 - 1}$.

The output θ_o does not settle to the required value of α_0 until (theoretically) an infinite time has elapsed. In practice, small amounts of Coulomb (dry) friction ensure that motion ceases reasonably quickly.

The viscous damper wastes power and cannot readily be constructed to give a precise amount of damping. There are other methods of introducing the first-derivative term ($CD\theta_o$) into the system equation 10.15, and one of these makes use of a d.c. device known as a tachogenerator, driven by the load. The voltage V_t produced by the tachogenerator is proportional to its angular velocity, so that

$$V_t = k_4 D\theta_o \qquad (10.18)$$

and the block-diagram form is shown in Fig. 10.14.

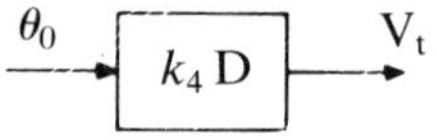

Figure 10.14

Consider the case of the undamped load with a tachogenerator attached. The tachogenerator is a relatively small device and applies a negligible torque to the load so that equations 10.9 are applicable. Assume that the voltage V_t is subtracted from the voltage V by an operational-amplifier system so that the voltage V_m applied to the motor is

$$V_m = V - V_t \qquad (10.19)$$

The system block diagram for this case is shown in Fig. 10.15 and we observe that the tachogenerator appears in an inner loop. The equations for the component parts of the system are listed below.

$$\theta_e = \theta_i - \theta_o \qquad (10.6)$$

$$V = k_2\theta_e \qquad (10.7)$$

$$V_m = V - V_t \qquad (10.19)$$

$$T_m = k_1 V_m$$

$$T_m = ID^2\theta_o \qquad (10.9b)$$

$$V_t = k_4 D\theta_o \qquad (10.18)$$

Eliminating T_m, V_m, V, V_t and θ_e, we obtain

$$ID^2\theta_o = k_1[k_2(\theta_i - \theta_o) - k_4 D\theta_o]$$

$$(ID^2 + k_1k_4D + k_1k_2)\theta_o = k_1k_2\theta_i \qquad (10.20)$$

The amount of damping in the system can be altered by regulating the techogenerator voltage by a potentiometer circuit. This method of introducing damping is known as *output velocity feedback*. Another common way of introducing damping is to use *proportional-plus-derivative* action (see problem 10.5).

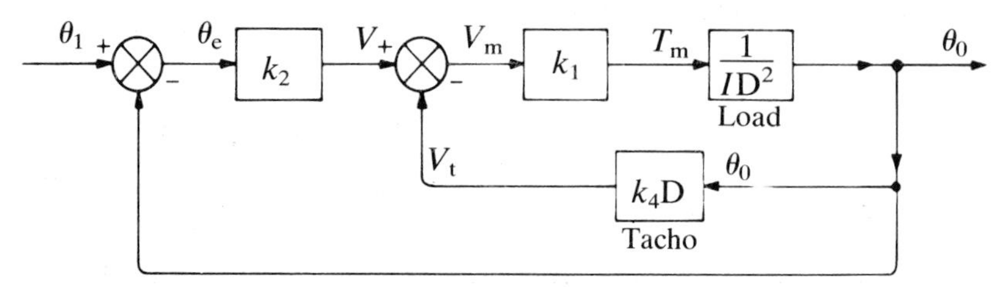

Figure 10.15

10.5 System errors

A system equation relates one of the loop variables to the input(s). It is conventional to have the loop variable on the left-hand side of the equation and the input(s) on the right. For example, in Fig. 10.13, θ_e, V, T_m and θ_o are loop variables and θ_i is the input; equation 10.15 is an output–input system equation. To obtain the error–input system equation we can replace θ_o in this equation by $\theta_i - \theta_e$ from equation 10.6, to obtain

$$(ID^2 + CD + K)(\theta_i - \theta_e) = K\theta_i$$
$$(ID^2 + CD + K)\theta_e = (ID^2 + CD)\theta_i \quad (10.21)$$

If θ_i has the constant value α_0 as shown in Fig. 10.9 then all its derivatives are zero and, for this input, equation 10.21 becomes, for $t>0$,

$$(ID^2 + CD + K)\theta_e = 0 \quad (10.22)$$

We already have the solution for θ_o, equations 10.17. Subtracting these functions of θ_o from θ_i we obtain

$$\left.\begin{aligned} \theta_e &= \alpha_0 e^{-\zeta\omega_n t}\{\cos\omega_d t \\ &\quad + [\zeta/\surd(1-\zeta^2)]\sin\omega_d t\} & \zeta<1 \\ &= \alpha_0 e^{-\omega_n t}\{1+\omega_n t\} & \zeta=1 \\ &= \alpha_0 e^{-\zeta\omega_n t}\{\cosh\omega_e t \\ &\quad + [\zeta/\surd(\zeta^2-1)]\sinh\omega_e t\} & \zeta>1 \end{aligned}\right\} \quad (10.23)$$

where $\omega_n{}^2 = K/I$ and $\zeta = \frac{1}{2}C/\surd(IK)$. Each of the above three equations contains the negative exponential term $e^{-\zeta\omega_n t}$ so that, as $t\to\infty$, $\theta_e \to 0$ and we say that the final or *steady-state error* is zero and write

$$[\theta_e]_{t\to\infty} = [\theta_e]_{ss} = 0$$

We do not need to solve equation 10.22 to find the steady-state value of θ_e since this is merely the particular-integral part of the solution, which is clearly zero. That the steady-state error is not always zero can be seen from the following example.

Consider the position-control system with viscously damped load which has already been described. Assume that the system is at rest with $\theta_i = 0$. What would be the steady-state error following the application of a constant external torque T_o to the load?

Equation 10.14 is replaced by

$$T_m - CD\theta_o + T_o = ID^2\theta_o \quad (10.24)$$

or

$$T_m + T_o = (ID^2 + CD)\theta_o \quad (10.25)$$

We could equally well have written $-T_o$, since the direction was unspecified. Putting $T = T_m + T_o$ and $T/(ID^2 + CD) = \theta_o$, we can draw the system block diagram (Fig. 10.16). Notice that the external torque T_o appears as an extra input to the system. Combining equations 10.25, 10.8, and 10.6 and putting $\theta_i = 0$, we have

$$k_1 k_2 \theta_e + T_o = (ID^2 + CD)(-\theta_e)$$
$$(ID^2 + CD + k_1 k_2)\theta_e = -T_o \quad (10.26)$$

This system equation is identical in form with equation 10.15 with θ_o and $K\theta_i$ replaced by θ_e and $-T_o$ respectively and so the solutions can be written down immediately from equation 10.17. The steady-state error can be obtained by letting $t\to\infty$ and is found to be

$$[\theta_e]_{t\to\infty} = [\theta_e]_{ss} = -T_o/(k_1 k_2)$$

which is independent of the amount of damping. (Note that for *zero* damping the system oscillates indefinitely with a mean error value of $-T_o/(k_1 k_2)$).

The complete solution of equation 10.26 consists of (a) the complementary function, which is the transient part of the solution and dies away with time, provided some positive damping is present, and (b) the particular integral or steady-state solution which remains after the transients have died away. For a constant forcing function, the steady-state solution must be a constant function.

Equation 10.26 describes the system for all time from 0 to ∞. In the steady-state, therefore, $D\theta_e = D^2\theta_e = 0$ and equation 10.26 becomes

$$k_1 k_2 [\theta_e]_{ss} = -T_o$$

and the steady-state error is

$$[\theta_e]_{ss} = -T_o/(k_1 k_2)$$

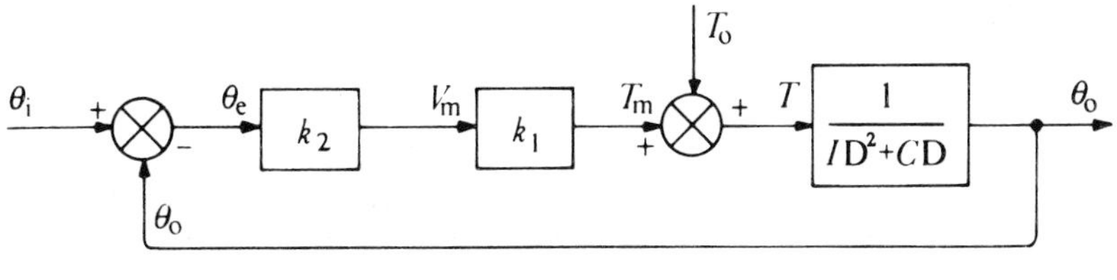

Figure 10.16

Consider once again Fig. 10.13. Assume that the system is initially at rest then, at time $t = 0$, it is required that the load have a constant angular velocity Ω_i. The desired position or input is therefore

$$\theta_i = 0, \quad t<0$$

and $\theta_i = \Omega_i t, \quad t \geq 0$

as shown in Fig. 10.17. This is known as a *ramp input*.

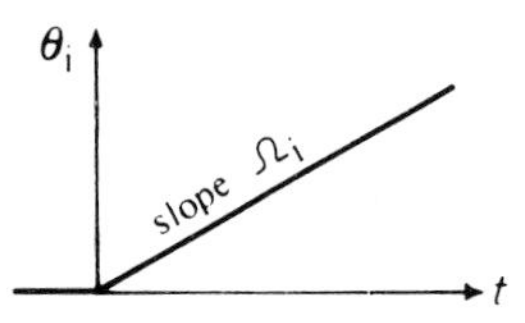

Figure 10.17

The error–input equation for this particular input is, from equation 10.21,

$$(ID^2 + CD + K)\theta_e = (ID^2 + CD)\Omega_i t = C\Omega_i \quad (10.27)$$

The steady-state error is equal to $C\Omega_i/K$ and the error response will be of the same form as equations 10.17. Since $\theta_o = \theta_i - \theta_e$, the output response can be obtained by subtracting the error response from the input function. The result is illustrated in Figs 10.18(a) and (b).

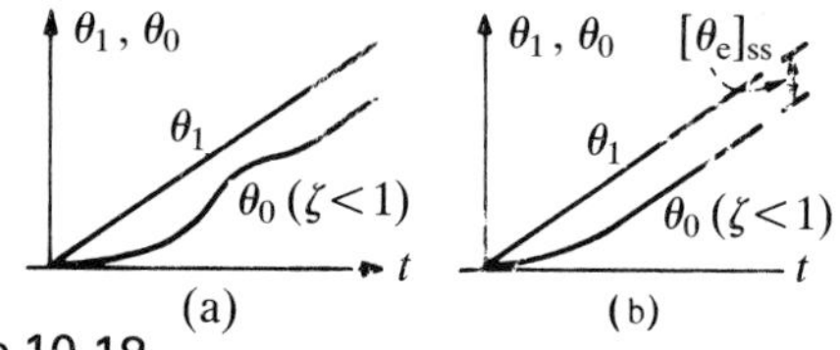

Figure 10.18

A control system with a residual error is normally unsatisfactory. Certain steady-state errors can be overcome by using a controller which incorporates *integral action*. Suppose that, in the above example, the voltage V_m applied to the motor, instead of being directly proportional to the error θ_e, is given by

$$V_m = k_2\theta_e + k_s \int_0^t \theta_e \, dt \quad (10.28)$$

In D-operator form this is written

$$V_m = \left(k_2 + \frac{k_5}{D}\right)\theta_e$$

and so the block diagram representation of this *proportional-plus-integral* controller is as shown in Fig. 10.19.

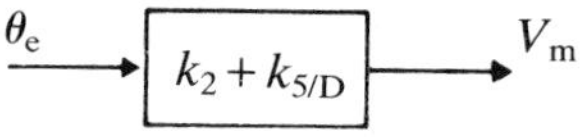

θ_e → $k_2 + k_{5/D}$ → V_m

Figure 10.19

The error–input equation for this system can be written down directly from equation 10.27, with $K = k_1 k_2$ replaced by $k_1(k_2 + k_5/D)$. Thus

$$[ID^2 + CD + k_1(k_2 + k_5/D)]\theta_e = C\Omega_i$$

To convert this to a purely differential equation we simply differentiate with respect to time by multiplying by D:

$$[ID^3 + CD^2 + k_1 k_2 D + k_1 k_5]\theta_e = DC\Omega_i = 0 \quad (10.29)$$

since $C\Omega_i$ is constant. The steady-state error is the particular integral of the above equation so that, for the ramp input,

$$[\theta_e]_{ss} = 0$$

10.6 Stability of control systems

The introduction of integral action in the above example had the effect of removing the steady-state error to a ramp input. It also had the effect of raising the *order* of the system. The order is defined as the highest power of D on the left-hand side of a system equation, and in the example it was raised from two to three.

For any particular control system, the system-equation loop variable, whichever one is chosen, will be preceded by the same polynomial in operator D (see problem 10.2). Thus the complementary functions (transient responses) for the loop variables will have different initial conditions but will otherwise be of the same form.

Before the concept of integral action was introduced in the previous section, all the system equations were of order two; that is, they were of the form

$$[a_2 D^2 + a_1 D + a_0]x = f(D)y \quad (10.30)$$

The transient response, and thus the stability of such a system, depends only on the coefficients a_0, a_1, and a_2. Assuming that $a_0 > 0$, then, provided that $a_1 > 0$ and $a_2 > 0$, the complementary function will not contain any positive time exponentials and the system will be stable. If $a_1 = 0$ (zero damping) the complementary function will oscillate indefinitely with constant amplitude and, although not strictly unstable, this represents unsatisfactory control. Such a system is described as *marginally stable*. If either $a_1<0$ (negative damping) or $a_2<0$ (negative mass), the transient will contain positive exponentials and the system will be unstable. Figure 10.20 illustrates the various types of stability.

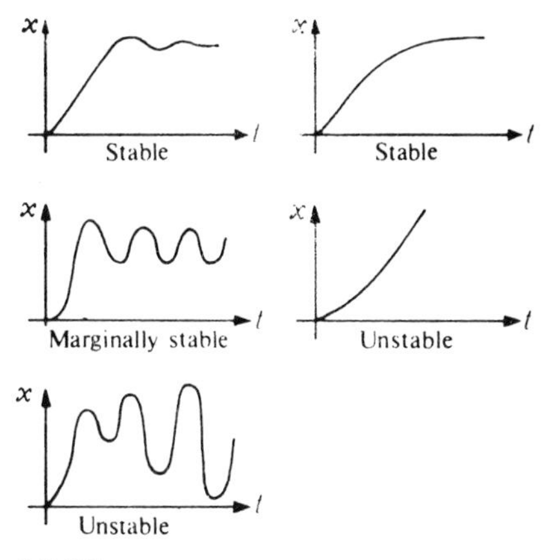

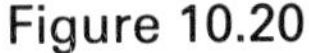
Figure 10.20

The system equation for the control system described by equation 10.29 is of third order. The general form of such an equation is

$$[a_3\mathrm{D}^3+a_2\mathrm{D}^2+a_1\mathrm{D}+a_0]x=f(\mathrm{D})y \quad (10.31)$$

If $a_0>0$, then, for such a system to be stable, it can be shown that the following conditions must hold: $a_1>0$, $a_2>0$, and $a_3>0$. Also it is necessary that

$$a_1a_2>a_0a_3 \quad (10.32)$$

We give below, without proof, the *Routh–Hurwitz* conditions for stability of a control system of order n described by a system equation of the form

$$[a_n\mathrm{D}^n+a_{n-1}\mathrm{D}^{n-1}+\ldots+a_1\mathrm{D}+a_0]x = f(\mathrm{D})y \quad (10.33)$$

where x is a loop variable. We assume throughout that a_0 is positive (if it is negative we multiply throughout by -1; if it is zero and $a_1>0$ then we rewrite the equation with $\dot{x}$ as the loop variable, and so on).

The first necessary condition is that all a_1, a_2, $\ldots$, a_{n-1}, a_n are positive.

Consider now the array

$$\begin{array}{cc|c|c|c|cc}
a_1 & a_0 & 0 & 0 & 0 & 0 & 0 & \cdot & \cdot & \cdot \\
a_3 & a_2 & a_1 & a_0 & 0 & 0 & 0 & \cdot & \cdot & \cdot \\
a_5 & a_4 & a_3 & a_2 & a_1 & a_0 & 0 & \cdot & \cdot & \cdot \\
a_7 & a_6 & a_5 & a_4 & a_3 & \cdot & \cdot & \cdot & \cdot & \cdot \\
a_9 & a_8 & \cdot & \cdot & \cdot & \cdot & \cdot & \cdot & \cdot & \cdot \\
\cdot & \cdot & \cdot & \cdot & \cdot & \cdot & \cdot & \cdot & \cdot & \cdot
\end{array}$$

For stability, it is necessary that the determinants

$$\begin{vmatrix} a_1 & a_0 \\ a_3 & a_2 \end{vmatrix}, \begin{vmatrix} a_1 & a_0 & 0 \\ a_3 & a_2 & a_1 \\ a_5 & a_4 & a_3 \end{vmatrix}, \begin{vmatrix} a_1 & a_0 & 0 & 0 \\ a_3 & a_2 & a_1 & a_0 \\ a_5 & a_4 & a_3 & a_2 \\ a_7 & a_6 & a_5 & a_4 \end{vmatrix}, \text{ etc.}$$

are all positive. For example, in a fourth-order system the necessary and sufficient conditions for stability are

$$\text{i)}\quad a_1>0, a_2>0, a_3>0, a_4>0 \quad (10.44)$$

$$\text{ii)}\quad \begin{vmatrix} a_1 & a_0 \\ a_3 & a_2 \end{vmatrix}>0 \quad (10.45)$$

$$\text{iii)}\quad \begin{vmatrix} a_1 & a_0 & 0 \\ a_3 & a_2 & a_1 \\ 0 & a_4 & a_3 \end{vmatrix}>0 \quad (10.46)$$

The coefficient a_4 appears first in the second of the string of determinants and it is unnecessary to proceed further since putting $a_5=a_6=a_7=\ldots=0$ in subsequent determinants gives identical results. (See example 10.2.) For a third order system we simply put $a_4=a_5=a_6=\ldots=0$ and obtain the result 10.32 given previously.

Consider now the case of a control system which is *marginally stable*, i.e. the output or any other loop variable x has a continuous sinusoidal oscillation of the form

$$x=A\cos(\omega_n t)$$

(this would be the complementary function part of the complete solution). It is convenient to use the complex exponential form so we write

$$x = \mathrm{Re}\,A\,e^{j\omega_n t} \qquad (10.47)$$

where ω_n is the natural circular frequency of the oscillation and A is the real amplitude.

We will use as an illustration the third-order system described by equation 10.31. With the right-hand side set equal to zero we have, for the complementary function,

$$[a_3 D^3 + a_2 D^2 a_1 D + a_0] A\,e^{j\omega_n t} = 0 \qquad (10.48)$$

where we assume that a_0, a_1, a_2 and a_3 are all real and positive.

Now

$$DA\,e^{j\omega_n t} = j\omega A\,e^{j\omega_n t} \qquad (10.49)$$

$$D^2 A\,e^{j\omega_n t} = (j\omega)^2 A\,e^{j\omega_n t} \qquad (10.50)$$

and it follows that

$$D^r A\,e^{j\omega_n t} = (j\omega_n)^r A\,e^{j\omega_n t} \qquad (10.51)$$

where r is any integer.

Equation 10.48 becomes

$$[a_3(j\omega_n)^3 + a_2(j\omega_n)^2 + a_1(j\omega_n) + a_0] A\,e^{j\omega_n t} = 0 \qquad (10.52)$$

so that

$$a_3(j\omega_n)^3 + a_2(j\omega_n)^2 + a_1(j\omega_n) + a_0 = 0 \qquad (10.53)$$

since $A\,e^{j\omega_n t} \neq 0$.

Hence $\quad -a_3 j\omega_n^{\ 3} - a_2\omega_n^{\ 2} + a_1 j\omega_n + a_0 = 0$

or $\quad (-a_2\omega_n^{\ 2} + a_0) + \omega_n(-a_3\omega_n^{\ 2} + a_1)j = 0$

The real and imaginary parts must separately be zero, hence

$$\omega_n^{\ 2} = \frac{a_0}{a_2} = \frac{a_1}{a_3}. \qquad (10.54)$$

We conclude that if $a_1 a_2 = a_0 a_3$ then the third-order system will be marginally stable and will oscillate at a circular frequency ω_n given by equation 10.54. We learned above (inequality 10.32) that if $a_1 a_2 > a_0 a_3$ the system will be stable. It is clear that if this inequality is reversed the system will be unstable.

A physical reason why this inequality determines the stability of the system can be found by considering a small applied sinusoidal forcing term, $Fe^{j\omega t}$, where F is a complex force amplitude and ω is close to ω_n.

The Argand diagram without the forcing term is as shown in Fig. 10.21(a) and that with the forcing term is shown in Fig. 10.21(b). For energy to be fed *into* the system the force must have a component which is in phase with the velocity (i.e. the imaginary part of the force must be positive). It follows that if energy is required to keep the system oscillating then the system must be stable. So we see that

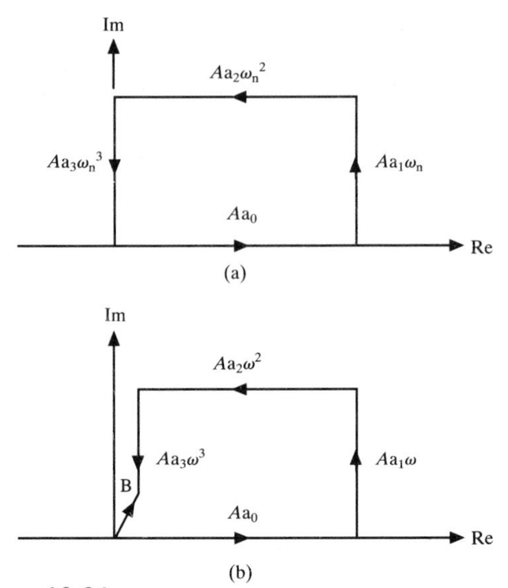

Figure 10.21

$$a_1\omega > a_3\omega^3$$

or $\quad a_1 > a_3\omega^2$.

Now since ω is close to ω_n we can write

$$a_0 = a_2\omega_n^{\ 2} = a_2(\omega \pm \varepsilon)^2$$

where ε is a small quantity. So as $\varepsilon \to 0$ then $\omega^2 \to a_0/a_2$.

Hence for a stable system

$$a_1 > a_3(a_0/a_2)$$

or $\quad a_1 a_2 > a_3 a_0$

Note that as previously mentioned all the a's must be positive because if any one a is negative the output will diverge for zero input.

10.7 Frequency response methods

An assessment of the behaviour of a closed-loop control system can be made from an examination of the frequency response of the open-loop system. Graphical methods are often employed in this work.

The main reasons for using open-loop system response methods are

(a) the overall open-loop system response can be built up quickly using standard response curves of the component parts of the system;

(b) in practice most open-loop systems are stable which is an advantage if experimental techniques are used!

Consider again a simple position-control system with proportional control driving an inertia load with viscous damping. The block diagram for the closed-loop system is shown in Fig. 10.22 which corresponds to Fig. 10.13 with $K = k_1 k_2$.

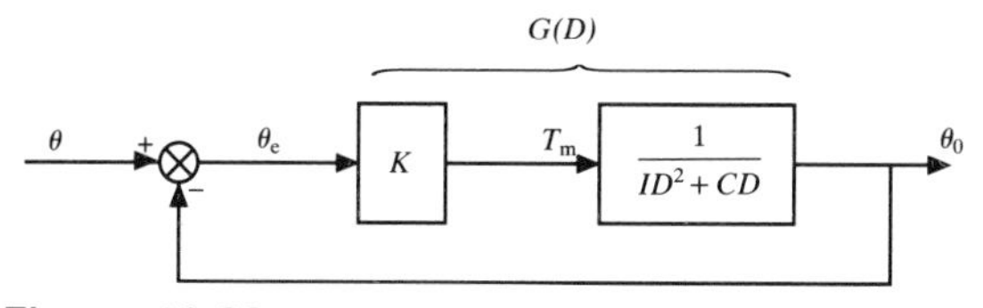

Figure 10.22

The *forward-path* operator $G(D)$ is given by

$$G(D) = \frac{\theta_o}{\theta_e} = \frac{K}{ID^2 + CD} \qquad (10.55)$$

If we disconnect the feedback loop we have the open-loop system of Fig. 10.23 and it can be seen that $G(D)$ is also the open-loop transfer operator. Here θ_i is simply the input to the open-loop system.

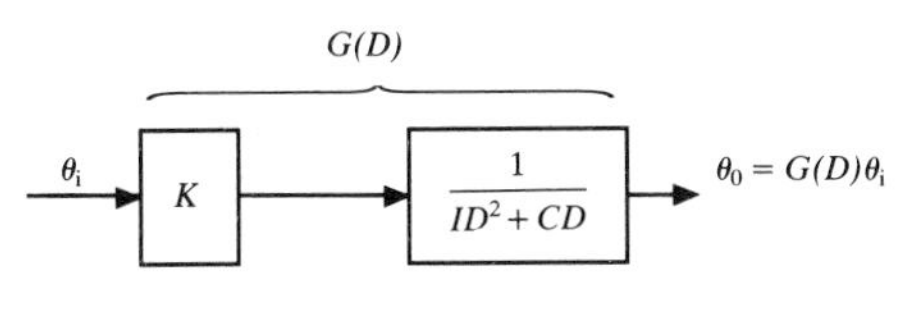

Figure 10.23

We can write $G(D)$ in standard form as

$$G(D) = \frac{K_o}{D(1 + \tau D)} \qquad (10.56)$$

where $K_o = K/C$ and $\tau = I/C$. (Note that τ has the dimensions of time.) So, for the open-loop

$$D(1 + \tau D)\,\theta_o = K_o \theta \qquad (10.57)$$

We wish to consider the frequency response of the open-loop so the input must be sinusoidal and we can write

$$\theta = A e^{j\omega t}$$

where A is complex. Equation 10.57 now becomes

$$D(1 + \tau D)\,\theta_o = K_o A\, e^{j\omega t} \qquad (10.58)$$

In frequency response methods we are only concerned with the steady-state oscillations (particular integral) part of the solution and so we ignore the transient (complementary function) response. Since the system is linear, the particular integral solution of equation 10.58 must be sinusoidal and at the same frequency ω as the input. The steady-state solution is therefore of the form

$$\theta_o = B e^{j\omega t}.$$

Substituting for θ_o in equation 10.58,

$$D(1 + \tau D)\, B e^{j\omega t} = K_o A\, e^{j\omega t} \qquad (10.59)$$

or, from equation 10.51

$$j\omega(1 + \tau j\omega)\, B e^{j\omega t} = K_o A\, e^{j\omega t} \qquad (10.60)$$

We see that, for sinusoidal inputs of frequency ω to the open loop system, the ratio of output to input is

$$\frac{\theta_o}{\theta} = \frac{B e^{j\omega t}}{A e^{j\omega t}} = \frac{K_o}{j\omega(1 + \tau j\omega)} = G(j\omega) \qquad (10.61)$$

which corresponds to the transfer operator $G(D)$ of equation 10.56 with D replaced by $j\omega$. $G(j\omega)$ is known as the *open-loop transfer function*.

We turn our attention now to the *closed-loop system* with unity feedback. A unity feedback system is, by definition, one for which the error $\theta_e = \theta_i - \theta_o$ and therefore, since

$$\theta_o = G(D)\,\theta_e \qquad (10.62)$$

eliminating θ_e we obtain

$$[1 + G(D)]\,\theta_o = G(D)\,\theta_i. \qquad (10.63)$$

(For a system with non-unity feedback see example 10.7.)

Suppose now, as was discussed at the end of section 10.6, that the closed-loop system is marginally stable, i.e. it oscillates continuously at frequency ω_n say, for no input. In this case the particular integral part of the solution is zero, but the complementary function, or 'transient' part is sinusoidal

$$\theta_o = C e^{j\omega_n t}.$$

Substituting into equation 10.63, with the right-hand side set equal to zero we have, for the complementary function

$$[1 + G(j\omega_n)]\, C e^{j\omega_n t} = 0$$

therefore

$$1+G(j\omega_n)=0$$

since $Ce^{j\omega_n t}\neq 0$.

What we have shown is that, for marginal stability of the closed-loop system,

$$G(j\omega_n)=-1 \tag{10.64}$$

In other words, if there exists a particular value of ω (i.e. $\omega=\omega_n$) which makes the open-loop transfer function, $G(j\omega)$, have the (real) value of -1, then *the closed-loop system will be marginally stable* and will oscillate continuously at the frequency $\omega=\omega_n$. Thus we can see that the open-loop transfer function $G(j\omega)$ can give us information about the closed-loop performance.

Returning to our example of the open-loop transfer function 10.61

$$G(j\omega)=\frac{K_0}{j\omega(1+\tau j\omega)}$$

we can check if a value of ω can be found which makes $G(j\omega)=-1$. In other words, does a value of ω exist which simultaneously makes the amplitude ratio have unity value and the phase angle have the value of $-180°$ or $-\pi$ radians?

The amplitude ratio is

$$\left|\frac{\theta_o}{\theta_i}\right|=\left|G(j\omega)\right|=\left|\frac{K_0}{j\omega(1+\tau j\omega)}\right|$$

$$=\left|K_0\right|\left|\frac{1}{\omega}\right|\left|\frac{1}{1+\tau j\omega}\right| \tag{10.65}$$

K_0 is real so that

$$\left|K_0\right|=K_0. \tag{10.66}$$

Further

$$\left|\frac{1}{j\omega}\right|=\frac{1}{\omega}\left|\frac{1}{j}\right|=\frac{1}{\omega}\left|-j\right|=\frac{1}{\omega} \tag{10.67}$$

and

$$\left|\frac{1}{1+\tau j\omega}\right|=\left|\frac{1-\tau j\omega}{(1+\tau j\omega)(1-\tau j\omega)}\right|=\left|\frac{1-\tau j\omega}{1+(\omega\tau)^2}\right|$$

$$=\frac{|1-\tau j\omega|}{1+(\omega\tau)^2}=\frac{\sqrt{[1+(\omega\tau)^2]}}{1+(\omega\tau)^2}$$

$$=\frac{1}{\sqrt{[1+(\omega\tau)^2]}} \tag{10.68}$$

Therefore

$$G(j\omega)=\frac{K_0}{\omega\sqrt{[1+(\omega\tau)^2]}} \tag{10.69}$$

A particular real, positive value of ω, say ω_1, can be found such that $|G(j\omega)|=1$ which is

$$\omega_1=\frac{1}{\tau}\sqrt{[-\tfrac{1}{2}+\sqrt{(\tfrac{1}{4}+K_0^2\tau^2)}]}.$$

The phase angle between input and output is

$$\phi=\arg G(j\omega)=\arg K_0+\arg\left(\frac{1}{j\omega}\right)+\arg\left(\frac{1}{1+\tau j\omega}\right)$$

$$=\arg K_0+\arg\left(\frac{-j}{\omega}\right)+\arg\left(\frac{1-j\tau\omega}{1+(\omega\tau)^2}\right)$$

$$=0-\frac{\pi}{2}-\arctan(\omega\tau). \tag{10.70}$$

Is there a value of ω which gives ϕ the value $-\pi$? There is only one, which is when ω is infinitely large. There is therefore no value of ω which makes $G(j\omega)=-1$ which shows that the closed-loop system of Fig. 10.22 can never be marginally stable. This is a result we already knew, since the second-order system is always stable.

The polar, or Argand, diagram of the open-loop frequency response is known as a Nyquist diagram (after H. Nyquist's work in the early 1930s). A sketch of the Nyquist diagram for the transfer function of equation 10.61 is shown in Fig. 10.24 where the arrow shows the direction of increasing frequency. It can be seen that $G(j\omega)$ never has the critical value of -1. The plotting of Nyquist diagrams and a logarithmic form of frequency response are discussed later in this chapter.

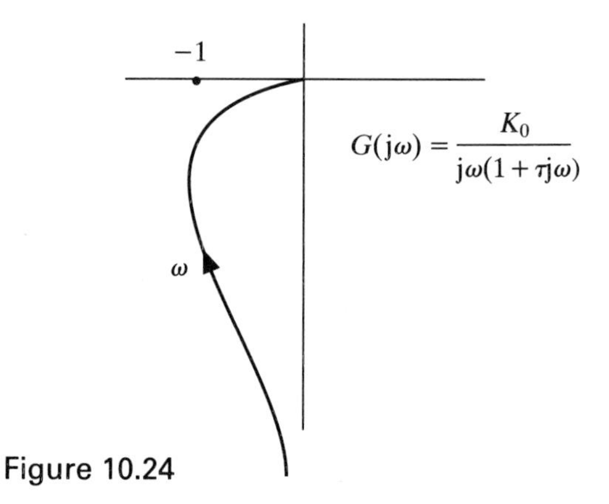

Figure 10.24

Assume now that the proportional controller of the above example is replaced by a controller with proportional-plus-integral action. The open-loop transfer operator of equation 10.55 is thus replaced by

$$G(D) = \frac{K + K_i/D}{ID^2 + CD} \qquad (10.71)$$

which we can write in the form

$$G(D) = \frac{K_0(1 + \tau_0 D)}{D^2(1 + \tau D)} \qquad (10.72)$$

where $K_0 = K_i/C$, $\tau_0 = K/K_i$ and $\tau = I/C$.

For sinusoidal inputs of circular frequency ω we can replace D by $j\omega$ as before to obtain the open-loop transfer function

$$G(j\omega) = \frac{K_0(1 + \tau_0 j\omega)}{(j\omega)^2(1 + \tau j\omega)} \qquad (10.73)$$

It is easy to show (see example 10.5) that the amplitude ratio is

$$|G(j\omega)| = \frac{K_0\sqrt{[1 + (\omega\tau_o)^2]}}{\omega^2\sqrt{[1 + (\omega\tau)^2]}} \qquad (10.74)$$

and the phase angle is

$$\phi = \arg G(j\omega) = \arctan(\omega\tau_o) - \pi - \arctan(\omega\tau) \qquad (10.75)$$

To check for marginal stability of the closed-loop system we look for the possibility of a value of ω which simultaneously gives $|G(j\omega)|$ the value of unity and ϕ the value of $-\pi$. We note that, from equation 10.75, ϕ will have the value of $-\pi$ if $\tau = \tau_0$ for any value of ω. If $|G(j\omega)| = 1$ then, from equation 10.74 with $\tau = \tau_0$

$$K_0/\omega^2 = 1$$

so, from equation 10.72

$$\omega^2 = K_0 = K_i/C = K/I.$$

We have shown that, for the open-loop frequency response, the amplitude ratio will be unity and simultaneously the phase lag will be π radians at the excitation frequency $\omega = \sqrt{(K/I)}$ provided that $K_i/C = K/I$ or $CK = IK_i$. This is equivalent to saying that the closed-loop system will be marginally stable provided that

$$CK = IK_i$$

and, if this is the case, the system will oscillate continuously at the frequency $\sqrt{(K/I)}$.

Sketches of the Nyquist diagrams for the open-loop response of the above system with proportional-plus-integral are shown in Fig. 10.25. It will be observed that the plot for $\tau = \tau_0$ passes through the critical point $G(j\omega) = -1$. The frequency response of equation 10.73 is discussed again in example 10.5.

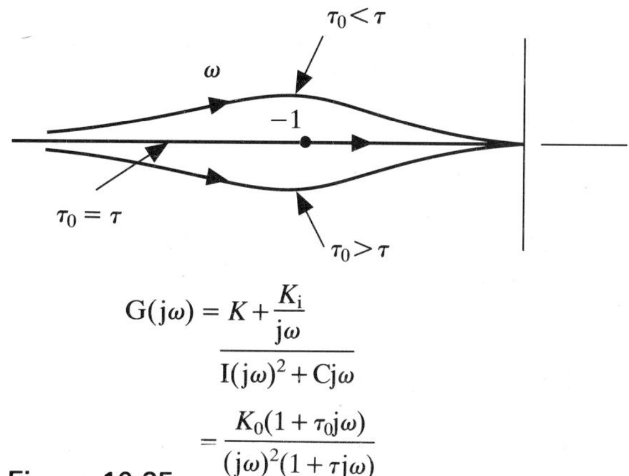

$$G(j\omega) = \frac{K + \dfrac{K_i}{j\omega}}{I(j\omega)^2 + Cj\omega} = \frac{K_0(1 + \tau_0 j\omega)}{(j\omega)^2(1 + \tau j\omega)}$$

Figure 10.25

Substituting for $G(D)$ from equation 10.72 into the closed-loop input/output system equation 10.63 we obtain

$$[\tau D^3 + D^2 + K_0\tau_o D + K_0]\theta_o = [K_0\tau_0 D + K_0]\theta_i$$

If the system is marginally stable, we have an equation of the form 10.48 for the complementary function, where $a_0 = K_0$, $a_1 = K_0\tau_0$, $a_2 = 1$ and $a_3 = \tau$. From equation 10.54, we find

$$\frac{K_0}{1} = \frac{K_0\tau_0}{\tau}$$

or $\qquad \tau_0 = \tau$

which confirms the result found from consideration of the open-loop frequency response.

Bode diagrams

The overall open-loop amplitude ratio is the product of the amplitude ratios of the component parts, and the overall phase angle is the sum of the phase angles of the component parts (see, for examples, equations 10.65 and 10.70).

When graphical techniques are employed it is convenient to plot the logarithm of the amplitude ratio $|G|$ (logarithms to the base 10 are always used). Traditionally, although not essentially, the logarithm of the amplitude is multiplied by 20 to give the ratio in the form of decibels (dB). When the frequency response information is presented in the form of two graphs, one $\log|G(j\omega)|$ or $20\log|G(j\omega)|$ plotted against $\log\omega$ and the other the phase angle ϕ plotted against $\log\omega$, the graphs are known as Bode diagrams after H. W. Bode who presented his work in the 1940s.

It is useful to build up a number of standard Bode diagrams of simple functions since knowledge of these enables (a) the rapid sketching of the overall frequency response plots and (b) the reduction of experimental results into component parts to assist with analysis. Below we use the notation $E(D)$ for the transfer operator of a component part and $G(D)$ for the transfer operator of the open-loop. $E(j\omega)$ and $G(j\omega)$ are the corresponding frequency response transfer functions.

(i) $E_1(D) = K$, a constant

The amplitude ratio $|E_1(i\omega)|$ of the frequency response is simply K and the phase angle ϕ is zero at all frequencies so that the Bode diagrams for this function are as shown in Fig. 10.26.

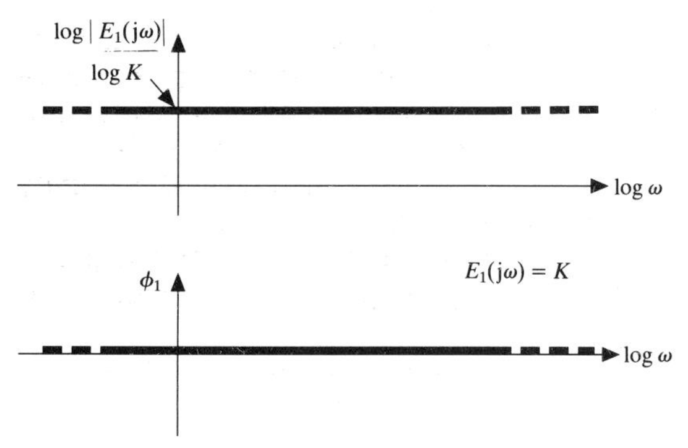

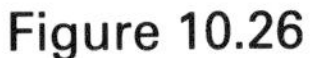
Figure 10.26

(ii) $E_2(D) = 1/D$, the integral operator

The amplitude ratio $|E_2(j\omega)| = 1/\omega$ from equation 10.67. The log of the amplitude ratio is $\log(1/\omega) = -\log\omega$ (or, in decibel form, $-20\log\omega$). The phase angle $\phi = \arg(-j/\omega) = -\pi/2$ at all values of the frequency ω. The Bode diagrams for this function are shown in Fig. 10.27. Each graph is a straight line with the log(amplitude ratio) graph having a slope of -1. (If decibels are used for this graph the gradient is -20. A tenfold increase in frequency is known as a decade and $\log 10 = 1$ so that this slope is often described as -20 dB/decade. A doubling in frequency is known as an octave and $\log 2 = 0.3010$ so the slope can also be described as -6 dB/octave since $20 \times 0.3010 = 6.02 \cong 6$).

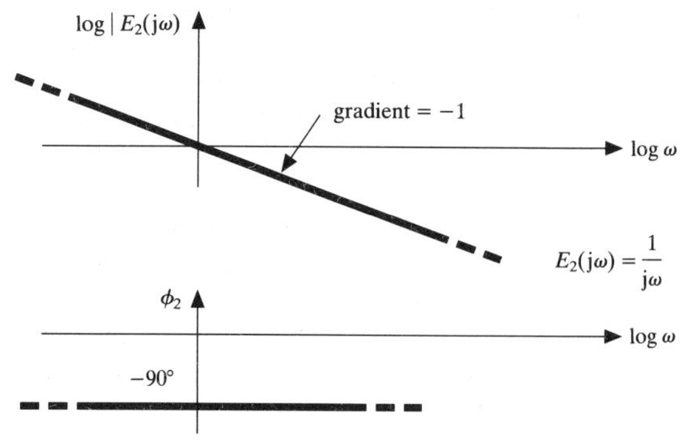

Figure 10.27

(iii) $E_3(D) = 1/(1+\tau D)$, the first order lag

The amplitude ratio $|E_3(j\omega)| = [1+(\omega\tau)^2]^{-1/2}$ from equation 10.68 and the phase angle is $\phi = \arg(1+\tau j\omega)^{-1} = -\arctan(\omega\tau)$ from equation 10.70. With regard to the overall shape of the Bode diagrams for this function we note that at low frequencies (small ω), $|E_3(j\omega)| \to 1$ and $\phi \to 0$ whereas at high frequencies (large ω), $|E_3(j\omega)| \to (\omega\tau)^{-1}$ and $\phi \to -\pi/2$. So at low frequencies $\log|E(j\omega)| \to -\log\omega - \log\tau$ or $-\log\omega + \log(1/\tau)$ which is a straight line of slope -1 on the graph of $\log|E(j\omega)|$ plotted against $\log\omega$ (or if decibels are used the slope is -20 dB/decade or -6 db/octave).

The log amplitude ratio and phase graphs are each therefore asymptotic to straight lines at both low and high frequencies. At the particular frequency $\omega = 1/\tau$, known as the break or corner frequency, $|E_3(j\omega)| = 2^{-1/2}$ and $\log|E_3(j\omega)| = -0.1505$ (and $20\log|E_3(j\omega)| \cong -3$ dB) also $\phi = \arctan(-1) = -\pi/4$ radians or $-45°$. The Bode diagrams for this function are shown in Fig. 10.28.

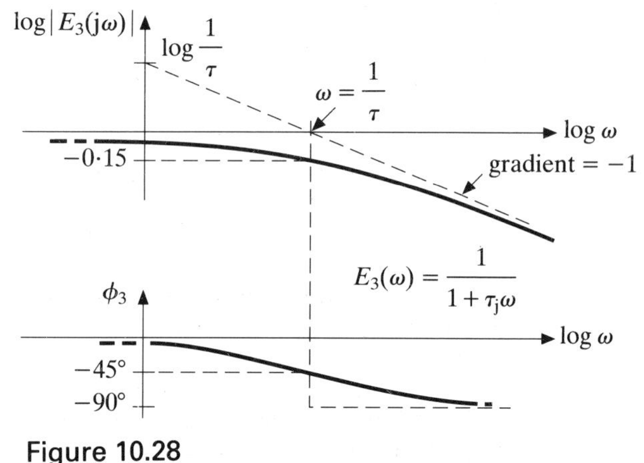

Figure 10.28

As an aid to sketching the phase graph of Fig. 10.28 it should be noted that the gradient of the graph at the break frequency $\omega = 1/\tau$ is $-(\ln 10)/2$ or -1.151 radians per decade $=$ $-66°$/decade (the proof of this is left for problem 10.25).

Accurate values of amplitude ratios in decibel form and phase angles for the function $E_3(j\omega)$ are listed in the table below and plotted to scale in Fig. 10.29 where a logarithmic scale is used for the frequency axis. In practice log graph paper is normally used for Bode diagrams.

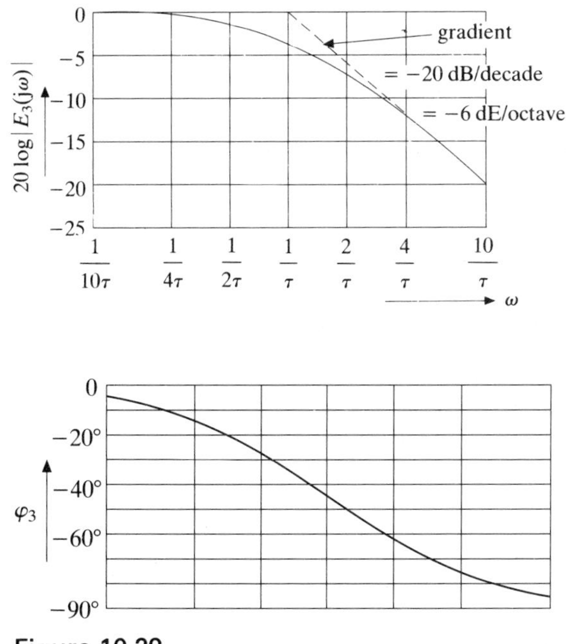

Figure 10.29

ω/(rad s^{-1})	$20\log\lvert E_3(j\omega)\rvert$/dB	ϕ/degrees
$1/(10\tau)$	−0.04	−5.71
$1/(4\tau)$	−0.26	−14.04
$1/(2\tau)$	−0.97	−26.57
$1/\tau$	−3.01	−45.00
$2/\tau$	−6.99	−63.43
$4/\tau$	−12.30	−75.96
$10/\tau$	−20.04	−84.29

Once scales for the graphs have been chosen, all of the graphs of $\log|E_3(j\omega)|$ will have the same shape, independent of the value of τ. A template can be made of the curve which has been drawn for a particular value of τ, the break point being $\omega = 1/\tau$, then the template can be moved along the $\log\omega$ axis for any other value of τ. The same applies to the phase graph $\phi = \arg[E_3(j\omega)]$. The proof of this phenomenon is left to the reader.

(iv) $E_4(D) = 1 + \tau D$, the first order lead

It is left to the reader to show that the log amplitude ratios and the phase curves are those for the transfer operator $E_3(D) = (1+\tau D)^{-1}$ rotated about the $\log\omega$ axis, as shown in Fig. 10.30.

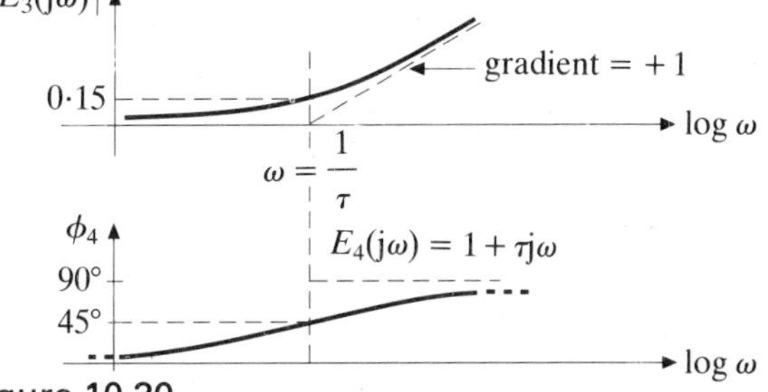

Figure 10.30

Assessment of closed-loop stability

We know that if, at a particular value of excitation frequency in the open-loop frequency response, the amplitude ratio is unity and simultaneously the phase angle is $-180°$ (i.e. $G(j\omega) = -1$) the loop, when closed by unity feedback, will be marginally stable. The closeness of the open-loop Nyquist plot to the critical point, $G(j\omega) = -1$, is a measure of the closed-loop stability.

Take, as an example, a control system with an open-loop transfer function of the form

$$G(D) = \frac{K}{D(1+\tau_1 D)(1+\tau_2 D)} \tag{10.76}$$

The Nyquist diagrams of $G(j\omega)$ for three particular values of K are sketched in Figs 10.31(a), (b) and (c).

In Fig. 10.31(b) a value of K has been chosen which makes the curve pass through the critical

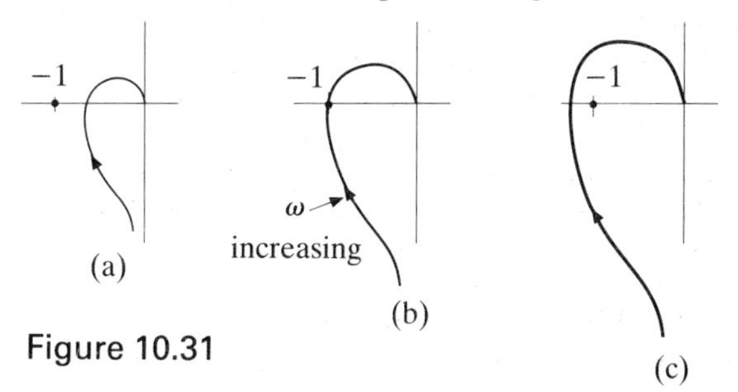

Figure 10.31

point (marginally stable closed-loop) while in Figs 10.31(a) and (c) smaller and larger values of K than that used in (b) are used respectively.

The *closed-loop* equation, from equation 10.63, is

$$\left[1+\frac{K}{D(1+\tau_1 D)(1+\tau_2 D)}\right]\theta_o$$

$$=\frac{K}{D(1+\tau_1 D)(1+\tau_2 D)}\theta_i$$

hence

$$[\tau_1\tau_2 D^3+(\tau_1+\tau_2)D^3+D+K]\theta_o=K\theta_i$$

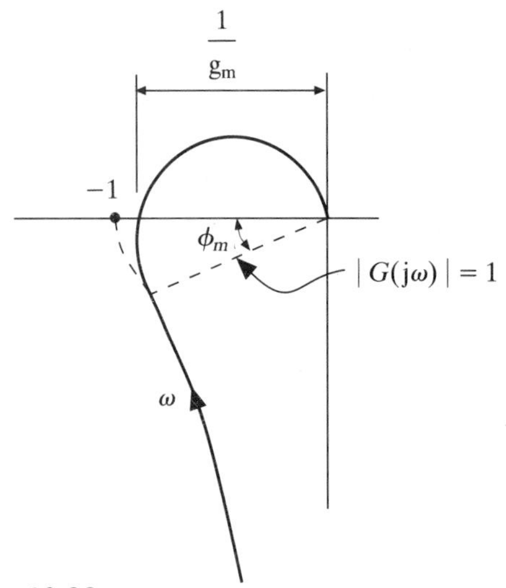

Figure 10.32

Comparing this with the standard form of equation 10.31 and applying inequality 10.32 we find, for a stable system,

$$K<(\tau_1+\tau_2)/(\tau_1\tau_2)$$

It follows that if $K=(\tau_1+\tau_2)/(\tau_1\tau_2)$ the closed-loop system will be marginally stable and if $K>(\tau_1+\tau_2)/(\tau_1\tau_2)$ the closed-loop system will be unstable. Figure 10.31(b) corresponds, as mentioned previously, to a marginally stable system while Figs 10.31(a) and (c) correspond to stable and unstable closed-loop systems respectively.

If one considers walking along the curve of Fig. 10.31(a) (stable closed-loop) in the direction of increasing frequency it will be observed that the critical point $G(j\omega)=-1$ falls to the left of the curve. Similarly for Fig. 10.31(c) (unstable closed-loop) the critical point falls to the right. This idea can be used as a rule of thumb for determining, from the open-loop response, whether or not the closed-loop system is stable and will suffice for all systems described in this chapter. There are a group of transfer functions where this rule does not apply, but they are outside the scope of this book and for these the reader is referred to more advanced texts on frequency response methods).

The closeness of the open-loop frequency response curve to the critical point is an indication of the performance of the closed-loop system and a measure of the closeness can be obtained from the *phase margin* ϕ_m and the *gain margin* g_m which are defined in the open-loop Nyquist diagram of Fig. 10.32.

The phase margin is given by $\phi_m=180°+\phi$ when the amplitude ratio is unity. It is the angle between the negative real axis and the $G(j\omega)$ vector when $|G(j\omega)|=1$.

The gain margin is the reciprocal of the amplitude ratio when the phase angle $\phi=-180°$.

If the gain K in, for example, equation 10.76 is multiplied by an amount equal to the gain margin, the open-loop will then pass through the critical point and the corresponding closed-loop will be marginally stable.

Typical values for satisfactory closed-loop performance are: ϕ_m should not be less than about 45° and g_m should not be less than about 2 or 6 dB.

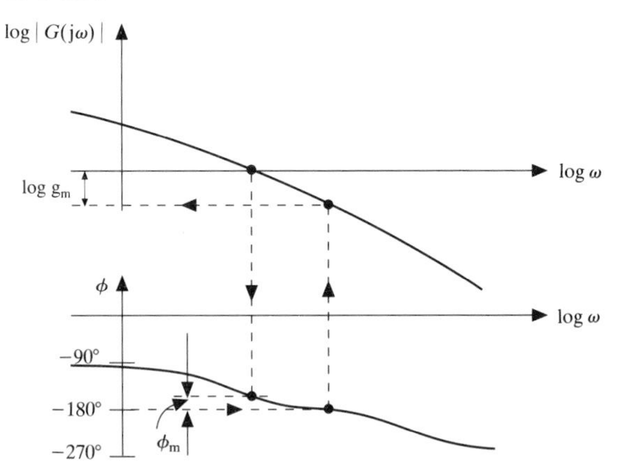

Figure 10.33

The phase margin and gain margin can of course be found from Bode diagrams and these are illustrated in Fig. 10.33.

Discussion examples

Example 10.1

Figure 10.34 shows a hydraulic relay with feedback used in a control system. Oil under pressure is supplied at P via the spool valve to the power ram and can exhaust to the drain at either Q or R. The value contains a sliding sleeve and the displacements of the spool and sleeve

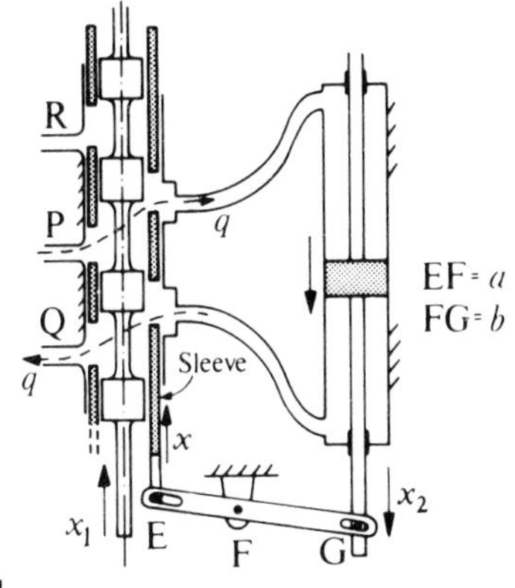

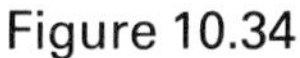
Figure 10.34

measured from the centralised position are x_1 and x respectively. Neglecting compressibility and inertias, the volumetric flow rate q through the valve can be taken to be proportional to the effective valve opening; that is,

$$q = k(x_1 - x)$$

The power ram, whose displacement is x_2, has an effective area A, and the sleeve is connected to the ram by the slotted lever EFG, which pivots about F.

Draw a block diagram for the relay and show that the transfer operator between x_2 and x_1 is of the form

$$\frac{x_2}{x_1} = \frac{K}{\tau \mathrm{D} + 1}$$

Find the values of the gain K and the first-order time-constant τ and determine the response of the relay to a step change in x_1 of magnitude X_1.

Solution The velocity v_2 of the ram downwards is equal to the product of the flow q and the area A:

$$v_2 = \mathrm{D}x_2 = qA = k(x_1 - x)A \qquad \text{(i)}$$

From the geometry of the feedback link (Fig. 10.35),

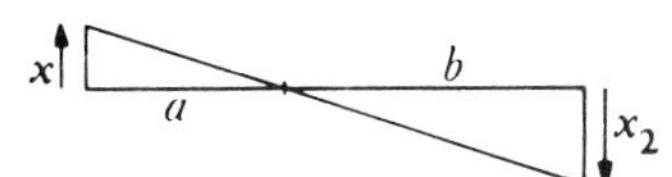

Figure 10.35

$$x/a = x_2/b \qquad \text{(ii)}$$

and equations (i) and (ii) are represented by the block diagram of Fig. 10.36.

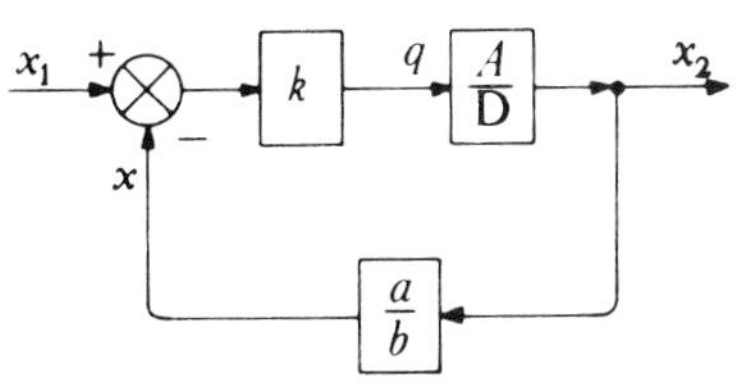

Figure 10.36

Eliminating x from equations (i) and (ii),

$$\mathrm{D}x_2 = k(x_1 - ax_2/b)A$$

$$(\mathrm{D} + kaA/b)x_2 = kAx_1$$

$$(b\mathrm{D}/(kaA) + 1)x_2 = (b/a)x_1$$

or

$$(\tau\mathrm{D} + 1)x_2 = Kx_1 \qquad \text{(iii)}$$

where $\tau = b/(kaA)$ and $K = b/a$.

The transfer operator is

$$\frac{x_2}{x_1} = \frac{K}{\tau\mathrm{D} + 1} \qquad \text{(iv)}$$

which can be represented by the block diagram of Fig. 10.37.

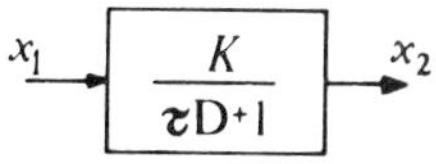

Figure 10.37

The complementary function (c.f.) of equation (iii) is the solution of $(\tau\mathrm{D} + 1)x_2 = 0$, which is of the form $x_2 = X_2\mathrm{e}^{mt}$; hence

$$\tau m X_2 \mathrm{e}^{mt} + X_2\mathrm{e}^{mt} = 0$$

and, dividing by $X_2\mathrm{e}^{mt}$,

$$m = -1/\tau$$

The c.f. is therefore

$$x_2 = X_2\mathrm{e}^{-t/\tau}$$

The particular integral is

$$x_2 = KX_1$$

and so the complete solution is

$$x_2 = X_2\mathrm{e}^{-t/\tau} + \mathrm{K}X_1 \qquad \text{(v)}$$

At $t = 0$, $x_2 = 0$, so that

$$0 = X_2(1) + KX_1$$
$$X_2 = -KX_1$$

and substituting for X_2 in equation (v) gives

$$x_2 = KX_1(1 - e^{-t/\tau})$$

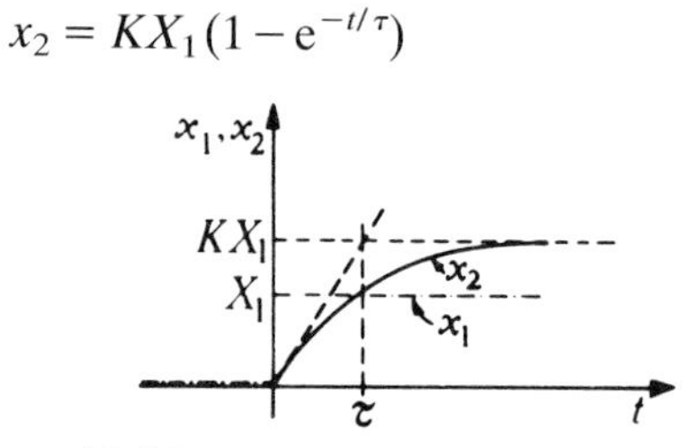

Figure 10.38

The response is sketched in Fig. 10.38. The transfer operator (iv) thus represents proportional control (the constant of proportionality being K) with a first-order lag of time-constant τ.

Example 10.2

Figure 10.39 shows a level-control rig. The tank T has a horizontal cross-sectional area A, and liquid flows into the tank at a rate q_i and out at a rate q_o. The height h_o of liquid in the tank is sensed by the float F, which displaces end G of link GHJ, where GH = HJ = l_1. The desired height h_i can be adjusted by altering the vertical position of end J. Link LMN is connected as shown; LM = MN = l_2. Point M, whose vertically downward displacement is y, is connected to the flapper U of a pneumatic flapper-nozzle valve

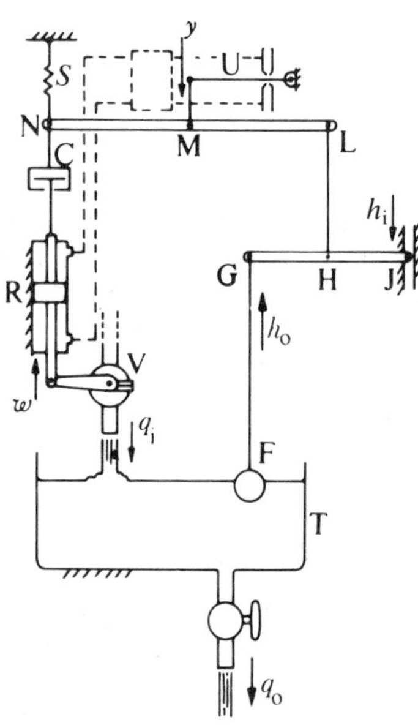

Figure 10.39

system which drives the power ram R whose vertically upward displacement is w. The velocity $\dot{w}$ of the ram is basically k_1 times the displacement y, but the pneumatic system introduces a first-order lag of time-constant τ. The ram adjusts valve V, and the flow q_i into the tank from a constant-head supply is k_2 times the ram displacement w. A dashpot of damping constant C and a spring of stiffness S connect end N to the ram and to ground respectively. The mass of all the links can be neglected.

a) Draw a block diagram for the system and show that it contains proportional-plus-integral action.
b) Show that there will be no steady-state error in level following a sudden change in the desired level h_i or demand q_o.
c) Determine the necessary and sufficient conditions for system stability.

Solution Considering small displacements of link GHJ, from similar triangles (Fig. 10.40),

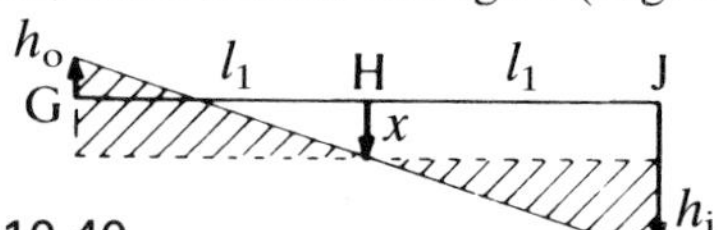

Figure 10.40

$$(h_i - x)/l_1 = (h_o + x)/l_1$$
$$x = \tfrac{1}{2}(h_i - h_o) = \tfrac{1}{2}h_e \qquad \text{(i)}$$

where x is the downward displacement of H (and L) and h_e is the error in level. Similarly, denoting the upward displacement of N by z,

$$y = \tfrac{1}{2}(x - z) \qquad \text{(ii)}$$

The motion of the ram R is given by

$$\mathrm{D}w = [k_1/(1 + \tau\mathrm{D})]y \qquad \text{(iii)}$$

The downward force acting on N due to the spring is Sz, and the upward force due to the dashpot is $C(\dot{w} - \dot{z})$. Since the links are light, the net force must be zero,

i.e. $Sz - C(\dot{w} - \dot{z}) = 0$

$$(C\mathrm{D} + S)z = C\mathrm{D}w \qquad \text{(iv)}$$

The flow into the tank is

$$q_i = k_2 w \qquad \text{(v)}$$

and the net inflow is equal to the area A times the rate of change of height h_o:

$$q_i - q_o = A\mathrm{D}h_o \qquad \text{(vi)}$$

Equations (i) to (vi) are represented in the

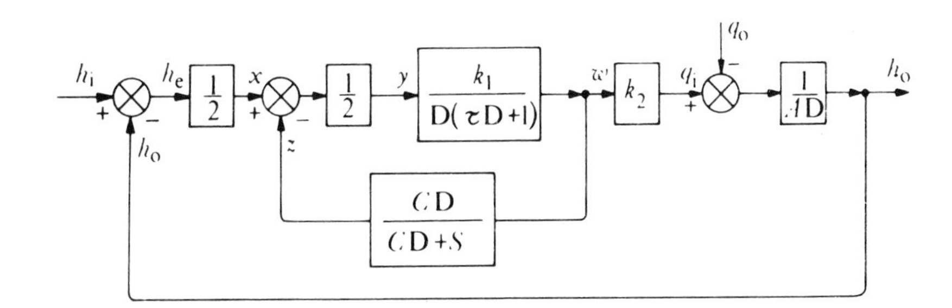

Figure 10.41

block diagram of Fig. 10.41. The transfer operator for the inner loop can be obtained by eliminating y and z from equations (ii), (iii), and (iv). The result is

$$\frac{w}{x}=\frac{\frac{1}{2}k(CD+S)}{D(\tau D+1)(CD+S)+\frac{1}{2}k_1CD}$$
$$=\frac{\frac{1}{2}k_1(C+S/D)}{\tau CD^2+(\tau S+C)D+(S+\frac{1}{2}k_1C)}$$

which indicates proportional-plus-integral action with a second-order lag. Eliminating x, w, and q_i and replacing h_o by h_i-h_e, we find

$$[A\tau CD^4+A(\tau S+C)D^3+A(S+\tfrac{1}{2}k_1C)D^2+\tfrac{1}{4}k_1k_2CD+\tfrac{1}{4}k_1k_2S]h_e$$
$$=AD^2[\tau CD^2+(\tau S+C)D+(S+\tfrac{1}{2}k_1C)]h_i+D[\tau CD^2+(\tau S+C)D+(S+\tfrac{1}{2}k_1C)]q_o \qquad \text{(vii)}$$

Note that, for step changes in desired level h_i or in demand q_o, the right-hand side of the system equation (vii) is zero and therefore the steady-state or particular-integral value of the error h_e is zero.

The system equation is of the fourth-order form

$$(a_4D^4+a_3D^3+a_2D^2+a_1D+a_0)h_e=\ldots$$

The first condition for stability is that all the coefficients a_0, a_1, a_2, a_3, a_4 be positive, which is satisfied. The next condition is

$$a_1a_2>a_3a_0$$
$$(k_1k_2C/4)(S+k_1C/2)A>A(\tau S+C)(k_1k_2S/4)$$

which reduces to

$$k_1C^2/2>\tau S^2 \qquad \text{(viii)}$$

The final condition for a fourth-order system is

$$a_1(a_2a_3)-a_0a_3^2-a_1^2a_4>0$$

which leads to

$$A(C+\tau S)(k_1C^2/2-\tau S^2)>k_1k_2C^3\tau/4 \qquad \text{(ix)}$$

We note that this inequality cannot be satisfied unless condition (viii) is satisfied.

Example 10.3

In a simple angular-position control system the driving torque on the load is k times the error θ_e. The load is a flywheel of moment of inertia I whose motion is opposed by a dry friction torque which can be assumed to have a constant magnitude T_f. Viscous damping is negligible. Numerical values are $k=0.2$ N m/rad, $I=1\times10^{-3}$ kg m^2, and $T_f=0.015$ N m.

Initially the system is at rest with zero error and then the double step input θ_i shown in Fig. 10.42 is applied. Find (a) the final position and (b) the time taken for all motion to cease.

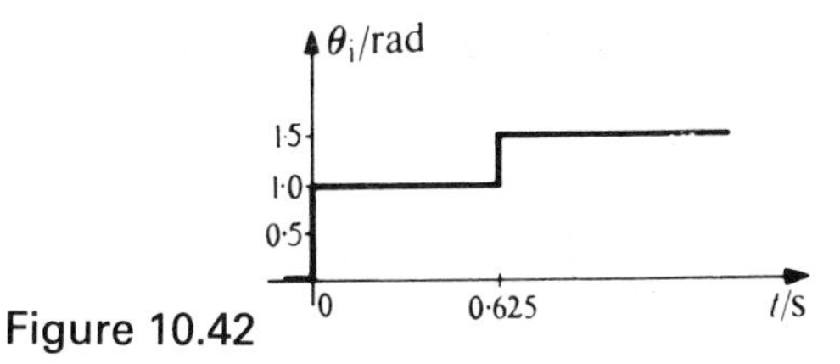

Figure 10.42

Solution If we choose to denote θ_o (and hence θ_i and θ_e) as positive in the anticlockwise sense, then the positive direction of its derivatives, $\dot{\theta}_o$ and $\ddot{\theta}_o$, is also anticlockwise. If the flywheel happens to be rotating anticlockwise ($\dot{\theta}_o>0$), then the friction torque T_f will be clockwise and vice versa (see Figs 10.43(a) and (b)).

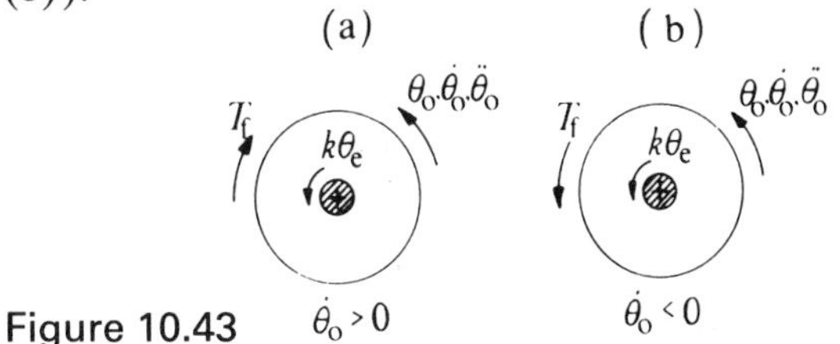

Figure 10.43

The two equations of motion are

$$k\theta_e - T_f = I\ddot{\theta}_o \quad (\text{for } \dot{\theta}_o > 0) \qquad \text{(i)}$$

$$k\theta_e + T_f = I\ddot{\theta}_o \quad (\text{for } \dot{\theta}_o < 0) \qquad \text{(ii)}$$

Replacing θ_e by $\theta_i - \theta_o$, we obtain

$$I\ddot{\theta}_o + k\theta_o = k\theta_i - T_f \quad (\dot{\theta}_o > 0)$$

$$I\ddot{\theta}_o + k\theta_o = k\theta_i + T_f \quad (\dot{\theta}_o < 0)$$

Substituting numerical values and dividing by $k = 0.2$,

$$\left.\begin{aligned}(5\times 10^{-3})\ddot{\theta}_0 + \theta_o &= 0.925 \quad (\dot{\theta}_o > 0) \qquad \text{(iii)}\\ (5\times 10^{-3})\ddot{\theta}_0 + \theta_o &= 1.075 \quad (\dot{\theta}_o < 0) \qquad \text{(iv)}\end{aligned}\right\} \theta_i = 1$$

$$\left.\begin{aligned}(5\times 10^{-3})\ddot{\theta}_0 + \theta_o &= 1.425 \quad (\dot{\theta}_o > 0) \qquad \text{(v)}\\ (5\times 10^{-3})\ddot{\theta}_0 + \theta_o &= 1.575 \quad (\dot{\theta}_o < 0) \qquad \text{(vi)}\end{aligned}\right\} \theta_i = 1.5$$

Each equation is a second-order linear differential equation with constant coefficients, no (viscous) damping term $\dot{\theta}_o$, and a constant forcing term. Thus the equations represent simple harmonic motion and we know (section 9.8) that if we draw a phase-plane plot of $\dot{\theta}_o/\omega_n$ against θ_o the result will be a series of circular arcs with centres on the θ_o axis corresponding to the particular integral or 'equilibrium' position for the relevant equation. Since the equations have been arranged to have a unity coefficient for θ_o, the right-hand sides of the equations are the particular integrals.

Each time the velocity $\dot{\theta}_o = 0$, we must check whether there is sufficient driving torque $k\theta_e$ to overcome the static friction torque T_f. So, whenever $\dot{\theta}_o = 0$, motion will not continue unless $0.2|\theta_e| > 0.015$ or $|\theta_e| > 0.075$. This corresponds to the range AB, $0.925 \leq \theta_o \leq 1.075$, when $\theta_i = 1$ and to the range CD, $1.425 \leq \theta_o \leq 1.575$, when $\theta_i = 1.5$ (see Fig. 10.44).

Immediately after the input θ_i is applied, $\theta_o = \dot{\theta}_o = 0$ and so the trajectory starts at point (a). The initial error is $\theta_i - \theta_o = 1 - 0 = 1$ and the initial driving torque $k\theta_e = (0.2)(1) = +0.2$ N m, causing the load velocity $\dot{\theta}_o$ to be positive. The appropriate centre for the trajectory is $\theta_o = 0.925$ (point A). We could have arrived at the same conclusion by noting that the trajectories always follow a clockwise pattern so that, when motion commences, the velocity $\dot{\theta}_o$ will be positive and the centre of the arc at point A.

The input $\theta_i = 1$ for 0.625 s. The angle swept by the radii which generate the trajectory is equal to $\omega_n t$ so that, while $\theta_i = 1$, the total angle of rotation will be $[1/(5\times 10^{-3})]^{1/2} \times 0.625$ radians or 506.4 degrees; that is, one revolution plus 146.4 degrees. After point (b), $\dot{\theta}_o$ becomes negative so we shift the centre to point B and so on until point (c) is reached. θ_i then takes on the value of 1.5 rad and so the appropriate centres become points C and D. The trajectory continues and then ceases at point (e), due to insufficient driving torque.

The final position is $\theta_o = 1.48$ rad, and the trajectory makes 2.5 revolutions so that the total time is $2.5(2\pi/\omega_n) = 2.5(2\pi)/(5\times 10^{-3})^{-1/2} = 0.89$ s.

It should be noted that this technique can be used for any input function θ_i if it is approximated by a series of small steps as shown in Fig. 10.45.

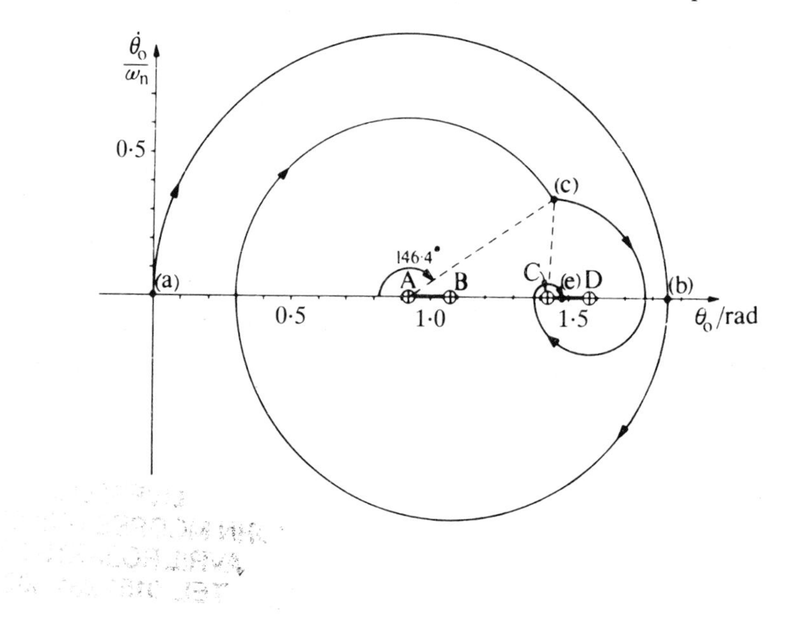

Figure 10.44

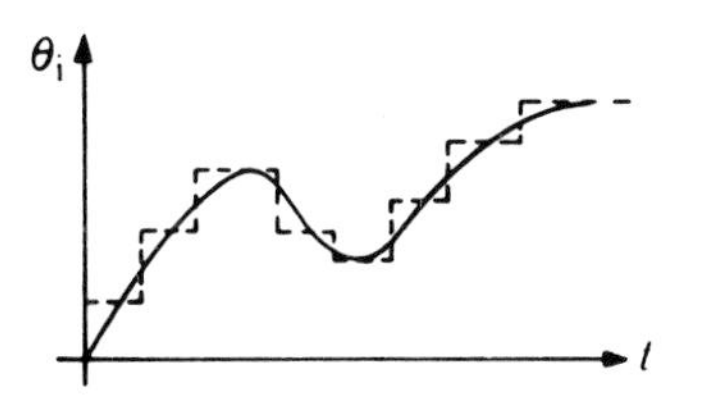

Figure 10.45

Example 10.4

A four-cylinder petrol engine is connected directly to a load. The total effective moment of inertia is I and the damping constant is C. A governor increases the throttle angle ϕ of the carburettor by N degrees for each rad/s of speed change. For steady-state operation over a particular operating range, the torque T_D driving the crankshaft increases by T_a for each degree increase in throttle angle and decreases by T_b for each rad/s increase in the engine speed ω_o. During speed variation, the carburettor introduces a first-order lag of time-constant τ. Numerical values are $I = 3$ kg m^2, $C = 0.2$ N m per rad/s, $N = 10$, $T_a = 4$ N m, $T_b = 0.35$ N m, and $\tau = 0.15$ s.

A brake applies a torque T_L of 50 N m to the load which is being driven by the engine at a constant speed of 200 rad/s. The brake is then removed. Assuming that the system is linear, obtain the differential equation which describes the engine speed (a) if the governor is disconnected and (b) if the governor is connected. Find the final speed for each case and draw a block diagram of the system for case (b).

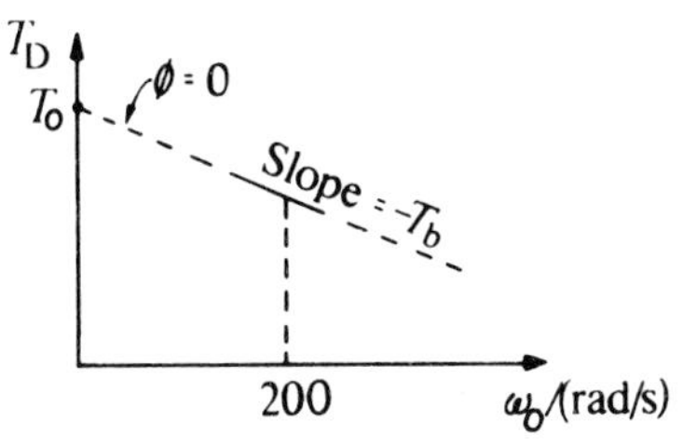

Figure 10.46

Solution Let us measure the throttle angle ϕ from its original position when the engine speed is steady at 200 rad/s. Figure 10.46 shows the engine torque T_D for $\phi = 0$ and steady-speed operation and is of the form

$$T_D = T_o - T_b\omega_o$$

The engine torque T_D for this case is simply equal to the damping torque $C\omega_o$ plus the braking torque T_L so that

$$T_o - T_b\omega_o - (C\omega_o + T_L) = 0$$

$$T_o = (C + T_b)\omega_o + T_L = (0.2) + 0.35)200 + 50 = 160 \text{ N m}$$

a) When the speed is varying, the equation of motion for the load with the brake removed and the governor disconnected (i.e. ϕ remaining constant at 0°), allowing for the time lag τ, is

$$T_D/(1 + \tau D) - C\omega_o = ID\omega_o$$

$$T_o - T_b\omega_o = (1 + \tau D)(ID + C)\omega_o$$

$$[I\tau D^2 + (I + C\tau)D + (C + T_b)]\omega_o = T_o$$

Substituting numerical values,

$$[0.45D^2 + 3.03D + 0.55]\omega_o = 160$$

The final value of ω_o is the particular integral of the above equation, which is

$$\omega_o = 160/0.55 = 290.9 \text{ rad/s}$$

b) When the governor is connected, it will operate in such a way as to increase the torque if the speed drops and vice versa. We can assume that, before the load was removed, the error in speed ω_e was zero; i.e. ω_i, the desired speed, is equal to 200 rad/s. The extra torque T_g provided by the governor is found from the two equations

$$\phi = N\omega_e$$

$$T_g = T_a\phi$$

so that

$$T_g = NT_a\omega_e$$

and T_D in part (a) is replaced by

$$T_D = T_o - T_b\omega_o + T_g$$

Substituting $\omega_i - \omega_o$ for ω_e, the resulting equation is

$$[I\tau D^2 + (I + C\tau)D + (C + T_b + NT_a)]\omega_o = T_o + NT_a\omega_i$$

$$[0.45D^2 + 3.03D + 40.55]\omega_o = 8160$$

The steady-state value of ω_o is

$$\omega_o = 8160/40.55 = 201.2 \text{ rad/s}$$

A block diagram of the control system is shown in Fig. 10.47.

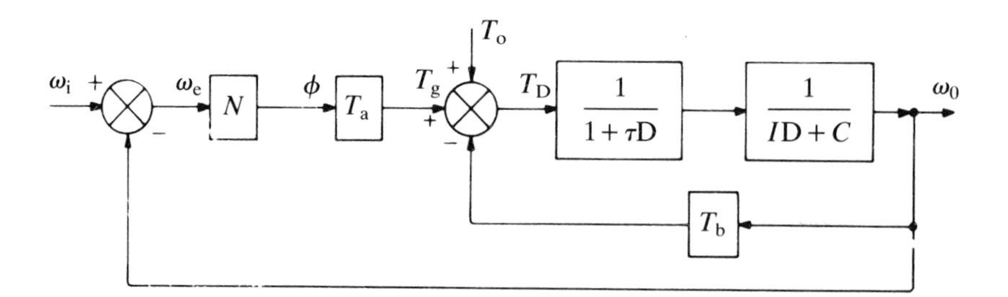

Figure 10.47

Example 10.5

Consider the control system with proportional-plus-integral action with the forward path transfer operator

$$G(D) = \frac{K + K_i/D}{ID^2 + cD} = \frac{K_0(1 + \tau_0 D)}{D^2(1 + \tau D)}$$

as in equations 10.71 and 10.2 and obtain sketches for $\tau_0 > \tau$, $\tau_0 = \tau$ and $\tau_0 < \tau$ of the overall open-loop frequency response in Bode and Nyquist form.

Solution The open-loop frequency response transfer function

$$G(j\omega) = \frac{K_0(1 + \tau_0 j\omega)}{(j\omega)^2(1 + \tau j\omega)} \qquad (10.77)$$

can be broken down into its individual components

$$E_a(j\omega) = K_0$$
$$E_b(j\omega) = 1 + \tau_0 j\omega$$
$$E_c(j\omega) = 1/(j\omega)^2$$
$$E_d(j\omega) = 1/(1 + \tau j\omega).$$

The Bode diagrams for E_a, E_b and E_d have been discussed above, see Figs 10.26, 10.30 and 10.28.

$$E_c = \left(\frac{1}{j\omega}\right)\left(\frac{1}{j\omega}\right)$$

so, combining two sets of Bode diagrams as in Fig. 10.27 the amplitude part for $E_c(j\omega)$ is a straight line of gradient -2 passing through the origin and the phase angle is constant at $-180°$.

The break frequencies for E_b and E_d are at $\omega = 1/\tau_0$ and $\omega = 1/\tau$ respectively.

Let us assume first that $\tau_0 > \tau$ (i.e. $1/\tau_0 < 1/\tau$) and initially, for simplicity, that $K_0 = 1$ and only the straight line approximations are required on the amplitude ratio graph. The component parts of the transfer function are shown with the dashed lines in Fig. 10.48 and the overall open-loop response with the full lines.

Assuming next that $\tau_0 < \tau$ (i.e. $1/\tau_0 > 1/\tau$) the break points of E_b and E_d are interchanged resulting in the Bode diagrams of Fig. 10.49.

Finally, if $\tau_0 = \tau$, the numerator and denominator terms $(1 + \tau_0 j\omega)$ and $(1 + \tau j\omega)$ in equation 10.77 cancel so that the transfer function reduces to

$$G(j\omega) = \frac{K_0}{(j\omega)^2}$$

Again letting $K_0 = 1$, the Bode diagrams for this system are simply the straight line graphs shown in Fig. 10.50.

For any positive value of K_0 not equal to unity an additional component of $\log K_0$ would be added to the amplitude ratio plots, so that the overall frequency response amplitude ratio curves obtained above would simply be moved upwards by an amount equal to $\log K_0$. If the above Bode diagrams are re-plotted in Nyquist form, the curves of Fig. 10.25 result.

Example 10.6

The forward path transfer function of a control system is given by

$$G(D) = \frac{C(D+5)}{D(D+2)(D+3)}$$

From the open-loop frequency response determine

a) the phase margin ϕ_m if $C = 10$,
b) the value of C if $\phi_m = 45°$, and
c) the gain margin for each case.

Solution We can rewrite the transfer operator in standard form as

$$G(D) = \frac{K_0(1 + \tau_1 D)}{D(1 + \tau_2 D)(1 + \tau_3 D)}$$

where $K_0 = 5C/6$, $\tau_1 = 1/5$, $\tau_2 = 1/2$ and $\tau_3 = 1/3$.

The Bode diagrams for each of the five

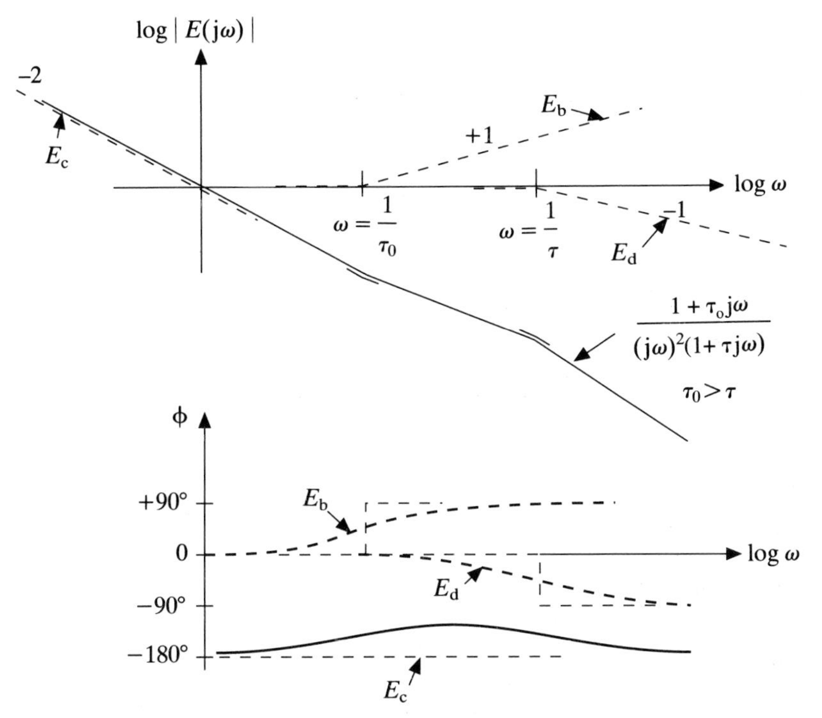

Figure 10.48

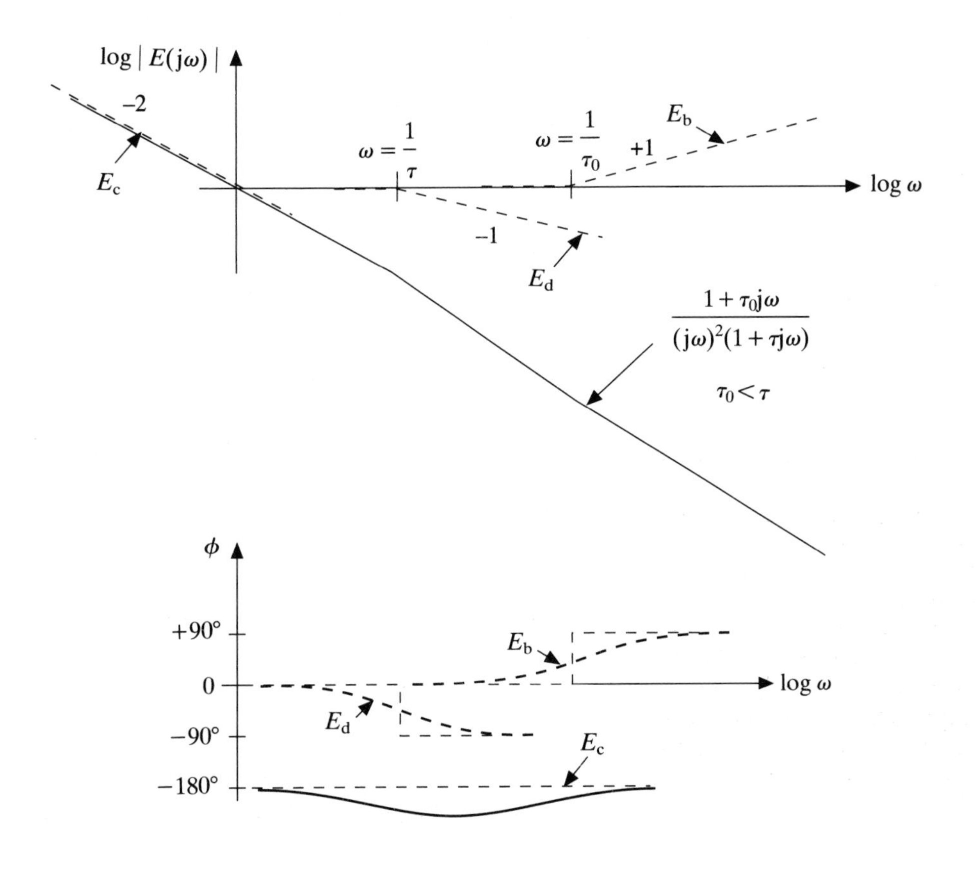

Figure 10.49

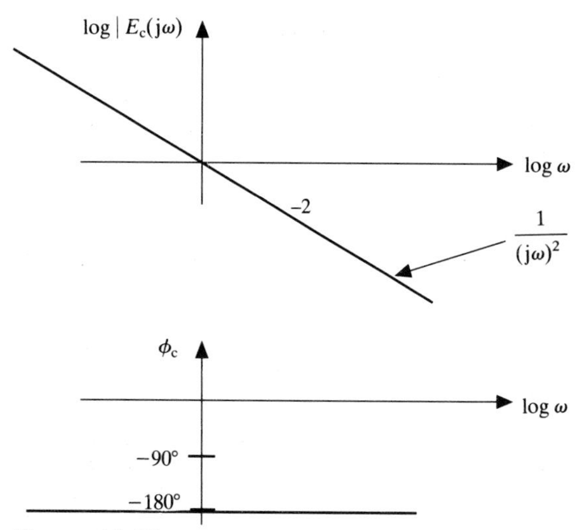

Figure 10.50

component parts of $G(j\omega)$ can be drawn, and the overall open-loop frequency response can be obtained by combining the components as in example 10.5. The phase margin and the gain margin can then be determined as in Fig. 10.33. This method is left as an exercise to the reader.

Alternatively we can work directly from the overall amplitude ratio and phase functions, as follows.

The overall open-loop frequency response function is

$$G(j\omega) = \frac{K_0(1+\tau_1 j\omega)}{j\omega(1+\tau_2 j\omega)(1+\tau_3 j\omega)}$$

The amplitude ratio is

$$|G(j\omega)| = \frac{K_0\sqrt{[1+(\tau_1\omega)^2]}}{\omega\sqrt{1+(\tau_2\omega)^2}\sqrt{[1+(\tau_3\omega)^2]}} \quad (10.78)$$

and the phase angle is

$$\phi = \arctan(\tau_1\omega) - \pi/2 - \arctan(\tau_2\omega) - \arctan(\tau_3\omega) \quad (10.79)$$

a) To find the phase margin we need to establish the value of ω at which $|G(j\omega)| = 1$. Substituting the numerical values into equations 10.78 and 10.79 and by trying a few values of ω, we obtain the following table

ω/(rad s^{-1})	$\lvert G(j\omega)\rvert$	ϕ/degrees
1	7.21	−123.7
2	2.64	−146.9
3	1.27	−160.3
4	0.72	−167.3

The amplitude ratio is seen to be unity somewhere between $\omega = 3$ and $\omega = 4$ rad/s. By trial and error we find that, at $\omega = 3.34$ rad/s, $|G(j\omega)| = 1.0001$ and the phase angle is $-163.8°$.

The phase margin is therefore

$$\phi_m = 180 - 163.8 = 16.2°.$$

(This would be too small in a practical system; the closed loop response would be too oscillatory).

b) If the phase margin ϕ_m is to be 45° we need to find the value of ω at which $\phi = -180° + 45° = -135°$. From the above table we see this occurs somewhere between $\omega = 1$ and $\omega = 2$ rad/s. By trial and error we find that at $\omega = 1.427$ rad/s the phase angle $\phi = -135.02°$. Using the value of $C = 10$ as in (a) above we find that the corresponding value of $|G(j\omega)|$ is 4.464. For a phase margin of 45° this value of $|G(j\omega)|$ should be unity so C needs to be reduced by a factor of 4.464. This gives $C = 10/4.464 = 2.24$.

Example 10.7

Figure 10.51(a) shows the block diagram for a system where the feedback is operated on by a transfer operator $H(D)$.

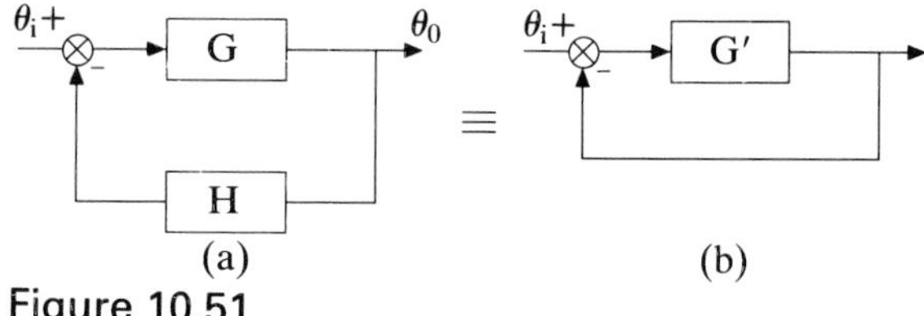

Figure 10.51

Determine (a) the closed-loop transfer operator and (b) equivalent unity feedback system.

Solution

a) The difference between the desired input and the signal which is now fed back is

$$\theta_e' = \theta_i - H(D)\,\theta_o$$

therefore the output

$$\theta_o = G(D)\,\theta_e' = G(D)\,\theta_i - G(D)H(D)\,\theta_o$$

thus

$$\theta_o = \frac{G(D)\,\theta_i}{1+G(D)H(D)}$$

so that the transfer operator is given by

$$\frac{\theta_o}{\theta_i} = \frac{G(D)}{1+G(D)H(D)}$$

b) Figure 10.51(b) shows the equivalent unity

feedback system such that the closed-loop transfer operator is the same as (a)

$$\text{i.e.}\quad \frac{G(D)}{1+G(D)H(D)} = \frac{G'(D)}{1+G'(D)}$$

$$\text{or}\quad G+GG' = G'+GHG'$$
$$G = G'(1+GH-G)$$

$$\text{finally}\quad G' = \frac{G}{(1+GH-G)}$$

Note that the open-loop transfer operator for (a) is GH and if $G(j\omega)H(j\omega)$ equals -1 then the closed-loop system is marginally stable. Also for (b) the open-loop transfer operator is G' and if $G'(j\omega)$ equals -1 then this system is marginally stable. A little algebra soon shows that $G'(j\omega) = -1$ implies that $G(j\omega)H(j\omega) = -1$, so that all the analysis carried out in this chapter considering unity feedback applies equally well to systems with non-unity feedback operators using the appropriate open-loop transfer operator.

Problems

10.1 For the system represented by the block diagram of Fig. 10.52, obtain the transfer operator for x/y.

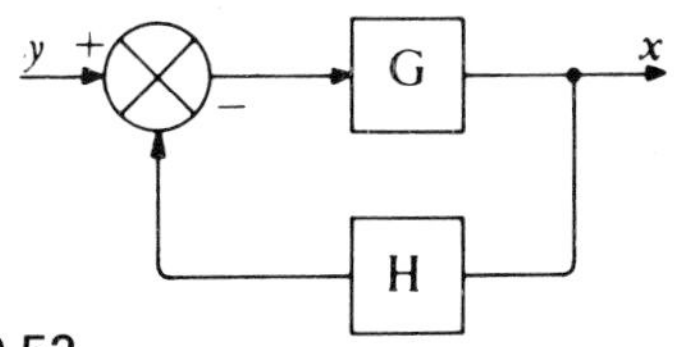

Figure 10.52

10.2 For the control system of Fig. 10.53, obtain the system equation for each of the loop variables.

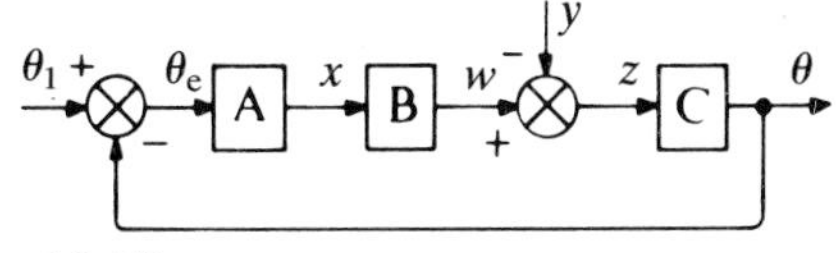

Figure 10.53

10.3 A motor used in a position-control system has its input voltage V_m, its output torque T_m, and its angular velocity ω_m related by the equation

$$T_m = K_1 V_m - K_2 \omega_m$$

The motor is connected directly to a load of moment of inertia I whose motion is opposed by a viscous damping torque equal to C times the angular velocity of the load. If the motor voltage V_m is given by $K_3\theta_e$, where θ_e is the position error, show that the output–input system equation is

$$[ID^2 + (C+K_2)D + K_1K_3]\theta_o = K_1K_3\theta_i$$

10.4 A voltage V is produced which is K_1 times the error in a position-control system. The load is a flywheel of moment of inertia I_L and the damping torque at the load is equal to C times the angular velocity of the load. The moment of inertia of the rotor of the motor which drives the load is I_m and the torque developed between the rotor and the stator is given by $T_m = K_2V$. Obtain the system equation for the output θ_o and also determine the damping factor for each of the following cases: (a) the motor is directly connected to the load; (b) as (a) with an external torque Q_L applied to the load; (c) a gearbox is placed between the motor and the load such that $\omega_M = n\omega_L$.

10.5 The amount of damping in a position-control system is increased by using *proportional-plus-derivative* action such that the driving torque T_D applied to the load is given by $T_D = k_1\theta_e + k_2\dot{\theta}_e$, where θ_e is the error. The moment of inertia of the load is I and the viscous damping constant is C. If the damping ratio of the system is $\frac{1}{2}$, show that $k_1 I = (C+k_2)^2$.

10.6 Figure 10.54 shows a hydraulic power ram B fed by a spool valve V. It can be assumed that the ram velocity is proportional to the spool displacement measured from the centralised position, the constant of proportionality being k. The slotted link PQR is connected to the spool and ram as shown.

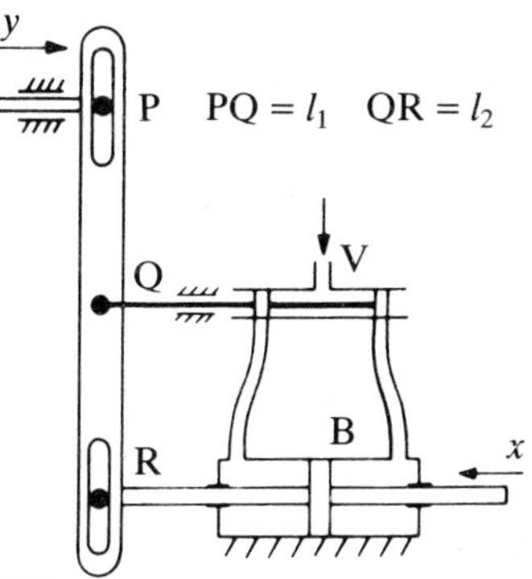

Figure 10.54

Show that the transfer operator for the arrangement is of the form $x/y = A/(1+\tau D)$ and write down expressions for the gain A and first-order time-constant τ.

10.7 The hydraulic relay of problem 10.6 is modified by the addition of a spring of stiffness S and a damper of damping constant C, as shown in Fig. 10.55. Show that the modified arrangement gives proportional-plus-integral action with a first-order lag of time-constant τ by obtaining the transfer operator in the form

$$\frac{x}{y} = A\frac{(1+1/(\tau_i D))}{(1+\tau D)}$$

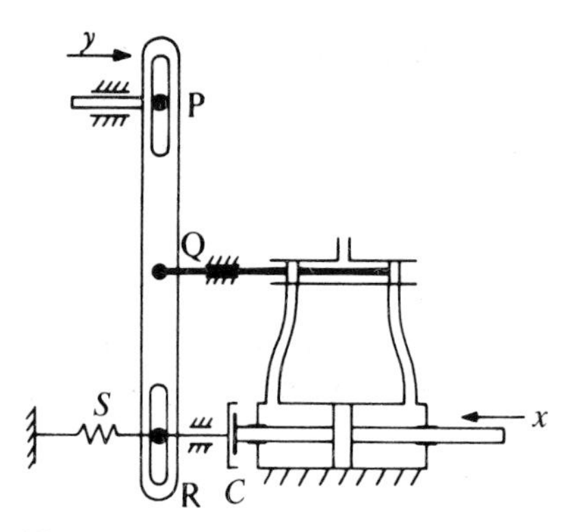

Figure 10.55

Evaluate the constant A, the lag constant τ, and the integral time constant τ_i.

10.8 Consider the level-control system of example 10.2 with the spring removed and the dashpot replaced by a rigid link. The system is steady, supplying a constant demand Q_o. Show that if the demand is increased by 10%, the level drops by $0.2Q_o/k_2$.

10.9 The load of a position-control system is an undamped flywheel of moment of inertia I. The driving torque on the load may be assumed to be K_o times the motor input voltage V. A three-term controller and amplifier whose output is V have a combined transfer operator

$$V/\theta_e = (K_1 + K_2\mathrm{D} + K_3/\mathrm{D})$$

where θ_e is the position error, and D the operator d/dt.

a) Show that the maximum value of K_3 for stability is $K_0K_1K_2/I$.

b) Show that the steady-state position error for each of the following inputs is zero: (i) step input, (ii) ramp input, and (iii) acceleration input.

10.10 A simple position-control system has a viscously damped load. The moment of inertia of the load is 4 kg m^2 and the damping constant is 8 N m per rad/s. The driving torque applied to the load is K times the position error and the system has a damping ratio of unity. (a) Find the value of K. (b) If the system is initially at rest and then at $t = 0$ the input shaft is rotated at 0.4 rad/s, find the steady-state position error.

10.11 For the previous problem, show that the position of the load is given by $\theta_o = 0.4t - 0.8[1 - \mathrm{e}^{-t}(1 + \frac{1}{2}t)]$ where t is the time in seconds. Find when the maximum acceleration of the load occurs and determine its value.

10.12 Derive all of equations 10.17.

10.13 See example 10.3. Rewrite equations (iii) to (vi) in terms of θ_e instead of θ_o. Draw the phase-plane plot of $\dot{\theta}_e/\omega_n$ against θ_e and hence show that the final error θ_e is 0.02 rad.

10.14 See problem 9.21. In a position-control system, the driving torque on the load is 0.2 N m/rad of error. The load is a flywheel of moment of inertia 5×10^{-4} kg m^2 whose motion is opposed by a dry-friction torque such that the torque required to initiate motion is 0.022 N m but once motion has started the resisting torque is 0.015 N m. Viscous damping is negligible. Initially the system is at rest and then a step input of 1 radian is applied. Find (a) the time taken for all motion to cease and (b) the steady-state error.

10.15 In an angular-position control system the load consists of a flywheel of moment of inertia I_L and the driving torque is K times the position error. Damping of the load is brought about by a viscous Lanchester damper in the form of a second flywheel of moment of inertia I_D mounted coaxially with the first and connected to it by a viscous damper. The torque transmitted through the damper is C times the relative angular velocity of the flywheels.

a) Show that the system is stable.

b) Determine the steady-state errors following inputs of the form (i) $Au(t)$, (ii) $Atu(t)$, and (iii) $At^2u(t)$ where A is constant and $u(t) = 0$ for $t < 0$, $u(t) = 1$ for $t \geq 0$.

10.16 Figure 10.56 shows a flywheel A which is driven by a motor M having an output torque Q. Flywheel A drives flywheel B by viscous action, the torque transmitted being C times the relative angular velocity. One end of a spring of torsional stiffness S is attached to B, the other end being fixed. The moments of inertia of A and B are I_A and I_B respectively; the inertia of M is negligible.

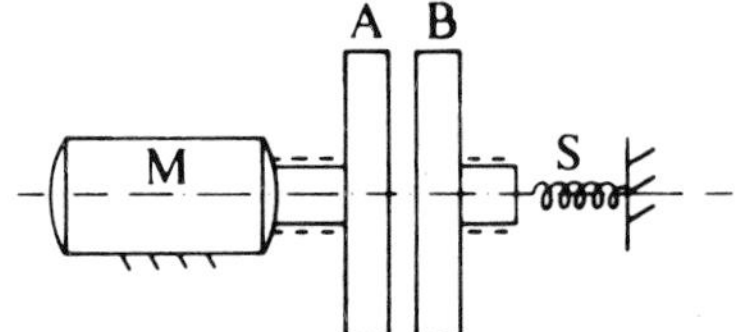

Figure 10.56

a) Derive a differential equation relating Q to the angular position θ_A of A.

b) If A is the load in a position-control system and Q is K times the error, obtain the fourth-order output–input system equation and show that the system is always stable.

10.17 In a speed-control system, the driving torque T_D which is applied to the load increases by 0.01 N m for each rad/s of error ω_e. The load consists of a flywheel of moment of inertia 0.5 kg m^2 with viscous damping amounting to 0.04 N m per rad/s of load speed.

a) If the load is running at a constant speed of 150 rad/s with no error, determine the equation relating T_D to ω_e and find the time-constant of the system.

b) If an input corresponding to a desired speed of 180 rad/s is applied to the system, what will be the final speed?

10.18 A petrol engine drives a load and the total effective moment of inertia and the viscous damping constant are 4 kg m^2 and 6 N m per rad/s respectively. The throttle of the carburettor is adjusted by a governor, and the basic control is that the engine torque increases by 30 N m for each rad/s of speed error but a first-order (exponential) lag of 0.15 s is introduced.

The engine is initially running at 250 rad/s and then a constant braking torque is applied, causing the speed to drop to 240 rad/s. Find the value of the braking torque and obtain a differential equation which defines the engine speed ω_o after the torque is applied.

10.19 A control system is represented by the equation

$$(a_3\mathrm{D}^3 + a_2\mathrm{D}^2 + a_1\mathrm{D} + a_0)\,\theta_o = (b_1\mathrm{D} + b_0)\,\theta_i$$

What is the condition which results in the system having a sustained oscillation, and what is the frequency of this oscillation?

10.20 A trolley T has pinned to it at O an inverted pendulum P as shown in Fig. 10.57. The pendulum is a uniform thin rigid rod of length $2a$ and mass m. The desired value of θ is zero, and, in order to balance the pendulum above the trolley, it is proposed to control the displacement x of the trolley such that $\mathrm{D}^2x = (A + B\mathrm{D})\,\theta$ where A and B are positive constants.

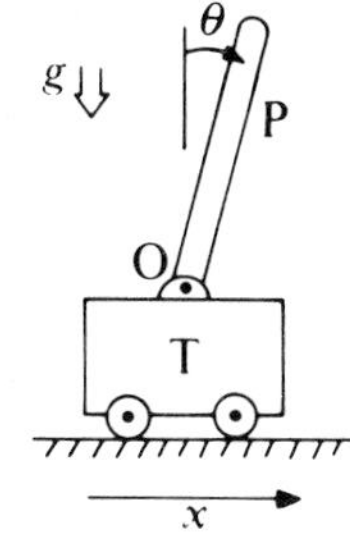

Figure 10.57

Assuming that θ remains small, show that the dynamics of the pendulum are represented by $(g - \frac{4}{3}a\mathrm{D}^2)\,\theta = \mathrm{D}^2x$ and that the control will be successful provided that $A > g$.

Initially the control is switched off and the pendulum held at an angle $\theta = \theta_1$. At time $t = 0$ the pendulum is released and simultaneously the control is brought into action. Show that, in the steady-state, the trolley has a constant velocity to the right and determine this velocity.

10.21 Obtain accurate Bode plots of the transfer function

a) $1/(1 + j\omega)$, and

b) $$\frac{1}{j\omega(1 + 0.025j\omega)(1 + 0.1j\omega)}$$

(*Hint*: use the results from (a) for each of the bracketed terms in (b).)

10.22 The open-loop transfer operator of a control system is

$$\frac{K}{D(1 + 0.025D)(1 + 0.1D)}$$

Using the Bode diagrams from problem 10.21(b), determine the value of K which will give, for open-loop frequency response,

a) a phase margin of 40°

b) a gain margin of 2.0.

10.23 Repeat problem 10.22 using the technique of example 10.6.

If the loop is closed by unity feedback, what value of K will make the system marginally stable?

10.24 In the tables given below, experimental frequency response results are given. Given that the component parts of the transfer functions are made up from such functions as K, $(1 + \tau j\omega)$, $1/j\omega$, and $(1 + \tau j\omega)^{-1}$, working graphically fit these types of function to the overall response results and hence determine the overall function for each case.

a)

Frequency (ω)/ (rad/s^{-1})	Amplitude ratio	Phase lag/ degrees
0.6	17.0	—
1	10	80
2	5.5	69
3	4.0	60
4	3.2	50
5	2.7	45
10	2.2	29
20	2.1	16
100	2.0	5

b)

Frequency (ω)/ (rad/s^{-1})	Amplitude ratio	Phase lag/ degrees
1	5.0	96
2	2.5	100
4	1.1	110
8	0.5	130
20	0.1	155
50	0.02	170
100	—	175

10.25 A simple rotary position servo is to have the load driving torque equal to K times the position error. The load is a viscously damped flywheel with a moment of inertia $I = 2.27$ kgm^2 and the damping constant $c = 75$ Nms^{-1}. The open-loop frequency response is to have a phase margin of 45°.

a) What is the appropriate value of K?

b) What is the gain margin?

c) What is the damping ratio of the closed-loop system?

10.26 The transfer function of a first-order lag is of the form $E(j\omega) = (1 + \tau j\omega)^{-1}$. Show that, at the break frequency $\omega = 1/\tau$, the slope of the phase-frequency plot is $-66°$/decade.

11
Dynamics of a body in three-dimensional motion

11.1 Introduction

A particle in three-dimensional motion requires three independent co-ordinates to specify its position and is said to have three degrees of freedom. For a rigid body the positions of three points specify the location and orientation of the body uniquely. The nine co-ordinates are not, however, independent because there are three equations of constraint expressing the fact that the distances between the three points are constant; thus there are only six independent co-ordinates. An unrestrained rigid body therefore has six degrees of freedom.

Another way of defining the position of a body is to locate one point of the body – three co-ordinates – then to specify the direction of a line fixed to the body, two co-ordinates, and finally a rotation about this line giving six co-ordinates in total.

In order to simplify the handling of three-dimensional problems it is frequently convenient to use translating and/or rotating axes. These axes may be regarded kinematically as a rigid body, so a study of the motion of a rigid body will be undertaken first.

11.2 Finite rotation

It has already been stated that finite rotation does not obey the laws of vector addition; this is easily demonstrated with reference to Fig. 11.1.

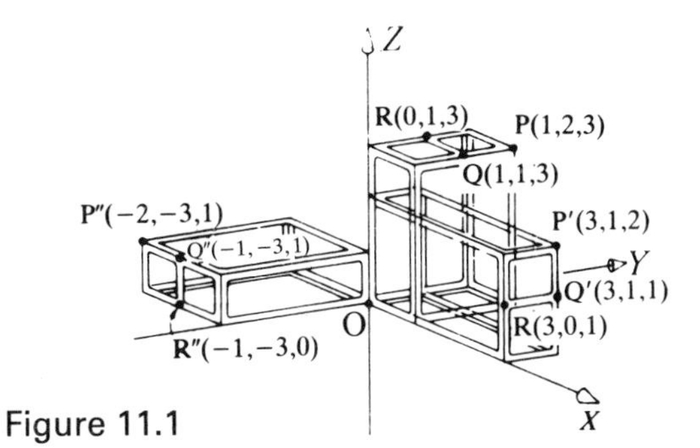

Figure 11.1

The displacement of point P to P′ has been achieved by a rotation of 90° about the X-axis followed by a rotation of 90° about Z-axis. If the order of rotation had been reversed, the point P would have been moved to P″, which is clearly a different position. If the rotations are defined relative to axes fixed to the body, it is found that a rotation of 90° about the X-axis followed by a 90° rotation about the new Z-axis transforms P to P″ and reversing the order transforms P to P′.

The change in position produced by a rotation about the X-axis followed by a rotation about the Z-axis can be effected by a single rotation about an axis through O. The direction of this axis is easily found since the displacements $\overrightarrow{PP'}$, $\overrightarrow{QQ'}$, and $\overrightarrow{RR'}$ are all normal to the axis of rotation; therefore the forming of the vector product of any two will give a vector parallel to the axis of rotation – see Fig. 11.2.

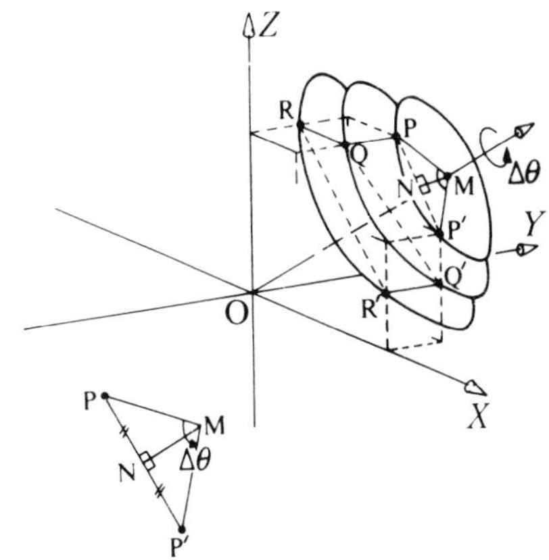

Figure 11.2

Two of the displacement vectors are

$$\overrightarrow{PP'} = \boldsymbol{i}(3-1)+\boldsymbol{j}(1-2)+\boldsymbol{k}(2-3) = 2\boldsymbol{i}-1\boldsymbol{j}-1\boldsymbol{k}$$

and

$$\overrightarrow{QQ'} = \boldsymbol{i}(3-1)+\boldsymbol{j}(1-1)+\boldsymbol{k}(1-3) = 2\boldsymbol{i}-2\boldsymbol{k}$$

The axis of the equivalent single rotation is perpendicular to both $\overrightarrow{PP'}$ and $\overrightarrow{QQ'}$; therefore it is parallel to

$$\overrightarrow{PP'} \times \overrightarrow{QQ'} = \begin{vmatrix} \boldsymbol{i} & \boldsymbol{j} & \boldsymbol{k} \\ 2 & -1 & -1 \\ 2 & 0 & -2 \end{vmatrix}$$
$$= 2\boldsymbol{i} + 2\boldsymbol{j} + 2\boldsymbol{k}$$

Since the point O does not change position, the axis of rotation must pass through the point O as well as through the point (2,2,2).

From Fig. 11.2 we can see that $|\overrightarrow{PP'}| = 2|\overrightarrow{MP}|\sin(\Delta\theta/2)$. The point M is on the axis of rotation, so

$$\overrightarrow{OM} = \lambda(2\boldsymbol{i} + 2\boldsymbol{j} + 2\boldsymbol{k})$$

where λ is a constant to be determined. It is clear that

$$\overrightarrow{MP} = \overrightarrow{OP} - \overrightarrow{OM}$$
$$= (\boldsymbol{i} + 2\boldsymbol{j} + 3\boldsymbol{k}) - \lambda(2\boldsymbol{i} + 2\boldsymbol{j} + 2\boldsymbol{k})$$

but $\overrightarrow{MP}$ is normal to $\overrightarrow{OM}$; hence $\overrightarrow{MP} \cdot \overrightarrow{OM} = 0$

or $\lambda(2+4+6) - \lambda^2(4+4+4) = 0$

giving $\lambda = 1$

So we have

$$\overrightarrow{MP} = -\boldsymbol{i} + \boldsymbol{k} \quad \text{hence} \quad |\overrightarrow{MP}| = \sqrt{2}$$

and since $\overrightarrow{PP'} = 2\boldsymbol{i} - \boldsymbol{j} - \boldsymbol{k}$

$$|\overrightarrow{PP'}| = \sqrt{6}$$

Therefore $\sin(\Delta\theta/2) = \dfrac{|\overrightarrow{PP'}|}{2|\overrightarrow{MP}|} = \dfrac{\sqrt{6}}{2\sqrt{2}} = 0.866$

giving $\Delta\theta = 2 \times 60° = 120°$

Had the point O not been a fixed point, the above calculation could still be performed by subtracting the displacement of O from all other displacements, thereby treating O as if it were a fixed point. The total displacement of the body is then the displacement of O plus a rotation about O. If the displacement of O is normal to the axis of rotation then a true fixed point can be found, but if a component of the displacement of O is along the axis of rotation then no such fixed point can be located. When the displacement of O is wholly along the axis of rotation through O, the motion is known as a screw displacement (see example 11.1).

In conclusion, we now state the following theorems.

i) Any finite displacement of a rigid body may be reduced to a single rotation about an axis plus a translation parallel to the same axis. This axis is known as Poinsot's central axis. (It should be noted that only the displacements are equivalent and not the paths taken by the points.)

ii) If a point on a rigid body does not change its position then any series of successive rotations can be compounded to a rotation about a single axis (Euler's theorem).

iii) Any displacement of a rigid body may be compounded from a single rotation about any given point plus a translation of that point (Chasles's theorem).

11.3 Angular velocity

First consider Fig. 11.3(a) which shows the surface of a sphere radius r. The finite displacement $\overrightarrow{PP'}$ has a magnitude $2\tan\frac{1}{2}\theta_x|\overrightarrow{NN'}|$ and is in a direction parallel to $\boldsymbol{i} \times \overrightarrow{NN'}$ or to $\boldsymbol{i} \times \overrightarrow{ON'}$.

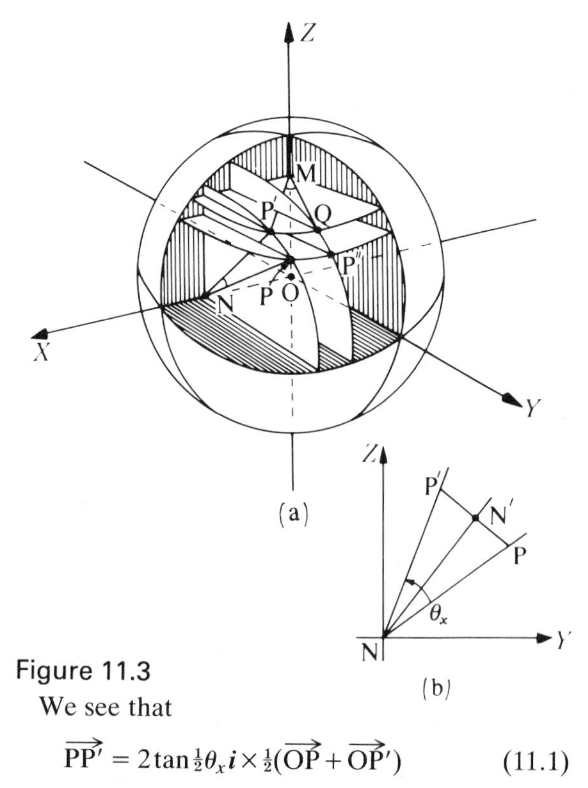

Figure 11.3

We see that

$$\overrightarrow{PP'} = 2\tan\tfrac{1}{2}\theta_x \boldsymbol{i} \times \tfrac{1}{2}(\overrightarrow{OP} + \overrightarrow{OP'}) \qquad (11.1)$$

since $\frac{1}{2}(\overrightarrow{OP} + \overrightarrow{OP'}) = \overrightarrow{ON'}$, see Fig. 11.3(b).

Similarly,

$$\overrightarrow{P'Q} = 2\tan\tfrac{1}{2}\theta_z \boldsymbol{k} \times \tfrac{1}{2}(\overrightarrow{OP'} + \overrightarrow{OQ})$$

For small angles,

$$2\tan\tfrac{1}{2}\Delta\theta_x \boldsymbol{i} \rightarrow \Delta\theta_x \boldsymbol{i}$$

and $\overrightarrow{OP'} \rightarrow \overrightarrow{OP} \rightarrow \boldsymbol{r}$

thus $\overrightarrow{PP'} = (\Delta\theta_x \boldsymbol{i}) \times \boldsymbol{r}$

and $\overrightarrow{P'Q} = (\Delta\theta_z \boldsymbol{k}) \times \boldsymbol{r}$

therefore $\overrightarrow{PQ} = \overrightarrow{PP'} + \overrightarrow{P'Q} = \Delta\boldsymbol{r}$

or $\qquad \Delta\boldsymbol{r} = (\Delta\theta_x \boldsymbol{i} + \Delta\theta_z \boldsymbol{k}) \times \boldsymbol{r}$

If this change takes place in a time Δt then

$$\boldsymbol{v} = \lim_{\Delta t \to 0} \frac{\Delta\boldsymbol{r}}{\Delta t} = \lim_{\Delta t \to 0} \left(\frac{\Delta\theta_x}{\Delta t} \boldsymbol{i} + \frac{\Delta\theta_z}{\Delta t} \boldsymbol{k} \right) \times \boldsymbol{r}$$

so $\boldsymbol{v} = (\omega_x \boldsymbol{i} + \omega_z \boldsymbol{k}) \times \boldsymbol{r}$

where $\omega = \lim_{\Delta t \to 0} \left(\frac{\Delta\theta}{\Delta t} \right)$

It is clear that if a third rotation about the y-axis is added then

$$\begin{aligned} \boldsymbol{v} &= (\omega_x \boldsymbol{i} + \omega_y \boldsymbol{j} + \omega_z \boldsymbol{k}) \times \boldsymbol{r} \\ &= \boldsymbol{\omega} \times \boldsymbol{r} \end{aligned} \tag{11.2}$$

where $\boldsymbol{\omega}$ is the angular velocity vector; therefore angular velocity is equal to the sum of its component parts in the same manner as any other vector quantity.

It is worth noting that a given angular velocity $\boldsymbol{\omega}$ gives rise to a specific velocity $\boldsymbol{v}$ of a point having a position vector $\boldsymbol{r}$. However the inverse is not unique because a given velocity $\boldsymbol{v}$ of a point at $\boldsymbol{r}$ can be produced by any angular velocity vector, of appropriate magnitude, which lies in a plane normal to $\boldsymbol{v}$. Because an angular velocity $\boldsymbol{\omega}_r$ about an axis along $\boldsymbol{r}$ does not alter $\boldsymbol{v}$, we see that

$$\boldsymbol{v} = \boldsymbol{\omega} \times \boldsymbol{r} = (\boldsymbol{\omega}_n + \boldsymbol{\omega}_r) \times \boldsymbol{r}$$

thus only $\boldsymbol{\omega}_n$, the component of $\boldsymbol{\omega}$ normal to $\boldsymbol{r}$, can be found.

It is obvious that if the three theorems previously quoted apply to finite displacements then they must apply to infinitesimal displacements and thus to angular velocities. Hence in terms of angular velocities we may state

i) any motion of a rigid body may be described by a single angular velocity plus a translational velocity parallel to the angular velocity vector;

ii) any motion of a body about a point may be represented by a single angular velocity about an axis through that point;

iii) any motion of a rigid body may be represented by the velocity of a point plus an angular velocity about an axis through that point.

So far we have discussed angular displacement and angular velocity, so a few words on angular acceleration will be timely. Angular acceleration, $d\boldsymbol{\omega}/dt$, is not as significant as the acceleration of the centre of mass of a body because, as we shall see, the moment of the forces acting externally on the body are related to the rate of change of the moment of momentum, which in many cases cannot be written as a constant times the angular acceleration (exceptions being fixed-axis rotation and cases where the inertial properties of the body do not depend on orientation).

11.4 Differentiation of a vector when expressed in terms of a moving set of axes

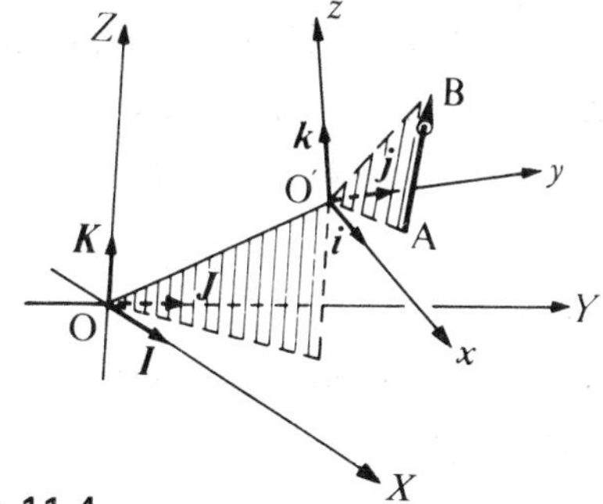

Figure 11.4

The vector $\overrightarrow{AB}$ shown in Fig. 11.4 may be expressed in terms of its components along a fixed set of X-, Y-, Z-axes as

$$\overrightarrow{AB} = C_X \boldsymbol{I} + C_Y \boldsymbol{J} + C_Z \boldsymbol{K} \tag{11.3}$$

or along a moving set of x-, y-, z-axes as

$$\overrightarrow{AB} = c_x \boldsymbol{i} + c_y \boldsymbol{j} + c_z \boldsymbol{k} \tag{11.4}$$

In all future work we must carefully distinguish between a vector expressed in terms of different base vectors and a vector *as seen from* a moving set of axes. In the first case we are merely expressing the same vector in different components, whereas in the second case the vector quantity may be different.

Imagine two observers, one attached to the fixed set of axes and the other attached to the moving x-, y-, z-axes. Both observers will agree on the magnitude and direction of $\overrightarrow{AB}$ although they will express the vector in terms of different base vectors, as in equations 11.3 and 11.4.

If $\overrightarrow{AB}$ is fixed with respect to the x-, y-, z-axes then $[d(\overrightarrow{AB})/dt]_{xyz} = 0$, the subscript xyz meaning 'as seen from the moving x-, y-, z-axes'. Our observer attached to the X-, Y-, Z-axes will detect a change in the vector $\overrightarrow{AB}$ if the x-, y, z-axes are rotating. Pure translation will not produce any change in the vector $\overrightarrow{AB}$ because its length and orientation will not be affected.

If O′ is moving relative to O with a velocity $\boldsymbol{v}_{O'/O}$ and in addition the x-, y-, z-axes are rotating at an angular velocity $\boldsymbol{\omega}$ relative to the X-, Y-, Z-axes, then the absolute velocity of point B, which is fixed in the xyz-frame, is given by

$$\begin{aligned} \boldsymbol{v}_{B/O} &= \boldsymbol{v}_{O'/O} + \boldsymbol{v}_{B/O'} \\ &= \boldsymbol{v}_{O'/O} + \boldsymbol{\omega} \times \boldsymbol{\rho}_{B/O'} \end{aligned}$$
(from equation 11.2)

The velocity of A is

$$\boldsymbol{v}_{A/O} = \boldsymbol{v}_{O'/O} + \boldsymbol{\omega} \times \boldsymbol{\rho}_{A/O'}$$

Thus
$$\begin{aligned} \boldsymbol{v}_{B/A} &= \boldsymbol{v}_{B/O} - \boldsymbol{v}_{A/O} = \boldsymbol{\omega} \times \boldsymbol{\rho}_{B/O'} - \boldsymbol{\omega} \times \boldsymbol{\rho}_{A/O'} \\ &= \boldsymbol{\omega} \times (\boldsymbol{\rho}_{B/O'} - \boldsymbol{\rho}_{A/O'}) = \boldsymbol{\omega} \times \overrightarrow{AB} \end{aligned}$$

or
$$\boldsymbol{v}_{B/A} = d(\overrightarrow{AB})/dt = \boldsymbol{\omega} \times \overrightarrow{AB} \qquad (11.5)$$

which, as stated earlier, is independent of $\boldsymbol{v}_{O'/O}$.

Although $\overrightarrow{AB}$ has been considered to be a displacement vector, it could represent any vector which is constant as seen from the xyz-frame.

A unit vector attached to the xyz-frame is a vector of the type just considered; hence, from equation 11.5,

$$d(\boldsymbol{i})/dt = \boldsymbol{\omega} \times \boldsymbol{i}$$
$$d(\boldsymbol{j})/dt = \boldsymbol{\omega} \times \boldsymbol{j}$$

and $d(\boldsymbol{k})/dt = \boldsymbol{\omega} \times \boldsymbol{k}$

Writing $\boldsymbol{\omega} = \omega_x \boldsymbol{i} + \omega_y \boldsymbol{j} + \omega_z \boldsymbol{k}$

we see that
$$\begin{aligned} d(\boldsymbol{i})dt &= -\omega_y \boldsymbol{k} + \omega_z \boldsymbol{j} \\ d(\boldsymbol{j})/dt &= \omega_x \boldsymbol{k} - \omega_z \boldsymbol{i} \\ d(\boldsymbol{k})/dt &= \omega_x \boldsymbol{j} + \omega_y \boldsymbol{i} \end{aligned}$$

Consider now a vector $\boldsymbol{A} = a_x \boldsymbol{i} + a_y \boldsymbol{j} + a_z \boldsymbol{k}$. By the usual rule for the differentiation of a product,

$$\begin{aligned} \frac{d\boldsymbol{A}}{dt} &= \left(\frac{da_x}{dt}\boldsymbol{i} + a_x\frac{d\boldsymbol{i}}{dt}\right) + \left(\frac{da_y}{dt}\boldsymbol{j} + a_y\frac{d\boldsymbol{j}}{dt}\right) + \left(\frac{da_z}{dt}\boldsymbol{k} + a_z\frac{d\boldsymbol{k}}{dt}\right) \\ &= \left(\frac{da_x}{dt}\boldsymbol{i} + \frac{da_y}{dt}\boldsymbol{j} + \frac{da_z}{dt}\boldsymbol{k}\right) \\ &\quad + (a_x \boldsymbol{\omega} \times \boldsymbol{i} + a_y \boldsymbol{\omega} \times \boldsymbol{j} + a_z \boldsymbol{\omega} \times \boldsymbol{k}) \qquad (11.6) \end{aligned}$$

The first group of three terms gives the rate of change of the vector *as seen from* the moving frame of reference, for which we shall use the notation $\partial \boldsymbol{A}/\partial t = [d\boldsymbol{A}/dt]_{xyz}$. The last group can be rearranged to give

$$\boldsymbol{\omega} \times (a_x \boldsymbol{i} + a_y \boldsymbol{j} + a_z \boldsymbol{k}) = \boldsymbol{\omega} \times \boldsymbol{A}$$

Thus equation 11.6 becomes

$$\frac{d\boldsymbol{A}}{dt} = \frac{\partial \boldsymbol{A}}{\partial t} + \boldsymbol{\omega} \times \boldsymbol{A} \qquad (11.7)$$

where $d\boldsymbol{A}/dt = [d\boldsymbol{A}/dt]_{XYZ}$ is the rate of change of a vector as seen from the fixed set of axes, $\partial \boldsymbol{A}/\partial t = [d\boldsymbol{A}/dt]_{xyz}$ is the rate of change of a vector as seen from the rotating set of axes, and $\boldsymbol{\omega}$ is the angular velocity of the moving set of axes relative to the fixed set of axes. Equation 11.7 is very important and will be used several times in the remaining part of this chapter.

11.5 Dynamics of a particle in three-dimensional motion

Cartesian co-ordinates

The equation of motion for the particle shown in Fig. 11.5 is simply

$$\boldsymbol{F} = m\ddot{\boldsymbol{r}} \qquad (11.8)$$

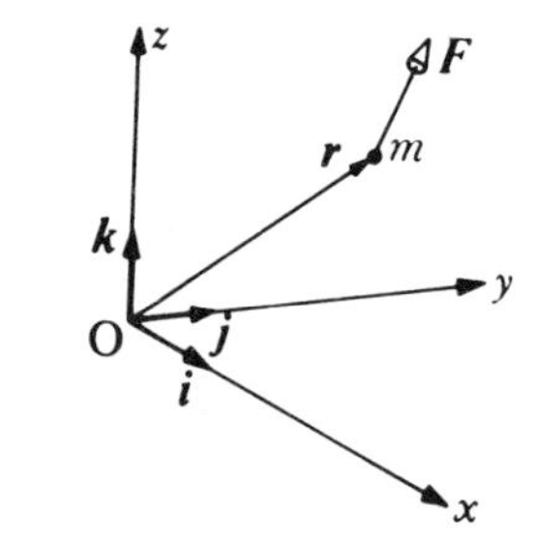

Figure 11.5

so the main task is to express the acceleration in a suitable co-ordinate system governed by the type of problem in hand. If the force is readily expressed in terms of Cartesian co-ordinates then it is convenient to express the acceleration in the same co-ordinates. This system poses no new problems for three-dimensional motion and the expressions for displacement, velocity, and acceleration are listed below for the sake of completeness:

$$\boldsymbol{r} = x\boldsymbol{i} + y\boldsymbol{j} + z\boldsymbol{k} \tag{11.9}$$

$$\boldsymbol{v} = \dot{\boldsymbol{r}} = \dot{x}\boldsymbol{j} + \dot{y}\boldsymbol{j} + \dot{z}\boldsymbol{k} \tag{11.10}$$

$$\boldsymbol{a} = \dot{\boldsymbol{v}} = \ddot{\boldsymbol{r}} = \ddot{x}\boldsymbol{i} + \ddot{y}\boldsymbol{j} + \ddot{z}\boldsymbol{k} \tag{11.11}$$

If the force is expressible in cylindrical or spherical co-ordinates then we shall need expressions for acceleration in these systems, or if the particle is constrained to move along a prescribed path then path co-ordinates may be required. These systems of co-ordinates will now be considered.

Cylindrical co-ordinates

Cylindrical co-ordinates are a simple extension of the polar co-ordinates encountered in Chapter 2: the position of a point is now defined by the co-ordinates R, θ, and z as shown in Fig. 11.6. The unit vectors $\boldsymbol{e}_{\mathrm{R}}$, $\boldsymbol{e}_\theta$ and $\boldsymbol{k}$ form an orthonormal triad, where $\boldsymbol{e}_{\mathrm{R}}$ is in the direction of $\mathbf{R}$, $\boldsymbol{e}_\theta$ is normal to $\boldsymbol{e}_{\mathrm{R}}$ and lies in the xy-plane, in the sense shown in the figure, and $\boldsymbol{k}$ is in the z-direction.

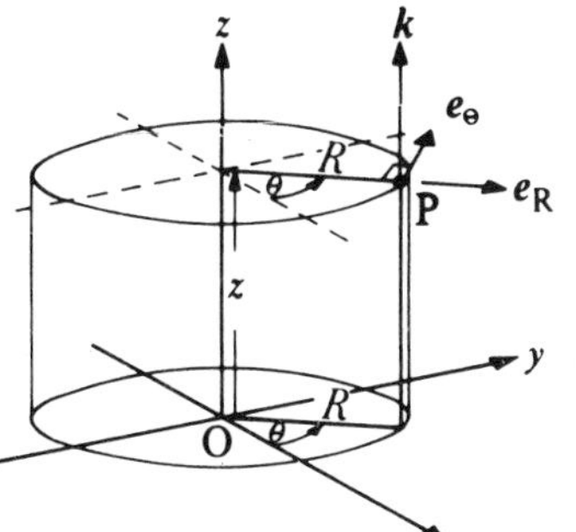

Figure 11.6

If the point P moves along any path then the triad will rotate about the z-axis at a rate $\dot{\theta}$, so that the angular velocity of the triad is $\boldsymbol{\omega} = \dot{\theta}\boldsymbol{k}$.

The position of P is given by

$$\boldsymbol{r} = R\boldsymbol{e}_R + z\boldsymbol{k} \tag{11.12}$$

the velocity $\boldsymbol{v} = \mathrm{d}\boldsymbol{r}/\mathrm{d}t = \dot{R}\boldsymbol{e}_R + R\dot{\boldsymbol{e}}_R + \dot{z}\boldsymbol{k}$

Using equation 11.7 to evaluate $\dot{\boldsymbol{e}}_R$,

$$\begin{aligned}\boldsymbol{v} &= \dot{R}\boldsymbol{e}_R + R\boldsymbol{\omega}\times\boldsymbol{e}_R + \dot{z}\boldsymbol{k}\\ &= \dot{R}\boldsymbol{e}_R + R\dot{\theta}\boldsymbol{k}\times\boldsymbol{e}_R + \dot{z}\boldsymbol{k}\\ &= \dot{R}\boldsymbol{e}_R + R\dot{\theta}\boldsymbol{e}_\theta + \dot{z}\boldsymbol{k}\end{aligned} \tag{11.13}$$

The acceleration is found by applying equation 11.7 again:

$$\begin{aligned}\boldsymbol{a} &= \partial\boldsymbol{v}/\partial t + \boldsymbol{\omega}\times\boldsymbol{v}\\ &= [\ddot{R}\boldsymbol{e}_R + (\dot{R}\dot{\theta} + R\ddot{\theta})\boldsymbol{e}_\theta + \ddot{z}\boldsymbol{k}]\\ &\quad + \dot{\theta}\boldsymbol{k}\times(\dot{R}\boldsymbol{e}_R + R\dot{\theta}\boldsymbol{e}_\theta + \dot{z}\boldsymbol{k})\\ &= \ddot{R}\boldsymbol{e}_R + (\dot{R}\dot{\theta} + R\ddot{\theta})\boldsymbol{e}_\theta + \ddot{z}\boldsymbol{k} + \dot{\theta}\dot{R}\boldsymbol{e}_\theta - R\dot{\theta}^2\boldsymbol{e}_R\\ &= (\ddot{R} - R\dot{\theta}^2)\boldsymbol{e}_R\\ &\quad + (2\dot{R}\dot{\theta} + R\ddot{\theta})\boldsymbol{e}_\theta + \ddot{z}\boldsymbol{k}\end{aligned} \tag{11.14}$$

Compare this derivation with those used in Chapter 2.

Path co-ordinates

In Fig. 11.7, $\boldsymbol{t}$ is the unit vector which is tangent to the path taken by the particle, $\boldsymbol{n}$ is the unit vector normal to the path and directed towards the centre of curvature, and $\boldsymbol{b}$ completes the orthonormal triad of unit vectors.

In this system,

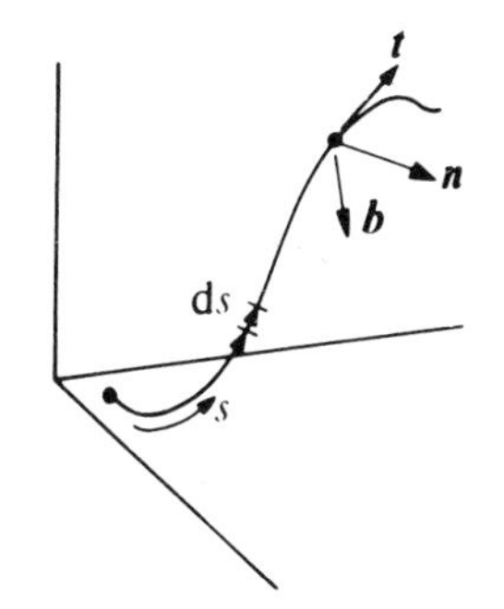

Figure 11.7

$$\boldsymbol{r} = \boldsymbol{r}_{\mathrm{o}} + \int(\mathrm{d}s\boldsymbol{t}) \tag{11.15}$$

and $\boldsymbol{v} = (\mathrm{d}s/\mathrm{d}t)\boldsymbol{t}$ (11.16)

where s is the distance measured along the path.

The angular velocity of the triad as the point moves along the path is $\omega\boldsymbol{b}$, since $\boldsymbol{t}$ and $\boldsymbol{n}$ both lie in the osculating plane of the curve; hence

$$\begin{aligned}\boldsymbol{a} = \mathrm{d}\boldsymbol{v}/\mathrm{d}t &= (\mathrm{d}^2s/\mathrm{d}t^2)\boldsymbol{t} + (\mathrm{d}s/\mathrm{d}t)(\mathrm{d}\boldsymbol{t}/\mathrm{d}t)\\ &= \ddot{s}\boldsymbol{t} + (\mathrm{d}s/\mathrm{d}t)\,\omega\boldsymbol{b}\times\boldsymbol{t}\\ &= \ddot{s}\boldsymbol{t} + \dot{s}\omega\boldsymbol{n}\end{aligned} \tag{11.17}$$

If the radius of curvature of the path at P is ρ, then

$$\boldsymbol{v} = \boldsymbol{\omega}\times\boldsymbol{\rho} = \omega\boldsymbol{b}\times\rho(-\boldsymbol{n}) = \omega\rho\boldsymbol{t}$$

therefore $\dot{s} = \omega\rho$

Again this should be compared with the approach in Chapter 2.

Spherical co-ordinates

The unit vectors shown in Fig. 11.8 are in directions such that $\boldsymbol{e}_r$ is in the direction of increasing radius, θ and ϕ constant. The direction of $\boldsymbol{e}_\theta$ is the same as the displacement of the point P if only θ varies, and $\boldsymbol{e}_\phi$ is similarly defined.

The angular velocity of the triad is

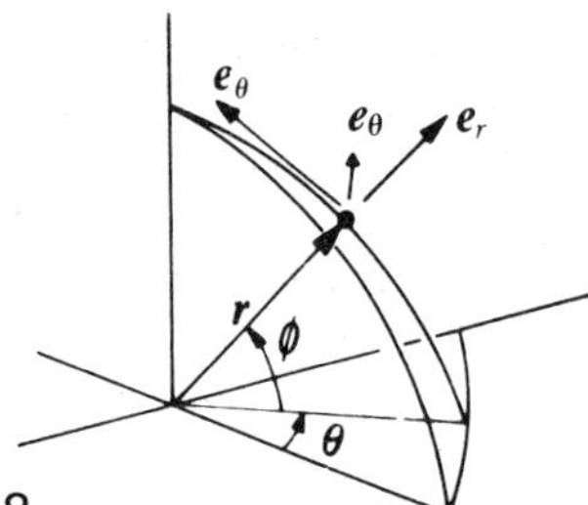

Figure 11.8

$$\boldsymbol{\omega} = \dot{\theta}\boldsymbol{k} - \dot{\phi}\boldsymbol{e}_\theta$$
$$= \dot{\theta}\sin\phi\boldsymbol{e}_r + \dot{\theta}\cos\phi\boldsymbol{e}_\phi - \dot{\phi}\boldsymbol{e}_\theta$$

Now $\boldsymbol{r} = r\boldsymbol{e}_r$ (11.18)

and, using equation 11.7,

$$\boldsymbol{v} = \dot{r}\boldsymbol{e}_r + r\boldsymbol{\omega}\times\boldsymbol{e}_r$$
$$= \dot{r}\boldsymbol{e}_r + r(\dot{\theta}\cos\phi\boldsymbol{e}_\theta + \dot{\phi}\boldsymbol{e}_\phi)$$
$$= \dot{r}\boldsymbol{e}_r + r\dot{\theta}\cos\phi\boldsymbol{e}_\theta + r\dot{\phi}\boldsymbol{e}_\phi \quad (11.19)$$

Differentiating again using equation 11.7, we have the acceleration:

$$\boldsymbol{a} = \ddot{r}\boldsymbol{e}_r + (\dot{r}\dot{\theta}\cos\phi + r\ddot{\theta}\cos\phi - r\dot{\theta}\sin\phi\dot{\phi})\boldsymbol{e}_\theta + (\dot{r}\dot{\phi} + r\ddot{\phi})\boldsymbol{e}_\phi$$
$$+\begin{vmatrix} \boldsymbol{e}_r & \boldsymbol{e}_\theta & \boldsymbol{e}_\phi \\ \dot{\theta}\sin\phi & -\dot{\phi} & \dot{\theta}\cos\phi \\ \dot{r} & r\dot{\theta}\cos\phi & r\dot{\phi} \end{vmatrix}$$

Expanding the determinant and collecting the terms gives

$$\boldsymbol{a} = (\ddot{r} - r\dot{\phi}^2 - r\dot{\theta}^2\cos^2\phi)\boldsymbol{e}_r$$
$$+ (r\ddot{\theta}\cos\phi - 2r\dot{\theta}\dot{\phi}\sin\phi + 2\dot{r}\dot{\theta}\cos\phi)\boldsymbol{e}_\theta$$
$$+ (r\ddot{\phi} + 2\dot{r}\dot{\phi} + r\dot{\theta}^2\sin\phi\cos\phi)\boldsymbol{e}_\phi \quad (11.20)$$

11.6 Motion relative to translating axes

In many problems it is often easier to express the motion of a system in terms of co-ordinate axes which are themselves in motion. Consider first the motion of a particle expressed in terms of co-ordinates which are moving, but not rotating, relative to a set of inertial axes. From Fig. 11.9 we have

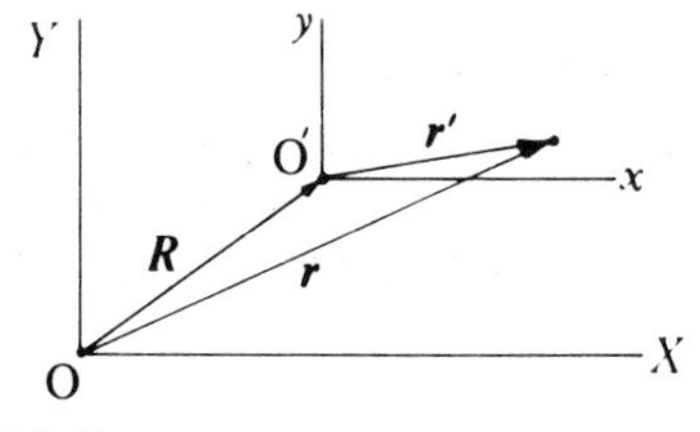

Figure 11.9

$$\boldsymbol{r} = \boldsymbol{R} + \boldsymbol{r}' \quad (11.21)$$
$$\dot{\boldsymbol{r}} = \dot{\boldsymbol{R}} + \dot{\boldsymbol{r}}' \quad (11.22)$$
$$\ddot{\boldsymbol{r}} = \ddot{\boldsymbol{R}} + \ddot{\boldsymbol{r}}' \quad (11.23)$$

If the force acting on the ith particle of a group is $\boldsymbol{F}_i$, then

$$\boldsymbol{F}_i = m_i\ddot{\boldsymbol{r}}_i = m_i\ddot{\boldsymbol{R}} + m_i\ddot{\boldsymbol{r}}_i' \quad (11.24)$$

Summing for the group,

$$\sum\boldsymbol{F}_i = \sum\text{external forces}$$
$$= M\ddot{\boldsymbol{R}} + \sum m_i\ddot{\boldsymbol{r}}_i' \quad (11.25)$$

Provided that $\ddot{\boldsymbol{R}}$ is zero then equation 11.25 is of the same form as that obtained using inertial axes, therefore any set of axes moving at a constant velocity, without rotation, relative to inertial axes are also inertial axes.

For cases when $\ddot{\boldsymbol{R}}$ is not zero, equation 11.25 may be written

$$\sum\boldsymbol{F}_i - M\ddot{\boldsymbol{R}} = \sum m_i\boldsymbol{r}_i'$$

thus from the point of view of an observer moving with the x-, y-, z-axes it appears that a body force, similar to weight, is acting on the system. Indeed, without the means of observing the rest of the universe it is impossible to distinguish between a real gravitational force and the apparent one which arises if the moving set of axes is taken as a frame of reference.

For example, imagine an observer descending in a lift which has a constant acceleration less than that due to gravity so that the observer does not leave the floor. There is no experiment which can be performed within the lift which will tell the observer whether the lift is accelerating or whether the strength of the gravitational field has been reduced.

11.7 Motion relative to rotating axes

We will now consider the case in which the x-, y-, z-axes shown in Fig. 11.10 are rotating at an angular velocity $\boldsymbol{\omega}$ relative to the inertial axes. We shall use primed symbols to indicate that the vector is as seen by an observer attached to the rotating frame.

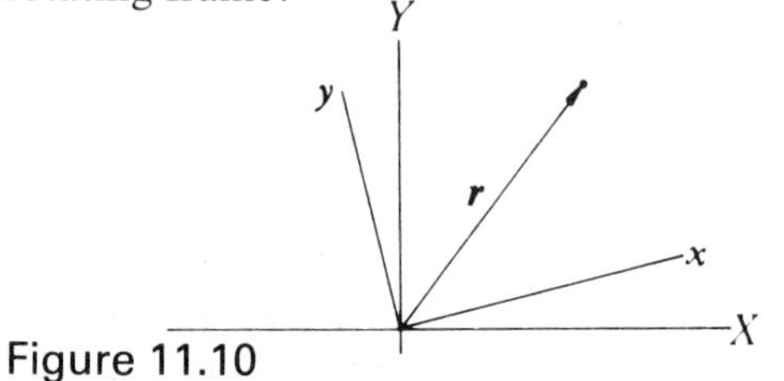

Figure 11.10

The displacement vector is the same when viewed from either frame, although each observer may use different base vectors. Therefore

$$\boldsymbol{r} = \boldsymbol{r}' \tag{11.26}$$

and by using equation 11.7 the velocity is given by

$$\mathrm{d}\boldsymbol{r}/\mathrm{d}t = \partial\boldsymbol{r}'/\partial t + \boldsymbol{\omega}\times\boldsymbol{r}'$$
$$\boldsymbol{v} = \boldsymbol{v}' + \boldsymbol{\omega}\times\boldsymbol{r}' \tag{11.27}$$

Similarly, the acceleration is given by

$$\begin{aligned}\boldsymbol{a} &= \mathrm{d}\boldsymbol{v}/\mathrm{d}t\\ &= \partial(\boldsymbol{v}' + \boldsymbol{\omega}\times\boldsymbol{r}')/\partial t + \boldsymbol{\omega}\times(\boldsymbol{v}' + \boldsymbol{\omega}\times\boldsymbol{r}')\\ &= \boldsymbol{a}' + (\partial\boldsymbol{\omega}/\partial t)\times\boldsymbol{r}' + \boldsymbol{\omega}\times\boldsymbol{v}' + \boldsymbol{\omega} + \boldsymbol{v}'\\ &\quad + \boldsymbol{\omega}\times(\boldsymbol{\omega}\times\boldsymbol{r}')\end{aligned}$$

Since $\mathrm{d}\boldsymbol{\omega}/\mathrm{d}t = \partial\boldsymbol{\omega}/\partial t + \boldsymbol{\omega}\times\boldsymbol{\omega} = \partial\boldsymbol{\omega}/\partial t$

we may without ambiguity use the dot notation to give

$$\boldsymbol{a} = \boldsymbol{a}' + \dot{\boldsymbol{\omega}}\times\boldsymbol{r} + 2\boldsymbol{\omega}\times\boldsymbol{v}' + \boldsymbol{\omega}\times(\boldsymbol{\omega}\times\boldsymbol{r}) \tag{11.28}$$

Coriolis's theorem

Equation 11.28 may easily be extended to cover the condition when the axes are also translating. Referring to Fig. 11.11, we have

$$\boldsymbol{a} = \mathrm{d}^2\boldsymbol{R}/\mathrm{d}t^2 + \boldsymbol{a}' + \dot{\boldsymbol{\omega}}\times\boldsymbol{r} + 2\boldsymbol{\omega}\times\boldsymbol{v}' + \boldsymbol{\omega}\times(\boldsymbol{\omega}\times\boldsymbol{r}) \tag{11.29}$$

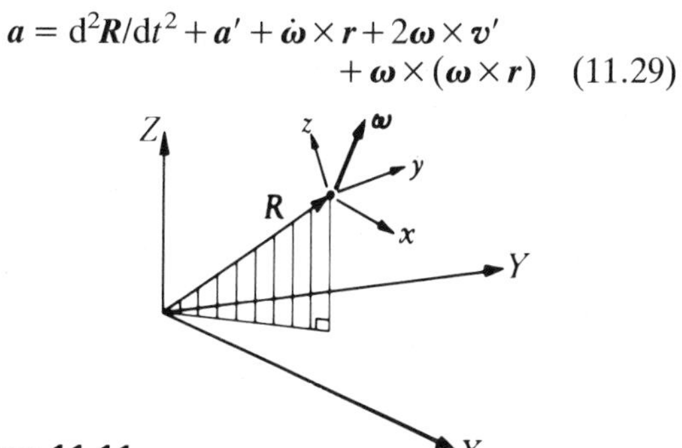

Figure 11.11

This result is known as Coriolis's theorem.

Now the equation of motion for a particle relative to the inertial frame of reference is

$$\boldsymbol{F} = m\boldsymbol{a}$$

but, if we choose to regard the moving frame of reference as an inertial frame, then

$$\boldsymbol{F} - m\ddot{\boldsymbol{R}} - m\dot{\boldsymbol{\omega}}\times\boldsymbol{r} - m2\boldsymbol{\omega}\times\boldsymbol{v}' - m\boldsymbol{\omega}\times(\boldsymbol{\omega}\times\boldsymbol{r}) = m\boldsymbol{a}' \tag{11.30}$$

The consequence of this choice of axes is that, in order to preserve Newton's laws of motion, four fictitious forces have to be introduced. We call them fictitious forces because we cannot locate the other body on which the equal and opposite force acts, whereas $\boldsymbol{F}$ is due to contact with another body or a body force of gravitational or electromagnetic origin, again due to the presence of some identifiable body.

The second term on the left-hand side of equation 11.30 is a 'force' due to the acceleration of the axes, and the third term is due to the angular acceleration of the axes. The fourth term is a 'force' acting on the particle in a direction mutually perpendicular to $\boldsymbol{v}'$ and $\boldsymbol{\omega}$ and is known as the Coriolis force. The fifth term is the 'centrifugal force', since it is always directed away from the origin and is normal to the axis of rotation of the axes.

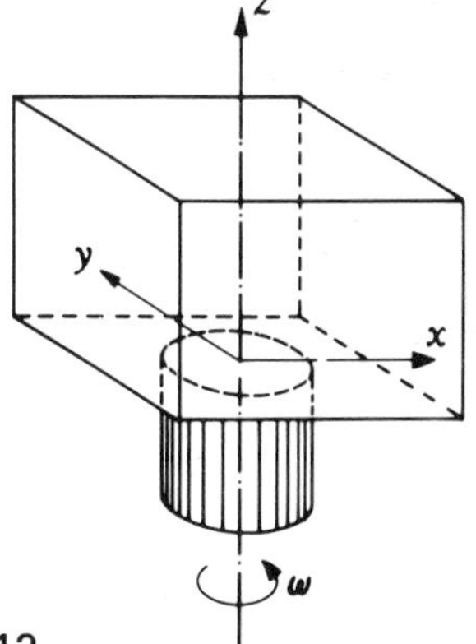

Figure 11.12

It is instructive to consider an experiment carried out in a rotating room as shown in Fig. 11.12. Let us assume that $\dot{\boldsymbol{\omega}} = 0$, $\ddot{\boldsymbol{R}} = 0$, and that rotation is about the z-axis. Equation 11.30 reduces to

$$\boldsymbol{F} - m2\boldsymbol{\omega}\times\boldsymbol{v}' - m\boldsymbol{\omega}\times(\boldsymbol{\omega}\times\boldsymbol{r}) = m\boldsymbol{a}'$$

and, using the expansion for the triple vector product (equation A1.16),

$$\boldsymbol{F} - 2m\boldsymbol{\omega}\times\boldsymbol{v}' - m\boldsymbol{\omega}(\boldsymbol{\omega}\cdot\boldsymbol{r}) + m\omega^2\boldsymbol{r} = m\boldsymbol{a}'$$

Further, if the motion is confined to the xy-plane,

$$\boldsymbol{\omega}\cdot\boldsymbol{r} = 0$$

thus $\boldsymbol{F} - m2\boldsymbol{\omega}\times\boldsymbol{v}' + m\omega^2\boldsymbol{r} = m\boldsymbol{a}'$ (11.31)

Let us consider a simple spring–mass system such that the force of the spring acting on the particle is $\boldsymbol{F} = -kr\boldsymbol{e}_r{}'$, where k is the stiffness of the spring. In terms of the rotating Cartesian co-ordinates,

$$\boldsymbol{r} = x\boldsymbol{i} + y\boldsymbol{j}$$

so that $\boldsymbol{F} = -k(x\boldsymbol{i}+y\boldsymbol{j})$

Equation 11.31 may now be written as

$$-kx\boldsymbol{i} - ky\boldsymbol{j} - 2m\omega\boldsymbol{k}\times(\dot{x}\boldsymbol{i}+\dot{y}\boldsymbol{j}) + m\omega^2(x\boldsymbol{i}+y\boldsymbol{j}) = m(\ddot{x}\boldsymbol{i}+\ddot{y}\boldsymbol{j})$$

giving the two scalar equations

$$-(k-m\omega^2)x + 2m\omega\dot{y} = m\ddot{x} \qquad (11.32)$$

$$-(k-m\omega^2)y - 2m\omega\dot{x} = m\ddot{y} \qquad (11.33)$$

If we assume solutions of the form

$$x = X\exp(\lambda t) \quad \text{and} \quad y = Y\exp(\lambda t)$$

equations 11.32 and 11.33 become

$$[m\lambda^2+(k-m\omega^2)]X = (2m\omega\lambda)Y$$
$$[m\lambda^2+(k-m\omega^2)]Y = -(2m\omega\lambda)X$$

therefore $$\frac{X}{Y} = \frac{2m\omega\lambda}{[m\lambda^2+(k-m\omega^2)]}$$

$$= \frac{-[m\lambda^2+(k-m\omega^2)]}{2m\omega\lambda} \qquad (11.34)$$

From the last equality,

$$[m\lambda^2+(k-m\omega^2)]^2+(2m\omega\lambda)^2 = 0$$

Expanding and collecting terms gives

$$\lambda^4+2[(k/m)+\omega^2]\lambda^2+[(k/m)-\omega^2]^2 = 0$$

Solving the quadratic in λ^2 yields

$$\lambda^2 = -[\surd(k/m)\pm\omega]^2$$
$$\lambda = \pm\mathrm{j}[\surd(k/m)\pm\omega]$$

and, since $\exp(\mathrm{j}\theta) = \cos\theta+\mathrm{j}\sin\theta$, we can write

$$x = X_1\cos[\surd(k/m)-\omega]t + X_2\sin[\surd(k/m)-\omega]t + X_3\cos[\surd(k/m)+\omega]t + X_4\sin[\surd(k/m)+\omega]t \qquad (11.35)$$

with a similar expression for y where Y is obtained from equation 11.34 with appropriate values of λ.

Equation 11.35 shows that the motion is stable except when $\omega = \surd(k/m)$; under these conditions the first two terms are X_1+X_2t. The four values of X depend on the initial conditions for x, $\dot{x}$, y, and $\dot{y}$.

It is interesting to note that, if the particle is constrained to move only in the x-direction, the motion is unstable for $\omega \geqslant \surd(k/m)$. Equations 11.32 and 11.33 are easily modified for $y = 0$, $\dot{y} = 0$, and $\ddot{y} = 0$; also, a constraining force F_y must be added to equation 11.33. Thus

$$m\ddot{x} = -(k-m\omega^2)x$$
$$0 = \pm 2m\omega\dot{x} + F_y$$

11.8 Kinematics of mechanisms

Introduction

Consider a link AB (Fig. 11.13) and denote $\overrightarrow{AB}$ by $\boldsymbol{l}$. Assume, for example, that $\boldsymbol{v}_A$ and $\boldsymbol{l}$ are known completely but $\boldsymbol{v}_B$ is known only in direction. If the link is of fixed length ($\dot{l} = 0$) then,

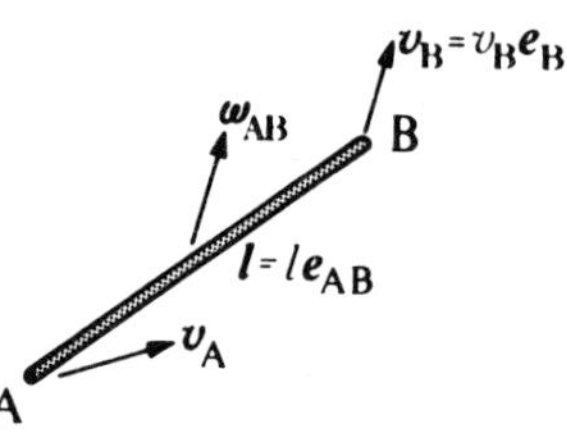

Figure 11.13

relative to A, B has no component of velocity along AB, so that

$$\boldsymbol{v}_{B/A}\cdot\boldsymbol{l} = 0 \qquad (11.36)$$

$$(v_B\boldsymbol{e}_B - \boldsymbol{v}_A)\cdot\boldsymbol{l} = 0 \qquad (11.37)$$

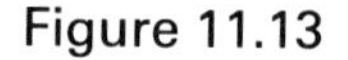

The only unknown in equation 11.37 is v_B, and performing the dot product leads to the value of v_B.

Consider now the case where the position of B is not known, although the path along which it travels is known. An example is the three-dimensional slider-crank chain shown in Fig. 11.14. Crank OA rotates about O with angular velocity $\omega\boldsymbol{j} = \dot{\theta}\boldsymbol{j}$, and slider B travels along

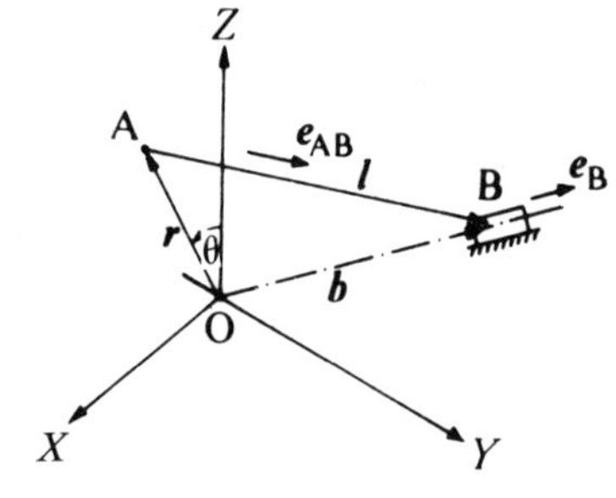

Figure 11.14

a known straight line passing through O, the unit vector for $\overrightarrow{OB}$ being denoted by $\boldsymbol{e}_B$ so that $\overrightarrow{OB} = b\boldsymbol{e}_B = \boldsymbol{b}$.

Suppose that r, l and θ are known and it is required to determine the position of B. From the triangle OAB we have

$$\boldsymbol{r}+\boldsymbol{l}=\boldsymbol{b}$$

$$r(\sin\theta\boldsymbol{i}+\cos\theta\boldsymbol{k})+l\boldsymbol{e}_{AB}=b\boldsymbol{e}_B \qquad (11.38)$$

In this equation, $\boldsymbol{e}_{AB}$ and b are unknown. Rewriting this as

$$l\boldsymbol{e}_{AB}=b\boldsymbol{e}_B-r(\sin\theta\boldsymbol{i}+\cos\theta\boldsymbol{k}) \qquad (11.39)$$

and taking the modulus of both sides eliminates $\boldsymbol{e}_{AB}$ and enables b to be found directly. Putting the known value of b back in equation 11.39 gives $\boldsymbol{e}_{AB}$.

A vector method for determining the position of a mechanism was given in example 5.1. For the slider-crank chain just considered, the position of point B can readily be determined by trigonometry.

Angular velocity of a link

It has already been pointed out that, if the relative velocity between two points on a body is known, *this information alone* does not permit the angular velocity $\boldsymbol{\omega}$ of the body to be found; only the component of $\boldsymbol{\omega}$ which is perpendicular to the line joining the two points can be determined.

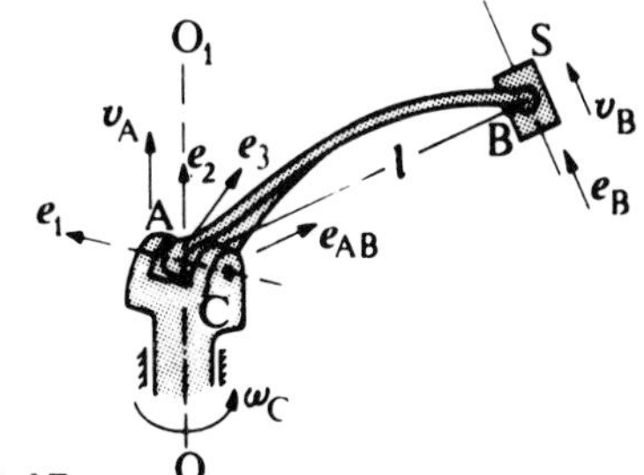

Figure 11.15

Consider (Fig. 11.15) the link AB which is pinned to the forked member C at A and connected by a ball-and-socket joint at B to slider S. The direction of the axis of the pin is denoted by the unit vector $\boldsymbol{e}_1$. The member C can rotate about the axis OO_1 and translate along it so that

$$\boldsymbol{\omega}_c=\omega_c\boldsymbol{e}_2$$

and $\quad \boldsymbol{v}_c=\boldsymbol{v}_A=v_A\boldsymbol{e}_2$

The velocity of point B is

$$\boldsymbol{v}_B=v_B\boldsymbol{e}_B$$

and we shall assume that $\boldsymbol{v}_A$ and $\overrightarrow{AB}=\boldsymbol{l}$ are known completely and that $\boldsymbol{v}_B$ is known only in direction.

We know that

$$\boldsymbol{v}_B-\boldsymbol{v}_A=\boldsymbol{v}_{B/A}=\boldsymbol{\omega}_{AB}\times\boldsymbol{l} \qquad (11.40)$$

so let us determine $\boldsymbol{\omega}_{AB}$. Writing equation 11.40 as

$$v_B\boldsymbol{e}_B-\boldsymbol{v}_A=(\omega_x\boldsymbol{i}+\omega_y\boldsymbol{j}+\omega_z\boldsymbol{k})\times\boldsymbol{l} \qquad (11.41)$$

we see that equation 11.41 contains four unknowns. Carrying out the cross product and comparing the coefficients of $\boldsymbol{i}$, $\boldsymbol{j}$, and $\boldsymbol{k}$ we find that the resulting three equations are not independent, although v_B can be found. Referring again to Fig. 11.15, we see that the angular velocity of AB can be represented by the angular velocity relative to C plus the angular velocity of C; thus

$$\boldsymbol{\omega}_{AB}=\omega_1\boldsymbol{e}_1+\omega_C\boldsymbol{e}_2 \qquad (11.42)$$

where ω_1 and ω_C are unknown. We observe that the link has no angular velocity component in the direction $\boldsymbol{e}_3=\boldsymbol{e}_1\times\boldsymbol{e}_2$, so

$$\boldsymbol{\omega}_{AB}\cdot\boldsymbol{e}_3=(\omega_x\boldsymbol{i}+\omega_y\boldsymbol{j}+\omega_z\boldsymbol{k})\cdot\boldsymbol{e}_3=0 \qquad (11.43)$$

If we perform the dot product and combine the resulting equation with any two of the three non-independent equations mentioned above, we shall have three independent equations containing the three unknowns ω_x, ω_y, and ω_z, which can thus be found. The components ω_1 and ω_C can then be found if required from equation 11.42.

The method for finding the angular velocity $\boldsymbol{\omega}_{AB}$ described above is rather tedious, and fortunately it is possible to determine $\boldsymbol{\omega}_{AB}$ from a single equation. If equation 11.40 is pre-cross-multiplied by the unit vector representing the direction for which $\boldsymbol{\omega}_{AB}$ has no component, namely $\boldsymbol{e}_3$, we obtain

$$\boldsymbol{e}_3\times\boldsymbol{v}_{B/A}=\boldsymbol{e}_3\times(\boldsymbol{\omega}_{AB}\times\boldsymbol{l})$$

and expanding the triple vector product (equation A1.16) gives

$$\boldsymbol{e}_3\times\boldsymbol{v}_{B/A}=\boldsymbol{\omega}_{AB}(\boldsymbol{e}_3\cdot\boldsymbol{l})-\boldsymbol{l}(\boldsymbol{e}_3\cdot\boldsymbol{\omega}_{AB}) \qquad (11.44)$$

From equation 11.43, the second product on the right is zero, so

$$\boldsymbol{\omega}_{AB}=(\boldsymbol{e}_3\times\boldsymbol{v}_{BA})/(\boldsymbol{e}_3\cdot\boldsymbol{l})$$

from which $\boldsymbol{\omega}_{AB}$ can be found directly.

Consider now the case where the link AB has ball-and-socket joints at each end, as shown in Fig. 11.16. In this case it is clear that any rotation the link may have about the line AB has no effect on the relative motion between A and B and in any case cannot be determined. We can thus assume that $\boldsymbol{\omega}_{AB}\cdot\boldsymbol{l}=0$, and pre-cross-multiplying equation 11.40 by $\boldsymbol{l}$ leads to

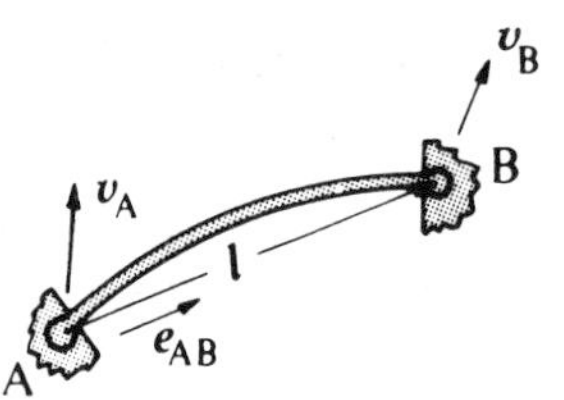

Figure 11.16

$$\boldsymbol{\omega}_{AB} = (\boldsymbol{l}\times\boldsymbol{v}_{B/A})/(\boldsymbol{l}\cdot\boldsymbol{l}) = (\boldsymbol{l}\times\boldsymbol{v}_{B/A})/l^2 \quad (11.45)$$

If there is a pinned joint at A or B but only the component of $\boldsymbol{\omega}_{AB}$ perpendicular to AB is of interest, we can use equation 11.45 to find this.

Angular acceleration of a link

Suppose now that $\boldsymbol{a}_A$, the acceleration of A in Fig. 11.15, is known completely and the acceleration of B is known apart from its magnitude a_B.

$$\boldsymbol{a}_{B/A} = \mathrm{d}(\boldsymbol{\omega}_{AB}\times\boldsymbol{l})/\mathrm{d}t \quad (11.46)$$

Using equation 11.7, where the moving axes are attached to the link AB, we can write equation 11.46 as

$$\boldsymbol{a}_B - \boldsymbol{a}_A = \dot{\boldsymbol{\omega}}_{AB}\times\boldsymbol{l} + \boldsymbol{\omega}_{AB}\times(\boldsymbol{\omega}_{AB}\times\boldsymbol{l}) \quad (11.47)$$

and equation 11.47 contains four unknowns, namely the magnitude of $\boldsymbol{a}_B$ and the three components of $\dot{\boldsymbol{\omega}}_{AB}$. The vector product $\dot{\boldsymbol{\omega}}_{AB}\times\boldsymbol{l}$ is perpendicular to both $\dot{\boldsymbol{\omega}}_{AB}$ and $\boldsymbol{l}$ so if we perform the dot product of $\boldsymbol{l}$ with $\dot{\boldsymbol{\omega}}_{AB}\times\boldsymbol{l}$ the result is zero. Thus if the dot product with $\boldsymbol{l}$ of each term in equation 11.47 is carried out, the term containing $\dot{\boldsymbol{\omega}}_{AB}$ is eliminated and the magnitude of $\boldsymbol{a}_B$ can be found.

If the component of $\dot{\boldsymbol{\omega}}_{AB}$ in the direction of AB is irrelevant, we can let $\dot{\boldsymbol{\omega}}_{AB}\cdot\boldsymbol{l} = 0$ and pre-cross-multiplying equation 11.47 by $\boldsymbol{l}$ we have after expanding the triple vector product containing $\dot{\boldsymbol{\omega}}_{AB}$,

$$\boldsymbol{l}\times\boldsymbol{a}_{B/A} = \dot{\boldsymbol{\omega}}_{AB}(\boldsymbol{l}\cdot\boldsymbol{l}) + \boldsymbol{l}\times[\boldsymbol{\omega}_{AB}\times(\boldsymbol{\omega}_{AB}\times\boldsymbol{l})] \quad (11.48)$$

and we thus find the angular acceleration of the link, perpendicular to the line AB.

Note that if the link has ball-and-socket joints at each end, so that we can write $\boldsymbol{\omega}_{AB}\cdot\boldsymbol{l} = 0$, then in equation 11.47,

$$\boldsymbol{\omega}_{AB}\times(\boldsymbol{\omega}_{AB}\times\boldsymbol{l}) = \boldsymbol{\omega}_{AB}(\boldsymbol{\omega}_{AB}\cdot\boldsymbol{l}) - \boldsymbol{l}(\boldsymbol{\omega}_{AB}\cdot\boldsymbol{\omega}_{AB}) = -\omega_{AB}{}^2\boldsymbol{l} \quad (11.49)$$

and in equation 11.48

$$\boldsymbol{l}\times[\boldsymbol{\omega}_{AB}\times(\boldsymbol{\omega}_{AB}\times\boldsymbol{l}) = \boldsymbol{l}\times[-\omega_{AB}{}^2\boldsymbol{l}] = 0 \quad (11.50)$$

11.9 Kinetics of a rigid body

Linear momentum

The linear momentum of a rigid body is the vector sum of all the individual momenta of its constituent particles, thus the total linear momentum is given by

$$\boldsymbol{p} = \sum \boldsymbol{p}_i = \sum m_i\boldsymbol{v}_i \quad (11.51)$$

and using the definition of the centre of mass

$$\sum m_i\boldsymbol{v}_i = \boldsymbol{v}_G(\sum m_i) = \boldsymbol{v}_G m$$

gives $\boldsymbol{p} = m\boldsymbol{v}_G$ (11.52)

This result is true for any group of particles, rigidly connected or not.

For a single particle $\boldsymbol{F}_i = \mathrm{d}\boldsymbol{p}_i/\mathrm{d}t$, where the force $\boldsymbol{F}_i$ may be due to other particles in the group or to external bodies. If we sum over the whole group, the contribution of the internal forces must be zero according to Newton's third law; hence

$$\sum \boldsymbol{F}_i = (\sum \boldsymbol{F}_i)_{\mathrm{external}} = \sum \mathrm{d}\boldsymbol{p}_i/\mathrm{d}t = \mathrm{d}\boldsymbol{p}/\mathrm{d}t = \mathrm{d}(m\boldsymbol{v}_G)/\mathrm{d}t = m\,\mathrm{d}\boldsymbol{v}_G/\mathrm{d}t \quad (11.53)$$

i.e. the sum of the external forces is equal to the total mass times the acceleration of the centre of mass, and is independent of any rotation.

Moment of momentum

The moment of momentum of a particle about a point O is defined to be $\boldsymbol{r}_i\times\boldsymbol{p}_i$, where $\boldsymbol{r}_i$ is the position of the particle relative to O. For a group of particles the total moment of momentum about O is

$$\boldsymbol{L}_O = \sum \boldsymbol{r}_i\times\boldsymbol{p}_i = \sum m_i\boldsymbol{r}_i\times\boldsymbol{v}_i \quad (11.54)$$

For a rigid body, the velocity of the particle can be written as the vector sum of the velocity of a specific point and the velocity of the particle relative to that point due to the rotation of the body.

We shall now consider two particular cases.

a) Motion, relative to fixed axes, referred to the centre of mass of the body (see Fig. 11.17).

In this case,

$$\boldsymbol{v}_i = \boldsymbol{v}_G + \boldsymbol{\omega} + \boldsymbol{\rho}_i$$

and $\boldsymbol{r}_i = \boldsymbol{r}_G + \boldsymbol{\rho}_i$

hence equation 11.54 becomes

$$\boldsymbol{L}_O = \sum m_i(\boldsymbol{r}_G + \boldsymbol{\rho}_i)\times(\boldsymbol{v}_G + \boldsymbol{\omega}\times\boldsymbol{\rho}_i)$$

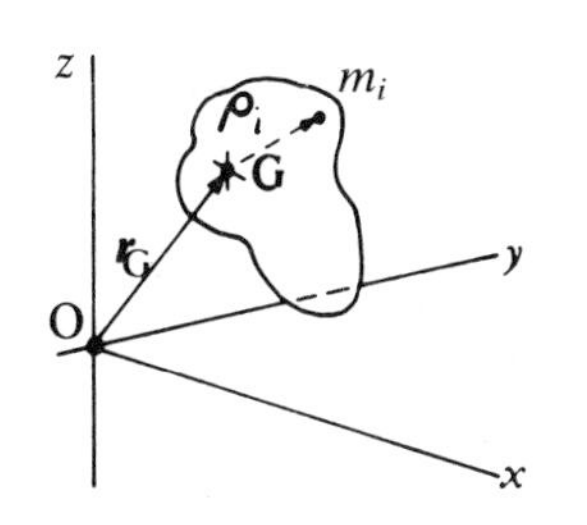

Figure 11.17

$$= r_G \times m v_G + r_G \times (\omega \times \sum m_i \rho_i) + (\sum m_i \rho_i) \times v_G + \sum m_i \rho_i \times (\omega \times \rho_i)$$

Since $\sum m_i = m$ and $\sum m_i \rho_i = 0$, we have

$$L_O = r_G \times m v_G + \sum m_i \rho_i \times (\omega \times \rho_i) \qquad (11.55)$$

If we choose O to be coincident with G, the centre of mass, then $r_G = 0$ so that the moment of momentum about G is

$$L_G = \sum m_i \rho_i \times (\omega \times \rho_i) \qquad (11.56)$$

b) Motion, relative to fixed axes, referred to a stationary point on the body (see Fig. 11.18).

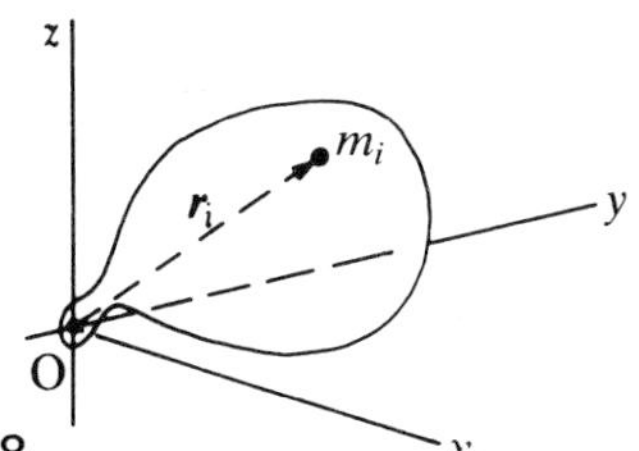

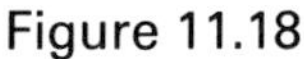
Figure 11.18

In this case, with O as the fixed point,

$$v_i = \omega \times r_i$$

thus equation 11.54 becomes

$$L_O = \sum m_i r_i \times (\omega \times r_i) \qquad (11.57)$$

It will be noticed that equation 11.57 is of the same form as equation 11.56.

Moment of momentum is an instantaneous quantity, therefore it is of no consequence whether or not we regard the point about which moments are taken to be fixed. However, it is important to state whether the velocities are relative to fixed or moving axes. If moments are taken about the centre of mass even this distinction is not required, as in equation 11.56 L_G is independent of v_G. Hence the concept of the centre of mass enables rotation and translation to be treated separately:

$$p = m v_G \qquad (11.58)$$

$$L_G = \sum m_i \rho_i \times (\omega \times \rho_i) \qquad (11.59)$$

Equations 11.56 and 11.57 are both of the form

$$L = \sum m_i r_i \times (\omega \times r_i)$$

Using the expansion for the triple vector product (equation A1.16),

$$L = \sum m_i \omega (r_i \cdot r_i) - \sum m_i r_i (\omega \cdot r_i)$$

Expressing ω and r in terms of Cartesian co-ordinates,

$$L = \sum m_i (\omega_x i + \omega_y j + \omega_z k)(x_i^2 + y_i^2 + z_i^2) - \sum m_i (x_i i + y_i j + z_i k)(\omega_x x_i + \omega_y y_i + \omega_z z_i)$$

Carrying out the multiplication and collecting the terms gives

$$\begin{bmatrix} L_x i \\ + L_y j \\ + L_z k \end{bmatrix} = \sum m_i \begin{bmatrix} i(\omega_x (y_i^2 + z_i^2) - \omega_y x_i y_i - \omega_z x_i z_i) \\ + j(\omega_y (z_i^2 + x_i^2) - \omega_z y_i z_i - \omega_x y_i x_i) \\ + k(\omega_z (x_i^2 + y_i^2) - \omega_x z_i x_i - \omega_y z_i y_i) \end{bmatrix} = \begin{bmatrix} i(\omega_x I_{xx} - \omega_y I_{xy} - \omega_z I_{xz}) \\ + j(\omega_y I_{yy} - \omega_z I_{yz} - \omega_x I_{yx}) \\ + k(\omega_z I_{zz} - \omega_x I_{zx} - \omega_y I_{zy}) \end{bmatrix} \qquad (11.60)$$

where $I_{xx} = \sum m_i (y_i^2 + z_i^2)$ is the moment of inertia about the x-axis and $I_{xy} = I_{yx} = \sum m_i x_i y_i$ is the product of inertia for the xy-plane. The other terms are similarly defined.

The three scalar equations may be written in matrix form as

$$\begin{bmatrix} L_x \\ L_y \\ L_z \end{bmatrix} = \begin{bmatrix} I_{xx} & -I_{xy} & -I_{xz} \\ -I_{yx} & I_{yy} & -I_{yz} \\ -I_{zx} & -I_{zy} & I_{zz} \end{bmatrix} \begin{bmatrix} \omega_x \\ \omega_y \\ \omega_z \end{bmatrix} \qquad (11.61)$$

The 3×3 symmetrical matrix is known as the inertia matrix.

11.10 Moment of force and rate of change of moment of momentum

Consider the moment of a force acting on a single particle. Since

$$F_i = \mathrm{d}p_i/\mathrm{d}t$$

then

$$r_i \times F_i = r_i \times (\mathrm{d}p_i/\mathrm{d}t) = \mathrm{d}(r_i \times p_i)/\mathrm{d}t = \mathrm{d}(\text{moment of momentum})/\mathrm{d}t \qquad (11.62)$$

the last statement being true since $(\mathrm{d}\boldsymbol{r}_i/\mathrm{d}t)\times\boldsymbol{p}_i = \dot{\boldsymbol{r}}_i\times m_i\dot{\boldsymbol{r}}_i = 0$.

Summing over the whole body and noting that the internal forces occur in equal, opposite, and collinear pairs,

$$\sum(\boldsymbol{r}_i\times\boldsymbol{F}_i)_{\text{external}} = \mathrm{d}[\sum(\boldsymbol{r}_i\times\boldsymbol{p}_i)]/\mathrm{d}t$$

or

$$\boldsymbol{M}_\mathrm{O} = \mathrm{d}\boldsymbol{L}_\mathrm{O}/\mathrm{d}t \tag{11.63}$$

where $\boldsymbol{M}_\mathrm{O}$ is the total moment of the externally applied forces about the point O.

Expressing $\boldsymbol{L}_\mathrm{O}$ as in equation 11.55 and differentiating with respect to time, we obtain

$$\boldsymbol{M}_\mathrm{O} = \dot{\boldsymbol{L}}_\mathrm{O} = \boldsymbol{r}_\mathrm{G}\times m\boldsymbol{a}_\mathrm{G} + \mathrm{d}[\sum m_i\boldsymbol{\rho}_i\times(\boldsymbol{\omega}\times\boldsymbol{\rho}_i)]/\mathrm{d}t$$
$$= \boldsymbol{r}_\mathrm{G}\times m\boldsymbol{a}_\mathrm{G} + \dot{\boldsymbol{L}}_\mathrm{G} \tag{11.64}$$

Moment of momentum referred to translating axes

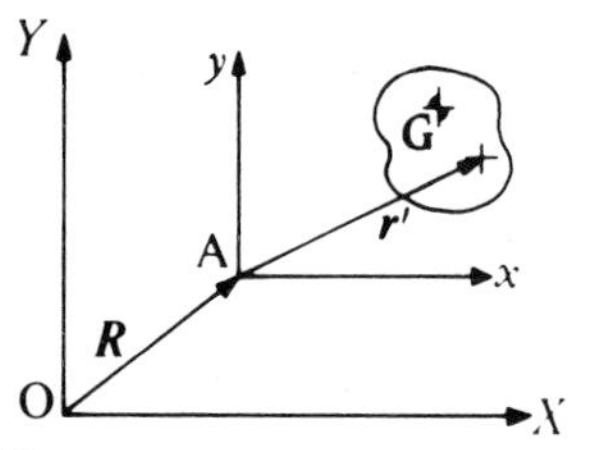

Figure 11.19

In Fig. 11.19 the X-, Y-, Z-axes are inertial and the x-, x-, z-axes are translating but not rotating. Taking moments about O we have for the total moment of the external forces

$$\boldsymbol{M}_\mathrm{O} = \mathrm{d}\boldsymbol{L}_\mathrm{O}/\mathrm{d}t = \sum\boldsymbol{r}_i\times m_i\ddot{\boldsymbol{r}}_i$$
$$= \mathrm{d}(\sum\boldsymbol{r}_i\times m_i\dot{\boldsymbol{r}}_i)/\mathrm{d}t$$

Replacing $\boldsymbol{r}_i$ by $\boldsymbol{R}+\boldsymbol{r}_i'$, where the prime indicates a vector as seen from the moving axes, we may write the moment about a fixed point in space coincident with A as

$$\boldsymbol{M}_\mathrm{A} = \sum(\boldsymbol{R}+\boldsymbol{r}_i')\times m_i(\ddot{\boldsymbol{R}}+\ddot{\boldsymbol{r}}_i')$$
$$\boldsymbol{R} = 0 \tag{11.65}$$

or

$$\boldsymbol{M}_\mathrm{A} = \mathrm{d}[\sum(\boldsymbol{R}_i+\boldsymbol{r}_i')\times m_i(\dot{\boldsymbol{R}}+\dot{\boldsymbol{r}}_i')]/\mathrm{d}t$$
$$= \mathrm{d}\boldsymbol{L}_\mathrm{A}/\mathrm{d}t \tag{11.66}$$

In equation 11.66 the vector $\boldsymbol{R}$ must be retained since $\dot{\boldsymbol{R}}\neq 0$.

Expanding equation 11.66 and putting $\boldsymbol{R} = 0$,

$$\boldsymbol{M}_\mathrm{A} = \sum m_i\boldsymbol{r}_i'\times\ddot{\boldsymbol{R}} + \sum\boldsymbol{r}_i'\times m_i\ddot{\boldsymbol{r}}_i'$$
$$= \boldsymbol{r}_\mathrm{G}'\times m\ddot{\boldsymbol{R}} + \dot{\boldsymbol{L}}_\mathrm{A}'$$

where $\boldsymbol{L}_\mathrm{A}' = \sum\boldsymbol{r}_i'\times m_i\dot{\boldsymbol{r}}_i'$ is the moment of momentum relative to the translating axes.

If the point A coincides with the centre of mass G then, since $\boldsymbol{r}_\mathrm{G}' = 0$,

$$\boldsymbol{M}_\mathrm{G} = \dot{\boldsymbol{L}}_\mathrm{G} = \dot{\boldsymbol{L}}_\mathrm{G}' \tag{11.67}$$

Differentiation of $\boldsymbol{L}$ is sometimes difficult because the moment of inertia changes if the body moves relative to the reference axes; however, the problem can be simplified if we choose a set of axes moving in such a way that the moment of inertia is constant with respect to these moving axes. One obvious set will be axes fixed to the body; also, if the body has an axis of symmetry, then the body may rotate relative to an axis coinciding with the axis of symmetry without altering the moment of inertia.

Using equation 11.7 for differentiating a vector with respect to a moving set of axes,

$$\boldsymbol{M} = \mathrm{d}\boldsymbol{L}/\mathrm{d}t = \partial\boldsymbol{L}/\partial t + \boldsymbol{\omega}\times\boldsymbol{L} \tag{11.68}$$

where $\boldsymbol{\omega}$ is the angular velocity of the moving axes.

In terms of Cartesian co-ordinates,

$$\boldsymbol{L} = L_x\boldsymbol{i} + L_y\boldsymbol{j} + L_z\boldsymbol{k}$$

and

$$\boldsymbol{\omega} = \omega_x\boldsymbol{i} + \omega_y\boldsymbol{j} + \omega_z\boldsymbol{k}$$

so

$$\boldsymbol{M} = \left(\frac{\partial L_x}{\partial t}\boldsymbol{i} + \frac{\partial L_y}{\partial t}\boldsymbol{j} + \frac{\partial L_z}{\partial t}\boldsymbol{k}\right)$$
$$+ (\omega_x\boldsymbol{i} + \omega_y\boldsymbol{j} + \omega_z\boldsymbol{k})\times(L_x\boldsymbol{i} + L_y\boldsymbol{j} + L_z\boldsymbol{k})$$

or

$$M_x\boldsymbol{i} + M_y\boldsymbol{j} + M_z\boldsymbol{k}$$
$$= \boldsymbol{i}\left(\frac{\partial L_x}{\partial t} + \omega_y L_z - \omega_z L_y\right)$$
$$+ \boldsymbol{j}\left(\frac{\partial L_y}{\partial t} + \omega_z L_x - \omega_x L_z\right)$$
$$+ \boldsymbol{k}\left(\frac{\partial L_z}{\partial t} + \omega_x L_y - \omega_y L_x\right) \tag{11.69}$$

From equation 11.60,

$$L_x = \Omega_x I_{xx} - \Omega_y I_{xy} - \Omega_z I_{xz}$$
$$L_y = -\Omega_x I_{yx} + \Omega_y I_{yy} - \Omega_z I_{yz}$$
$$L_z = -\Omega_x I_{zx} - \Omega_y I_{zy} + \Omega_z I_{zz} \tag{11.70}$$

where $\boldsymbol{\Omega} = \Omega_x\boldsymbol{i} + \Omega_y\boldsymbol{j} + \Omega_z\boldsymbol{k}$ is the angular velocity of body when different to $\boldsymbol{\omega}$, the angular velocity of the reference axes. Substituting equation 11.70 into 11.69 will give the full set of general equations. These equations are rarely used in this form because some form of simplification is generally possible by choosing axes to coincide either with a fixed axis of rotation or with the principal axes of the body. These are defined below.

An important fact is that it is always possible to find, through any point in a body, a set of axes such that the products of inertia are all zero; such sets are called principal axes. If a body has a plane of symmetry then any axis perpendicular to this plane is a principal axis. If a body has an axis of symmetry then this axis is a principal axis, and any axis perpendicular to it is also a principal axis.

In the former case, the two other principal axes are found by a simple transformation of axes as shown on Fig. 11.20. From the figure,

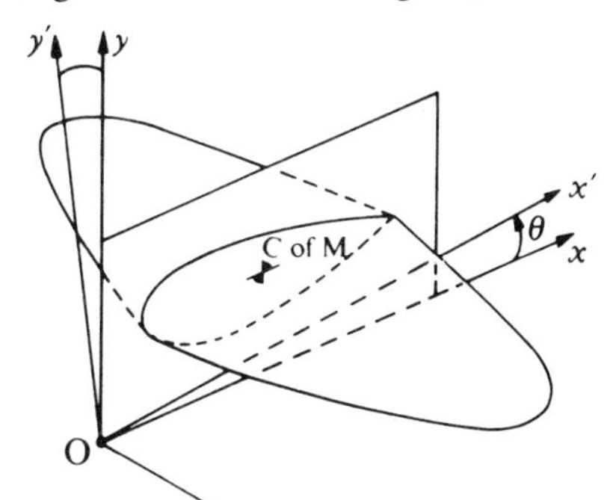

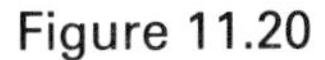
Figure 11.20

$$x' = x\cos\theta + y\sin\theta + 0$$
$$y' = -x\sin\theta + y\cos\theta + 0$$
$$z' = +0 + 0 + z$$

so $I_{xy}' = \sum m_i x_i' y_i' = \sum m_i (x_i \cos\theta + y_i \sin\theta)$

$$\times(-x_i \sin\theta + y_i \cos\theta)$$
$$= (\cos^2\theta - \sin^2\theta)\sum m_i x_i y_i$$
$$+ (\sum m_i y_i^2 - \sum m_i x_i^2)\sin\theta\cos\theta$$

But $(\sum m_i y_i^2 - \sum m_i x_i^2) = [\sum m_i (y_i^2 + z_i^2)$

$$- \sum m_i (x_i^2 + z_i^2)]$$
$$= I_{xx} - I_{yy}$$

therefore

$$I_{xy}' = I_{xy}\cos 2\theta + (I_{xx} - I_{yy})\tfrac{1}{2}\sin 2\theta$$

If the x'-, y'-, z'-axes are to be principal axes, then $I_{xy}' = 0$; thus

$$\tan 2\theta = 2I_{xy}/(I_{yy} - I_{xx}) \qquad (11.71)$$

If principal axes are chosen, equation 11.61 becomes

$$\begin{bmatrix} L_x \\ L_y \\ L_z \end{bmatrix} = \begin{bmatrix} I_{xx} & 0 & 0 \\ 0 & I_{yy} & 0 \\ 0 & 0 & I_{zz} \end{bmatrix} \begin{bmatrix} \omega_x \\ \omega_y \\ \omega_z \end{bmatrix}$$
$$= \begin{bmatrix} I_{xx}\omega_x \\ I_{yy}\omega_y \\ I_{zz}\omega_z \end{bmatrix}$$

or $\boldsymbol{L} = (I_{xx}\omega_x)\boldsymbol{i} + (I_{yy}\omega_y)\boldsymbol{j} + (I_{zz}\omega_z)\boldsymbol{k}$ (11.72)

So, substituting equation 11.72 into equation 11.68, we have that the moment of the external forces (about either a fixed point on the body or the centre of mass) is, with the appropriate principal moments of inertia,

$$\boldsymbol{M} = \boldsymbol{i}(I_{xx}\partial\omega_x/\partial t) + \boldsymbol{j}(I_{yy}\partial\omega_y/\partial t) + \boldsymbol{k}(I_{zz}\partial\omega_z/\partial t)$$
$$+ \begin{vmatrix} \boldsymbol{i} & \boldsymbol{j} & \boldsymbol{k} \\ \omega_x & \omega_y & \omega_z \\ I_{xx}\omega_x & I_{yy}\omega_y & I_{zz}\omega_z \end{vmatrix}$$

or $M_x\boldsymbol{i} + M_y\boldsymbol{j} + M_z\boldsymbol{k}$

$$= \boldsymbol{i}(I_{xx}\partial\omega_x/\partial t + \omega_y I_{zz}\omega_z - \omega_z I_{yy}\omega_y)$$
$$+ \boldsymbol{j}(I_{yy}\partial\omega_y/\partial t + \omega_z I_{xx}\omega_x - \omega_x I_{zz}\omega_z)$$
$$+ \boldsymbol{k}(I_{zz}\partial\omega_z/\partial t + \omega_x I_{yy}\omega_y - \omega_y I_{xx}\omega_x)$$

The three scalar equations are

$$M_x = I_{xx}\dot{\omega}_x + (I_{zz} - I_{yy})\omega_y\omega_z$$
$$M_y = I_{yy}\dot{\omega}_y + (I_{xx} - I_{zz})\omega_z\omega_x$$
$$M_z = I_{zz}\dot{\omega}_z + (I_{yy} - I_{xx})\omega_x\omega_y \qquad (11.73)$$

These equations are known as Euler's equations. It will be noted that $\partial\boldsymbol{\omega}/\partial t$ has been replaced, without ambiguity, by $\dot{\boldsymbol{\omega}}$ since

$$\mathrm{d}\boldsymbol{\omega}/\mathrm{d}t = \partial\boldsymbol{\omega}/\partial t + \boldsymbol{\omega}\times\boldsymbol{\omega} = \partial\boldsymbol{\omega}/\partial t$$

Thus for this case $(\dot{\omega}_x, \dot{\omega}_y, \dot{\omega}_z)$ are indeed the components of the angular acceleration of the body, but this is not true for the modified Euler's equations which follow.

When the angular velocity of the body $\boldsymbol{\Omega}$ is different from the angular velocity of the reference axes $\boldsymbol{\omega}$, then

$$\boldsymbol{M} = \boldsymbol{i}(I_{xx}\partial\Omega_x/\partial t) + \boldsymbol{j}(I_{yy}\partial\Omega_y/\partial t) + \boldsymbol{k}(I_{zz}\partial\Omega_z/\partial t)$$
$$+ \begin{vmatrix} \boldsymbol{i} & \boldsymbol{j} & \boldsymbol{k} \\ \omega_x & \omega_y & \omega_z \\ I_{xx}\Omega_x & I_{yy}\Omega_y & I_{zz}\Omega_z \end{vmatrix}$$

giving the three scalar equations

$$M_x = I_{xx}\partial\Omega_x/\partial t + \omega_y I_{zz}\Omega_z - \omega_z I_{yy}\Omega_y$$
$$M_y = I_{yy}\partial\Omega_y/\partial t + \omega_z I_{xx}\Omega_x - \omega_x I_{zz}\Omega_z$$
$$M_z = I_{zz}\partial\Omega_z/\partial t + \omega_x I_{yy}\Omega_y - \omega_y I_{xx}\Omega_x \quad (11.74)$$

It must be emphasised that the above equations apply equally for moments about either a fixed point on the body or the centre of mass. The moments of inertia must, of course, be evaluated for the point considered.

11.11 Rotation about a fixed axis

This case is of practical interest in connection with the forces appearing at the bearings of an imperfect rigid motor. By imperfect we mean that the centre of mass of the rotor is not on the axis of rotation and/or the axis of rotation does not coincide with a principal axis.

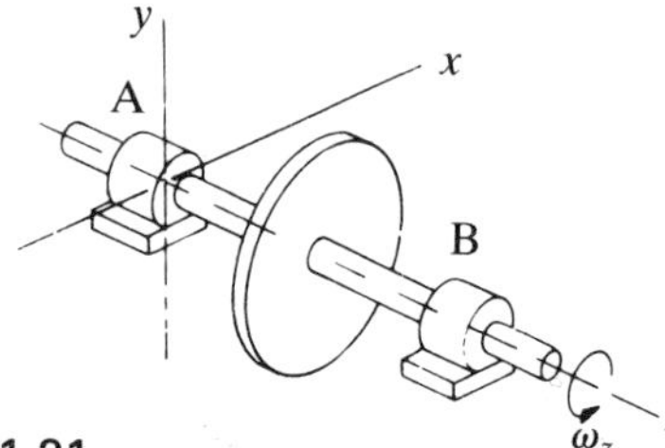

Figure 11.21

For the rotor shown in Fig. 11.21, assume that the moment of inertia is known relative to the x-, y-, z-axes which are attached to the body. From equation 11.60, with $\omega_x = \omega_y = 0$,

$$L_x = -I_{xz}\omega_z$$
$$L_y = -I_{yz}\omega_z$$
$$L_z = I_{zz}\omega_z$$

The angular velocity of the body and of the axes $\boldsymbol{\omega} = \omega_z \boldsymbol{k}$ and is assumed to be constant. Thus

$$\boldsymbol{M}_A = \boldsymbol{\omega}\times\boldsymbol{L} = \begin{vmatrix} \boldsymbol{i} & \boldsymbol{j} & \boldsymbol{k} \\ 0 & 0 & \omega_z \\ -I_{xz}\omega_z & -I_{yz}\omega_z & I_{zz}\omega_z \end{vmatrix}$$
$$= \boldsymbol{i}I_{yz}\omega_z^2 - \boldsymbol{j}I_{xz}\omega_z^2 \quad (11.75)$$

In practice, the products of inertia are small and may be considered to be due to the addition, or subtraction, of point masses to an otherwise perfect rotor, in which case I_{xz} and I_{yz} are easily calculated.

A variation of the same problem occurs when a perfect rotor is misaligned as shown in Fig. 11.22.

The angular velocity of rotor and axes, referred

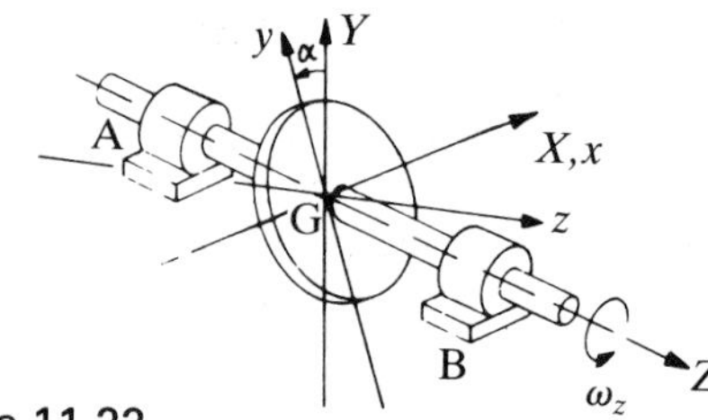

Figure 11.22

to the rotating axes which are coincident with the principal axes, is

$$\boldsymbol{\omega} = \omega_z\cos\alpha\boldsymbol{k} - \omega_z\sin\alpha\boldsymbol{j}$$

so $\quad \boldsymbol{L}_G = I_{zz}\omega_z\cos\alpha\boldsymbol{k} - I_{yy}\omega_z\sin\alpha\boldsymbol{j}$

Hence, assuming ω_z to be constant,

$$\boldsymbol{M}_G = \boldsymbol{\omega}\times\boldsymbol{L}_G$$
$$= \begin{vmatrix} \boldsymbol{i} & \boldsymbol{j} & \boldsymbol{k} \\ 0 & -\omega_z\sin\alpha & \omega_z\cos\alpha \\ 0 & -I_{yy}\omega_z\sin\alpha & I_{zz}\omega_z\cos\alpha \end{vmatrix}$$
$$= \boldsymbol{i}\omega_z^2(I_{yy} - I_{zz})\tfrac{1}{2}\sin 2\alpha \quad (11.76)$$

See example 11.8 and refer to problem 11.13 for the case of non-constant angular velocity.

11.12 Euler's angles

A convenient set of co-ordinates for describing the position of a rigid body are the Euler angles. Referring to Fig. 11.23, the x-, y-, z-axes are attached to the body and the X-, Y-, Z-axes are fixed in space. The position is defined by a

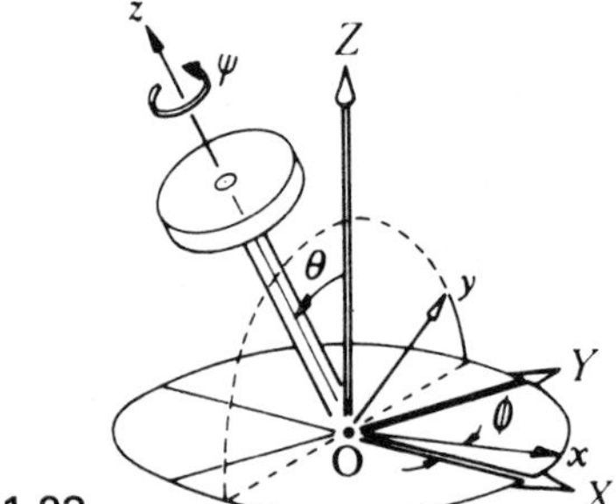

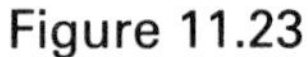

Figure 11.23

rotation ϕ about the Z-axis followed by a rotation θ about the displaced x-axis. Finally a rotation ψ about the z-axis completes the displacement. This last rotation will lift the x-axis out of the XY-plane in which it is shown in the figure. It is seen that the z-axis is given some prominence and is chosen to coincide with the axis of symmetry of the body, should the body possess one.

11.13 Rotation about a fixed point of a body with an axis of symmetry

In the context of this section, the property of symmetry means that the moment of inertia about an axis through the fixed point and normal to the axis of symmetry Oz is independent of the orientation of the axis. From this it follows that $I_{xx} = I_{yy}$.

This form of symmetry is found in any body which is formed from an integer number of segments radiating from the axis of symmetry. For example, a prismatic body with a cross-section in the form of a regular polygon satisfies these conditions of symmetry.

In this situation it is possible to allow the body to rotate relative to the z-axis, as this is the axis of symmetry, thereby keeping the x-axis in the XY-plane. For this case, $\dot{\psi}$ is the angular velocity of the body relative to the z-axis, but the absolute angular velocity component along the z-axis is $(\dot{\psi} + \dot{\phi}\cos\theta)$.

In terms of the Eulerian angles, the angular velocity of the body is

$$\begin{aligned}\boldsymbol{\Omega} &= \dot{\phi}\boldsymbol{K} + \dot{\theta}\boldsymbol{i} + \dot{\psi}\boldsymbol{k} \\ &= \dot{\phi}\sin\theta\boldsymbol{j} + \dot{\phi}\cos\theta\boldsymbol{k} + \dot{\theta}\boldsymbol{i} + \dot{\psi}\boldsymbol{k} \\ &= \dot{\theta}\boldsymbol{i} + \dot{\phi}\sin\theta\boldsymbol{j} + (\dot{\phi}\cos\theta + \dot{\psi})\boldsymbol{k} \qquad (11.77) \\ &= \Omega_x\boldsymbol{i} + \Omega_y\boldsymbol{j} + \Omega_z\boldsymbol{k}\end{aligned}$$

but the angular velocity of the reference axes is now

$$\boldsymbol{\omega} = \dot{\theta}\boldsymbol{i} + \dot{\phi}\sin\theta\boldsymbol{j} + \dot{\phi}\cos\theta\boldsymbol{k}$$

Since the body has an axis of symmetry, let $I_{xx} = I_{yy} = I_{\mathrm{O}}$ and let $I_{zz} = I_z$; then

$$\boldsymbol{L}_{\mathrm{O}} = I_{\mathrm{O}}\dot{\theta}\boldsymbol{i} + I_{\mathrm{O}}\dot{\phi}\sin\theta\boldsymbol{j} + I_z(\dot{\psi} + \dot{\phi}\cos\theta)\boldsymbol{k}$$

Using equation 11.7,

$$\begin{aligned}\mathrm{d}\boldsymbol{L}_{\mathrm{O}}/\mathrm{d}t &= I_{\mathrm{O}}\ddot{\theta}\boldsymbol{i} + I_{\mathrm{O}}(\ddot{\phi}\sin\theta + \dot{\phi}\dot{\theta}\cos\theta)\boldsymbol{j} \\ &\quad + I_z[\partial(\dot{\psi} + \dot{\phi}\cos\theta)\partial t]\boldsymbol{k} \\ &\quad + \begin{vmatrix} \boldsymbol{i} & \boldsymbol{j} & \boldsymbol{k} \\ \dot{\theta} & \dot{\phi}\sin\theta & \dot{\phi}\cos\theta \\ I_{\mathrm{O}}\dot{\theta} & I_{\mathrm{O}}\dot{\phi}\sin\theta & I_z(\dot{\psi} + \dot{\phi}\cos\theta) \end{vmatrix} \\ &= \boldsymbol{i}[I_{\mathrm{O}}\ddot{\theta} + I_z(\dot{\psi} + \dot{\phi}\cos\theta)\dot{\phi}\sin\theta \\ &\qquad - I_{\mathrm{O}}\dot{\phi}^2\sin\theta\cos\theta] \\ &\quad + \boldsymbol{j}[I_{\mathrm{O}}(\ddot{\phi}\sin\theta + \dot{\phi}\dot{\theta}\cos\theta) \\ &\qquad + I_{\mathrm{O}}\dot{\phi}\dot{\theta}\cos\theta - I_z\dot{\theta}(\dot{\psi} + \dot{\phi}\cos\theta)] \\ &\quad + \boldsymbol{k}\{\partial[I_z(\dot{\psi} + \dot{\phi}\cos\theta)]/\partial t\} \qquad (11.78)\end{aligned}$$

The absolute angular velocity of the body resolved along the z-axis, $\Omega_z = (\dot{\psi} + \dot{\phi}\cos\theta)$, is known as the spin velocity, and the angular velocity, $\dot{\phi}$, about the Z-axis, is known as the precessional velocity, while $\dot{\theta}$ is known as the nutational velocity.

For steady precession, $\dot{\phi} = \text{constant}$ and $\theta = \text{constant}$, so equation 11.78 reduces to

$$\begin{aligned}\mathrm{d}\boldsymbol{L}_{\mathrm{O}}/\mathrm{d}t = \boldsymbol{i}[I_z(\dot{\psi} + \dot{\phi}\cos\theta)\dot{\phi}\sin\theta \\ - I_{\mathrm{O}}\dot{\phi}^2\sin\theta\cos\theta] \qquad (11.79)\end{aligned}$$

so we see that if steady precession is achieved then only a moment of force about the x-axis is required.

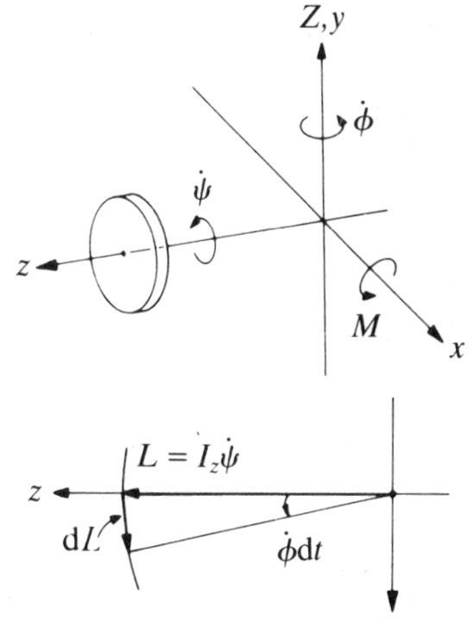

Figure 11.24

If $\theta = 90°$ then $\mathrm{d}\boldsymbol{L}_{\mathrm{O}}/\mathrm{d}t = \boldsymbol{i}I_z\dot{\psi}\dot{\phi}$. This can be seen from the simple vector diagram shown in Fig. 11.24 from which, for the small angle $\dot{\phi}\mathrm{d}t$,

$$\mathrm{d}\boldsymbol{L} = \boldsymbol{i}(I_z\dot{\psi})(\dot{\phi}\mathrm{d}t)$$

and, dividing through by dt,

$$\mathrm{d}\boldsymbol{L}_{\mathrm{O}}/\mathrm{d}t = \boldsymbol{i}I_z\dot{\psi}\dot{\phi}$$

Since $\boldsymbol{M}_{\mathrm{O}} = \mathrm{d}\boldsymbol{L}_{\mathrm{O}}/\mathrm{d}t$, the torque which must be applied to the body to produce the motion has a magnitude of $I_z\dot{\psi}\dot{\phi}$ and is about the x-axis.

A well-known problem covered by equation 11.78 is that of the freely spinning top. From Fig. 11.25 we see that

$$\boldsymbol{M}_{\mathrm{O}} = Wl\sin\theta\boldsymbol{i}$$

where W is the weight of the body and l is the distance of the centre of gravity G from the fixed pivot O.

Equation 11.78 shows that

$$M_z = 0 = \partial(\dot{\psi} + \dot{\phi}\cos\theta)/\partial t = \partial\Omega_z/\partial t$$

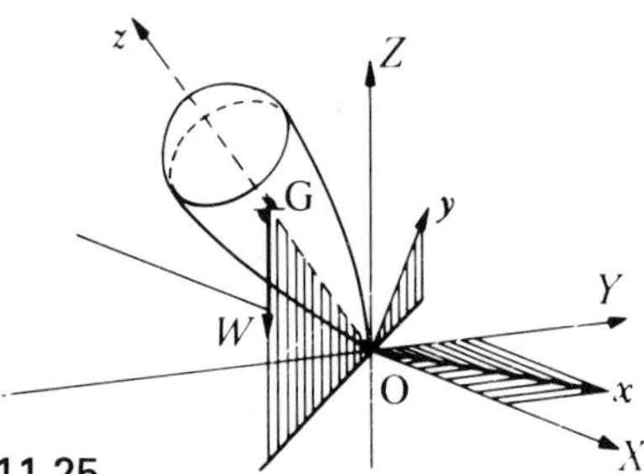

Figure 11.25

thus $(\dot{\psi}+\dot{\phi}\cos\theta)=\Omega_z=$ constant

For steady precession, equation 11.79 gives

$$Wl\sin\theta = I_z\Omega_z\dot{\phi}\sin\theta - I_O\dot{\phi}^2\sin\theta\cos\theta$$

or $$Wl = I_z\Omega_z\dot{\phi} + I_O\dot{\phi}^2\cos\theta$$

hence $$\cos\theta = I_z\Omega_z/I_O\dot{\phi} - Wl/I_O\dot{\phi}^2 \quad (11.80)$$

We must take $I_z\Omega_z$ as a constant since whatever torques are applied in order to establish steady precession they cannot produce a torque on the body about the z-axis as we are assuming frictionless conditions.

Letting $1/\dot{\phi}=T/2\pi$, where T is the periodic time of the precession,

$$\cos\theta = \left(\frac{I_z\Omega_z}{I_O}\right)\left(\frac{T}{2\pi}\right) - \left(\frac{Wl}{I_O}\right)\left(\frac{T}{2\pi}\right)^2 \quad (11.81)$$

A plot of $\cos\theta$ against $T/2\pi$ is shown in Fig. 11.26.

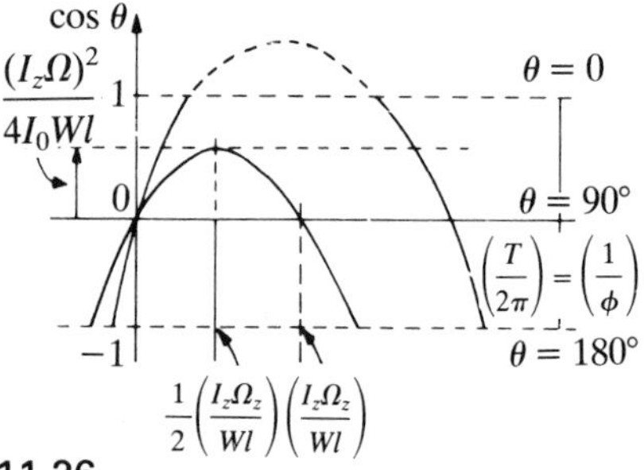

Figure 11.26

From the graph it is seen that, if $(I_z\Omega_z)^2 \geq 4I_O Wl$, it is possible for the top to 'sleep'; that is, for θ to be zero. In this condition precession has no meaning. For a given value of θ between 0 and 90°, if steady precession occurs there are two precession rates both with $\dot{\phi}>0$; however, for $90°<\theta<180°$ there are always two precession rates, one positive and one negative. In practice it is the slower of the precession rates which is observed.

The components of the moment of momentum are shown in Fig. 11.27. Since there is no applied couple about the Z-axis, the moment of momentum about this axis must be constant. Thus

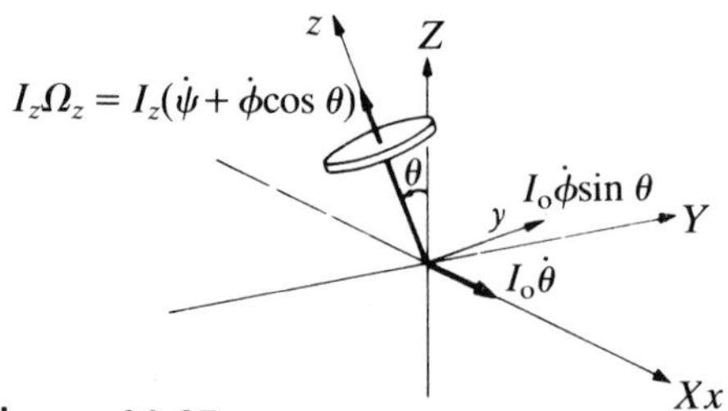

Figure 11.27

$$L_Z = I_z\Omega_z\cos\theta + I_O\dot{\phi}\sin^2\theta = \text{constant} \quad (11.82)$$

This relationship is true for all conditions of free motion.

11.14 Kinetic energy of a rigid body

The total kinetic energy $T=\sum\frac{1}{2}m_i\boldsymbol{v}_i\cdot\boldsymbol{v}_i$. If the body is rotating about a fixed point O, then $\boldsymbol{v}_i=\boldsymbol{\omega}\times\boldsymbol{r}_i$, so

$$T=\sum\tfrac{1}{2}m_i(\boldsymbol{\omega}\times\boldsymbol{r}_i)\cdot(\boldsymbol{\omega}\times\boldsymbol{r}_i)$$

when expressed in Cartesian co-ordinates is

$$T=\sum\tfrac{1}{2}m_i[\omega_y z_i-\omega_z y_i)^2+(\omega_z x_i-\omega_x z_i)^2 + (\omega_x y_i-\omega_y x_i)^2]$$

or $$2T=\omega_x^2\sum m_i(z_i^2+y_i^2)+\omega_z^2\sum m_i(x_i^2+z_i^2) + \omega_z^2\sum m_i(x_i^2+y_i^2)-\omega_y\omega_z 2\sum m_i z_i y_i - \omega_z\omega_x 2\sum m_i x_i z_i - \omega_x\omega_y 2\sum m_i y_i x_i$$

and, using the definitions of the moments of inertia,

$$T=\tfrac{1}{2}\omega_x^2 I_{xx}+\tfrac{1}{2}\omega_y^2 I_{yy}+\tfrac{1}{2}\omega_z^2 I_{zz} - \omega_y\omega_z I_{yz}-\omega_z\omega_x I_{xz}-\omega_x\omega_y I_{xy}$$

By direct matrix multiplication it is seen that

$$T=\tfrac{1}{2}(\omega_x,\omega_y,\omega_z) \times \begin{bmatrix} I_{xx} & -I_{xy} & -I_{xz} \\ -I_{yx} & I_{yy} & -I_{yz} \\ -I_{zx} & -I_{zy} & I_{zz} \end{bmatrix} \begin{bmatrix} \omega_x \\ \omega_y \\ \omega_z \end{bmatrix} \quad (11.83)$$

If principal axes are used, then

$$T = \tfrac{1}{2}(\omega_x, \omega_y, \omega_z) \times \begin{bmatrix} I_{xx} & 0 & 0 \\ 0 & I_{yy} & 0 \\ 0 & 0 & I_{zz} \end{bmatrix} \begin{bmatrix} \omega_x \\ \omega_y \\ \omega_z \end{bmatrix}$$

$$= \tfrac{1}{2}(I_{xx}\omega_x^2 + I_{yy}\omega_y^2 + I_{zz}\omega_z^2) \qquad (11.84)$$

In the special case when rotation is about a fixed x-axis, equation 11.83 reduces to $T = \tfrac{1}{2}I_{xx}\omega_x^2$, as given in Chapter 7.

If the body is in general motion, then

$$\boldsymbol{v}_i = \boldsymbol{v}_G + \boldsymbol{\omega} \times \boldsymbol{\rho}_i$$

thus $2T = \sum m_i(\boldsymbol{v}_G + \boldsymbol{\omega} \times \boldsymbol{\rho}_i)\cdot(\boldsymbol{v}_G + \boldsymbol{\omega} \times \boldsymbol{\rho}_i)$

$$= (\textstyle\sum m_i)\boldsymbol{v}_G \cdot \boldsymbol{v}_G + \sum m_i(\boldsymbol{\omega} \times \boldsymbol{\rho}_i)\cdot(\boldsymbol{\omega} \times \boldsymbol{\rho}_i) + 2\sum m_i \boldsymbol{v}_G \cdot (\boldsymbol{\omega} \times \boldsymbol{\rho}_i)$$

From Appendix 3 we know that

$$\textstyle\sum m_i \boldsymbol{v}_G \cdot (\boldsymbol{\omega} \times \boldsymbol{\rho}_i) = \sum m_i (\boldsymbol{v}_G \times \boldsymbol{\omega}) \cdot \boldsymbol{\rho}_i = (\boldsymbol{v}_G \times \boldsymbol{\omega}) \cdot (\sum m_i \boldsymbol{\rho}_i)$$

hence $T = \tfrac{1}{2}mv_G^2 + \sum \tfrac{1}{2}m_i(\boldsymbol{\omega} \times \boldsymbol{\rho}_i)\cdot(\boldsymbol{\omega} \times \boldsymbol{\rho}_i)$

because by definition of the centre of mass, $\sum m_i \boldsymbol{\rho}_i = 0$.

If principal axes are used,

$$T = \tfrac{1}{2}mv_G^2 + \tfrac{1}{2}(I_{xx}\omega_x^2 + I_{yy}\omega_y^2 + I_{zz}\omega_z^2) \qquad (11.85)$$

where the moments of inertia are calculated relative to the principal axes through the centre of mass.

Discussion examples

Example 11.1

An access panel OPQ, shown in Fig. 11.28, is to be moved to the position O″P″Q″. It is desired that this shall be accomplished by a rotation about a fixed hinge line. If this is possible, find the position of the hinge and the angle of rotation.

Solution We will consider the displacement in two stages: rotation about a fixed point followed by a translation.

Consider one point, say O, to be fixed, in which case P goes to P′ and Q goes to Q′, where

$$\overrightarrow{OP}' = \overrightarrow{OP}'' - \overrightarrow{OO}'' = (2\boldsymbol{i}+\boldsymbol{j}) - (\boldsymbol{i}+\boldsymbol{j}) = \boldsymbol{i}$$

and $\overrightarrow{OQ}' = \overrightarrow{OQ}'' - \overrightarrow{OO}'' = (\boldsymbol{i}+\boldsymbol{j}-\boldsymbol{k}) - (\boldsymbol{i}+\boldsymbol{j}) = -\boldsymbol{k}$

The relative displacements are

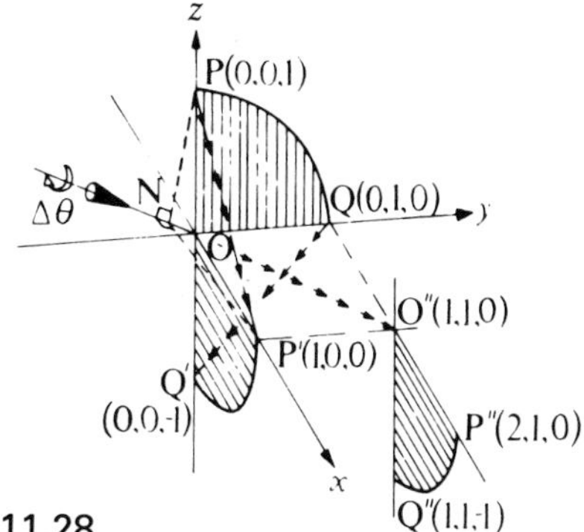

Figure 11.28

$$\overrightarrow{PP'} = \overrightarrow{OP}' - \overrightarrow{OP} = (\boldsymbol{i}) - (\boldsymbol{k}) = \boldsymbol{i} - \boldsymbol{k}$$

and $\overrightarrow{QQ}' = \overrightarrow{OQ}' - \overrightarrow{OQ} = (-\boldsymbol{k}) - (\boldsymbol{j}) = -\boldsymbol{j} - \boldsymbol{k}$

The axis of rotation for O fixed passes through O and is parallel to

$$\overrightarrow{QQ}' \times \overrightarrow{PP'} = \begin{vmatrix} \boldsymbol{i} & \boldsymbol{j} & \boldsymbol{k} \\ 0 & -1 & -1 \\ 1 & 0 & -1 \end{vmatrix} = \boldsymbol{i} - \boldsymbol{j} + \boldsymbol{k}$$

Thus the axis passes through the point $(1,-1,1)$.

Point N is located on the axis of rotation such that NP is normal to the axis; thus

$$\overrightarrow{ON} = b(\boldsymbol{i} - \boldsymbol{j} + \boldsymbol{k})$$

where b is a constant to be determined from $\overrightarrow{ON}\cdot\overrightarrow{NP} = 0$.

But $\overrightarrow{NP} = \overrightarrow{OP} - \overrightarrow{ON}$; thus

$$\overrightarrow{ON}\cdot(\overrightarrow{OP} - \overrightarrow{ON}) = b(\boldsymbol{i} - \boldsymbol{j} + \boldsymbol{k}) \cdot [\boldsymbol{k} - b(\boldsymbol{i} - \boldsymbol{j} + \boldsymbol{k})] = b - b^2 3 = 0$$

therefore $b = \tfrac{1}{3}$

The angle of rotation is found from

$|\overrightarrow{PP'}| = 2|\overrightarrow{NP}|\sin\tfrac{1}{2}\Delta\theta$, as ΔNPP′ is isosceles,

Now $|\overrightarrow{PP'}| = \sqrt{2}$

and $|\overrightarrow{NP}| = |\boldsymbol{k} - \tfrac{1}{3}(\boldsymbol{i} - \boldsymbol{j} + \boldsymbol{k})| = |-\tfrac{1}{3}\boldsymbol{i} + \tfrac{1}{3}\boldsymbol{j} + \tfrac{2}{3}\boldsymbol{k}| = \sqrt{6}/3$

hence $\sin\tfrac{1}{2}\Delta\theta = \dfrac{3\sqrt{2}}{2\sqrt{6}} = \sqrt{3}/2$

giving $\Delta\theta = 120°$

In order to determine whether a change in the

position of the axis of rotation will move OPQ to O″P″Q″, it is necessary to show that the displacement $\overrightarrow{OO''}$ (or $\overrightarrow{P'P''}$) is wholly normal to the axis of rotation.

We see that $\overrightarrow{ON}\cdot\overrightarrow{OO''} = \frac{1}{3}(\boldsymbol{i}-\boldsymbol{j}+\boldsymbol{k})\cdot(\boldsymbol{i}+\boldsymbol{j}) = 0$; therefore there is no component of displacement along the axis, so a single fixed hinge line is possible.

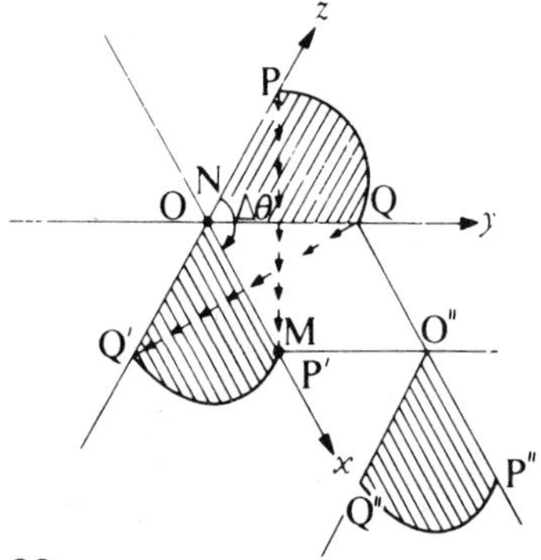

Figure 11.29

Figure 11.29 is a view looking along the axis of rotation, i.e. along the normal to the plane containing NPP′. Because all displacements take place in parallel planes, the displacements are the same as the displacements of their projections on to the NPP′ plane.

The intersection of the new axis of rotation with the NPP plane, point M, must lie on the perpendicular bisector of OO″ projected and be such that the angle OMO″ projected is 120°. For this case, M is clearly the point P′; thus the hinge line passes through (1,0,0) and is parallel to vector $(\boldsymbol{i}-\boldsymbol{j}+\boldsymbol{k})$. The angle of rotation is 120°.

Example 11.2

Figure 11.30 shows a set of x-, y-, z-axes rotating at a constant angular velocity $3\boldsymbol{j}$ rad/s about the Y-axis of a fixed set of X-, Y-, Z-axes. A vector $\overrightarrow{OA}$ is given as $\boldsymbol{A} = (2\boldsymbol{i}+t\boldsymbol{j}+3t^2\boldsymbol{k})$m, where t is the time in seconds and $\boldsymbol{i}$, $\boldsymbol{j}$, $\boldsymbol{k}$ are the unit vectors associated with the moving x-, y-, z-axes.

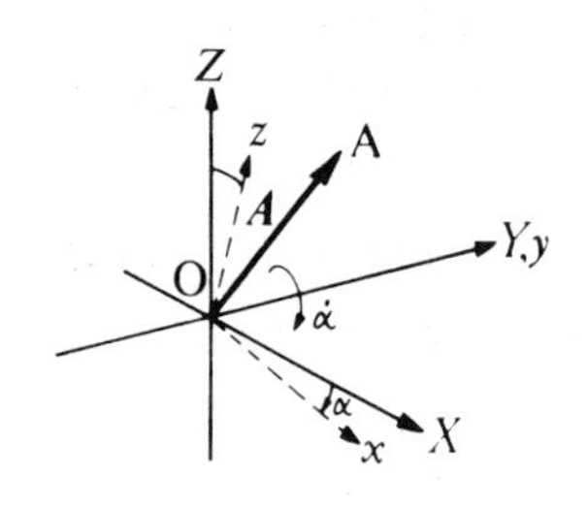

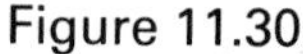
Figure 11.30

Express $\boldsymbol{A}$ and $\mathrm{d}\boldsymbol{A}/\mathrm{d}t$ in terms of both sets of axes for the conditions $t = 2$ s, $\alpha \doteq \arcsin(3/5)$ and $\dot{\alpha} = 3$ rad/s.

Solution

a) In terms of $\boldsymbol{i}, \boldsymbol{j}, \boldsymbol{k}$

$$\boldsymbol{A} = 2\boldsymbol{i}+t\boldsymbol{j}+3t^2\boldsymbol{k} \quad \text{(i)}$$

From equation 11.7,

$$\mathrm{d}\boldsymbol{A}/\mathrm{d}t = \partial\boldsymbol{A}/\partial t + \boldsymbol{\omega}\times\boldsymbol{A}$$

where $\boldsymbol{\omega}$ (the angular velocity of the moving axes) $= 3\boldsymbol{j}$

$$\text{so} \quad \mathrm{d}\boldsymbol{A}/\mathrm{d}t = (\boldsymbol{j}+6t\boldsymbol{k})+3\boldsymbol{j}\times(2\boldsymbol{i}+t\boldsymbol{j}+3t^2\boldsymbol{k})$$
$$= 9t^2\boldsymbol{i}+\boldsymbol{j}+(6t-6)\boldsymbol{k} \quad \text{(ii)}$$

Thus, when $t = 2$ s,

$$\boldsymbol{A} = (2\boldsymbol{i}+2\boldsymbol{j}+12\boldsymbol{k})\text{m} \quad \text{(iii)}$$
$$\text{and} \quad \mathrm{d}\boldsymbol{A}/\mathrm{d}t = (36\boldsymbol{i}+\boldsymbol{j}+6\boldsymbol{k})\text{m/s} \quad \text{(iv)}$$

b) In terms of $\boldsymbol{I}, \boldsymbol{J}, \boldsymbol{K}$

To express the vectors in terms of the unit vectors $\boldsymbol{I}, \boldsymbol{J}, \boldsymbol{K}$ associated with the fixed axes, we must first establish the transformation relationships between $\boldsymbol{i}, \boldsymbol{j}, \boldsymbol{k}$ and $\boldsymbol{I}, \boldsymbol{J}, \boldsymbol{K}$.

Referring to Fig. 11.31 we see that

$$\boldsymbol{i} = \cos\alpha\boldsymbol{I} - \sin\alpha\boldsymbol{K}$$
$$\boldsymbol{j} = \boldsymbol{J}$$
$$\boldsymbol{k} = \sin\alpha\boldsymbol{I} + \cos\alpha\boldsymbol{K}$$

$$\text{or} \quad \begin{bmatrix}\boldsymbol{i}\\ \boldsymbol{j}\\ \boldsymbol{k}\end{bmatrix} = \begin{bmatrix}\cos\alpha & 0 & -\sin\alpha\\ 0 & 1 & 0\\ \sin\alpha & 0 & \cos\alpha\end{bmatrix}\begin{bmatrix}\boldsymbol{I}\\ \boldsymbol{J}\\ \boldsymbol{K}\end{bmatrix} \quad \text{(v)}$$

$$\text{Now} \quad \boldsymbol{A} = (\boldsymbol{i},\boldsymbol{j},\boldsymbol{k})\begin{pmatrix}2\\ t\\ 3t^2\end{pmatrix} = (2,t,3t^2)\begin{pmatrix}\boldsymbol{i}\\ \boldsymbol{j}\\ \boldsymbol{k}\end{pmatrix}$$

$$= (2,t,3t^2)\begin{bmatrix}\cos\alpha\\ 0\\ \sin\alpha\end{bmatrix}\begin{matrix}0\\ 1\\ 0\end{matrix}\begin{bmatrix}-\sin\alpha\boldsymbol{I}\\ 0 \quad \boldsymbol{J}\\ \cos\alpha\boldsymbol{K}\end{bmatrix}$$

$$= [(2\cos\alpha+3t^2\sin\alpha),t,$$
$$(-2\sin\alpha-3t^2\cos\alpha)]\begin{bmatrix}\boldsymbol{I}\\ \boldsymbol{J}\\ \boldsymbol{K}\end{bmatrix} \quad \text{(vi)}$$

When $t = 2$, $\sin\alpha = 3/5$, and $\cos\alpha = 4/5$,

$$\boldsymbol{A} = [(44/5)\boldsymbol{I}+2\boldsymbol{J}+(42/5)\boldsymbol{K}]\text{m} \quad \text{(vii)}$$

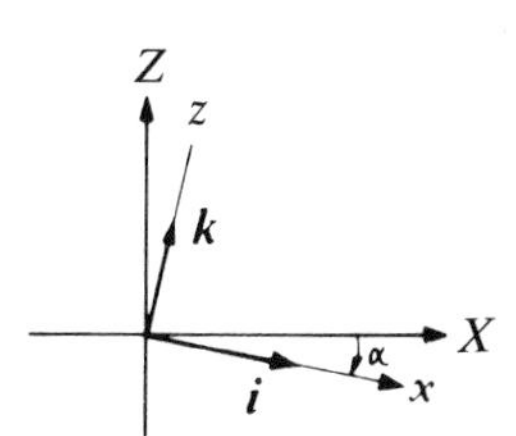

Figure 11.31

Similarly, using equations (iv) and (v),

$$\frac{dA}{dt} = (36,1,6)\begin{bmatrix} 4/5 & 0 & -3/5 \\ 0 & 1 & 0 \\ 3/5 & 0 & 4/5 \end{bmatrix}\begin{bmatrix} I \\ J \\ K \end{bmatrix}$$

$$= [(162/5)I + J - (84/5)K]\text{m/s} \qquad \text{(viii)}$$

c) It is possible to obtain equation (viii) from equation (vi) by direct differentiation:

$$\frac{dA}{dt} = [(-2\sin\alpha\dot{\alpha} + 6t\sin\alpha + 3t^2\cos\alpha\dot{\alpha}),1,$$

$$(-2\cos\alpha\dot{\alpha} + 6t\cos\alpha - 3t^2\sin\alpha\dot{\alpha})]\begin{bmatrix} I \\ J \\ K \end{bmatrix}$$

Substituting appropriate values gives

$$\frac{dA}{dt} = \left[\left(-\frac{18}{5} + \frac{36}{5} + \frac{144}{5}\right)I + J\right.$$

$$\left. + \left(-\frac{24}{5} + \frac{48}{5} - \frac{108}{5}\right)K\right]$$

$$= [(162/5)I + J - (84/5)K]\text{m/s}$$

Example 11.3

Figure 11.32(a) shows a wheel APB free to rotate on an axle OC which is free to rotate about the point O. The wheel is constrained to roll without slipping on the *XY*-plane while the axle OC is rotating at a constant angular velocity of 15 rad/s about the *Z*-axis. Determine the velocity and acceleration of the point P when the angle ACP is 120°. The radius *r* of the wheel is 0.3 m and the length *b* of OC is 0.4 m.

Solution The *X*-, *Y*-, *Z*-axes are fixed and the *x*-, *y*-, *z*-axes are attached to the axle such that the *x*-axis is along the axle, the *z*-axis is in the *xOZ*-plane and the *y*-axis completes the triad, being parallel to the *XY*-plane.

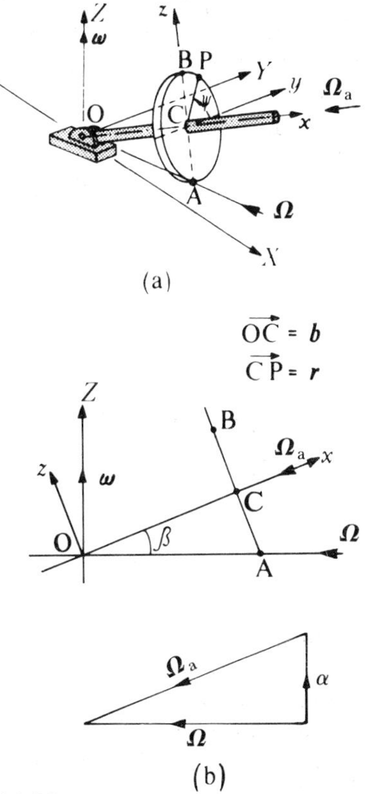

Figure 11.32

Let $\boldsymbol{\omega}$ = the angular velocity of the bearing housing about the *Z*-axis

$\boldsymbol{\Omega}_a$ = the angular velocity of the wheel relative to the axle OC

and $\boldsymbol{\Omega}$ = the absolute angular velocity of the wheel

If no slip is to take place then OA must be the instantaneous axis of rotation.

Now $\boldsymbol{\Omega} = \boldsymbol{\Omega}_a + \boldsymbol{\omega}$

and this relationship is shown on the vector diagram, Fig. 11.32(b), from which

$$\boldsymbol{\omega} = \omega\sin\beta\boldsymbol{i} + \omega\cos\beta\boldsymbol{k}$$
$$= 15\times0.6\boldsymbol{i} + 15\times0.8\boldsymbol{k}$$
$$= (9\boldsymbol{i} + 12\boldsymbol{k})\text{ rad/s}$$
$$\boldsymbol{\Omega}_a = (-\omega/\sin\beta)\boldsymbol{i} = (-15/0.6)\boldsymbol{i}$$
$$= -25\boldsymbol{i}\text{ rad/s}$$

and $\boldsymbol{\Omega} = \boldsymbol{\Omega}_a + \boldsymbol{\omega} = (-16\boldsymbol{i} + 12\boldsymbol{k})$ rad/s

The velocity of P is given by

$$v_{P/O} = \boldsymbol{\Omega}\times(\boldsymbol{b}+\boldsymbol{r})$$
$$= (-16\boldsymbol{i} + 12\boldsymbol{k})$$
$$\times(0.4\boldsymbol{i} + 0.3(0.866\boldsymbol{j} + 0.50\boldsymbol{k}))$$

$$= (-3.12\boldsymbol{i} + 7.2\boldsymbol{j} - 4.16\boldsymbol{k})\ \text{m/s}$$

The acceleration of P is given by

$$\boldsymbol{a}_{P/O} = d\boldsymbol{v}_{P/O}/dt = \partial(\boldsymbol{v}_{P/O})/\partial t + \boldsymbol{\omega} \times \boldsymbol{v}_{P/O}$$
$$= \partial[\boldsymbol{\Omega} \times (\boldsymbol{b} + \boldsymbol{r})]/\partial t + \boldsymbol{\omega} \times \boldsymbol{v}_{P/O}$$

Now, as seen from the x-, y-, z-axes, $\boldsymbol{\Omega}$ and $\boldsymbol{b}$ are constant vectors; thus

$$\partial\boldsymbol{\Omega}/\partial t = 0 \quad \partial\boldsymbol{b}/\partial t = 0$$

But $\partial\boldsymbol{r}/\partial t = \boldsymbol{\Omega}_a \times \boldsymbol{r}$

hence

$$\boldsymbol{a}_{P/O} = \boldsymbol{\Omega} \times (\boldsymbol{\Omega}_a \times \boldsymbol{r}) + \boldsymbol{\omega} \times \boldsymbol{v}_{P/O}$$

$$\boldsymbol{\Omega}_a \times \boldsymbol{r} = -25\boldsymbol{i} \times (0.3 \times 0.866\boldsymbol{j} + 0.3 \times 0.50\boldsymbol{k})$$
$$= -6.50\boldsymbol{k} + 3.75\boldsymbol{j}$$

$$\boldsymbol{a}_{P/O} = \begin{vmatrix} \boldsymbol{i} & \boldsymbol{j} & \boldsymbol{k} \\ -16 & 0 & 12 \\ 0 & 3.75 & -6.50 \end{vmatrix} + \begin{vmatrix} \boldsymbol{i} & \boldsymbol{j} & \boldsymbol{k} \\ 9 & 0 & 12 \\ -3.12 & 7.2 & -4.16 \end{vmatrix}$$

$$= (-45\boldsymbol{i} + 104\boldsymbol{j} - 60\boldsymbol{k}) + (-86.4\boldsymbol{i} + 0\boldsymbol{j} + 64.8\boldsymbol{k})$$
$$= (-131.4\boldsymbol{i} - 104\boldsymbol{j} + 4.8\boldsymbol{k})\ \text{m/s}^2$$

Example 11.4

Figure 11.33 shows part of a mechanical handling system. The cylindrical cannister C is held firmly by grab G which, relative to the arm AB, is rotating with angular velocity $\dot{\psi}\boldsymbol{k}$ and angular acceleration $\ddot{\psi}\boldsymbol{k}$. The arm AB has an angular velocity $\dot{\theta}\boldsymbol{k}$ and angular acceleration $\ddot{\theta}\boldsymbol{k}$ and is extending at a constant rate $\dot{R}$, where R = AB. At the same time, the arm has a vertical velocity $v\boldsymbol{k}$ and a vertical acceleration $a\boldsymbol{k}$.

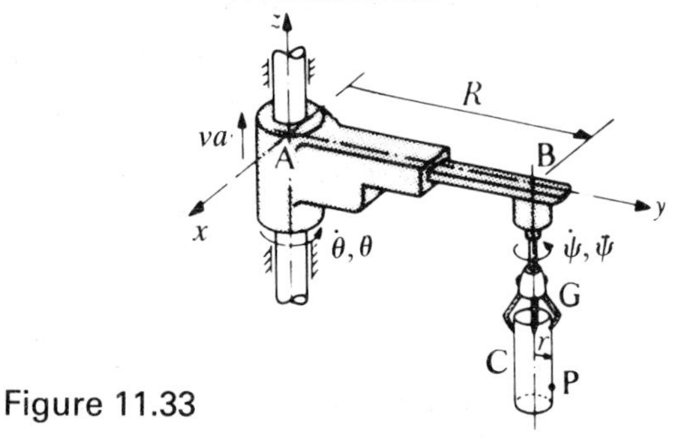

Figure 11.33

Determine the velocity and acceleration of a point P on the cannister which has a y-component of $R + r$.

Solution We shall find the velocity of P from the relative-velocity equation

$$\boldsymbol{v}_P = \boldsymbol{v}_B + \boldsymbol{v}_{P/B} \tag{i}$$

and denote the absolute angular velocity of the cannister by

$$\dot{\phi}\boldsymbol{k} = (\dot{\theta} + \dot{\psi})\boldsymbol{k} \tag{ii}$$

For the arm AB, from equation 11.13, replacing $\boldsymbol{e}_R$ by $\boldsymbol{j}$ and $\boldsymbol{e}_\theta$ by $(-\boldsymbol{i})$ we have

$$\boldsymbol{v}_B = \dot{R}\boldsymbol{j} + R\dot{\theta}(-\boldsymbol{i}) + v\boldsymbol{k} \tag{iii}$$

Similarly, noting that the cannister has no vertical velocity relative to the arm, and that r = constant,

$$\boldsymbol{v}_{P/B} = r\dot{\phi}(-\boldsymbol{i}) \tag{iv}$$

Combining equations (i), (ii), (iii) and (iv), the velocity of P is

$$\boldsymbol{v}_P = -[(R + r)\dot{\theta} + r\dot{\psi}]\boldsymbol{i} + \dot{R}\boldsymbol{j} + v\boldsymbol{k}$$

The relative-acceleration equation is

$$\boldsymbol{a}_P = \boldsymbol{a}_B + \boldsymbol{a}_{P/B} \tag{v}$$

From equation 11.14, noting that $\ddot{R} = 0$, the acceleration of B is

$$\boldsymbol{a}_B = -R\dot{\theta}^2\boldsymbol{j} + (R\ddot{\theta} + 2\dot{R}\dot{\theta})(-\boldsymbol{i}) + a\boldsymbol{k} \tag{vi}$$

and the relative acceleration between P and B is

$$\boldsymbol{a}_{P/B} = -r\dot{\phi}^2\boldsymbol{j} + r\ddot{\phi}(-\boldsymbol{i}) \tag{vii}$$

where

$$\ddot{\phi} = \ddot{\theta} + \ddot{\psi} \tag{viii}$$

Combining equations (v), (vi), (vii) and (viii), the acceleration of P is

$$\boldsymbol{a}_P = -[(R + r)\ddot{\theta} + r\ddot{\psi} + 2\dot{R}\dot{\theta}]\boldsymbol{i} - [(R + r)\dot{\theta}^2 + r\dot{\psi}^2 + 2r\dot{\theta}\dot{\psi}]\boldsymbol{j} + a\boldsymbol{k}$$

Example 11.5

See Fig. 11.34. A radar station A continuously measures, relative to the fixed X-, Y-, Z-axes, the spherical co-ordinates r, θ, and ϕ of an aircraft B. The derivatives of these co-ordinates are computed and, at a particular instant, the numerical values are

$$r = 20\,370\ \text{m} \qquad \theta = 22.35°$$
$$\dot{r} = -288.7\ \text{m/s} \qquad \dot{\theta} = 0.025\,56\ \text{rad/s}$$

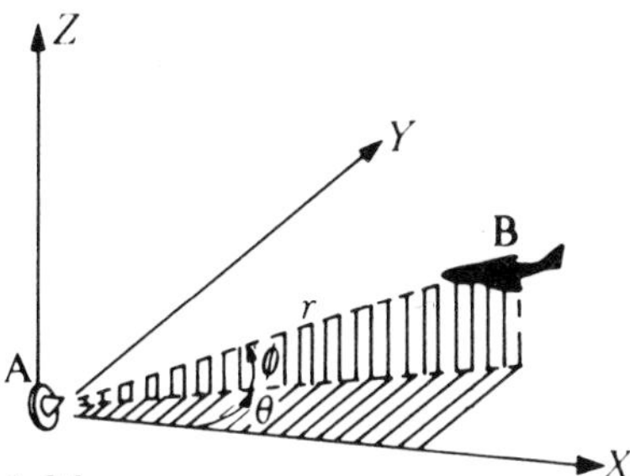

Figure 11.34

$$\ddot{r} = 15.74 \text{ m/s}^2 \qquad \ddot{\theta} = 7.518 \times 10^{-4} \text{ rad/s}$$
$$\phi = 11.69°$$
$$\dot{\phi} = 0.006223 \text{ rad/s}$$
$$\ddot{\phi} = 5.197 \times 10^{-4} \text{ rad/s}^2$$

Determine the velocity of the aircraft relative to the X-, Y-, Z-axes, the speed of the aircraft, and the angle of inclination of the velocity to the horizontal XY-plane. Also find the acceleration of the aircraft relative to the X-, Y-, Z-axes.

Solution The velocity and acceleration in spherical co-ordinates (see equations 11.19 and 11.20) are given in terms of the unit vectors $\boldsymbol{e}_r$, $\boldsymbol{e}_\theta$, and $\boldsymbol{e}_\phi$ corresponding to the x-, y-, z-axes in Fig. 11.35.

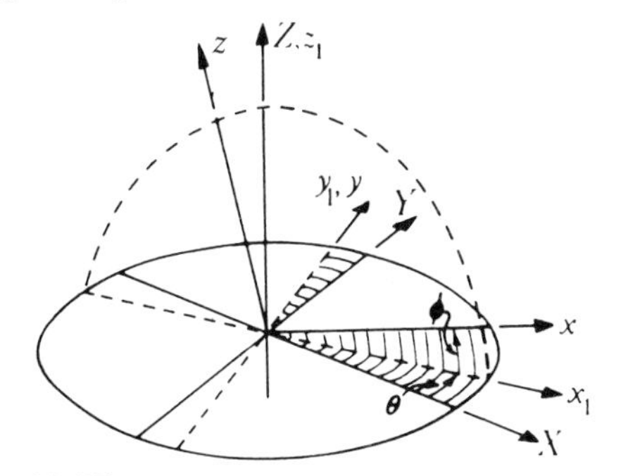

Figure 11.35

Substituting the given numerical values into equation 11.19, we obtain

$$\boldsymbol{v} = (-288.7\boldsymbol{e}_r + 509.9\boldsymbol{e}_\theta + 126.8\boldsymbol{e}_\phi) \text{ m/s} \qquad \text{(i)}$$

The directions of the x-, y-, z-axes can be obtained by considering the rotation of a set of axes, originally coincident with the X-, Y-, Z-axes. The first rotation is about the Z-axis through an angle θ to the orientation x_1, y_1, z_1 with associated unit vectors $\boldsymbol{i}_1$, $\boldsymbol{j}_1$, $\boldsymbol{k}_1$. The second rotation is about the y_1 axis through an angle ϕ to the orientation x, y, z with the associated unit vectors $\boldsymbol{e}_r$, $\boldsymbol{e}_\theta$, $\boldsymbol{e}_\phi$. From the figure we observe that

$$\boldsymbol{i}_1 = \cos\theta\boldsymbol{I} + \sin\theta\boldsymbol{J}$$
$$\boldsymbol{j}_1 = -\sin\theta\boldsymbol{I} + \cos\theta\boldsymbol{J}$$
$$\boldsymbol{k}_1 = \boldsymbol{K}$$

or, in matrix notation,

$$\begin{bmatrix} \boldsymbol{i}_1 \\ \boldsymbol{j}_1 \\ \boldsymbol{k}_1 \end{bmatrix} = \begin{bmatrix} \cos\theta & \sin\theta & 0 \\ -\sin\theta & \cos\theta & 0 \\ 0 & 0 & 1 \end{bmatrix} \begin{bmatrix} \boldsymbol{I} \\ \boldsymbol{J} \\ \boldsymbol{K} \end{bmatrix} \qquad \text{(ii)}$$

Also $\boldsymbol{e}_r = \cos\phi\boldsymbol{i}_1 + \sin\phi\boldsymbol{k}_1$

$$\boldsymbol{e}_\theta = \boldsymbol{j}_1$$
$$\boldsymbol{e}_\phi = -\sin\phi\boldsymbol{i}_1 + \cos\phi\boldsymbol{k}_1$$

or
$$\begin{bmatrix} \boldsymbol{e}_r \\ \boldsymbol{e}_\theta \\ \boldsymbol{e}_\phi \end{bmatrix} = \begin{bmatrix} \cos\phi & 0 & \sin\phi \\ 0 & 1 & 0 \\ -\sin\phi & 0 & \cos\phi \end{bmatrix} \begin{bmatrix} \boldsymbol{i}_1 \\ \boldsymbol{j}_1 \\ \boldsymbol{k}_1 \end{bmatrix} \qquad \text{(iii)}$$

Eliminating $\boldsymbol{i}_1$, $\boldsymbol{j}_1$, and $\boldsymbol{k}_1$, we obtain

$$\begin{bmatrix} \boldsymbol{e}_r \\ \boldsymbol{e}_\theta \\ \boldsymbol{e}_\phi \end{bmatrix} = \begin{bmatrix} \cos\phi & 0 & \sin\phi \\ 0 & 1 & 0 \\ -\sin\phi & 0 & \cos\phi \end{bmatrix} \times \begin{bmatrix} \cos\theta & \sin\theta & 0 \\ -\sin\theta & \cos\theta & 0 \\ 0 & 0 & 1 \end{bmatrix} \begin{bmatrix} \boldsymbol{I} \\ \boldsymbol{J} \\ \boldsymbol{K} \end{bmatrix}$$

$$= \begin{bmatrix} \cos\phi\cos\theta & \cos\phi\sin\theta & \sin\phi \\ -\sin\theta & \cos\theta & 0 \\ -\sin\phi\cos\theta & -\sin\phi\sin\theta & \cos\phi \end{bmatrix} \begin{bmatrix} \boldsymbol{I} \\ \boldsymbol{J} \\ \boldsymbol{K} \end{bmatrix} \qquad \text{(iv)}$$

or
$$\left.\begin{aligned} \boldsymbol{e}_r &= \cos\phi\cos\theta\boldsymbol{I} + \cos\phi\sin\theta\boldsymbol{J} + \sin\phi\boldsymbol{K} \\ \boldsymbol{e}_\theta &= -\sin\theta\boldsymbol{I} + \cos\theta\boldsymbol{J} \\ \boldsymbol{e}_\phi &= -\sin\phi\cos\theta\boldsymbol{I} - \sin\phi\sin\theta\boldsymbol{J} + \cos\phi\boldsymbol{K} \end{aligned}\right\} \qquad \text{(v)}$$

The 3×3 matrices which convert one set of unit vectors to another set are known as transformation matrices. The inverse of such matrices is obtained simply by exchanging the appropriate elements across the leading diagonal:

$$\begin{bmatrix} \boldsymbol{I} \\ \boldsymbol{J} \\ \boldsymbol{K} \end{bmatrix} = \begin{bmatrix} \cos\phi\cos\theta & -\sin\theta & -\sin\phi\cos\theta \\ \cos\phi\sin\theta & \cos\theta & -\sin\phi\sin\theta \\ \sin\phi & 0 & \cos\phi \end{bmatrix} \times \begin{bmatrix} \boldsymbol{e}_r \\ \boldsymbol{e}_\theta \\ \boldsymbol{e}_\phi \end{bmatrix} \qquad \text{(vi)}$$

Substituting equations (v) into equation (i), we

obtain the velocity of the aircraft:

$$v = (-479.1\boldsymbol{I} + 354.3\boldsymbol{J} + 65.64\boldsymbol{K})\ \text{m/s}$$

The speed is the magnitude of $\boldsymbol{v}$, which is 599.5 m/s.

Denoting the angle which $\boldsymbol{v}$ makes with the horizontal by λ (Fig. 11.36) we have

$$\tan\lambda = v_Z/\sqrt{(v_X{}^2 + v_Y{}^2)}$$

$$\lambda = \arctan[65.64/\sqrt{(479.1^2 + 354.3^2)}]$$

$$= 6.286°$$

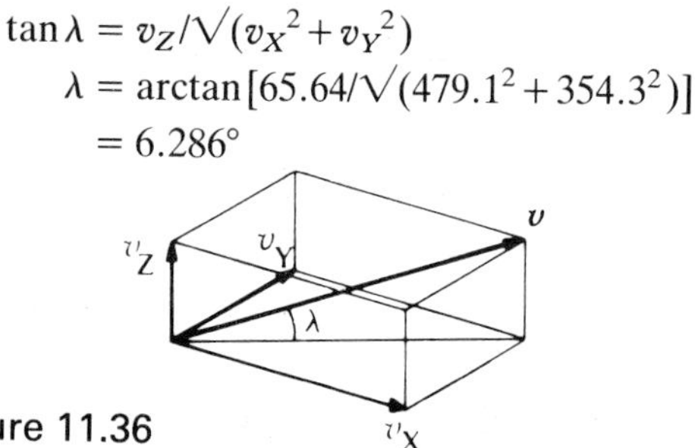

Figure 11.36

Substituting equations (v) into equation 11.20, we obtain the acceleration of the aircraft:

$$\boldsymbol{a} = (0.4675\boldsymbol{I} - 0.6398\boldsymbol{J} + 9.877\boldsymbol{K})\ \text{m/s}^2$$

Example 11.6

A three-dimensional four-bar chain is shown in Fig. 11.37. Crank DC is driven by crank AB via the connecting link BC which has ball-and-socket joints at each end. A and D are located at points (0, 0, 0) mm and (20, 40, 100) mm respectively. At the instant under consideration, AB and DC are parallel and crank AB has an angular velocity of $1\boldsymbol{k}$ rad/s and an angular acceleration of $2\boldsymbol{k}$ rad/s^2.

AB = 50 mm
DC = 30 mm

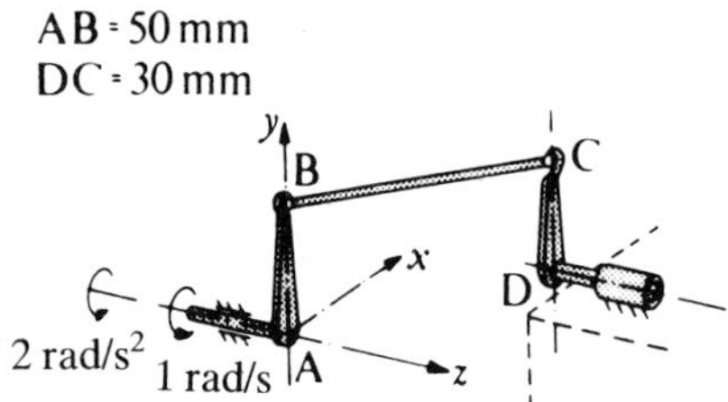

Figure 11.37

Determine the velocity and acceleration of C and the angular velocity and angular acceleration of DC.

Solution The velocity of B is given by

$$\boldsymbol{v}_B = \boldsymbol{\omega}_{AB} \times \boldsymbol{r}_{AB}$$

$$= 1\boldsymbol{k} \times 50\boldsymbol{j} = -50\boldsymbol{i}\ \text{mm/s}$$

and the velocity of C by

$$\boldsymbol{v}_C = \boldsymbol{\omega}_{DC} \times \boldsymbol{r}_{DC}$$

$$= \omega_{DC}\boldsymbol{k} \times 30\boldsymbol{j} = -30\omega_{DC}\boldsymbol{i}\ \text{mm/s} \qquad \text{(i)}$$

Since the relative velocity $\boldsymbol{v}_{C/B}$ is perpendicular to $\overrightarrow{BC}$,

$$(\boldsymbol{v}_{C/B} \cdot \overrightarrow{BC}) = 0 \qquad \text{(ii)}$$

where $\overrightarrow{BC} = \overrightarrow{AC} - \overrightarrow{AB}$

$$= (20\boldsymbol{i} + 70\boldsymbol{j} + 100\boldsymbol{k}) - 50\boldsymbol{j}$$

$$= (20\boldsymbol{i} + 20\boldsymbol{j} + 100\boldsymbol{k})\ \text{mm}$$

Substituting in equation (ii) we have

$$(-30\omega_{DC}\boldsymbol{i} + 50\boldsymbol{i}) \cdot (20\boldsymbol{i} + 20\boldsymbol{j} + 100\boldsymbol{k} = 0$$

$$-600\omega_{CD} + 1000 = 0$$

$$\omega_{CD} = 1\tfrac{2}{3}\ \text{rad/s}$$

and $\boldsymbol{\omega}_{CD} = 1\tfrac{2}{3}\boldsymbol{k}$ rad/s

From equation (i),

$$\boldsymbol{v}_C = -30(1\tfrac{2}{3})\boldsymbol{i} = -50\boldsymbol{i}\ \text{mm/s}$$

We note that, for the given position, $\boldsymbol{v}_C = \boldsymbol{v}_B$ so that $\boldsymbol{v}_{C/B} = 0$ and hence $\boldsymbol{\omega}_{BC} = 0$, from equation 11.45.

Now the acceleration of C is given by

$$\boldsymbol{a}_{C/A} = \boldsymbol{a}_{C/B} + \boldsymbol{a}_{B/A} = \boldsymbol{a}_{C/D} \qquad \text{(iii)}$$

since both A and D are fixed points. The acceleration components of B in cylindrical co-ordinates are shown in Fig. 11.38.

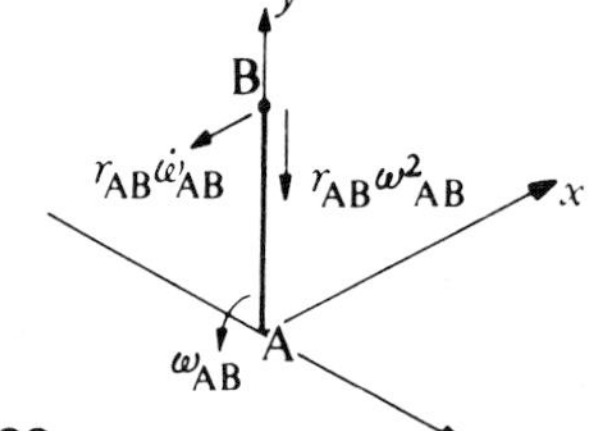

Figure 11.38

$$\boldsymbol{a}_B = \boldsymbol{a}_{B/A} = -r_{AB}\,\omega_{AB}{}^2\boldsymbol{j} - r_{AB}\,\dot{\omega}_{AB}\,\boldsymbol{i}$$

$$= -50(1)^2\boldsymbol{j} - 50(2)\,\boldsymbol{i}$$

$$= (-50\boldsymbol{j} - 100\boldsymbol{i})\ \text{mm/s}^2$$

Similarly,

$$\boldsymbol{a}_C = \boldsymbol{a}_{C/D} = -r_{DC}\,\omega_{DC}{}^2\boldsymbol{j} - r_{DC}\,\dot{\omega}_{DC}\boldsymbol{i}$$

$$= -30(1\tfrac{2}{3})^2\boldsymbol{j} - 30\dot{\omega}_{DC}\boldsymbol{i}$$

$$= -(250/3)\boldsymbol{j} - 30\dot{\omega}_{DC}\boldsymbol{i} \qquad \text{(iv)}$$

The relative acceleration is given by

$$\boldsymbol{a}_{C/B} = \boldsymbol{a}_C - \boldsymbol{a}_B$$

$$= -(100/3)\boldsymbol{j} - (30\dot{\omega}_{DC} - 100)\,\boldsymbol{i} \qquad \text{(v)}$$

Now

$$\boldsymbol{a}_{C/B} = \text{d}(\boldsymbol{\omega}_{BC} \times \boldsymbol{r}_{BC})/\text{d}t$$

$$= \dot{\omega}_{BC} \times r_{BC} + \omega_{BC} \times (\omega_{BC} \times r_{BC})$$

but $\quad \omega_{BC} = 0$

so that $\quad a_{C/B} = \dot{\omega}_{BC} \times r_{BC} \qquad$ (vi)

To eliminate the unknown and unwanted vector $\dot{\omega}_{BC}$, we carry out the dot product with r_{BC} of each side of equation (vi) and obtain

$$r_{BC} \cdot a_{C/B} = 0 \qquad \text{(vii)}$$

since r_{BC} is perpendicular to $\dot{\omega}_{BC} \times r_{BC}$.

Combining equations (v), (vi), and (vii), we obtain

$$\begin{aligned}
&(20i + 20j + 100k) \\
&\cdot[-(30\dot{\omega}_{DC} - 100)i - (100/3)j] = 0 \\
&-600\dot{\omega}_{DC} + 2000 - 2000/3 = 0 \\
&\dot{\omega}_{DC} = 20/9 = 2.222 \text{ rad/s}^2
\end{aligned}$$

and $\qquad \dot{\omega}_{DC} = 2.222k \text{ rad/s}^2$

The acceleration of C is now obtained from equation (iv):

$$\begin{aligned}
a_C &= -[30(20/9)]i - (250/3)j \\
&= -(66.67i + 83.33j) \text{ mm/s}^2
\end{aligned}$$

Example 11.7

The cylinder C shown in Fig. 11.39 is mounted on the cranked arm OAB which is rotating about the fixed Y-axis with angular velocity qJ and angular acceleration $\dot{q}J$. Relative to the arm the cylinder has angular velocity pk and angular acceleration $\dot{p}k$. The cylinder is uniform, its mass is m, its centre of mass is at G, its length is b, and its radius is r. Determine the moments, about the axes Gx, Gy, and Gz, of the forces being applied to the cylinder by assuming that (a) the x-, y-, z-axes are fixed to the arm and (b) the x-, y-, z-axes are fixed to the cylinder.

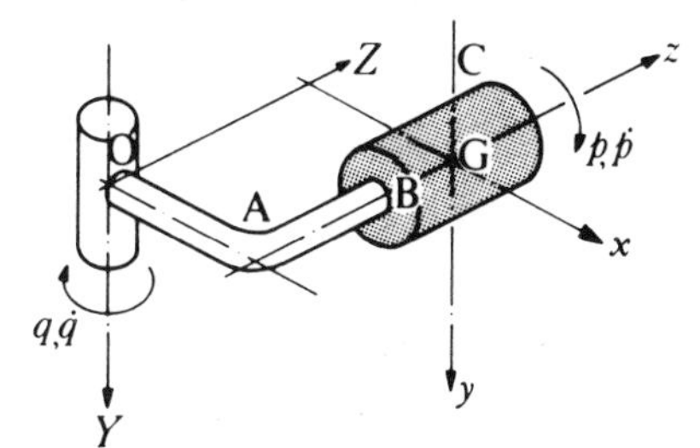

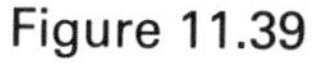
Figure 11.39

Solution

a) The angular velocity of the axes is $\omega = qJ = qj$. These are principal axes and so, from equation 11.72, the moment of momentum about G is

$$\begin{aligned}
L_G &= I_{xx}\Omega_x i + I_{yy}\Omega_y j + I_{zz}\Omega_z k \\
&= 0 + I_{yy}qj + I_{zz}pk \qquad \text{(i)}
\end{aligned}$$

The moment of the forces about G is given by equation 11.68:

$$M_G = \dot{L}_G = \partial(L_G)_{xyz}/\partial t + \omega \times L_G \qquad \text{(ii)}$$

Now $\quad \partial(L_G)_{xyz}/\partial t = I_{yy}\dot{q}j + I_{zz}\dot{p}k \qquad$ (iii)

and
$$\begin{aligned}
\omega \times L_g &= qj \times (I_{yy}qj + I_{zz}pk) \\
&= I_{zz}pqi \qquad \text{(iv)}
\end{aligned}$$

Substituting equations (iii) and (iv) into equation (ii) we find

$$M_G = I_{zz}pqi + I_{yy}\dot{q}j + I_{zz}\dot{p}k \qquad \text{(v)}$$

Evaluation of the moments of inertia gives the results $I_{yy} = m(3r^2 + b^2)/12$ and $I_{zz} = mr^2/2$.

b) The angular velocity of the axes is now $\omega = qj + pk$. The moment of momentum L_G is the same as that given in equation (i) but this is *not* a general expression in terms of the axes fixed to the cylinder; it is the particular value when the axes are aligned as shown in Fig. 11.39. We can obtain a general expression for L_G by allowing the axes to rotate through an angle ψ as shown in Fig. 11.40, and putting $\psi = 0$ *after* carrying out the partial derivative.

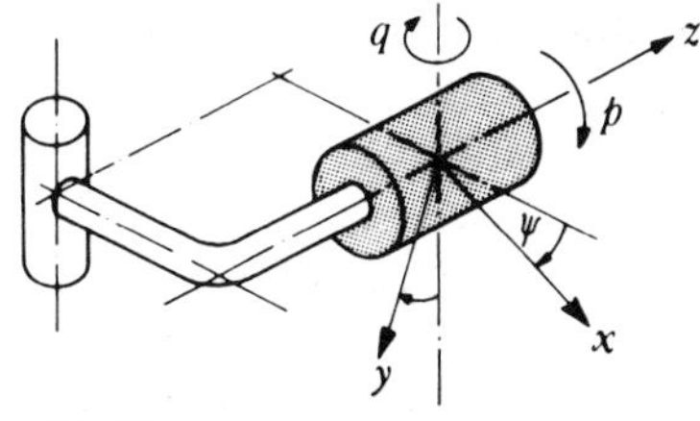

Figure 11.40

The angular velocity of the axes is

$$\begin{aligned}
\omega = \Omega &= \dot{\psi}k + q(\sin\psi i + \cos\psi j) \\
L_G &= I_{xx}\Omega_x i + I_{yy}\Omega_y j + I_{zz}\Omega_z k \\
&= I_{xx}q\sin\psi i + I_{yy}q\cos\psi j + I_{zz}\dot{\psi}k
\end{aligned}$$

$$\begin{aligned}
\partial(L_G)_{xyz}/\partial t = {} & I_{xx}(q\cos\psi\dot{\psi} + \dot{q}\sin\psi)i \\
& + I_{yy}(-q\sin\psi\dot{\psi} + \dot{q}\cos\psi)j + I_{zz}\ddot{\psi}k
\end{aligned}$$

Putting $\psi = 0$, $\dot{\psi} = p$ and $\ddot{\psi} = \dot{p}$, we have

$$\partial(L_G)_{xyz}/\partial t = I_{xx}qpi + I_{yy}\dot{q}j + I_{zz}\dot{p}k \qquad \text{(vi)}$$

Now
$$\begin{aligned}
\omega \times L_G &= (qj + pk) \times (I_{yy}qj + I_{zz}pk) \\
&= (I_{zz} - I_{yy})pqi \qquad \text{(vii)}
\end{aligned}$$

Substituting equations (vi) and (vii) into equation (ii) we find the same result as equation (v), since $I_{xx} = I_{yy}$.

Example 11.8

Figure 11.41 shows an electric motor fixed to a plate P which is mounted in bearings A and B. The motor drives a thin uniform disc of diameter 200 mm and of mass 3.0 kg.

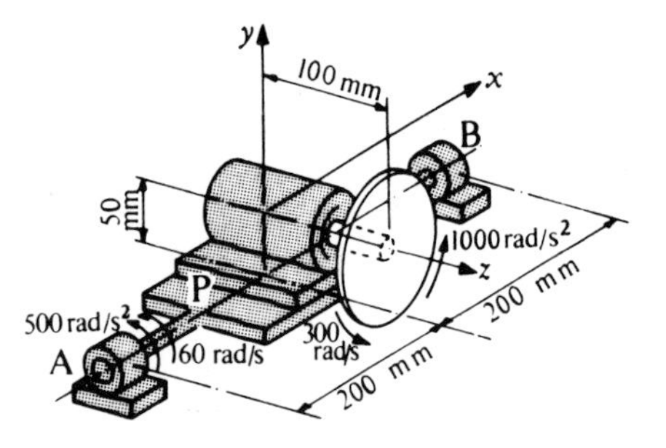

Figure 11.41

When the system is in the position shown, the angular velocity and acceleration of the disc, *measured relative to the plate P*, are $300\boldsymbol{k}$ rad/s and $1000\boldsymbol{k}$ rad/s^2 respectively. At the same time the angular velocity and acceleration of the plate are $-60\boldsymbol{i}$ rad/s and $-500\boldsymbol{i}$ rad/s^2 respectively. Find the anticlockwise couple which is being applied to the shaft AB and the components, in the y- and z-directions, of the forces acting on the bearings at A and B due to the inertial effects of the disc.

Solution Section 11.13 referred to the rotation about a fixed point O of a body with an axis of symmetry. Expressions were obtained for $\boldsymbol{L}_O$, the moment of momentum about O, and $\boldsymbol{M}_O$, the moment of the forces about O. In the present example the disc does not rotate about a fixed point. We have emphasised the interchangeability of expressions for $\boldsymbol{L}_O$ and $\boldsymbol{M}_O$ for a fixed point O with expressions for $\boldsymbol{L}_G$ and $\boldsymbol{M}_G$, and so expressions similar to those in section 11.13 can be used with O replaced by G, the centre of mass.

In Fig. 11.42 the x-, y-, z-axes are attached to the motor frame and the disc rotates relative to these axes with an angular velocity $\dot{\psi}\boldsymbol{k}$. The angular velocity of the axes is $\boldsymbol{\omega} = \dot{\theta}\boldsymbol{i}$. We note that, although the disc rotates relative to the axes, the moments of inertia relative to the axes do not change, due to symmetry. The angular velocity of the body is

$$\boldsymbol{\Omega} = \dot{\theta}\boldsymbol{i} + \dot{\psi}\boldsymbol{k}$$

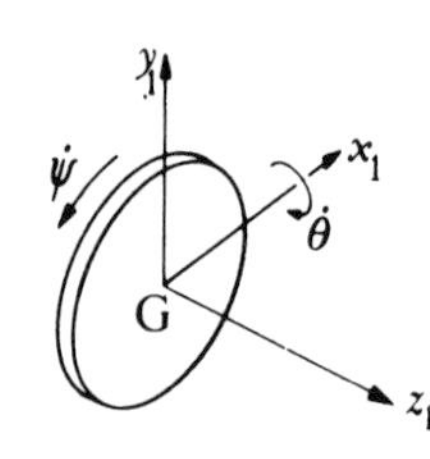

Figure 11.42

The moment of momentum about G is

$$\boldsymbol{L}_G = I_O\dot{\theta}\boldsymbol{i} + I_z\dot{\psi}\boldsymbol{k}$$

The moment of the forces applied to the disc is

$$\begin{aligned}\boldsymbol{M}_G = \dot{\mathbf{H}}_G &= \partial\mathbf{H}_G/\partial t + \boldsymbol{\omega}\times\boldsymbol{L}_G \\ &= I_O\ddot{\theta}\boldsymbol{i} + I_z\ddot{\psi}\boldsymbol{k} + \dot{\theta}\boldsymbol{i}\times(I_O\dot{\theta}\boldsymbol{i} + I_z\dot{\psi}\boldsymbol{k}) \\ &= I_O\ddot{\theta}\boldsymbol{i} - I_z\dot{\theta}\dot{\psi}\boldsymbol{j} + I_z\ddot{\psi}\boldsymbol{k} \qquad \text{(i)}\end{aligned}$$

The moment of inertia I_z about the disc axle is $mr^2/2$ and, since the disc is thin, we can use the perpendicular-axis theorem to show that the moment of inertia I_O about a diameter is $mr^2/4$.

$$\begin{aligned}\boldsymbol{M}_G &= [3\times(0.1)^2/4](-500)\boldsymbol{i} - [(3\times(0.1)^2/2] \\ &\quad \times(-60)(300)\boldsymbol{j} + [3\times(0.1)^2/2](1000)\boldsymbol{k} \\ &= (-3.75\boldsymbol{i} + 270\boldsymbol{j} + 15\boldsymbol{k})\text{ N m} \qquad \text{(ii)}\end{aligned}$$

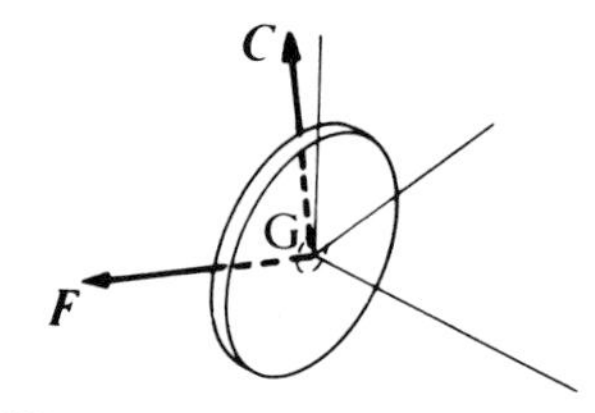

Figure 11.43

A free-body diagram for the disc is shown in Fig. 11.43. The motor shaft applies the couple $\boldsymbol{C}$ to the disc and $\boldsymbol{C} = \boldsymbol{M}_G$. To determine the force $\boldsymbol{F}$ applied to the disc by the shaft, we must first determine the acceleration of G. Referring to Fig. 11.44, we note that G is rotating about the fixed point O and the components of the acceleration of G in cylindrical co-ordinates are as shown.

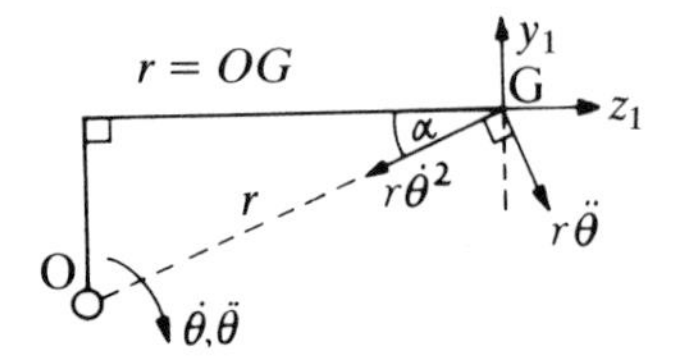

Figure 11.44

$$\begin{aligned}\boldsymbol{a}_G = &-(r\ddot{\theta}\cos\alpha + r\dot{\theta}^2\sin\alpha)\boldsymbol{j} \\ &+ (r\ddot{\theta}\sin\alpha - r\dot{\theta}^2\cos\alpha)\boldsymbol{k}\end{aligned}$$

$$= -[0.1(-500) + 0.05(60)^2]\boldsymbol{j} + [0.05(-500) - 0.1(60)^2]\boldsymbol{k}$$

$$\boldsymbol{a}_G = -(130\boldsymbol{j} + 385\boldsymbol{k}) \text{ m/s}^2$$

The force $\boldsymbol{F}$ is obtained from

$$\boldsymbol{F} = m\boldsymbol{a}_G$$

where $m = 3$ kg; hence

$$\boldsymbol{F} = -(390\boldsymbol{j} + 1155\boldsymbol{k}) \text{ N} \qquad \text{(iii)}$$

The free-body diagram of the system excluding the disc is shown in Fig. 11.45. $\boldsymbol{Q}$ is the required couple, the forces $\boldsymbol{F}_A$ and $\boldsymbol{F}_B$ applied to the shaft AB by the bearings at A and B are shown in their y- and z-components. The force and couple at G are equal and opposite to those on the disc. Note that we shall treat this part of the system as inertialess since we are concerned with the inertial effects of the disc only.

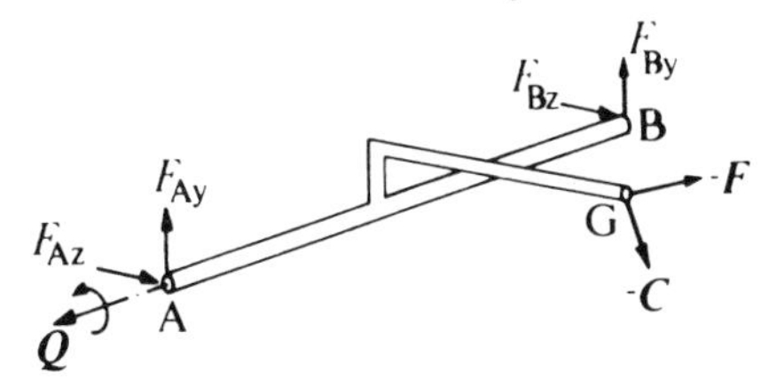

Figure 11.45

Taking moments about A we have

$$\boldsymbol{Q} + \boldsymbol{r}_{AG} \times (-\boldsymbol{F}) + \boldsymbol{r}_{AB} \times (F_{By}\boldsymbol{j} + F_{Bz}\boldsymbol{k}) + (-\boldsymbol{C}) = 0$$

where $\boldsymbol{Q} = -Q\boldsymbol{i}$

$$\boldsymbol{r}_{AG} = \overrightarrow{AG} = (0.2\boldsymbol{i} + 0.05\boldsymbol{j} + 0.1\boldsymbol{k}) \text{ m}$$

$$\boldsymbol{r}_{AB} = \overrightarrow{AB} = (0.4\boldsymbol{i}) \text{ m}$$

hence $(-Q + 18.75 + 3.75)\boldsymbol{i}$

$$-(231 + 0.4F_{Bz} + 270)\boldsymbol{j} + (78 + 0.4F_{By} - 15)\boldsymbol{k} = 0$$

Equating each of the coefficients of $\boldsymbol{i}$, $\boldsymbol{j}$, and $\boldsymbol{k}$ to zero, we find

$$Q = 22.5 \text{ N m} \quad F_{Bz} = -1252.5 \text{ N}$$

and $F_{By} = -157.5$ N

Then, either by taking moments about B or, more simply, by summing the forces to zero, we find

$$F_{Ay} = -232.5 \text{ N} \quad F_{Az} = 97.5 \text{ N}$$

The forces applied to the bearings at A and B are equal and opposite to those on the shaft.

Example 11.9

Figure 11.46 shows a three-dimensional mechanism. Collar C can rotate about the Z-axis and slide along it. Slider S is constrained to a path parallel to the X-axis. Connecting link AB is pinned to the collar at A and attached to the slider at B by a ball-and-socket joint. The path of B intersects the Y-axis and is 150 mm from the X-axis as shown. Collar C is travelling towards O at a constant speed of 600 mm/s, pushed by component E. A bearing between E and C ensures that the torque transmitted about the Z-axis between E and C is negligible. Link AB is 200 mm long and may be considered as a thin uniform rod of mass 0.12 kg. The mass of slider S is 0.5 kg and that of collar C is negligible.

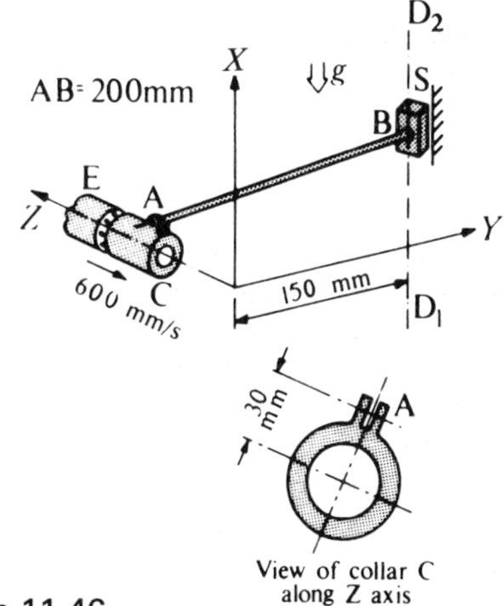

Figure 11.46

When the pin at A is 100 mm from the XY-plane determine the force at B applied to the slider by link AB, neglecting friction.

Solution The geometry of the mechanism is such that the simplest way of obtaining the necessary velocities and accelerations is generally by differentiation of trigonometric and algebraic equations which define the configuration. The techniques of section 11.8 and example 11.6 can be used, but these are left as problems for the end of the chapter.

In Fig. 11.47 the shaded plane ADBE is attached to the collar C and the link AB rotates in this plane. From the figure,

$$0.2^2 = q^2 + z^2 \qquad \text{(i)}$$

$$p = q + 0.03 \qquad \text{(ii)}$$

$$p^2 = 0.15^2 + x^2 \qquad \text{(iii)}$$

$$\tan\alpha = x/0.15 \qquad \text{(iv)}$$

$$\sin\beta = z/0.20 \qquad \text{(v)}$$

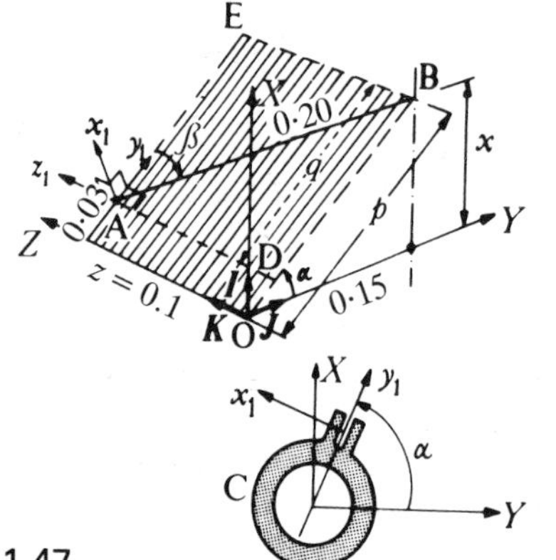

Figure 11.47

The angle α defines the rotation of collar C and angle β defines the rotation of AB in the plane ADBE. From equations (i) to (v), with $z = 0.1$ m, $\dot{z} = -0.6$ m/s, and $\ddot{z} = 0$, the following results are obtained:

$$\alpha = 42.42° \qquad \dot{\alpha} = 1.865 \text{ rad/s}$$
$$\beta = 30° \qquad \dot{\beta} = -3.464 \text{ rad/s}$$
$$x = 0.1371 \text{ m} \qquad \dot{x} = 0.5135 \text{ m/s}$$
$$\ddot{\alpha} = -46.05 \text{ rad/s}^2$$
$$\ddot{\beta} = 6.928 \text{ rad/s}^2$$
$$\ddot{x} = -5.156 \text{ m/s}^2 \qquad \text{(vi)}$$

If we attach axes x_1, y_1, z_1 with unit vectors $\boldsymbol{i}_1$, $\boldsymbol{j}_1$, and $\boldsymbol{k}_1$ to the plane ADBE as shown, then the angular velocity of these axes is

$$\boldsymbol{\omega}_{x_1y_1z_1} = -\dot{\alpha}\boldsymbol{K} = \boldsymbol{\omega}_{\text{C}} \qquad \text{(vii)}$$

The total angular velocity of link AB is its angular velocity relative to the axes plus the angular velocity of the axes:

$$\boldsymbol{\omega}_{\text{AB}} = -\dot{\alpha}\boldsymbol{K} + (-\dot{\beta})\boldsymbol{i}_1$$
$$= -\dot{\alpha}\boldsymbol{K} - \dot{\beta}(\cos\alpha\boldsymbol{I} - \sin\alpha\boldsymbol{J}) \qquad \text{(viii)}$$

Equation (viii) is a completely general expression for $\boldsymbol{\omega}_{\text{AB}}$, and $\boldsymbol{I}$, $\boldsymbol{J}$, and $\boldsymbol{K}$ are fixed vectors. The total or absolute angular acceleration of AB is therefore

$$\dot{\boldsymbol{\omega}}_{\text{AB}} = -\ddot{\alpha}\boldsymbol{K} - \ddot{\beta}(\cos\alpha\boldsymbol{I} - \sin\alpha\boldsymbol{J})$$
$$-\dot{\beta}(-\sin\alpha\dot{\alpha}\boldsymbol{I} - \cos\alpha\dot{\alpha}\boldsymbol{J}) \qquad \text{(ix)}$$

Alternatively, we can use equation 11.7, using the x_1-, y_1-, z_1-axes as the rotating set of axes:

$$\dot{\boldsymbol{\omega}}_{\text{AB}} = \partial[\boldsymbol{\omega}_{\text{AB}}]/\partial t + \boldsymbol{\omega}_{\text{C}} \times \boldsymbol{\omega}_{\text{AB}}$$
$$= -\ddot{\alpha}\boldsymbol{K} - \ddot{\beta}\boldsymbol{i}_1 + [-\dot{\alpha}\boldsymbol{K} \times (-\dot{\alpha}\boldsymbol{K} - \dot{\beta}\boldsymbol{i}_1)]$$
$$= -\ddot{\alpha}\boldsymbol{K} - \ddot{\beta}\boldsymbol{i}_1 + \dot{\alpha}\dot{\beta}\boldsymbol{j}_1$$

which is the same as equation (ix), since $\boldsymbol{i}_1 = \cos\alpha\boldsymbol{I} - \sin\alpha\boldsymbol{J}$ and $\boldsymbol{j}_1 = \sin\alpha\boldsymbol{I} + \cos\alpha\boldsymbol{J}$. Substituting numerical values from equations (vi), we find

$$\boldsymbol{\omega}_{\text{AB}} = 2.557\boldsymbol{I} - 2.337\boldsymbol{J} - 1.865\boldsymbol{K} \qquad \text{(x)}$$

and $$\dot{\boldsymbol{\omega}}_{\text{AB}} = -9.473\boldsymbol{I} - 0.096\boldsymbol{J} + 46.05\boldsymbol{K} \qquad \text{(xi)}$$

The acceleration of B is

$$\boldsymbol{a}_{\text{B}} = \ddot{x}\boldsymbol{I} = -5.156\boldsymbol{I} \text{ m/s}^2 \qquad \text{(xii)}$$

Using cylindrical co-ordinates (equation 11.14), the acceleration of A (see Fig. 11.48) is

$$\boldsymbol{a}_{\text{A}} = r\ddot{\alpha}\boldsymbol{i}_1 - r\dot{\alpha}^2\boldsymbol{j}_1$$
$$= r\ddot{\alpha}(\cos\alpha\boldsymbol{I} - \sin\alpha\boldsymbol{J})$$
$$- r\dot{\alpha}^2(\sin\alpha\boldsymbol{I} + \cos\alpha\boldsymbol{J})$$
$$= (-1.090\boldsymbol{I} + 0.8548\boldsymbol{J}) \text{ m/s}^2 \qquad \text{(xiii)}$$

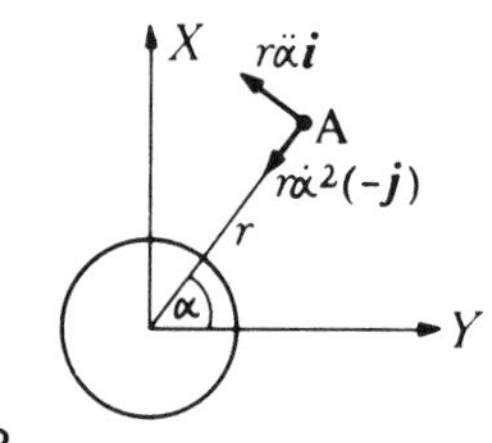

Figure 11.48

The acceleration of G, the centre of mass of the link AB, is given by

$$\boldsymbol{a}_{\text{G}} = \boldsymbol{a}_{\text{A}} + \boldsymbol{a}_{\text{G/A}} \qquad \text{(xiv)}$$

where $\boldsymbol{a}_{\text{G/A}} = 0.5\boldsymbol{a}_{\text{B/A}} = 0.5(\boldsymbol{a}_{\text{B}} - \boldsymbol{a}_{\text{A}})$

since G is at the mid-point of AB. Hence

$$\boldsymbol{a}_{\text{G}} = (-3.123\boldsymbol{I} + 0.4274\boldsymbol{J}) \text{ m/s}^2 \qquad \text{(xv)}$$

We have now determined all the necessary kinematics. To determine the force at B it is necessary to write equations of motion for the link AB, the slider S, and the collar C, making use of free-body diagrams.

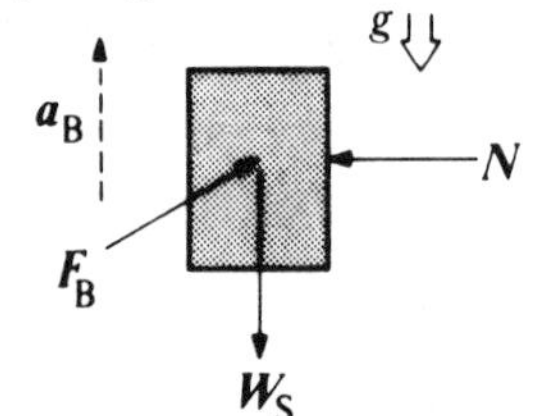

Figure 11.49

Figure 11.49 shows the forces acting on slider S.

The force $\boldsymbol{N}$ is the action of the guide on the slider and is horizontal because of the lack of friction. The equation of motion is

$$\boldsymbol{W}_{\mathrm{S}}+\boldsymbol{N}+\boldsymbol{F}_{\mathrm{B}}=m_{\mathrm{S}}\boldsymbol{a}_{\mathrm{B}} \quad \text{(xvi)}$$

where $\boldsymbol{W}_{\mathrm{S}}=m_{\mathrm{S}}g(-\boldsymbol{I})$

$$=-(0.5)(9.81)\boldsymbol{I}=-4.905\boldsymbol{I}\ \mathrm{N}$$

Let $\boldsymbol{F}_{\mathrm{B}}=F_{\mathrm{B}X}\boldsymbol{I}+F_{\mathrm{B}Y}\boldsymbol{J}+F_{\mathrm{B}Z}\boldsymbol{K}$ (xvii)

Taking the X-components of equation (xvi) to eliminate $\boldsymbol{N}$,

$$-4.905+F_{\mathrm{B}X}=(0.5)(-5.156)$$

$$F_{\mathrm{B}X}=2.327\ \mathrm{N} \quad \text{(xviii)}$$

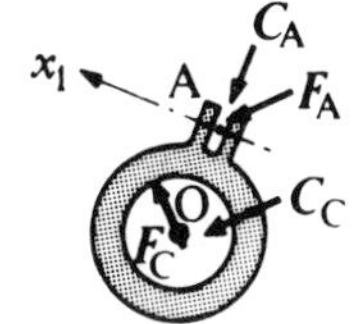

Figure 11.50

Figure 11.50 shows the free-body diagram for the collar C. $\boldsymbol{F}_{\mathrm{A}}$ and $\boldsymbol{F}_{\mathrm{C}}$ are, respectively, the forces applied by the link AB and by the shaft along which the collar slides. There will also be a contact force (not shown) in the Z-direction applied by component E. The couple $\boldsymbol{C}_{\mathrm{A}}$ applied by link AB has no component in the x_1-direction so we can write $\boldsymbol{C}_{\mathrm{A}}$ as

$$\boldsymbol{C}_{\mathrm{A}}=C_{\mathrm{A}y}\boldsymbol{j}_1+C_{\mathrm{A}Z}\boldsymbol{K}$$

$$=C_{\mathrm{A}y}(0.6746\boldsymbol{I}+0.7382\boldsymbol{J})+C_{\mathrm{A}Z}\boldsymbol{K}$$

The couple $\boldsymbol{C}_{\mathrm{C}}$ has no component in the Z-direction. Taking moments about O we have

$$\boldsymbol{C}_{\mathrm{C}}+\boldsymbol{C}_{\mathrm{A}}+\boldsymbol{r}_{\mathrm{OA}}\times\boldsymbol{F}_{\mathrm{A}}=0 \quad \text{(xix)}$$

since the collar is massless. Taking the dot product with $\boldsymbol{K}$ of equation (xix) to eliminate $\boldsymbol{C}_{\mathrm{C}}$,

$$C_{\mathrm{A}Z}+\boldsymbol{K}\cdot(\boldsymbol{r}_{\mathrm{OA}}\times\boldsymbol{F}_{\mathrm{A}})=0 \quad \text{(xx)}$$

Now $\boldsymbol{r}_{\mathrm{OA}}=r_{\mathrm{OA}}(\sin\alpha\boldsymbol{I}+\cos\alpha\boldsymbol{J})$

where $r_{\mathrm{OA}}=0.03\ \mathrm{m}$

hence $\boldsymbol{r}_{\mathrm{OA}}=(0.02024\boldsymbol{I}+0.02215\boldsymbol{J})\ \mathrm{m}$

Writing $\boldsymbol{F}_{\mathrm{A}}=F_{\mathrm{A}X}\boldsymbol{I}+F_{\mathrm{A}Y}\boldsymbol{J}+F_{\mathrm{A}Z}\boldsymbol{K}$

and substituting in equation (xx) we find

$$C_{\mathrm{A}Z}-0.02215F_{\mathrm{A}X}+0.02024F_{\mathrm{A}Y}=0 \quad \text{(xxi)}$$

To determine the kinetic relationships for the link, let us attach a set of x-, y-, z-axes to the link as shown in Fig. 11.51. The x-axis is parallel to the x_1-axis, the y-axis lies along the link, and the z-axis lies in the ADBE-plane. The moment of momentum about G is (equation 11.61)

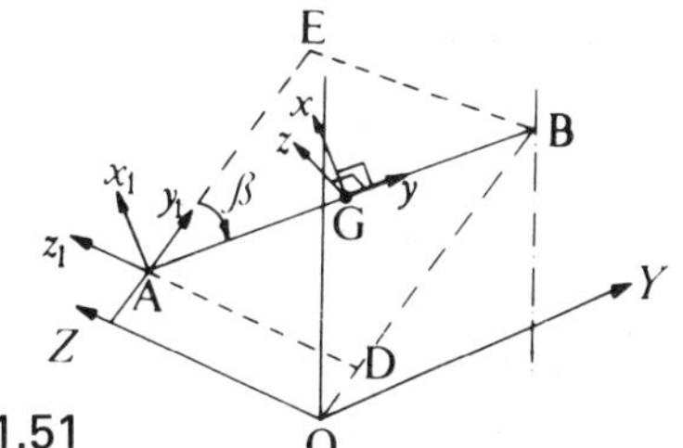

Figure 11.51

$$\boldsymbol{L}_{\mathrm{G}}=I_{xx}\omega_x\boldsymbol{i}+I_{zz}\omega_z\boldsymbol{k} \quad \text{(xxii)}$$

Note that $I_{yy}=0$ since the rod is thin, and that x, y, z are principal axes.

Denoting the unit vectors for the x-, y-, z-axes by $\boldsymbol{i}$, $\boldsymbol{j}$, and $\boldsymbol{k}$, we find

$$\left.\begin{aligned}\boldsymbol{i}&=\boldsymbol{i}_1\\ \boldsymbol{j}&=\cos\beta\boldsymbol{j}_1-\sin\beta\boldsymbol{k}_1\\ \boldsymbol{k}&=\sin\beta\boldsymbol{j}_1+\cos\beta\boldsymbol{k}_1\end{aligned}\right\} \quad \text{(xxiii)}$$

hence

$$\left.\begin{aligned}\boldsymbol{i}&=0.7382\boldsymbol{I}-0.6746\boldsymbol{J}\\ \boldsymbol{j}&=0.5842\boldsymbol{I}+0.6393\boldsymbol{J}-0.5\boldsymbol{K}\\ \boldsymbol{k}&=0.3373\boldsymbol{I}+0.3691\boldsymbol{J}+0.8660\boldsymbol{K}\end{aligned}\right\} \quad \text{(xxiv)}$$

Now $I_{xx}=ml^2/12=(0.12)(0.2)^2/12$

$$=4\times10^{-4}\ \mathrm{kg\ m^2}$$

and, from equations (x) and (xxiv),

$$\omega_x=\boldsymbol{\omega}_{\mathrm{AB}}\cdot\boldsymbol{i}=3.464\ \mathrm{rad/s}$$

$$\omega_z=\boldsymbol{\omega}_{\mathrm{AB}}\cdot\boldsymbol{k}=-1.615\ \mathrm{rad/s}$$

Note that, as a check, $\omega_x=\omega_{x1}=-\dot{\beta}$.

Substituting into equation (xxii), we obtain

$$\boldsymbol{L}_{\mathrm{G}}=10^{-4}(6.453\boldsymbol{I}-13.479\boldsymbol{J}+3.221\boldsymbol{K})\ \mathrm{kg\ m^2/s} \quad \text{(xxv)}$$

The rate change of the moment of momentum is, from equation 11.68,

$$\dot{\boldsymbol{L}}_{\mathrm{G}}=\partial(\boldsymbol{L}_{\mathrm{G}})_{xyz}/\partial t+\boldsymbol{\omega}_{xyz}\times\boldsymbol{L}_{\mathrm{G}}$$

$$=I_{xx}\dot{\omega}_x\boldsymbol{i}+I_{zz}\dot{\omega}_z\boldsymbol{j}+\boldsymbol{\omega}_{\mathrm{AB}}\times\boldsymbol{L}_{\mathrm{G}} \quad \text{(xxvi)}$$

Note that the derivative of the angular velocity of the body with respect to the moving x-, y-, z-axes is the same as the absolute angular acceleration since $\boldsymbol{\omega}_{xyz}$ and $\boldsymbol{\omega}_{\mathrm{AB}}$ are one and the same. From equations (xi) and (xxiv),

$$\dot{\omega}_x=\dot{\boldsymbol{\omega}}_{\mathrm{AB}}\cdot\boldsymbol{i}=-6.928\ \mathrm{rad/s^2}$$

$$\dot{\omega}_z=\dot{\boldsymbol{\omega}}_{\mathrm{AB}}\cdot\boldsymbol{k}=36.645\ \mathrm{rad/s^2}$$

Substituting the numerical values into equation (xxvi) gives

$$\dot{\boldsymbol{L}}_{\mathrm{G}}=(-3.68\times10^{-4}\boldsymbol{I}+5.252\times10^{-3}\boldsymbol{J}+1.0756\times10^{-2}\boldsymbol{K})\ \mathrm{N\ m} \quad \text{(xxvii)}$$

Now, $\dot{\boldsymbol{L}}_G = \boldsymbol{M}_G$, the sum of the moments of forces about G. If we take moments of forces about G, our equation will involve seven unknowns. We can eliminate the three components of $\boldsymbol{F}_A$ by taking moments about A, using equation 11.64:

$$\boldsymbol{M}_A = \dot{\boldsymbol{L}}_G + \boldsymbol{r}_{AG} \times m\boldsymbol{a}_G \qquad \text{(xxviii)}$$

Now $\boldsymbol{r}_{AB} = 0.20\boldsymbol{j} = 0.1168\boldsymbol{I} + 0.1279\boldsymbol{J} - 0.1\boldsymbol{K}$

and $\boldsymbol{r}_{AG} = 0.5\boldsymbol{r}_{AB} = 0.0584\boldsymbol{I} + 0.0639 - 0.05\boldsymbol{K}$

and, using equation (xv),

$$\boldsymbol{r}_{AG} \times m\boldsymbol{a}_G = 0.12(0.02137\boldsymbol{I} + 0.1562\boldsymbol{J} + 0.2245\boldsymbol{K})$$

hence $$\boldsymbol{M}_A = (2.196 \times 10^{-3}\boldsymbol{I} + 2.399 \times 10^{-2}\boldsymbol{J} + 3.770 \times 10^{-2}\boldsymbol{K})\ \text{N m} \qquad \text{(xxix)}$$

Figure 11.52 shows the free-body diagram for link AB, where

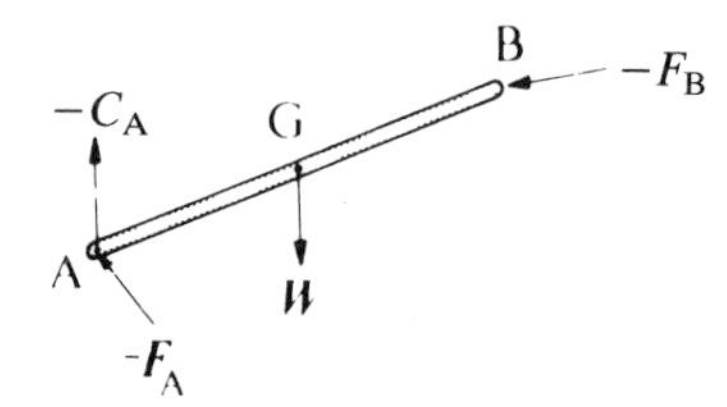

Figure 11.52

$$\boldsymbol{W} = mg(-\boldsymbol{I}) = (0.12)(9.81)(-\boldsymbol{I}) = -1.177\boldsymbol{I}\ \text{N}$$

Summing the forces, we have

$$-\boldsymbol{F}_A + \boldsymbol{W} - \boldsymbol{F}_B = m\boldsymbol{a}_G$$

$$-(F_{AX}\boldsymbol{I} + F_{AY}\boldsymbol{J} + F_{AZ}\boldsymbol{K}) + 1.177\boldsymbol{I} - (2.327\boldsymbol{I} + F_{BY}\boldsymbol{J} + F_{BZ}\boldsymbol{K}) = 0.12(-3.1231\boldsymbol{I} + 0.4274\boldsymbol{J})$$

hence $$-F_{AX} = 3.129 \qquad \text{(xxx)}$$

$$-F_{AY} - F_{BY} = 0.05129 \qquad \text{(xxxi)}$$

$$-F_{AZ} - F_{BZ} = 0 \qquad \text{(xxxii)}$$

Taking moments about A,

$$\boldsymbol{M}_A = (-\boldsymbol{C}_A) + \boldsymbol{r}_{AG} \times \boldsymbol{W} + \boldsymbol{r}_{AB} \times (-\boldsymbol{F}_B)$$
$$= (-0.6746C_{Ay} - 7.522 \times 10^{-2} + 0.1279F_{BZ} + 0.1F_{BY})\boldsymbol{I}$$
$$+ (-0.7382C_{Ay} + 6.877 \times 10^{-2} - 0.1168F_{BZ} - 0.2327F_{BY})\boldsymbol{J}$$
$$+ (-C_{AZ} + 0.1168F_{BY} - 0.2976)\boldsymbol{K} \qquad \text{(xxxiii)}$$

We can obtain three further scalar equations by comparing the coefficients of $\boldsymbol{I}$, $\boldsymbol{J}$, and $\boldsymbol{K}$ in equations (xxxiii) and (xxix). Combining these with equations (xxi), (xxx), (xxxi), and (xxxii) we can determine, after some labour, all the unknown scalars. For the force at B the results are

$$F_{BY} = 2.765\ \text{N} \quad \text{and} \quad F_{BZ} = -3.181\ \text{N}$$

Problems

11.1 A packing case is to be moved from the ground to a platform as shown in Fig. 11.53, with points P, Q, and R moving to points P′, Q′ and R′ respectively. The movement is to be in two stages: (i) a translation of point Q to Q′ without rotation of the case and (ii) a rotation about a fixed hinge line.

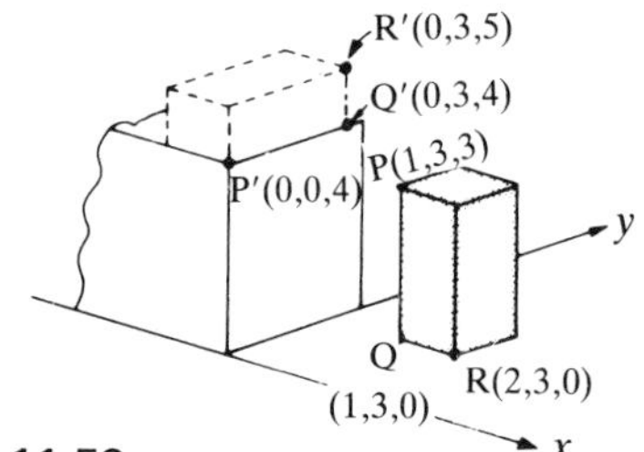

Figure 11.53

a) Find the direction of the hinge line and the required angle of rotation.
b) Could the movement have been accomplished by a single rotation about a fixed axis?

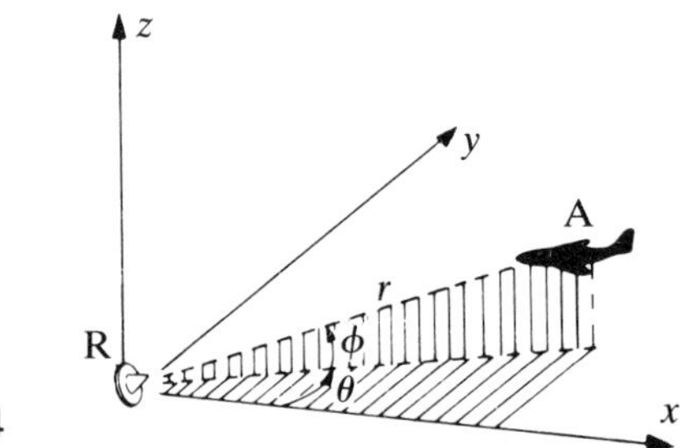

Figure 11.54

11.2 See Fig. 11.54. A radar station R is tracking an aircraft A which is flying horizontally at a constant velocity at an altitude of 2.0 km. The Cartesian co-ordinates of the aircraft relative to the radar station are (4.3, 2.7, 2.0) km. The velocity of the aircraft in the same co-ordinates is $(-700\boldsymbol{i} + 400\boldsymbol{j})$ km/h. Determine the numerical values of r, ϕ, θ, $\dot{r}$, $\dot{\phi}$, $\dot{\theta}$, $\ddot{r}$, $\ddot{\phi}$, and $\ddot{\theta}$.

11.3 In Fig. 11.55, rod AB, which is 400 mm long, pivots about pin B. The angular velocity of arm CB is $10\boldsymbol{k}$ rad/s and its angular acceleration is $50\boldsymbol{k}$ rad/s². The angle CBA is denoted by θ, and $\theta = 60°$, $\dot{\theta} = 20$ rad/s, $\ddot{\theta} = -100$ rad/s². Determine the velocity and acceleration of point A.

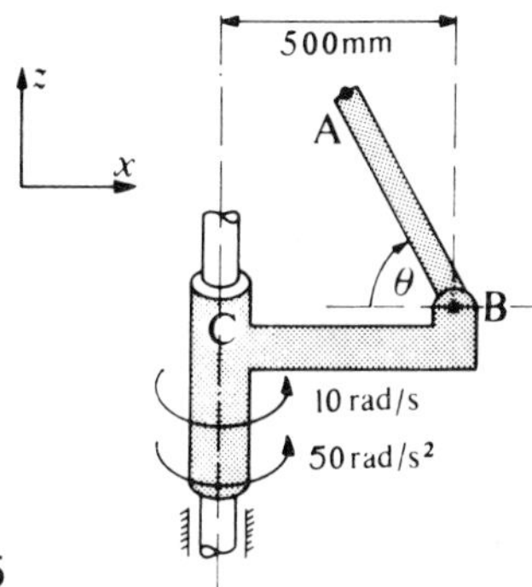

Figure 11.55

11.4 A cranked arm PQRS carries at S a rotating wheel W of radius r as shown in Fig. 11.56. The x-, y-, z-axes are attached to the arm. The angular velocity and angular acceleration of the wheel relative to the arm are $\omega_W \boldsymbol{j}$ and $\dot{\omega}_W \boldsymbol{j}$ respectively. The angular velocity and angular acceleration of the arm are $\omega_A \boldsymbol{k}$ and $\dot{\omega}_A \boldsymbol{k}$ respectively. B is a point attached to the wheel and B′ is a point fixed in the xyz-frame coincident with B.

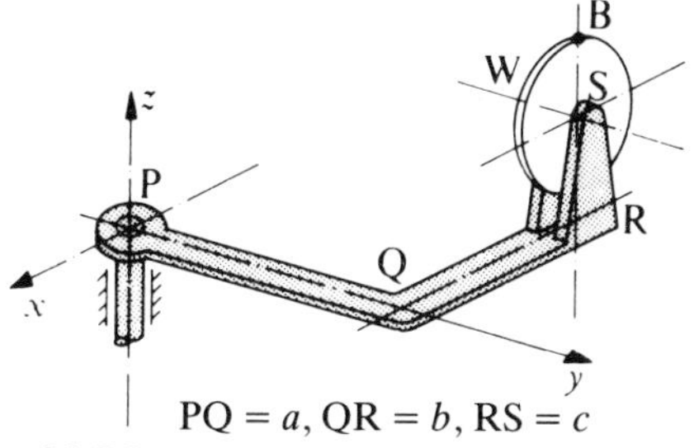

Figure 11.56

a) Find $\boldsymbol{v}_{B'}$ and $\boldsymbol{v}_{B/B'}$, and hence find $\boldsymbol{v}_B$.
b) Find $\boldsymbol{a}_{B'}$ and $\boldsymbol{a}_{B/B'}$, and hence find $\boldsymbol{a}_B$.
c) Find the angular velocity and angular acceleration of the wheel.

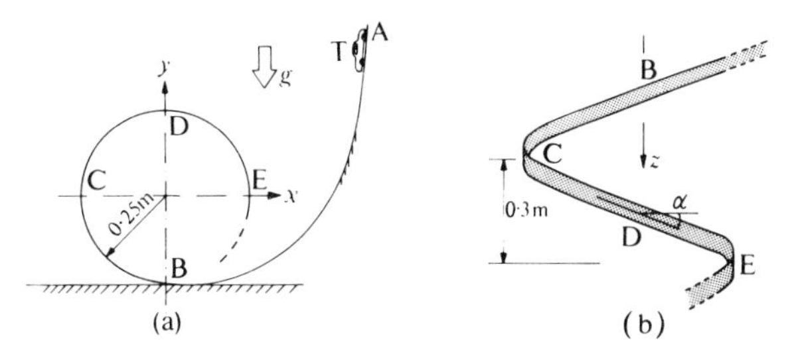

Figure 11.57

11.5 Figure 11.57(a) shows the elevation of the track ABCDE for a small toy motor car T; the plan view is shown at (b). BCDE is a helix of radius 0.25 m and the helix angle α is such that the pitch is 0.6 m. The track rests on the ground and is released from rest. Determine the velocity and acceleration of the toy as it passes point C. (If necessary it may be assumed that the radius of curvature of a helix of radius r is $r \sec^2 \alpha$.)

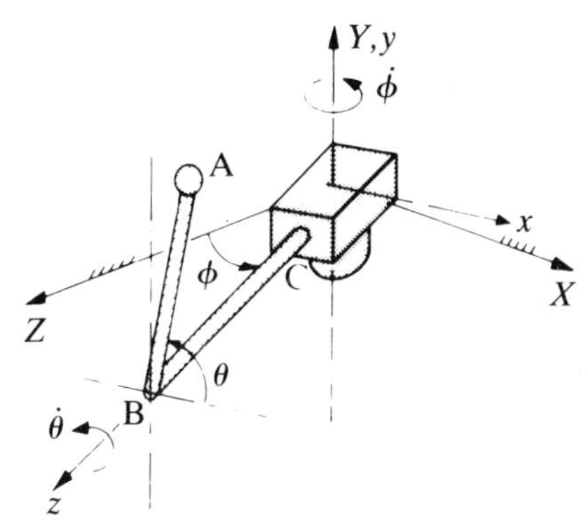

Figure 11.58

11.6 See Fig. 11.58. ABC is a cranked arm, with angle ABC a right angle. BC is in the XZ-plane, AB = r, and BC = a. The arm rotates at a constant speed $\dot{\theta}$ relative to block C, which is itself rotating at constant speed $\dot{\phi}$ about the fixed Y-axis. Neglecting all inertias apart from a mass m at A, find the force exerted by the cranked arm on the block. Express the force in terms of unit vectors (a) fixed to the block and (b) fixed in space.

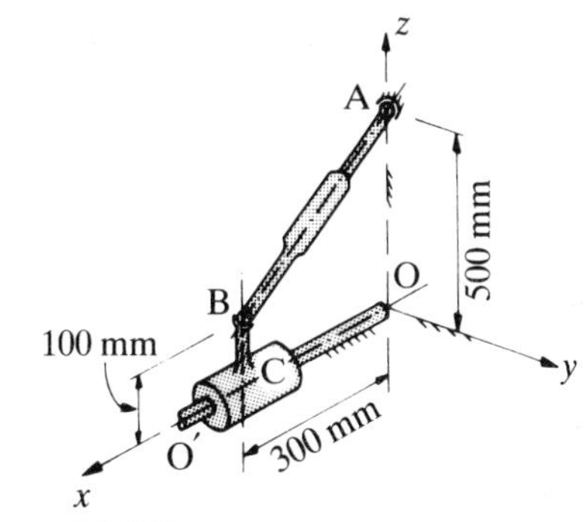

Figure 11.59

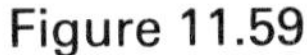

11.7 In Fig. 11.59, collar C can slide along the fixed rod OO′ and also rotate about it. AB is a ball-and-socket-jointed telescopic link. When the mechanism is in the configuration shown, $\boldsymbol{v}_C = 4\boldsymbol{i}$ m/s and $\boldsymbol{\omega}_C = 20\boldsymbol{i}$ rad/s. Find the velocity of B, the rate of extension of link AB, and the angular velocity of the line AB.

11.8 Use the methods of section 11.8 to determine the velocity of B and the angular velocities of the collar C and the link AB in the mechanism of example 11.9.

11.9 See Fig. 11.60. Crank OA rotates in the xy-plane at a constant angular velocity $\omega \boldsymbol{k}$. Slider B moves along path PQ, which is parallel to the z-axis and lies in the yz-plane. The connecting rod AB has ball-and-socket joints at each end. OA = r and AB = l. Show that the velocity and acceleration of B are given by

$$\boldsymbol{v}_B = -[br\omega(\cos\theta)/z]\boldsymbol{k}$$
$$\boldsymbol{a}_B = (br\omega^2/z^3)(z^2 \sin\theta - br\cos^2\theta)\boldsymbol{k}$$

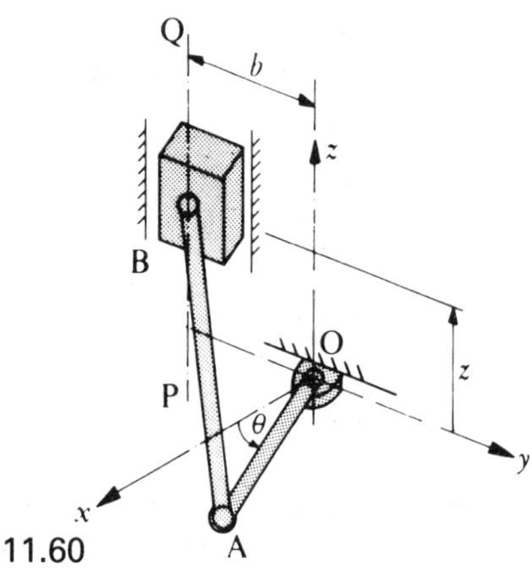

Figure 11.60

where $z = (l^2 - b^2 - r^2 - 2br\sin\theta)^{1/2}$. Also, find expressions for the angular velocity and angular acceleration of AB, neglecting their components along AB.

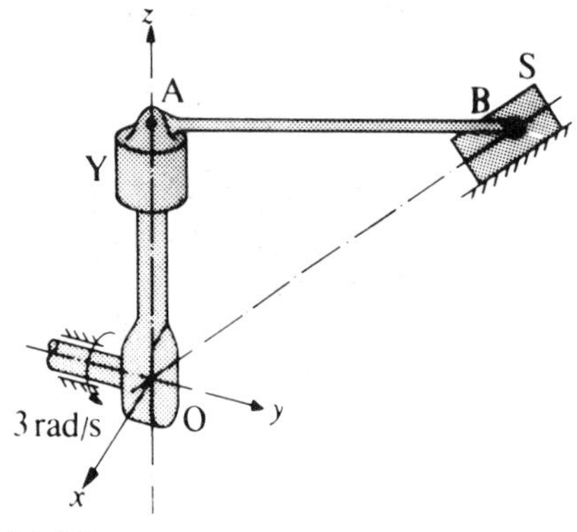

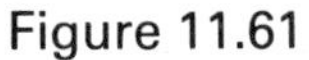

Figure 11.61

11.10 A three-dimensional slider-crank chain is shown in Fig. 11.61. Crank OA rotates about the fixed y-axis with a constant angular velocity of $3\boldsymbol{j}$ rad/s. Connecting link AB is pinned to the yoke Y which is free to rotate about the axis OA. The pin axis is perpendicular to OA and to AB. A ball-and-socket joint at B connects the link to slider S. The motion of the slider is such that the path of B is a straight line passing through O. At the instant under consideration A is located at point (0, 0, 1) m and B is at point (−1, 2, 1.5) m. Find the velocity of B and the angular velocity of AB. Also find the acceleration of B.

11.11 Figure 11.62 shows a uniform rectangular prism of sides a, b, and c and mass M. Determine the following moments and products of inertia: I_{xx}, I_{yy}, I_{zz}, I_{xy}, I_{xz}, I_{yz}.

11.12 A thin uniform rod is bent into the shape shown in Fig. 11.63. BC = CD = DE = a and the mass per unit length of the rod is ρ. Determine the moment of inertia I_{zz} and the products of inertia I_{xz} and I_{yz}.

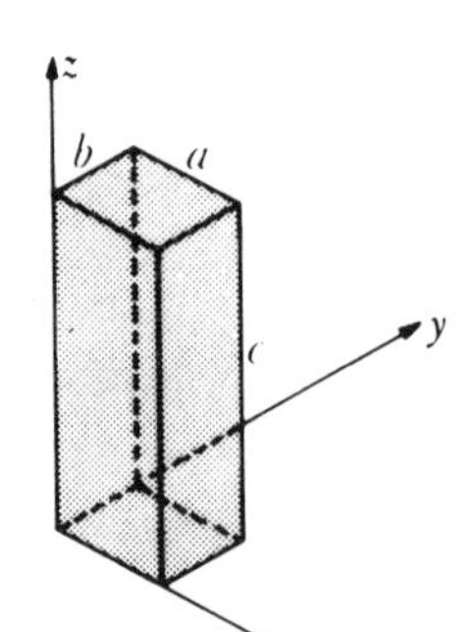

Figure 11.62

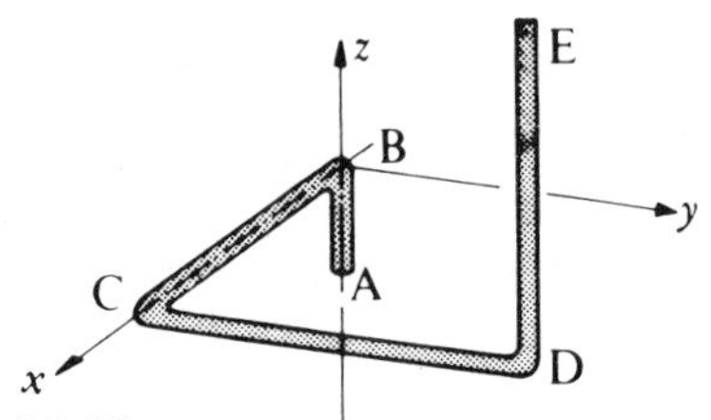

Figure 11.63

11.13 Show that, if ω is not constant, the moment equation for rotation about a fixed axis (see section 11.11) becomes

$$\boldsymbol{M} = (-I_{xz}\dot{\omega}_z + I_{yz}\omega_z^2)\boldsymbol{i} - (I_{yz}\dot{\omega}_z + I_{xz}\omega_z^2)\boldsymbol{j} + I_{zz}\dot{\omega}_z\boldsymbol{k}$$

Also show that this equation applies to the more general case for moments about G where the axis is not necessarily fixed, but the angular velocity vector is constant in direction.

11.14 Refer to problem 11.12. The cranked rod is rotating about the fixed z-axis with an angular velocity $\Omega_z\boldsymbol{k}$ and an angular acceleration $\dot{\Omega}_z\boldsymbol{k}$. Determine the twisting and bending moments in AB at B.

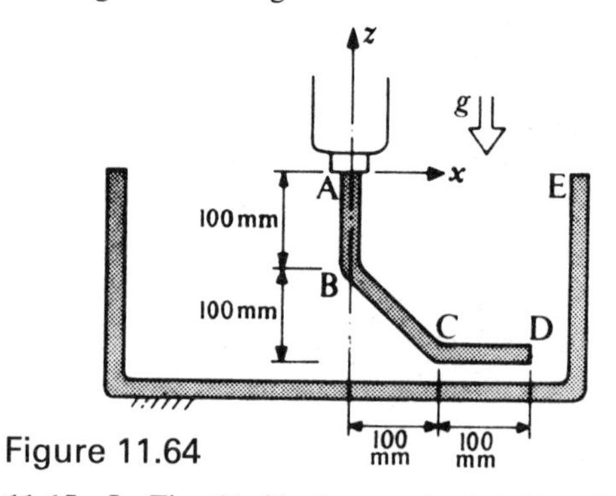

Figure 11.64

11.15 In Fig. 11.64, the cranked rigid rod ABCD is used to stir the contents of container E. The mass per unit length of the rod is 4 kg/m. Determine for the rod the moment of inertia I_{zz} and the product of inertia I_{xz}. If the rod has an angular velocity of $10\boldsymbol{k}$ rad/s and an angular acceleration of $-100\boldsymbol{k}$ rad/s^2, determine the magnitudes of the twisting and bending couples in the rod at A due to the inertia and weight of the rod.

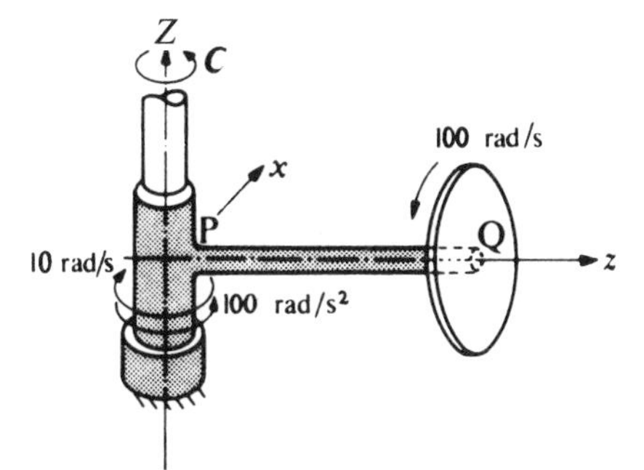

Figure 11.65

11.16 The freely spinning thin disc shown in Fig. 11.65 is rotating at an angular velocity $100\boldsymbol{k}$ rad/s relative to the arm PQ which is 100 mm long. At the same time, the arm has an angular velocity of $-10\boldsymbol{K}$ rad/s and an angular acceleration of $100\boldsymbol{K}$ rad/s^2. The disc has a radius of 80 mm and its mass is 1.0 kg. The moment of inertia about the Z-axis of arm PQ is 2.5×10^{-3} kg m^2. Determine the external couple $\boldsymbol{C}$ which is being applied to the arm and the bending moment at P.

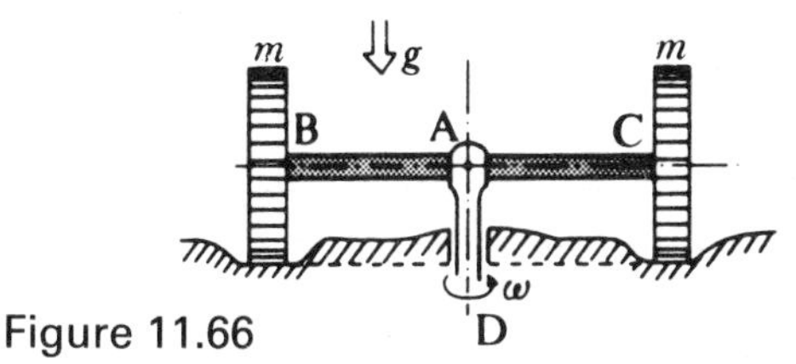

Figure 11.66

11.17 Figure 11.66 shows part of a machine used for compacting sand. Arms AB and AC are of equal length and are each connected to the vertical post AD by a horizontal pin at A. Rollers B and C, each of mass m and radius r, are mounted on bearings at the ends of the arms and roll over a horizontal surface. The constant angular speed of the post AD is ω. Show that, if $\omega = 2\sqrt{(g/r)}$, the contact force between the rollers and the surface will be three times greater than when the post is not rotating. Neglect the horizontal component of the contact force and the mass of the arms.

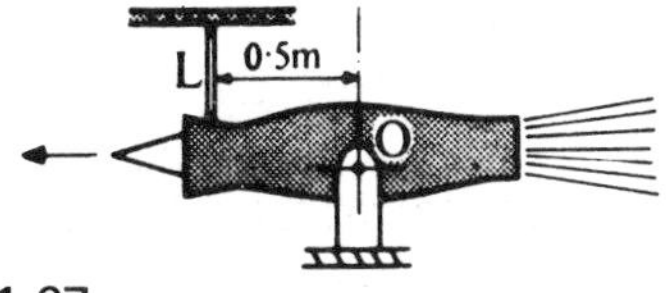

Figure 11.67

11.18 An aircraft has a single gas turbine, the rotor of which rotates clockwise when viewed from the front. The moment of inertia of the rotor is 15 kg m^2. The engine is mounted on trunnions which would allow it to pitch about the axis O, but this motion is prevented by the provision of the link L between the engine and the airframe as shown in Fig. 11.67. Axis O passes through the centre of mass of the engine. If the rotor speed is 10 000 rev/min and the aircraft is turning to port at 3°/s and banked at 30°, find the load in the link.

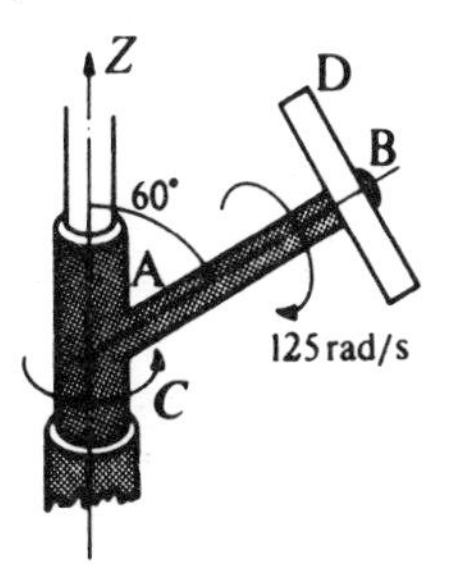

Figure 11.68

11.19 In Fig. 11.68 arm AB, of length 0.2 m, is free to rotate about the Z-axis and its moment of inertia about this axis is 5×10^{-3} kg m^2. A thin disc of radius 0.1 m and mass 1 kg is mounted in a bearing at the end of the arm and is set spinning at 125 rad/s in the direction indicated, the arm AB being stationary. A couple $\boldsymbol{C} = 0.2\boldsymbol{K}$ N m is then applied to the arm. Neglecting friction, determine the time taken until the disc reverses its direction of rotation relative to the arm.

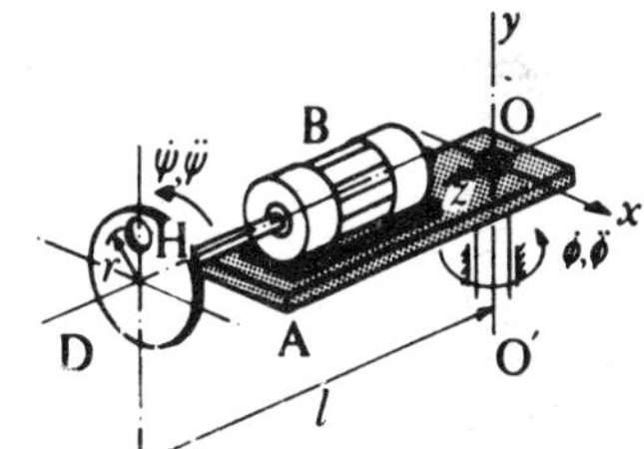

Figure 11.69

11.20 In Fig. 11.69, platform OA rotates about the fixed axis OO′ with angular velocity $\dot{\phi}\boldsymbol{j}$ and angular acceleration $\ddot{\phi}\boldsymbol{j}$. Motor B is fixed to the platform and carries disc D on its output shaft. The radius of the disc is R and its mass per unit area is ρ. A small hole H of diameter d is drilled in the disc at a radius r from the axle. The angular velocity and angular acceleration of the disc, relative to the platform, are $\dot{\psi}\boldsymbol{k}$ and $\ddot{\psi}\boldsymbol{k}$ respectively. When the hole is in the position shown, determine the couple being applied to the disc by the motor shaft.

11.21 An aircraft is flying at a constant speed v_0 in a horizontal circle of radius R. The angular speed of the propeller relative to the aircraft is ω_0 and the centre of the propeller is at C. By treating the propeller as a thin uniform rod of mass m and length l, show that, due to the inertia of the propeller, the magnitude of the bending moment in the propeller shaft at C is equal to $(\cos\theta)ml^2 v_0 \omega_0/(6R)$ and the magnitude of the twisting moment is equal to $ml^2 v_0^{\,2} \sin 2\theta/(24R^2)$, where θ is the angle the propeller makes with the vertical.

11.22 In the three-dimensional mechanism of Fig. 11.70, sliders A, B, and C move in their respective guides and are connected by the light rigid links AB

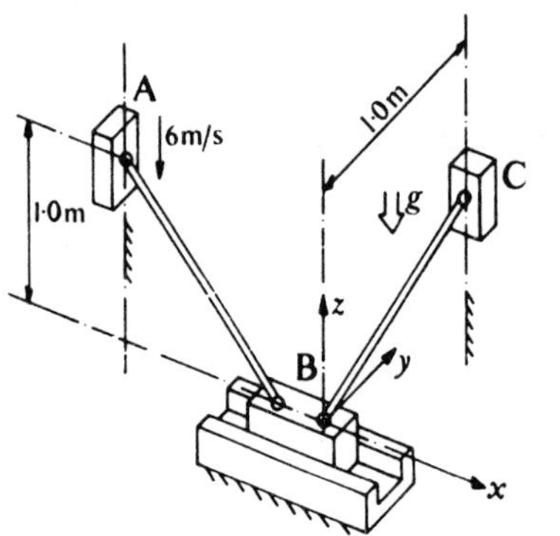

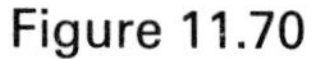
Figure 11.70

and BC. Link AB moves in the xz-plane. Each link has a length of $\sqrt{2}$ m and the connections to the sliders are by ball-and-socket joints. In the configuration shown, $\boldsymbol{v}_A = -6\boldsymbol{k}$ m/s and BC is in the yz-plane.

a) Determine the velocities of B and C and the angular velocity of the line BC.

b) Find the acceleration of C and show that it is independent of the acceleration of B for this configuration.

c) If the mass of slider C is 4 kg, determine the magnitude of the force in link BC, neglecting friction, and state whether this link is in tension or compression.

12
Introduction to continuum mechanics

SECTION A
One-dimensional continuum

12.1 Introduction

In the previous chapters physical objects have been modelled mainly as rigid bodies which may translate and rotate but not deform. The concept of an ideal spring was used to represent a body having no mass but capable of being deformed, the deformation being proportional to the applied load.

Any real body consists of a finite number of molecules each of which can translate and rotate, so that the number of degrees of freedom a body will have, in the mechanical sense, is six times (that is three translational and three rotational) the number of molecules. This is a very large number even for a particle 1 micron in diameter. Because of this the concept of a continuous distribution of matter is postulated.

Matter will be regarded here as being either solid or fluid. Both states require forces to be applied to change their volumes, but a fluid requires no force to change the shape of an element in the static condition.

Fluids are further divided into liquids and gases. A gas expands into any vessel but liquids will form a free surface with another fluid, also gases are much more compressible than liquids.

These descriptions are idealisations but many engineering materials behave in a manner which approximates to these ideal substances.

Internal friction can usually be ignored for most structural metals, but some materials are designed to have high internal damping characteristics for use when vibration is a problem. Fluid friction, viscosity, is an important consideration when dealing with oils, but for air or water the effects of viscosity are confined to a thin layer adjacent to a solid boundary known as the boundary layer. As internal friction complicates the governing equations they will not be included in this chapter. However such an approximation is quite common in a first approach to many engineering problems.

The reader is advised that the sections on fluid dynamics (12.9 to 12.15) are not intended to replace a text in fluid dynamics but are included to draw attention to the similarities and the differences in setting-up the basic equations. They may be omitted if solid mechanics only is required.

12.2 Density

The average mass density of a substance is defined as the quotient of mass to volume; and for a continuum it is assumed that no matter how small a volume is considered, the ratio remains finite so that the point mass density, or simply density, is defined by

$$\rho = \text{limit}\,(\Delta\text{mass}/\Delta\text{volume})$$

as the volume tends to zero.

Conversely we may write that for a region of space having a volume V

$$\text{mass} = \int \rho \,\mathrm{d}V \qquad (12.1)$$

where the integration is over the whole volume.

12.3 One-dimensional continuum

The term one-dimensional here means that only one spatial co-ordinate is required to describe the position of an element.

A simple one-dimensional continuum is a straight uniform solid bar or a fluid contained in a straight pipe of constant bore.

The co-ordinate system used to define the location and movement of an element of a substance is usually different when dealing with fluids from that used for solids. It is very important that these methods are understood, so to this end we shall consider a pipe and fluid as shown in Fig. 12.1.

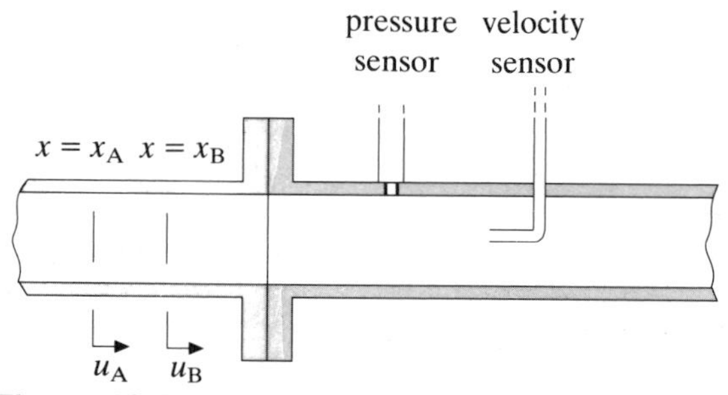

Figure 12.1

Consider first how measurements of deformation are made for the pipe itself. To determine how much the material has been stretched we could measure the relative movement of two marks, one at $x = x_A$ and the other at $x = x_B$. Note that if the pipe moves as a rigid body the marks will move with the pipe so the marks will not be at their original locations. We must regard x_A and x_B as being the 'names' of the marks: that is they define the original positions of the marks. In this context x does not vary, so we must use a different symbol to denote the displacement of the marks from their original positions. The symbols u_A and u_B will be used.

12.4 Elementary strain

The longitudinal, or axial, strain is defined to be the change in length per unit length

thus the strain $$\varepsilon = \frac{u_B - u_A}{x_B - x_A} \tag{12.2}$$

As the distance between the marks approaches to zero

$$\varepsilon = \frac{\partial u}{\partial x} \tag{12.3}$$

The partial differential is required since strain could vary with time as well as with position.

12.5 Particle velocity

The velocity of a particle at a given value of x, say x_A, is simply

$$v = \frac{\partial u}{\partial t} \tag{12.4}$$

Again the partial derivative is used to indicate that x is held constant.

The above co-ordinate system is known as Lagrangian.

If we are concerned with the fluid in the pipe then a pressure measuring device would be fixed to the pipe and, assuming that the pipe is rigid, the device would give a reading for a specific point in space and does not follow a particular particle of fluid. A flow velocity device would also be attached to the pipe.

A system of co-ordinates which relates properties to a specific point in space is known as Eulerian. Thus pressure p is a function of x and time t.

The particle velocity is here defined by

$$v = \frac{\mathrm{d}x}{\mathrm{d}t} \tag{12.5}$$

note here that the partial derivative is not required, however the velocity will be a function of both x and t.

In summary, Lagrangian co-ordinates refer to a particular particle whilst Eulerian co-ordinates refer to a particular point in space.

12.6 Ideal continuum

An ideal solid is defined as one which is homogeneous and isotropic, by which we mean that the properties are uniform throughout the region and so not depend on orientation. In addition we will assume that the material only undergoes small deformation and that this deformation is proportional to the applied loading system. This last statement is known as Hooke's law.

An ideal fluid is also homogeneous and isotropic and the term is usually restricted to incompressible, inviscid fluids. This is clearly a good approximation to the properties of water in conditions where the compressibility is negligible and the effects of viscosity are confined to a thin layer close to a solid surface known as the boundary layer. For gases such as air, which are very compressible, it is found that the effects of compressibility in flow processes are not significant until relative velocities approaching the speed of sound are reached.

12.7 Simple tension

Figure 12.2 shows a straight uniform bar length L, cross-section area A and under the action of a

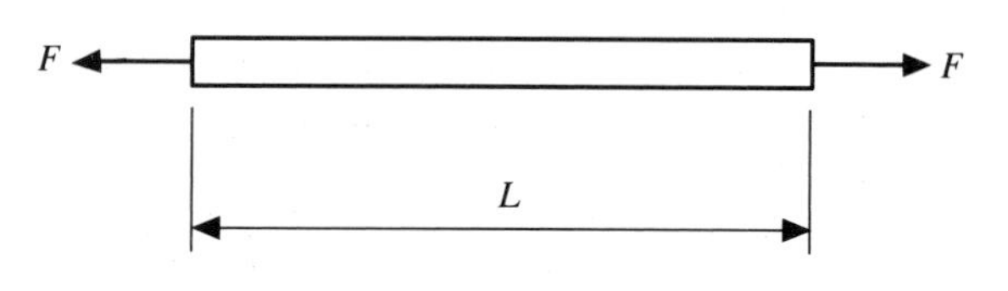

Figure 12.2

tensile force F. The state of tension along the bar is constant, this means that if a cut is made anywhere along the bar the force on a right-facing surface is F to the right and on a left-facing surface it is F to the left.

It follows that at any point the tensile load divided by the original cross-section area, F/A, is constant and this quantity is called the stress σ. A negative stress implies that the load is compressive. If the extension under this load is δ then the strain $\varepsilon = \delta/L$.

By Hooke's law $\delta \propto F$ so $\varepsilon \propto \sigma$ or

$$\sigma = E\varepsilon \tag{12.6}$$

where the constant of proportionality E is a property of the material known as Young's modulus.

Re-arranging the above equations gives

$$\delta = \frac{FL}{AE} \tag{12.7}$$

A state of tension resulting in an extension is regarded as being associated with a positive stress and a positive strain. (*See Appendix 8 for a discussion of material properties.*)

12.8 Equation of motion for a one-dimensional solid

Figure 12.3 shows an element of a uniform bar which has no external loads applied along its length, the external loads or constraints occurring only at the ends. The material is homogeneous with a density ρ, Young's modulus E and a constant cross-section area A.

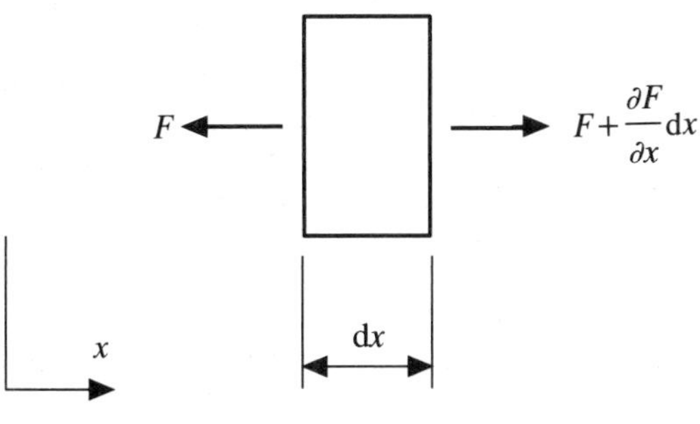

Figure 12.3

The mass of an element of length dx is $\rho A\, dx$ and this is constant as these quantities refer to the original values.

Resolving the forces in the x direction and equating the net force to the mass of the element times its acceleration gives

$$\left(F + \frac{\partial F}{\partial x} dx\right) - F = (\rho A\, dx)\frac{\partial^2 u}{\partial t^2}$$

or

$$\frac{\partial F}{\partial x} = \rho A \frac{\partial^2 u}{\partial t^2} \tag{12.8}$$

Since nominal or engineering stress is defined as force/original area, then dividing both sides of equation 12.8 by A gives

$$\frac{\partial \sigma}{\partial x} = \rho \frac{\partial^2 u}{\partial t^2} \tag{12.9}$$

We have already shown that the strain

$$\varepsilon = \frac{\partial u}{\partial x} \tag{12.10}$$

and also $\sigma = E\varepsilon$ (12.11)

so substituting 12.11 into 12.9 and using 12.10 we finally obtain

$$E\frac{\partial^2 u}{\partial x^2} = \rho \frac{\partial^2 u}{\partial t^2} \tag{12.12}$$

This is a very common equation in applied physics and is known as the wave equation.

In a statics case the right-hand side of equation 12.12 is zero so that u is a function of x only, giving

$$\frac{d^2 u}{dx^2} = 0$$

the solution of which is $u = a + bx$ where a and b are constants depending on the boundary conditions.

Now strain

$$\varepsilon = \frac{du}{dx} = b = \sigma/E = (F/A)/E$$

so if at $x = 0$ $u = 0$ then $a = 0$, $u = Fx/(AE)$. At $x = L$ the displacement is equal to $FL/(AE)$, as expected.

12.9 The control volume

The equations of motion developed for rigid bodies and commonly used for a solid continuum refer to a fixed amount of matter. However for fluids it is usually more convenient to concentrate on a fixed region of space with a volume V and a surface S. The properties of the fluid are expressed as functions of spatial position and of time, it being noted that different particles will occupy a given location at different times.

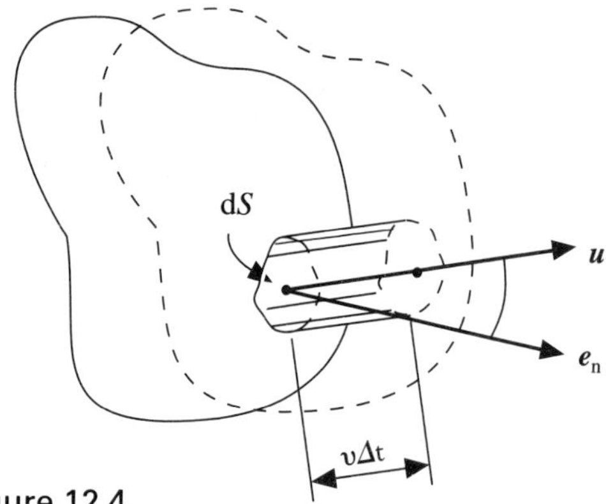

Figure 12.4

At time t the control volume is shown in Fig. 12.4 by the solid boundary. At time $t+\Delta t$ the position of the set of particles originally within the control volume is indicated by the dashed boundary.

The velocity of the fluid at an elemental part of the surface is $\boldsymbol{v}$ and the outward normal to the surface is $\mathbf{e}_n$.

At the elemental surface area, dS, the increase in mass in the time Δt is

$$\rho(dS\,v\Delta t\cos\alpha) = \rho\boldsymbol{v}\cdot\mathbf{e}_n dS\,\Delta t = \rho\boldsymbol{v}\cdot d\boldsymbol{S}\,\Delta t.$$

Note that the area vector ($d\boldsymbol{S} = \boldsymbol{e}_n dS$) is defined as having a magnitude equal to the elemental surface area and a direction defined by the outward normal unit vector, $\mathbf{e}_n$.

Integrating over the whole surface we obtain the net total mass *gained* by the original group due to the velocity at the surface. In addition to this there is a further increase in mass due to the density over the whole volume changing with time.

Thus the change in mass,

$$\Delta m = \left[\int_S \rho\boldsymbol{v}\cdot d\boldsymbol{S} + \int_V \frac{\partial\rho}{\partial t}dV\right]\Delta t$$

12.10 Continuity

Since the mass must remain constant

$$\frac{\Delta m}{\Delta t} = \int_S \rho\boldsymbol{v}\cdot d\boldsymbol{S} + \int_V \frac{\partial\rho}{\partial t}dV = 0 \qquad (12.13)$$

this is known as the continuity equation.

12.11 Equation of motion for a fluid

To obtain the equations of motion we need to consider the time rate of change in linear momentum. This is achieved by simply replacing ρ by $\rho\boldsymbol{v}$ in the development of the continuity equation. This is possible since ρ is the mass per unit volume and $\rho\boldsymbol{v}$ is the momentum per unit volume. Thus the change in momentum in time Δt is

$$\Delta G = \left[\int_S \rho\boldsymbol{v}(\boldsymbol{v}\cdot d\boldsymbol{S}) + \int_V \frac{\partial(\rho\boldsymbol{v})}{\partial t}dV\right]$$

Now force $\mathbf{F} = \lim_{\Delta t\to 0}\dfrac{\Delta G}{\Delta t}$

$$= \int_S \rho\boldsymbol{v}(\boldsymbol{v}\cdot d\boldsymbol{S}) + \int_V \frac{\partial(\rho\boldsymbol{v})}{\partial x}dV \qquad (12.14)$$

12.12 Streamlines

A streamline is a line drawn in space at a specific time such that the velocity of the fluid at that instant is, at all points, tangent to the streamline.

The distance along the streamline is s and, as in path co-ordinates, $\mathbf{e}_t$ is the unit tangent vector and $\mathbf{e}_n$ is the unit normal vector; as shown in Fig. 12.5.

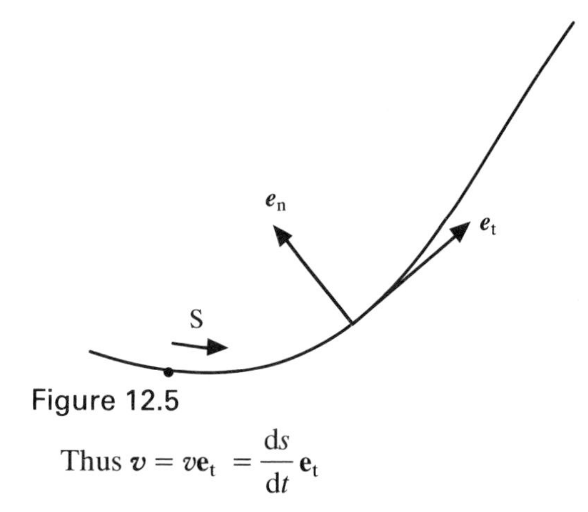

Figure 12.5

Thus $\boldsymbol{v} = v\mathbf{e}_t = \dfrac{ds}{dt}\mathbf{e}_t$

If the flow is steady, that is the velocity at any point does not vary with time, a streamline is also a path line.

12.13 Continuity for an elemental volume

The continuity equation, 12.13, is in the form for a finite volume. We now wish to obtain an expression for an elemental volume corresponding to that derived for the solid.

Figure 12.6 shows a rectangular element with sides dx, dy and dz. Considering the continuity equation we first evaluate the surface integral

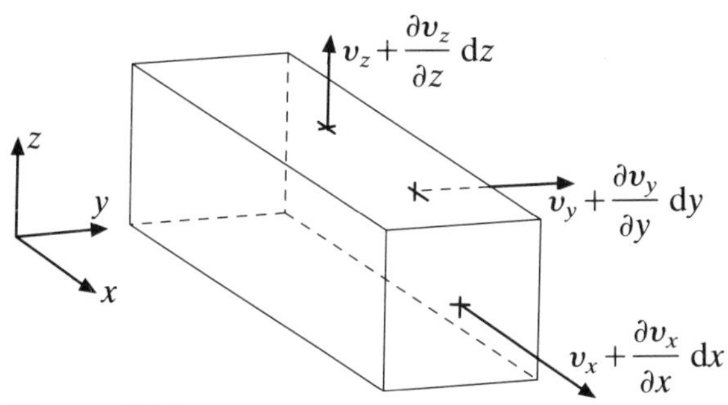

Figure 12.6

$$\int_{s} \rho \boldsymbol{v} \cdot d\boldsymbol{S} = \left[\rho v_x + \frac{\partial \rho v_x}{\partial x}\,dx\right] dy\,dz - \rho v_x\,dy\,dz$$

$$+ \left[\rho v_y + \frac{\partial \rho v_y}{\partial y}\,dy\right] dz\,dx - \rho v_y\,dz\,dx$$

$$+ \left[\rho v_z + \frac{\partial \rho v_z}{\partial z}\,dz\right] dx\,dy - \rho v_z\,dx\,dy$$

$$= \left[\frac{\partial \rho v_x}{\partial x} + \frac{\partial \rho v_y}{\partial y} + \frac{\partial \rho v_z}{\partial z}\right] dx\,dy\,dz.$$

The vector operator ∇ is defined, in Cartesian co-ordinates, to be

$$\nabla = \boldsymbol{i}\frac{\partial}{\partial x} + \boldsymbol{j}\frac{\partial}{\partial y} + \boldsymbol{k}\frac{\partial}{\partial z}$$

so with $\rho \boldsymbol{v} = \boldsymbol{i}\rho v_x + \boldsymbol{j}\rho v_y + \boldsymbol{k}\rho v_z$

$$\int_{s} \rho \boldsymbol{v} \cdot d\boldsymbol{S} = \nabla \cdot \rho \boldsymbol{v}\,dx\,dy\,dz.$$

The operation $\nabla \cdot (\rho \boldsymbol{v})$ is said to be the divergence of the $\rho \boldsymbol{v}$ field and is often written as $\operatorname{div}(\rho \boldsymbol{v})$.

Also $\displaystyle\int_{v} \frac{\partial \rho}{\partial t}\,dV = \frac{\partial \rho}{\partial t}\,dx\,dy\,dz$

so the complete continuity equation is

$$\left[\nabla \cdot \rho \boldsymbol{v} + \frac{\partial \rho}{\partial t}\right] dx\,dy\,dz = 0$$

or

$$\nabla \cdot \rho \boldsymbol{v} + \frac{\partial \rho}{\partial t} = 0 \qquad (12.15)$$

12.14 Euler's equation for fluid flow

In applying the momentum equation we shall choose a small cylindrical element with its axis along a streamline. However at the curved surface, of area dS', there could be a small radial component of velocity u since the streamlines at this surface may be diverging.

A stream tube could have been used where the curved surface is composed of streamlines, but this means that the cross-section area would be a variable and the effect of pressure on this surface would have to be considered.

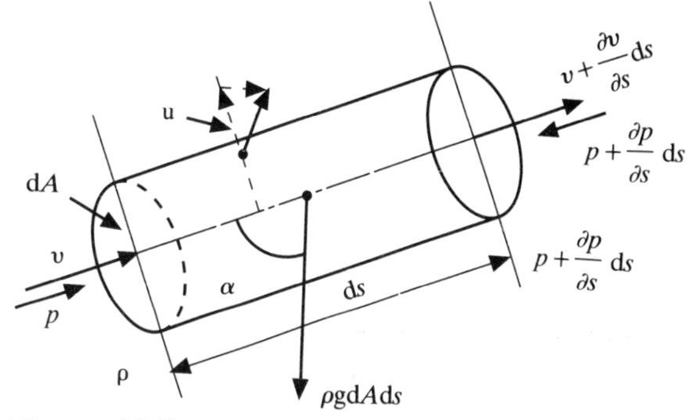

Figure 12.7

First we need to apply the continuity equation so with reference to Fig. 12.7

$$\left(\rho + \frac{\partial \rho}{\partial s}\,ds\right)\left(v + \frac{\partial v}{\partial s}\,ds\right) dA$$

$$- \rho v\,dA + \rho(u\,dS') + \frac{\partial \rho}{\partial t}\,dA\,ds = 0$$

Neglecting second order terms

$$\rho\frac{\partial v}{\partial s}\,dV + v\frac{\partial \rho}{\partial s}\,dV + \rho u\,dS + \frac{\partial \rho}{\partial t}\,dV = 0 \quad (12.16)$$

where $dV = ds\,dA$.

In applying the force equation we are going to include a body force, in this case gravity, in addition to the pressure difference. Resolving forces along the streamline

$$dF = p\,dA - \left(p + \frac{\partial p}{\partial s}\,ds\right) dA$$

$$- \rho g\,ds\,dA \cos\alpha$$

or

$$dF = \left(\frac{\partial p}{\partial s} - \rho g \cos\alpha\right) ds\,dA. \qquad (12.17)$$

The rate of change of momentum is, from equation 12.14,

$$dG = \left(\rho + \frac{\partial \rho}{\partial s}\,ds\right)\left(v + \frac{\partial v}{\partial s}\,ds\right)^2 dA - \rho v v\,dA$$

$$+\rho v(udS')+\frac{\partial(\rho v)}{\partial t}\,ds\,dA$$

$$=2\rho v\frac{\partial v}{\partial s}\,ds\,dA+v^2\frac{\partial \rho}{\partial s}\,ds\,dA+\rho v u\,dS'$$

$$+\frac{\partial(\rho v)}{\partial t}\,ds\,dA$$

The right-hand side of this expression can be simplified by subtracting v times equation 12.16 to give

$$dG=\rho v\frac{\partial v}{\partial s}\,ds\,dA-v^2\frac{\partial \rho}{\partial t}\,ds\,dA+\frac{\partial(\rho v)}{\partial t}\,ds\,dA$$

combining with 12.17 and dividing through by $ds\,dA$

$$-g\rho\cos\alpha-\frac{\partial p}{\partial s}=\rho v\frac{\partial v}{\partial s}-v\frac{\partial \rho}{\partial t}+\left(\rho\frac{\partial v}{\partial t}+v\frac{\partial \rho}{\partial t}\right)$$

and finally re-arranging gives

$$-g\cos\alpha-\frac{1}{\rho}\frac{\partial p}{\partial s}=v\frac{\partial v}{\partial s}+\frac{\partial v}{\partial t} \qquad (12.18)$$

This is known as Euler's equation for fluid flow. Since $v=v(s,t)$,

$$\frac{dv}{dt}=\frac{\partial v}{\partial s}\frac{ds}{dt}+\frac{\partial v}{\partial t}\frac{dt}{dt}=\frac{\partial v}{\partial s}v+\frac{\partial v}{\partial t}$$

the right-hand side of 12.18 may be written as

$$\frac{dv}{dt}.$$

12.15 Bernoulli's equation

If we consider the case for steady flow where the velocity at a given point does not change with time, Euler's equation may be written

$$-g\cos\alpha-\frac{1}{\rho}\frac{dp}{ds}=v\frac{dv}{ds}$$

the partial differentials have been replaced by total differentials because v is defined to be a function of s only. Multiplying through by ds and integrating gives

$$-\int g\cos\alpha\,ds-\int\frac{dp}{\rho}=\frac{v^2}{2}+\text{constant}$$

now $\cos\alpha\,ds=dz$ thus

$$\int\frac{dp}{\rho}+\frac{v^2}{2}+gz=\text{constant}$$

If ρ is a known function of p then the integral can be determined but if we take ρ to be constant we have

$$\frac{p}{\rho}+\frac{v^2}{2}+gz=\text{constant} \qquad (12.19)$$

this is known as Bernoulli's equation.

This equation is strictly applicable to steady flow of a non-viscous, incompressible fluid; it is, however, often used in cases where the flow is changing slowly. The effects of friction are usually accounted for by the inclusion of experimentally determined coefficients. As has already been mentioned, the effects of compressibility can often be neglected in flow cases where the relative speeds are small compared with the speed of sound in the fluid.

SECTION B Two- and three-dimensional continua

12.16 Introduction

We are now going to extend our study of solid continua to include more than one dimension. In our treatment of one-dimensional tension or compression we did not consider any changes in the lateral dimensions. Although we are going to use three dimensions we shall restrict the analysis to plane strain conditions. By plane strain we mean that any group of particles which lie in a plane will, after deformation, remain in a plane. It is possible that the plane will be displaced from the original plane but will still be parallel to it.

It is an experimental fact that a stress applied in one direction only will produce strain in that direction and also at right angles to the stress axis. If a specimen is strained within the x–y plane then, if the strain in the z direction is to be zero, there must be a stress in the z direction as well as in the x and y directions. Conversely, if stresses are applied in the x and y directions with a zero stress in the z direction, there will be a resulting strain in the z direction as well as those in the x and y directions. The two-dimensional analyses presented later are based on the latter case.

12.17 Poisson's ratio

If Hooke's law is obeyed, then the transverse strain produced in axial tension will also be proportional to the applied load; thus it follows that the lateral strain will be proportional to the axial strain. The ratio

$$\frac{\text{transverse strain}}{\text{axial strain}} = -\nu$$

where ν is known as Poisson's ratio.

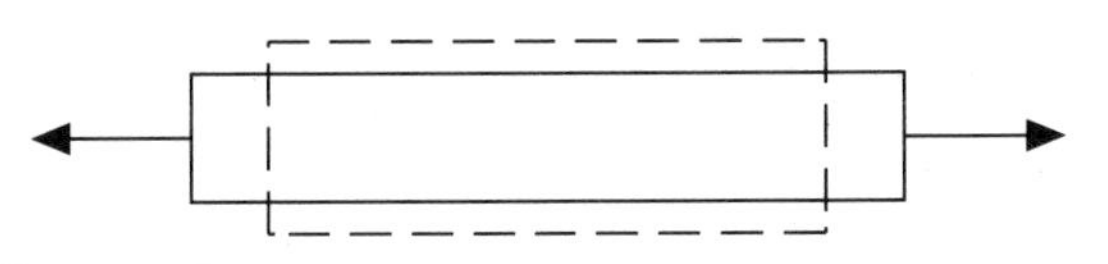

Figure 12.8

If a uniform rectangular bar, as shown in Fig. 12.8, is loaded along the x axis then

$$\varepsilon_x = \sigma_x/E$$

$$\varepsilon_y = -\nu\sigma_x/E$$

and $\varepsilon_z = -\nu\sigma_x/E.$

12.18 Pure shear

Figure 12.9 shows a rectangular element which is deformed by a change in shape such that the length of the sides remain unaltered. The shear strain γ_{xy} is defined as the change in angle (measured in radians) of the right angle between adjacent edges. This is a small angle consistent with our discussion of small strains.

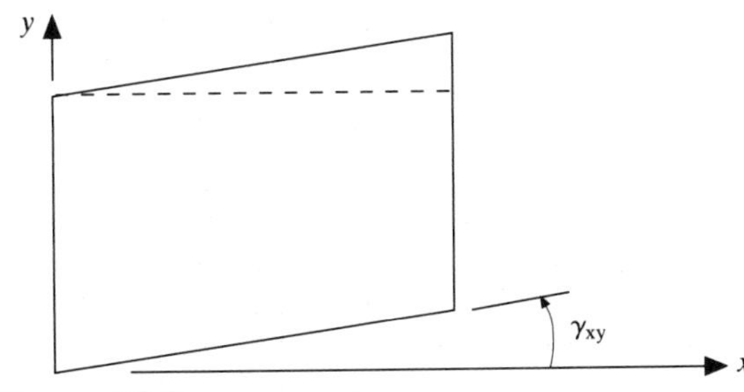

Figure 12.9

The loading applied to the element to produce pure shear is as shown in Fig. 12.10. This set of forces is in equilibrium, so by considering the sum of the moments of the two couples in the xy plane

$$F_y\,\mathrm{d}x - F_x\,\mathrm{d}y = 0 \tag{12.20}$$

The shear stress is defined as

$$\tau_{xy} = F_x/(\mathrm{d}y\,\mathrm{d}z)$$

and $\tau_{yx} = F_y/(\mathrm{d}x\,\mathrm{d}z)$

Substitution into equation 12.20 gives

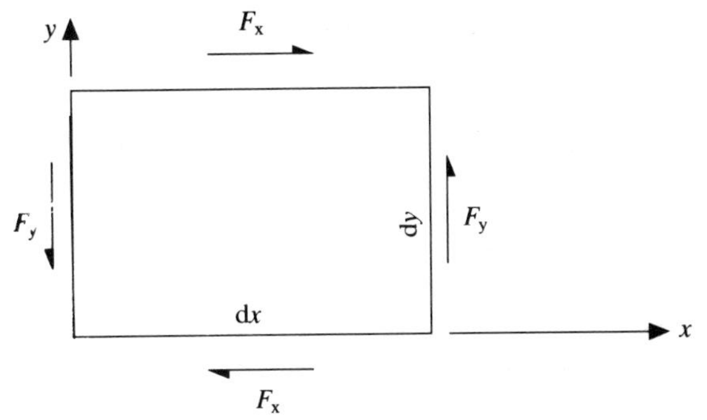

Figure 12.10

$$\tau_{xy} = \tau_{yx} \tag{12.21}$$

This shows the equivalence of the complementary shear stresses.

Again by Hooke's law, shear stress is proportional to the shear strain

$$\tau_{xy} = G\gamma_{xy} \tag{12.22}$$

where G is known as the Shear Modulus or as the Modulus of Rigidity.

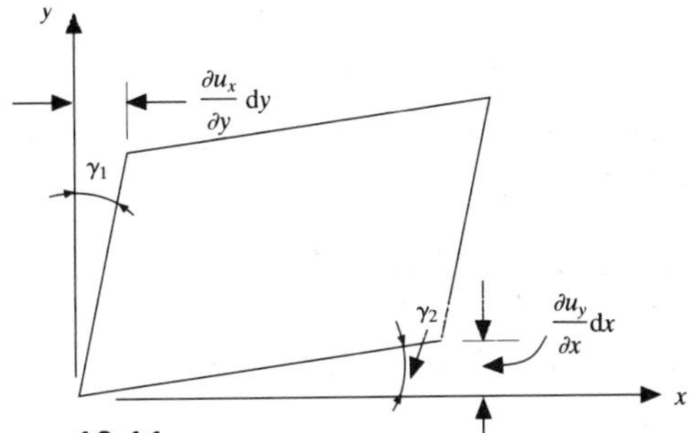

Figure 12.11

Referring to Fig. 12.11 it is seen that the shear strain can be expressed in terms of partial differential coefficients as

$$\gamma_{xy} = \gamma_1 + \gamma_2$$

$$\gamma_{xy} = \frac{\partial u_y}{\partial x} + \frac{\partial u_x}{\partial y} \tag{12.23}$$

12.19 Plane strain

The rectangular element, shown in Fig. 12.12, has one face in the xy plane and is distorted such that the corner points A, B, C and D move in the xy plane only.

The translation of point A is $\boldsymbol{u}$ and that of point C is $\boldsymbol{u} + \mathrm{d}\boldsymbol{u}$. For small displacements

$$\mathrm{d}\boldsymbol{u} = \left[\frac{\partial u_x}{\partial x}\mathrm{d}x + \frac{\partial u_x}{\partial y}\mathrm{d}y\right]\boldsymbol{i} + \left[\frac{\partial u_y}{\partial x}\mathrm{d}x + \frac{\partial u_y}{\partial y}\mathrm{d}y\right]\boldsymbol{j}$$

or in matrix form

$$\begin{bmatrix} du_x \\ du_y \end{bmatrix} = \begin{bmatrix} \dfrac{\partial u_x}{\partial x} & \dfrac{\partial u_x}{\partial y} \\ \dfrac{\partial u_y}{\partial x} & \dfrac{\partial u_y}{\partial y} \end{bmatrix} \begin{bmatrix} dx \\ dy \end{bmatrix} \quad (12.24)$$

Let us now introduce the notation

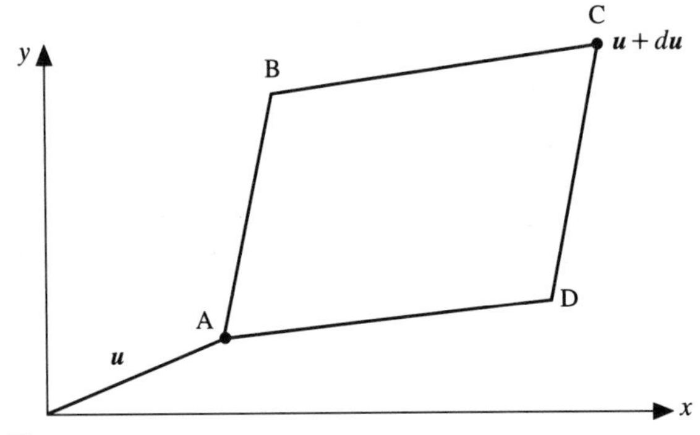

Figure 12.12

$$\frac{\partial u_x}{\partial y} = u_{x,y} \text{ etc.}$$

In this notation the strain in the x direction

$$\varepsilon_{xx} = u_{x,x}$$

similarly $\quad \varepsilon_{yy} = u_{y,y}$

and the shear strain $\quad \gamma_{xy} = u_{y,x} + u_{x,y}$

and equation 12.24 becomes

$$\begin{bmatrix} du_x \\ du_y \end{bmatrix} = \begin{bmatrix} u_{x,x} & u_{x,y} \\ u_{y,x} & u_{y,y} \end{bmatrix} \begin{bmatrix} dx \\ dy \end{bmatrix}$$

The square matrix can be written as the sum of a symmetrical and an anti-symmetrical matrix. By this means the shear strain can be introduced.

$$\begin{bmatrix} u_{x,x} & u_{x,y} \\ u_{y,x} & u_{y,y} \end{bmatrix} = \begin{bmatrix} u_{x,x} & \frac{1}{2}(u_{x,y} + u_{y,x}) \\ \frac{1}{2}(u_{x,y} + u_{y,x}) & u_{y,y} \end{bmatrix} + \begin{bmatrix} 0 & -\frac{1}{2}(u_{y,x} - u_{x,y}) \\ \frac{1}{2}(u_{x,y} - u_{y,x}) & 0 \end{bmatrix}$$

therefore

$$\begin{bmatrix} du_x \\ du_y \end{bmatrix} = \left\{ \begin{bmatrix} \varepsilon_{xx} & \frac{1}{2}\gamma_{xy} \\ \frac{1}{2}\gamma_{xy} & \varepsilon_{yy} \end{bmatrix} + \begin{bmatrix} 0 & -\Omega_{xy} \\ \Omega_{xy} & 0 \end{bmatrix} \right\} \begin{bmatrix} dx \\ dy \end{bmatrix} \quad (12.25)$$

where Ω_{xy} is the rigid body rotation in the xy plane given by

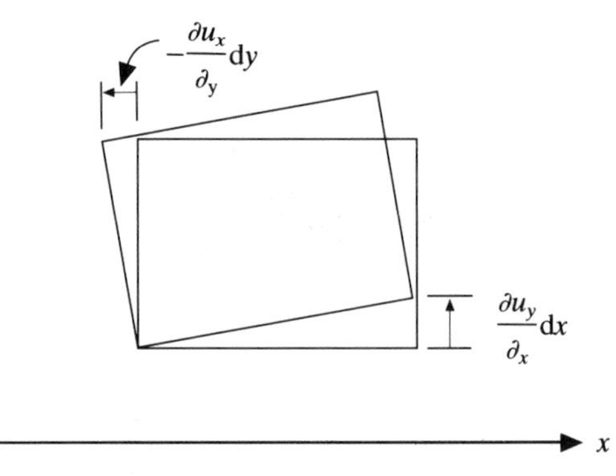

Figure 12.13

$$\Omega_{xy} = \frac{1}{2}\left(\frac{\partial u_y}{\partial x} - \frac{\partial u_x}{\partial y}\right)$$

(see Fig. 12.13).

The 1/2 in the strain matrix spoils the simplicity of the notation therefore it is common to replace $\frac{1}{2}\gamma_{xy}$ by ε_{xy}.

12.20 Plane stress

The triangular elements shown in Fig. 12.14 are in equilibrium under the action of forces which have components in the x and y directions but not in the z direction. Note that the surface *abcd* has area $dy\,dz\,\boldsymbol{i}$ and area *abef* has an area $dx\,dz\,\boldsymbol{j}$; these are the vector components of the area $e'f'c'd'$. The sense of the stress component, shown on the diagram, is such that when multiplied by the area vector it gives the force vector.

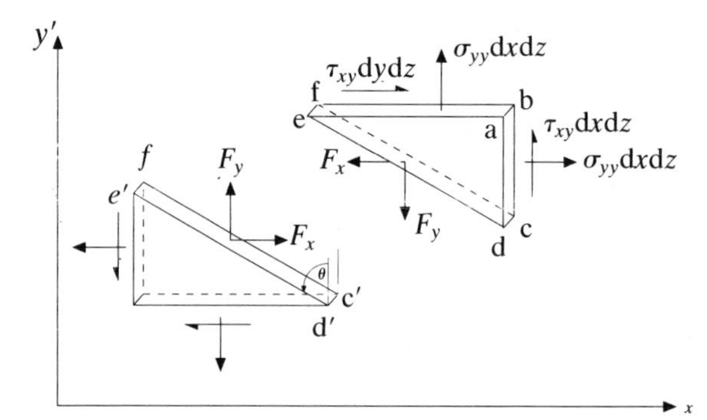

Figure 12.14

Resolving in the x direction we obtain

$$F_x = \sigma_{xx}\,dy\,dz + \tau_{xy}\,dx\,dz$$
$$F_y = \sigma_{yy}\,dx\,dz + \tau_{xy}\,dy\,dz$$

or, in matrix form,

$$\begin{bmatrix} F_x \\ F_y \end{bmatrix} = \begin{bmatrix} \sigma_{xx} & \tau_{xy} \\ \tau_{xy} & \sigma_{yy} \end{bmatrix} \begin{bmatrix} dy\,dz \\ dx\,dz \end{bmatrix}$$

Letting $dy\,dz = S_x$ and $dx\,dz = S_y$

$$\begin{bmatrix} F_x \\ F_y \end{bmatrix} = \begin{bmatrix} \sigma_{xx} & \tau_{xy} \\ \tau_{xy} & \sigma_{yy} \end{bmatrix} \begin{bmatrix} S_x \\ S_y \end{bmatrix} \tag{12.26}$$

In many texts τ_{xy} is replaced by σ_{xy}.

12.21 Rotation of reference axes

The values of the components of stress and strain depend on the orientation of the reference axes. In Fig. 12.15 the axes have been rotated by an angle θ about the z axis.

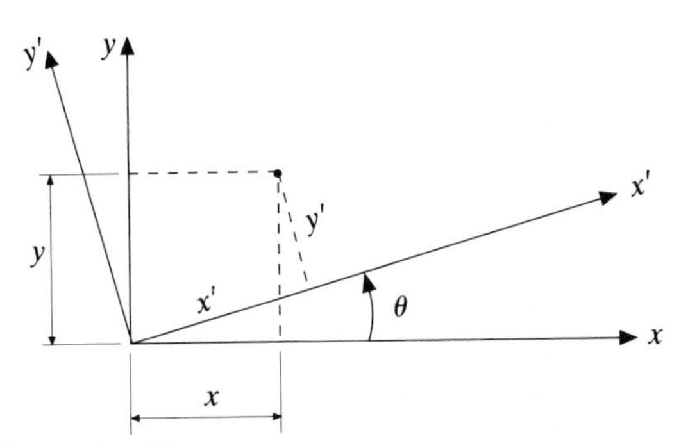

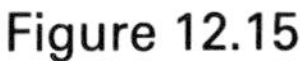
Figure 12.15

From the figure we have

$$x = x'\cos\theta - y'\sin\theta$$
$$y = y'\cos\theta + x'\sin\theta$$

which, in matrix form, becomes

$$\begin{bmatrix} x \\ y \end{bmatrix} = \begin{bmatrix} \cos\theta & -\sin\theta \\ \sin\theta & \cos\theta \end{bmatrix} \begin{bmatrix} x' \\ y' \end{bmatrix} \tag{12.27}$$

or, in abbreviated form

$$(\boldsymbol{x}) = [\boldsymbol{T}](\boldsymbol{x}').$$

The matrix $[\boldsymbol{T}]$ is a transformation matrix. It is easily shown from the geometry or by matrix inversion that the inverse of this matrix is the same as its transpose.

$$[\boldsymbol{T}]^{-1} = [\boldsymbol{T}]^{\mathrm{T}} = \begin{bmatrix} \cos\theta & \sin\theta \\ -\sin\theta & \cos\theta \end{bmatrix}$$

Writing equations 12.25 and 12.26 in abbreviated form as

$$(\mathbf{d}\boldsymbol{u}) = \{[\boldsymbol{\varepsilon}] + [\boldsymbol{\Omega}]\}(\mathbf{d}\boldsymbol{x})$$

and $\quad (\boldsymbol{F}) = [\boldsymbol{\sigma}](\boldsymbol{S})$

they may now be transformed by use of the transformation matrix.

12.22 Principal strain

Since $(\mathrm{d}\boldsymbol{u}) = [\boldsymbol{T}](\mathrm{d}\boldsymbol{u}')$ and $(\mathrm{d}\boldsymbol{x}) = [\boldsymbol{T}](\mathrm{d}\boldsymbol{x}')$ we can write

$$[\boldsymbol{T}](\mathrm{d}\boldsymbol{u}') = \{[\boldsymbol{\varepsilon}] + [\boldsymbol{\Omega}]\}[\boldsymbol{T}](\mathrm{d}\boldsymbol{x}')$$

and pre-multiplying by $[\boldsymbol{T}]^{\mathrm{T}}$ we obtain

$$(\mathrm{d}\boldsymbol{u}') = [\boldsymbol{T}]^{\mathrm{T}}\{[\boldsymbol{\varepsilon}] + [\boldsymbol{\Omega}]\}[\boldsymbol{T}](\mathrm{d}\boldsymbol{x}').$$

The rotation $[\boldsymbol{\Omega}]$ is not affected by the change in axes because they are rotated in the xy plane. The transformed strain matrix is

$$[\boldsymbol{\varepsilon}'] = [\boldsymbol{T}]^{\mathrm{T}}[\boldsymbol{\varepsilon}][\boldsymbol{T}]$$

$$= \begin{bmatrix} \cos\theta & \sin\theta \\ -\sin\theta & \cos\theta \end{bmatrix} \begin{bmatrix} \varepsilon_{xx} & \varepsilon_{xy} \\ \varepsilon_{xy} & \varepsilon_{yy} \end{bmatrix} \times \begin{bmatrix} \cos\theta & -\sin\theta \\ \sin\theta & \cos\theta \end{bmatrix}$$

$$= \begin{bmatrix} \varepsilon'_{xx} & \varepsilon'_{xy} \\ \varepsilon'_{xy} & \varepsilon'_{yy} \end{bmatrix}$$

where

$$\varepsilon'_{xx} = \varepsilon_{xx}\cos^2\theta + \varepsilon_{yy}\sin^2\theta + \varepsilon_{xy}2\cos\theta\sin\theta \tag{12.28}$$

$$\varepsilon'_{yy} = \varepsilon_{yy}\cos^2\theta + \varepsilon_{xx}\sin^2\theta - \varepsilon_{xy}2\cos\theta\sin\theta \tag{12.29}$$

$$\varepsilon'_{xy} = (\varepsilon_{yy} - \varepsilon_{xx})\sin\theta\cos\theta + \varepsilon_{xy}(\cos^2\theta - \sin^2\theta)$$

$$= \frac{(\varepsilon_{yy} - \varepsilon_{xx})}{2}\sin 2\theta + \varepsilon_{xy}\cos 2\theta \tag{12.30}$$

also $\quad (\varepsilon'_{xx} + \varepsilon'_{yy}) = (\varepsilon_{xx} + \varepsilon_{yy}) \qquad (12.31)$

From equation 12.30 it is seen that it is possible to choose a value for θ such that $\varepsilon'_{xy} = 0$. The value of θ is found from

$$\tan 2\theta = \frac{2\varepsilon_{xy}}{(\varepsilon_{xx} - \varepsilon_{yy})} \tag{12.32}$$

The axes for which the shear strain is zero are known as the principal strain axes. Let us therefore take our original axes as the principal axes, that is $\varepsilon_{xy} = 0$. The longitudinal strains are now the principal strains and will be denoted by ε_1 in the x direction and by ε_2 in the y.

From equations 12.28 and 12.29 we now have

$$\frac{(\varepsilon'_{xx} - \varepsilon'_{yy})}{2} = \frac{(\varepsilon_1 - \varepsilon_2)\cos 2\theta}{2}$$

and from equation 12.30

$$-\varepsilon'_{xy} = \frac{(\varepsilon_1 - \varepsilon_2)\sin 2\theta}{2}$$

A simple geometric construction, known as Mohr's circle, gives the relationship between the strains and the angle θ. Figure 12.16 shows a circle plotted with its centre on the normal strain axis, the ordinate being the negative shear strain. The location of the centre is given by the average strain, and the radius of the circle is half the difference between the principal strains. It is seen that this diagram satisfies the above equations.

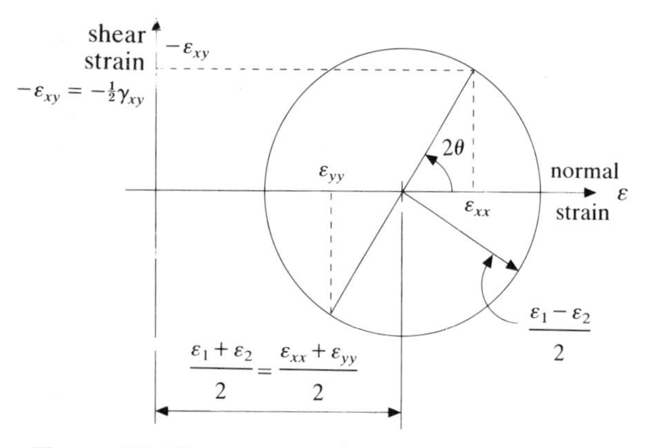

Figure 12.16

It can be seen that when $\theta = \pi/4$ the shear strain is maximum and the normal strains are equal. If the circle has its centre at the origin then for $\theta = \pi/4$ the normal strains are zero. So for the case of pure shear the principal strains are equal and opposite with a magnitude $\varepsilon_{xy} = \gamma_{xy}/2$.

In the case of uniaxial loading $\varepsilon_2 = -v\varepsilon_1$ hence the radius of the circle is $(\varepsilon_1 + v\varepsilon_1)/2$ which also equals the maximum shear strain at $\theta = \pi/4$.

So $\gamma_{xy} = \varepsilon_1(1 + v) = \sigma_1(1 + v)/E$.

12.23 Principal stress

Equation 12.26 can also be written in abbreviated form as

$$(\boldsymbol{F}) = [\boldsymbol{\sigma}](\boldsymbol{S})$$

and since the components of any vector can be expressed in terms of rotated co-ordinates we may write

$$(\boldsymbol{F}) = [\boldsymbol{T}](\boldsymbol{F}') \text{ and } (\boldsymbol{S}) = [\boldsymbol{T}](\boldsymbol{S}')$$

thus $[\boldsymbol{T}](\boldsymbol{F}') = [\boldsymbol{\sigma}][\boldsymbol{T}](\boldsymbol{S}')$

so pre-multiplying by $[\boldsymbol{T}]^{\mathrm{T}}$ gives

$$(\boldsymbol{F}') = [\boldsymbol{T}]^{\mathrm{T}}[\boldsymbol{\sigma}][\boldsymbol{T}](\boldsymbol{S}')$$

therefore

$$[\boldsymbol{\sigma}'] = [\boldsymbol{T}]^{\mathrm{T}}[\boldsymbol{\sigma}][\boldsymbol{T}]$$

$$= \begin{bmatrix} \cos\theta & \sin\theta \\ -\sin\theta & \cos\theta \end{bmatrix} \begin{bmatrix} \sigma_{xx} & \sigma_{xy} \\ \sigma_{xy} & \sigma_{yy} \end{bmatrix} \times \begin{bmatrix} \cos\theta & -\sin\theta \\ \sin\theta & \cos\theta \end{bmatrix}$$

$$= \begin{bmatrix} \sigma'_{xx} & \sigma'_{xy} \\ \sigma'_{xy} & \sigma'_{yy} \end{bmatrix}$$

where

$$\sigma'_{xx} = \sigma_{xx}\cos^2\theta + \sigma_{yy}\sin^2\theta + \sigma_{xy}2\cos\theta\sin\theta \quad (12.33)$$

$$\sigma'_{yy} = \sigma_{yy}\cos^2\theta + \sigma_{xx}\sin^2\theta - \sigma_{xy}2\cos\theta\sin\theta \quad (12.34)$$

$$\sigma'_{xy} = (\sigma_{yy} - \sigma_{xx})\sin\theta\cos\theta + \sigma_{xy}(\cos^2\theta - \sin^2\theta)$$

$$= \frac{(\sigma_{yy} - \sigma_{xx})}{2}\sin 2\theta + \sigma_{xy}\cos 2\theta \quad (12.35)$$

also $(\sigma'_{xx} + \sigma'_{yy}) = (\sigma_{xx} + \sigma_{yy})$

From equations 12.33 and 12.34 we now have

$$\frac{(\sigma'_{xx} - \sigma'_{yy})}{2} = \frac{(\sigma_1 - \sigma_2)\cos 2\theta}{2}$$

and from equation 12.35

$$-\sigma'_{xy} = \frac{(\sigma_1 - \sigma_2)\sin 2\theta}{2}$$

The form of these equations is the same as those for strain therefore a similar geometrical construction can be made, which is Mohr's circle for stress as shown in Fig. 12.17.

Because we have taken the material to be isotropic it follows that the principal axes for stress coincide with those for strain. This is because normal stresses cannot produce shear strain in a material which shows no preferred directions.

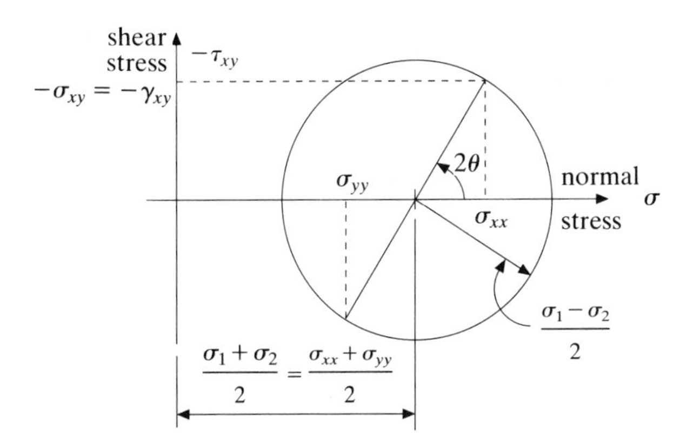

Figure 12.17

12.24 The elastic constants

So far we have encountered three elastic constants namely Young's modulus (E), the shear modulus (G) and Poisson's ratio (ν). There are three others which are of importance, the first of which is the bulk modulus.

For small strains the change in volume of a rectangular element with sides dx, dy and dz is

$$(\varepsilon_{xx}\,dx)\,dy\,dz + (\varepsilon_{yy}\,dy)\,dz\,dx + (\varepsilon_{zz}\,dz)\,dx\,dy.$$

The volumetric strain, also known as the dilatation, is the ratio of the change in volume to the original volume; thus the dilatation

$$\Delta = \varepsilon_{xx} + \varepsilon_{yy} + \varepsilon_{zz}$$

It should be remembered that shear strain has no effect on the volume.

The average stress

$$\sigma_{ave.} = (\sigma_{xx} + \sigma_{yy} + \sigma_{zz})/3$$

and the bulk modulus K is defined by

$$\sigma_{ave.} = K\Delta \tag{12.36}$$

(For fluids the average stress is the negative of the pressure p).

The two other constants are the Lamé constants and they will be defined during the following discussion.

In general every component of stress depends linearly on each component of strain. If we consider an element which is aligned with the principal axes of stress and strain, then each principal stress will be a function of each principal strain, thus

$$\sigma_1 = a\varepsilon_1 + b\varepsilon_1 + c\varepsilon_3.$$

Because of the symmetry b must be equal to c so we can write

$$\sigma_1 = (b + (a-b))\,\varepsilon_1 + b\varepsilon_2 + b\sigma_3$$

or $\sigma_1 = b(\varepsilon_1 + \varepsilon_2 + \varepsilon_3) + (a-b)\,\varepsilon_1.$

Let $b = \lambda$ and $(a-b) = 2\mu$ where λ and μ are the Lamé constants, and introducing dilatation Δ, the sum of the strains, we have

$$\sigma_1 = \lambda\Delta + 2\mu\varepsilon_1 \tag{12.37}$$

and again because of symmetry

$$\sigma_2 = \lambda\Delta + 2\mu\varepsilon_2 \tag{12.38}$$

$$\sigma_3 = \lambda\Delta + 2\mu\varepsilon_3. \tag{12.39}$$

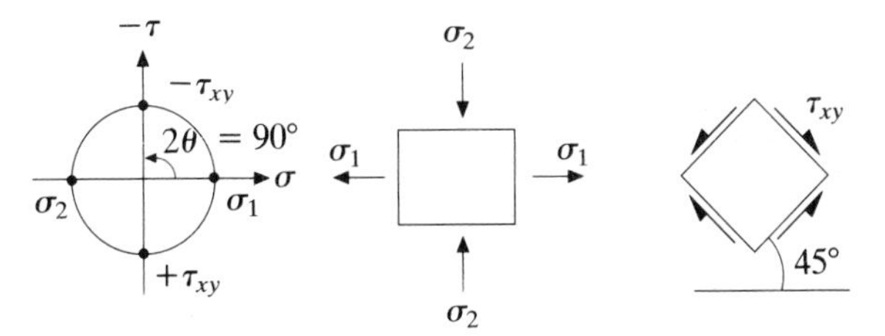

Figure 12.18

Let us now consider the case of pure shear, see Fig. 12.18. We have already seen that $\sigma_1 = -\tau_{xy}$, $\sigma_2 = \tau_{xy}$, $\varepsilon_1 = -\varepsilon_{xy}$ and $\varepsilon_2 = \varepsilon_{xy}$ so substituting into equations 12.37 and 12.38 we have

$$-\tau_{xy} = \lambda\Delta + 2\mu(-\varepsilon_{xy})$$

and $\tau_{xy} = \lambda\Delta + 2\mu\varepsilon_{xy}.$

Solving the last two equations shows that $\Delta = 0$ and $\tau_{xy} = 2\mu\varepsilon_{xy}$ giving

$$\mu = \frac{\tau_{xy}}{2\varepsilon_{xy}} = \frac{\tau_{xy}}{\gamma_{xy}} = G \tag{12.40}$$

Now consider the case of pure tension, see Fig. 12.19, such that $\sigma_2 = 0$ and $\varepsilon_2 = -\nu\varepsilon_1$. Substitution into equations 12.38 and 12.39 gives

$$\sigma_1 = \lambda\Delta + 2\mu\varepsilon_1$$
$$0 = \lambda\Delta - 2\mu\nu\varepsilon_1$$

from which $\sigma_1 = 2\mu(1+\nu)\,\varepsilon_1$

so $\sigma_1/\varepsilon_1 = E = 2\mu(1+\nu) = 2G(1+\nu)$ (12.41)

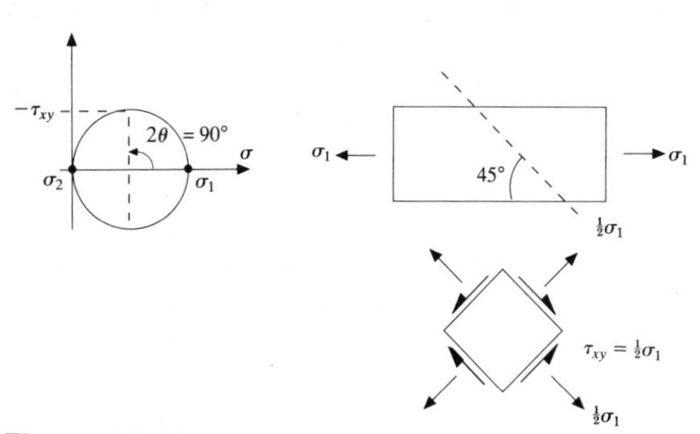

Figure 12.19

If we add together the three equations 12.37 to 12.39 we obtain

$$3\sigma_{\text{ave.}} = 3\lambda\Delta + \mu\Delta = (3\lambda + 2\mu)\Delta$$

so $K = \dfrac{\sigma_{\text{ave.}}}{\Delta} = \lambda + 2\mu/3.$ (12.42)

(For an ideal fluid $\mu = 0$ and $\lambda = K$).

Using equation 12.28 it is seen that taking the $OxOy$ axes as principal axes

$$\varepsilon_{xx} = \varepsilon_1 \cos^2\theta + \varepsilon_2 \sin^2\theta$$

and using equation 12.33

$$\sigma_{xx} = \sigma_1 \cos^2\theta + \sigma_2 \sin^2\theta.$$

Substituting from equations 12.37 and 12.38 leads to

$$\begin{aligned}\sigma_{xx} &= [\lambda\Delta + 2\mu\varepsilon_1]\cos^2\theta + [\lambda\Delta + 2\mu\varepsilon_2]\sin^2\theta \\ &= \lambda\Delta + 2\mu[\varepsilon_1\cos^2\theta + \varepsilon_2\sin^2\theta] \\ &= \lambda\Delta + 2\mu\varepsilon_{xx}.\end{aligned}$$

In general we may write

$$\sigma_{ii} = \lambda\Delta + 2\mu\varepsilon_{ii} \tag{12.43}$$

and $\tau_{ij} = 2\mu\varepsilon_{ij}\ (i \neq j)$ (12.44)

This can also be written in matrix form as

$$[\boldsymbol{\sigma}] = \lambda\Delta[\boldsymbol{I}] + 2\mu[\boldsymbol{\varepsilon}] \qquad 12.45)$$

where $[\boldsymbol{I}]$ is the identity matrix.

Note that for a homogeneous isotropic elastic material there are only two independent elastic moduli.

12.25 Strain energy

If a body is strained then work is done on that body and if the body is elastic then, by definition of the term elastic, the process is reversible. Consider a unit cube of material so that the force on a face is numerically equal to the stress, and the extension is equal to the strain. For the case where only normal stresses are acting the increase in work done is

$$dU = \sigma_{xx}\,d\varepsilon_{xx} + \sigma_{yy}\,d\varepsilon_{yy} + \sigma_{zz}\,d\varepsilon_{zz}$$

For a linearly elastic material obeying Hooke's law where stress is proportional to strain, the total energy may be found by applying the load in each direction sequentially rather than simultaneously. Applying the load in the x direction first the work done is the area under the stress–strain graph, so since the strain is due to σ_{xx} only

$$U_x = \frac{\sigma_{xx}}{2}\frac{\sigma_{xx}}{E}$$

we now apply σ_{yy} slowly whilst σ_{xx} remains constant

thus $U_y = \dfrac{\sigma_{yy}}{2}\dfrac{\sigma_{yy}}{E} + \sigma_{xx}(-\nu)\dfrac{\sigma_{yy}}{E}$

and $U_z = \dfrac{\sigma_{zz}}{2}\dfrac{\sigma_{zz}}{E} + \sigma_{xx}(-\nu)\dfrac{\sigma_{zz}}{E} + \sigma_{yy}(-\nu)\dfrac{\sigma_{zz}}{E}$

The total energy due to normal stresses is

$$\begin{aligned}U &= U_x + U_y + U_z \\ &= \frac{\sigma_{xx}}{2}\left(\frac{\sigma_{xx}}{E} - \frac{\nu}{E}(\sigma_{yy} + \sigma_{zz})\right) \\ &\quad + \frac{\sigma_{yy}}{2}\left(\frac{\sigma_{yy}}{E} - \frac{\nu}{E}(\sigma_{zz} + \sigma_{xx})\right) \\ &\quad + \frac{\sigma_{zz}}{2}\left(\frac{\sigma_{zz}}{E} - \frac{\nu}{E}(\sigma_{xx} + \sigma_{yy})\right) \\ &= \frac{\sigma_{xx}\,\varepsilon_{xx}}{2} + \frac{\sigma_{yy}\,\varepsilon_{yy}}{2} + \frac{\sigma_{zz}\,\varepsilon_{zz}}{2}\end{aligned}$$

In the case of pure shear strain the strain energy is simply

$$\frac{\tau_{xy}\,\gamma_{xy}}{2}$$

and since the shear strains are independent the total strain energy can be written

$$U = \frac{\sigma_{xx}\,\varepsilon_{xx}}{2} + \frac{\sigma_{yy}\,\varepsilon_{yy}}{2} + \frac{\sigma_{zz}\,\varepsilon_{zz}}{2} + \frac{\tau_{xy}\,\gamma_{xy}}{2} + \frac{\tau_{yz}\,\gamma_{yz}}{2} + \frac{\tau_{zx}\,\gamma_{zx}}{2} \qquad (12.46)$$

or in indicial notation

$$U = \frac{\sigma_{ij}\,\varepsilon_{ij}}{2}$$

where summation is taken over all values of i and j. (Remember that $\varepsilon = \gamma/2$, $\varepsilon_{ij} = \varepsilon_{ji}$ and $\sigma_{ij} = \sigma_{ji}$.)

SECTION C
Applications to bars and beams

12.26 Introduction

The exact solution to the three-dimensional stress strain relationships are known for only a small number of special cases. So for the common engineering problems – involving prismatic bars under the action of tension, torsion and bending – certain simplying assumptions are made. The most important of these is that any cross-section of the bar remains plane when under load. This assumption provides very good solutions except for very short bars or ones which have a high degree of initial curvature.

12.27 Compound column

To illustrate the use of the simple tension/compression formulae we shall consider a compound column as shown in Fig. 12.20.

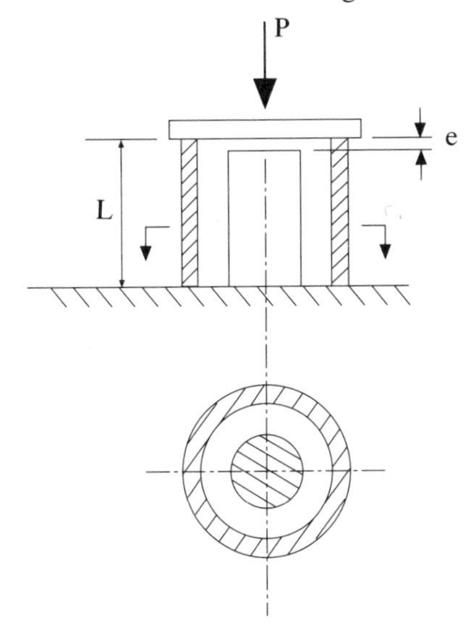

Figure 12.20

We assume that a light, rigid plate is resting on top of a tube which is concentric with a solid rod. The rod is slightly shorter than the tube by an amount e which is very small compared with the length L.

The problem is to find the stresses in the component parts when the plate is axially loaded with a sufficiently large compressive force that the gap is closed and the rod further compressed.

The solution is to consider equilibrium, compatibility and the elastic relationship.

Equilibrium of the plate is considered with reference to the free body diagram depicted in Fig. 12.21 where P is the applied load and P_R and P_T are the compressive forces in the rod and the tube respectively.

$$P - P_R - P_T = 0 \qquad (12.47)$$

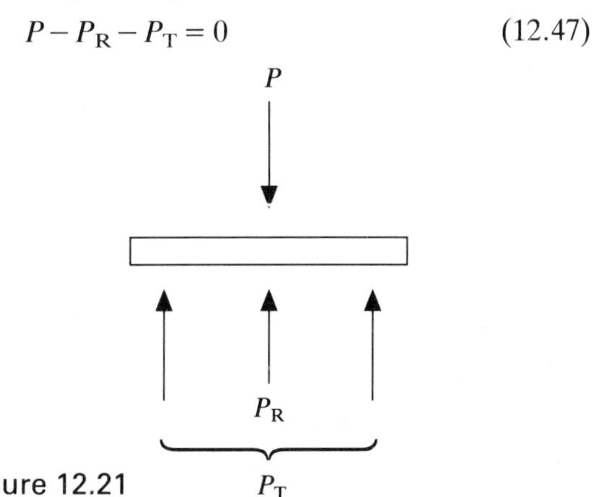

Figure 12.21

The compatibility condition is that the final length of the tube shall be the same as that of the rod. So with reference to Fig. 12.22 we see that the compression of the tube is equal to the initial lack of fit plus the compression of the rod.

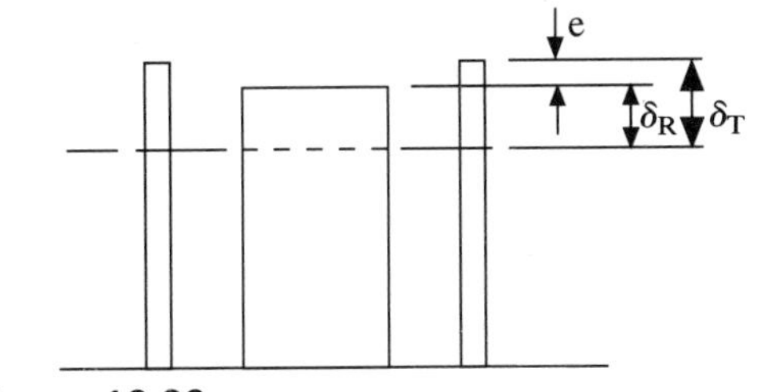

Figure 12.22

$$\delta_T = e + \delta_R \qquad (12.48)$$

The application of Hooke's law to the tube and rod in turn gives

$$-P_T/A_T = -E_T(\delta_T/L) \qquad (12.49)$$

and $$-P_R/A_R = -E_R(\delta_R/L) \qquad (12.50)$$

Substituting these last three equations into equation 12.47 gives

$$P - \delta_T E_T A_T / L - (\delta_T - e) E_R A_R / L = 0$$

or
$$\delta_T = \frac{LP - eE_R A_R}{E_T A_T + E_R A_R} \qquad (12.51)$$

and from 12.48

$$\delta_R = \delta_T - e \qquad (12.52)$$

From equations 12.49 and 12.50 the forces in each component can be found and hence the stresses.

12.28 Torsion of circular cross-section shafts

As an example of the use of shear stress and strain we now develop the standard formulae for describing the torsion of a uniform circular cross-section shaft. Other forms of cross-section lead to more difficult solutions and will not be covered in this book.

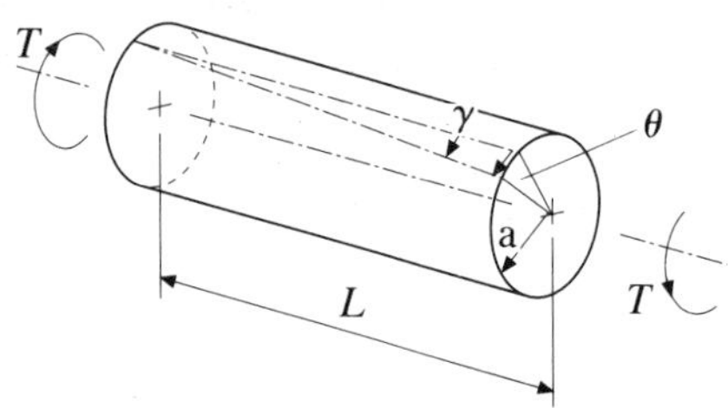

Figure 12.23

Figure 12.23 shows a length of shaft, radius a and length L, under the action of a couple in a plane normal to the shaft axis. This couple is known as the torque.

The following assumptions are to be made

a) the material is elastic,
b) plane cross-sections remain plane and
c) the shear strain varies linearly with radius.

From Fig. 12.23 and the definition of shear strain the shear strain at the surface $\gamma_a = a\theta/L$ and the shear stress at the surface $\tau_a = G\gamma_a = Ga\theta/L$. Therefore at a radius r

$$\tau = Gr\theta/L \qquad (12.53)$$

We can now form an expression for the torque carried by the shaft. Consider an elemental area of cross-section as shown in Fig. 12.24. The elemental shear force is

$$dF = \tau(r\,d\theta\,dr) = (Gr\theta/L)(r\,d\theta\,dr)$$

and the torque due to this is

$$r\,dF = r^3 G(\theta/L)\,d\theta\,dr.$$

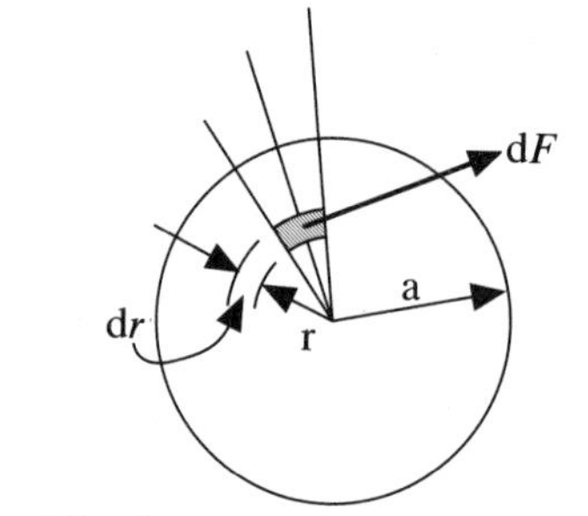

Figure 12.24

For an annulus $d\theta$ is replaced by 2π thus integrating over the radius from 0 to a gives the total torque

$$T = G(\theta/L)\int_0^a r^3 2\pi\,dr = G(\theta/L)\left(\frac{\pi a^4}{2}\right).$$

The integral $\int r^3 2\pi\,dr = \int r^2\,dA$, where dA is the elemental area, is known as the second polar moment of area and the usual symbol is J.

The above expression for torque may now be written

$$T = G(\theta/L)J \qquad (12.54)$$

Combining this with equation 12.53 we have

$$\boxed{\frac{T}{J} = \frac{G\theta}{L} = \frac{\tau}{r}} \qquad (12.55)$$

where

T = torque
J = second polar moment of area
G = shear modulus
θ = angle of twist
L = length of shaft
τ = shear stress
r = radius at which stress is required.

For a hollow shaft with outside radius a and inside radius b the second polar moment of area is $\pi(a^4 - b^4)/2$.

12.29 Shear force and bending moment in beams

In the case of rods, ties and columns the load is axial, and for shafts we considered a couple applied in a plane normal to the axis of the shaft. In the case of beams the loading is transverse to the axis of the beam. In practice the applied loading may well be a combination of the three standard types, in which case for elastic materials undergoing small deflections the effects are simply additive.

A beam is a prismatic bar with its unstrained axis taken to be coincident with the x direction and usually loaded in the y direction. Figure 12.25 shows an element of such a beam.

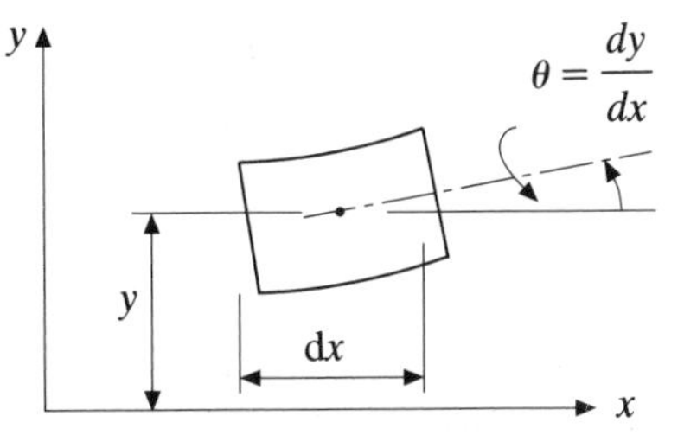

Figure 12.25

It is assumed that the angle that the axis of the beam makes with the x axis is always small. The lateral load intensity is w and is a measure of the load per unit length of the beam. The resultant force acting on the cross-section is expressed as a shear force V and a couple M known as the bending moment. The convention for a positive bending moment is that which gives rise to a positive curvature: concave upwards. Note that this does not follow a vector sign convention since the moments at the ends of the element are of opposite signs.

Figure 12.26 shows the free body diagram for the element, note that the x dimension has been exaggerated.

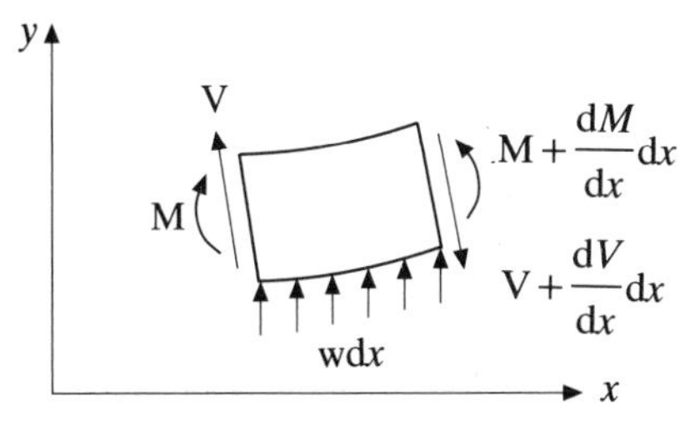

Figure 12.26

We resolve forces in the y direction and equate to zero since in this analysis dynamic effects are not to be included.

So $$w\,\mathrm{d}x + V - \left(V + \frac{\mathrm{d}V}{\mathrm{d}x}\right) = 0$$

leads to $$\frac{\mathrm{d}V}{\mathrm{d}x} = w \qquad (12.56)$$

By taking moments about the centre of the element and again equating to zero

$$M + \frac{\mathrm{d}M}{\mathrm{d}x}\mathrm{d}x - M - V\,\mathrm{d}x/2 - \left(V + \frac{\mathrm{d}v}{\mathrm{d}x}\mathrm{d}x\right)\frac{\mathrm{d}x}{2} = 0$$

or $$\frac{\mathrm{d}M}{\mathrm{d}x} = V \qquad (12.57)$$

Substituting equation 12.57 into 12.56 gives

$$\frac{\mathrm{d}^2M}{\mathrm{d}x^2} = \frac{\mathrm{d}V}{\mathrm{d}x} = w \qquad (12.58)$$

If the loading $w(x)$ is a given function of x, then by integration

$$V = \int w\,\mathrm{d}x \qquad (12.59)$$

and $$M = \iint w\,\mathrm{d}x\,\mathrm{d}x = \int V\,\mathrm{d}x \qquad (12.60)$$

However in the majority of practical problems the loading is not of a continuous nature but frequently consists of loads concentrated at discrete points. In these cases it is often advantageous to use a graphical or semi-graphical method. These methods are especially useful when only maximum values of shear force and bending moment are required.

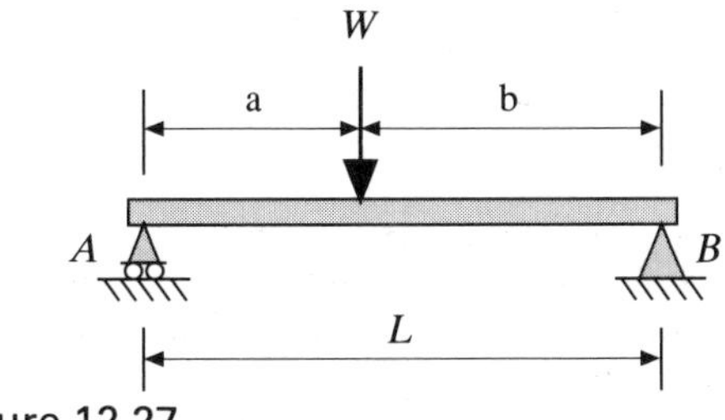

Figure 12.27

As an example of the use of graphical techniques we will consider the case of a simply supported beam as shown in Fig. 12.27.

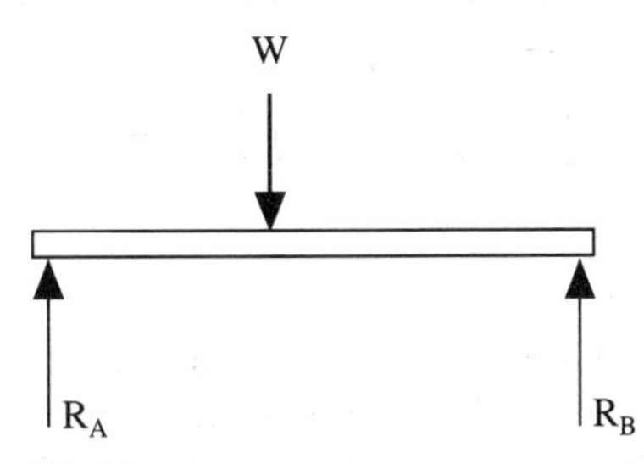

Figure 12.28

The free body diagram for the beam is given in Fig. 12.28 from which, resolving in the y direction,

$$R_A + R_B - W = 0$$

and by moments about A (anticlockwise positive)

$$R_B L - Wa = 0$$

therefore $R_A = Wb/L$ and $R_B = Wa/L$

Figure 12.29 is the shear force and bending moment diagram for the beam and is constructed in the following way.

The shear force just to the right of A is positive and equal to R_A. The value remains constant until the concentrated load W is reached, the shear force is now reduced by W to $R_A - W$ which

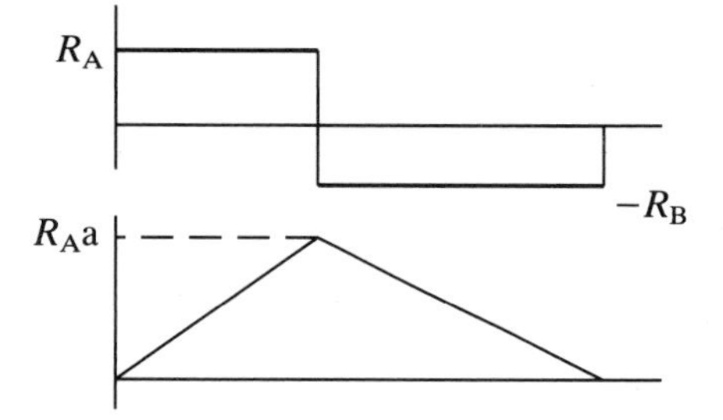

Figure 12.29

is equal to $-R_B$. This value remains constant until reduced to zero by the reaction of point B.

The bending moment is found by integrating the shear force which is, of course, just the area under the shear force diagram. Since the shear force is constant between A and C it follows that the bending moment will be linear. Because point A is a pin joint the bending moment is, by definition, zero. The rest of the diagram can be constructed by continuing the integration or by starting from end B. The maximum bending moment is

$$R_A a = -R_B(-b) = Wab/L$$

12.30 Stress and strain distribution within the beam

Consider the element of the beam, shown in Fig. 12.30, under the action of a pure bending moment (i.e. no shear force). The beam cross-section is symmetrical about the yy axis and its area is A. It is clear that the upper fibres will be in compression and the lower fibres will be in tension, so there must be a layer of fibres which are unstrained. This is called the neutral layer and the z axis is defined to run through this layer.

We shall now assume that plane cross-sections remain plane so that the strain in a layer y from the neutral layer (which retains its original length)

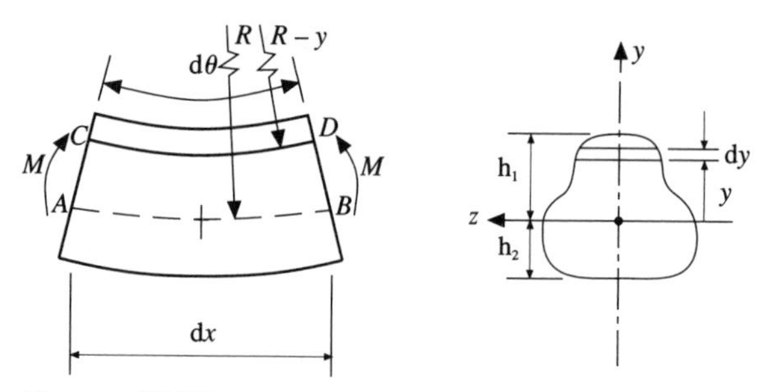

Figure 12.30

can be expressed as

$$\varepsilon = \frac{CD - AB}{AB} = \frac{(R-y)\,d\theta - R\,d\theta}{R\,d\theta} = -y/R$$

therefore the stress $\sigma = E\varepsilon = -Ey/R$ (see Fig. 12.31)

$$\text{or} \quad -\frac{\sigma}{y} = \frac{E}{R} \tag{12.61}$$

where R is the radius of curvature of the beam. Note that in many texts, due to a choice of different sign convention, the above equation appears without the minus sign.

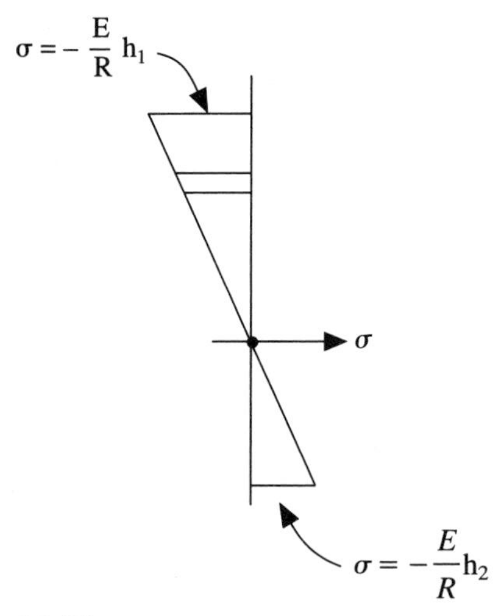

Figure 12.31

The resultant load acting on the section normal to the surface is

$$\int_{y=-h_1}^{y=h_2} \sigma b\,dy = -\int_{y=-h_1}^{y=h_2} \frac{Eyb}{R}\,dy$$

$$= -\frac{E}{R}\int_{y=-h_1}^{y=h_2} yb\,dy$$

Since this must equate to zero as a pure couple has been applied

$$\int_{y=-h_1}^{y=h_2} yb\,\mathrm{d}y = 0$$

This is the first moment of area so by definition the centroid of the cross-section area lies in the neutral layer.

If we now take moments about the z axis we obtain an expression for the bending moment

$$M = -\int_{y=-h_2}^{y=h_1} y\sigma b\,\mathrm{d}y = \frac{E}{R}\int_{y=-h_2}^{y=h_1} y^2 b\,\mathrm{d}y \qquad (12.62)$$

The integral $\int y^2 b\,\mathrm{d}y = \int y^2 \mathrm{d}A$ is known as the second moment of area and denoted by I. Similar to moment of inertia, the second moment of area is often written as $I = Ak^2$ where A is the cross-section area and k is known as the radius of gyration.

The parallel axes theorem relates the second moment of area about an arbitrary axis to that about an axis through the centroid, by the formula

$$I_{zz} = I_{GG} + Ah^2 \qquad (12.63)$$

where h is the distance between the xx and the GG axes.

The perpendicular axes theorem states that for a lamina in the yz plane

$$I_{xx} = I_{yy} = I_{zz} \qquad (12.64)$$

The proofs of these two theorems are similar to those given for moments of inertia in section 6.3.

Using the definition of second moment of area equation 12.62 becomes

$$M = \frac{E}{R} I$$

or

$$\frac{M}{I} = \frac{E}{R} \qquad (12.65)$$

and combining this with equation 12.61 we obtain

$$\boxed{-\frac{\sigma}{y} = \frac{M}{I} = \frac{E}{R}} \qquad (12.66)$$

Where σ is the stress at a fibre at a distance y from the neutral layer, M is the bending moment, I is the second moment of area, E is Young's modulus and R is the radius of curvature produced in the beam.

Equation 12.66 is sometimes referred to as the engineer's theory of bending, and is widely used even for cases where the shear force is not zero as the effect of shear has little effect on the stresses as defined above. However the bending does have a significant effect on the distribution of shear stress over the cross-section.

12.31 Deflection of beams

The governing equation for beam deflection is

$$\frac{M}{I} = \frac{E}{R}$$

For small slope (i.e. $\mathrm{d}y/\mathrm{d}x \ll 1$) the curvature

$$\frac{1}{R} = \frac{\mathrm{d}^2 y}{\mathrm{d}x^2}$$

so

$$\frac{M}{EI} = \frac{\mathrm{d}^2 y}{\mathrm{d}x^2} \qquad (12.67)$$

Integrating with respect to x we have

$$\frac{\mathrm{d}y}{\mathrm{d}x} = \int \frac{M}{EI}\mathrm{d}x \qquad (12.68)$$

and

$$y = \iint \frac{M}{EI}\mathrm{d}x\,\mathrm{d}x. \qquad (12.69)$$

As an example of calculating the deflection of a beam we will consider the cantilever shown in Fig. 12.32. The loading is uniformly distributed with an intensity of w.

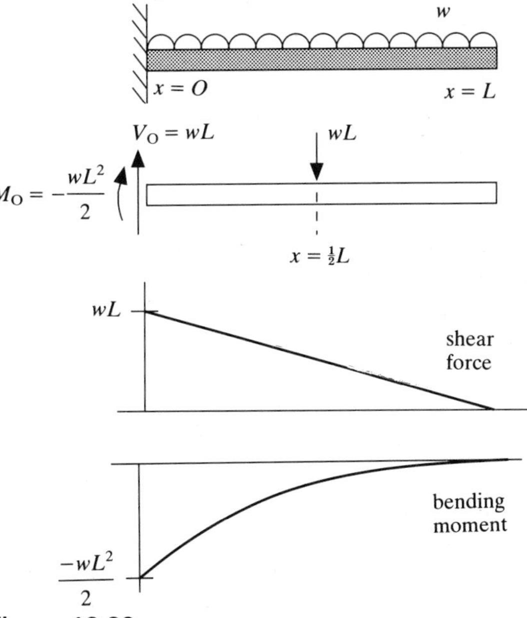

Figure 12.32

From the free body diagram the shear and bending at the fixed end are found to be wL and $-wL^2/2$ respectively. We now use equations 12.59 and 12.60 to evaluate the shear force and bending moment as functions of x.

$$V = \int (-w)\,\mathrm{d}x + \text{constant} = -wx + wL$$

$$M = \int (-wx + wL)\,\mathrm{d}x + \text{constant}$$

$$= -\frac{wx^2}{2} + wLx + (-wL^2/2).$$

Now using equations 12.68 and 12.69

$$\frac{\mathrm{d}y}{\mathrm{d}x} = \frac{1}{EI}\int\left(-\frac{wx^2}{2} + wxL - wL^2/2\right)\mathrm{d}x$$

$$+ \text{constant}$$

$$= \frac{1}{EI}\left(-wx^3/6 + wx^2L/2 - wL^2x/2 + 0\right)$$

the constant is zero since the slope is zero at the fixed end.

$$y = \frac{1}{EI}\int(-wx^3/6 + wx^2L/2 - wL^2x/2)\mathrm{d}x$$

$$+ \text{constant}$$

$$= \frac{1}{EI}\left(-wx^4/24 + wx^3L/6 - wL^2x^2/4 + 0\right).$$

The maximum deflection clearly is at the right-hand end of the beam where $x = L$

$$y_{\max} = \frac{1}{EI}wL^4(-1/24 + 1/6 - 1/4)$$

$$= -\frac{wL^4}{8EI}$$

12.32 Area moment method

The double integration of $M/(EI)$ can be performed in a semi-graphical way by a technique known as the area moment method. Integrating equation 12.67 between the limits x_1 and x_2 gives

$$\frac{\mathrm{d}y}{\mathrm{d}x_2} - \frac{\mathrm{d}y}{\mathrm{d}x_1} = \theta_2 - \theta_1 = \int_{x_1}^{x_2}\frac{M}{EI}\mathrm{d}x \qquad (12.70)$$

or $\theta_2 - \theta_1 =$ area under the $M/(EI)$ diagram as shown in Fig. 12.33.

Now by definition

$$\frac{\mathrm{d}y}{\mathrm{d}x} = \theta$$

so integrating between limits

$$y_2 - y_1 = \int_{x_1}^{x_2}\theta\,\mathrm{d}x$$

and integrating by parts we obtain

$$y_2 - y_1 = \theta x\Big|_{x_1}^{x_2} - \int_{x_1}^{x_2} x\frac{\mathrm{d}\theta}{\mathrm{d}x}\mathrm{d}x$$

We know that

$$\frac{1}{R} = \frac{\mathrm{d}}{\mathrm{d}x}\left(\frac{\mathrm{d}y}{\mathrm{d}x}\right) = \frac{\mathrm{d}\theta}{\mathrm{d}x} = \frac{M}{EI}$$

and by choosing x_1 as the origin we may write

$$y_2 - y_1 = \theta_2(x_2 - x_1) - \int_0^{(x_2-x_1)} x\frac{M}{EI}\mathrm{d}x \qquad (12.71)$$

The interpretation of the last equation can be seen in Figs 12.33 and 12.34. The difference in deflection between positions 1 and 2, relative to the tangent at point 2, is the moment of the area under the M/EI diagram, between points 1 and 2, about the point 1.

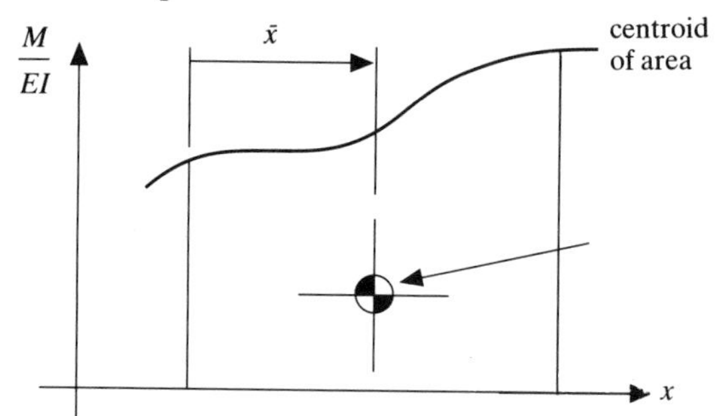

Figure 12.33

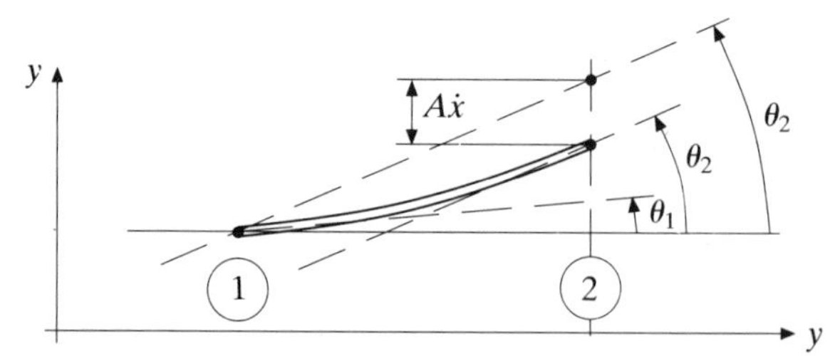

Figure 12.34

As a simple example of the use of the area moment method we will consider the case of a cantilever, length L, with a concentrated load at a distance a from the fixed end, as shown in Fig.

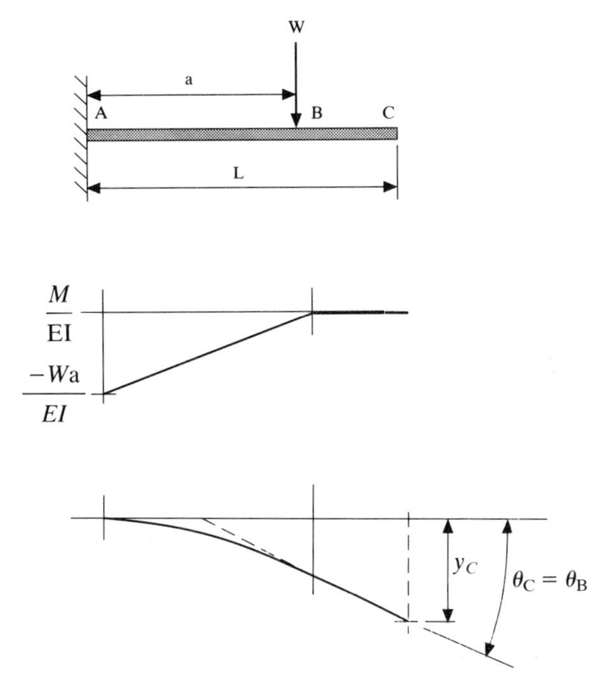

Figure 12.35

12.35. We wish to find the slope and deflection at the free end.

The first step is to sketch the bending moment diagram which is linear from B to A and has a maximum value of $-Wa$ at A.

The change in slope between A and C is the area of the M/EI diagram thus

$$\theta_c - 0 = -\frac{Wa}{EI}\frac{a}{2}, \text{ since } \theta_a = 0.$$

Applying equation 12.71

$$y_c - y_a = \theta_a x_a - \left(-\frac{Wa}{EI}\frac{a}{2}\right)\left(-(L - a/3)\right).$$

As both y_a and θ_a are zero

$$y_c = -\frac{Wa^2(L - a/3)}{2EI}$$

Discussion examples

Example 12.1

A circular cross-section rod, made from steel, has a length L and tapers linearly from a diameter D_2 at one end to a diameter of D_1 at the other. The rod is subjected to a constant tension F.

Assuming that the taper is slight so that the stress distribution across the cross-section is uniform, derive an expression for the change in length of the rod.

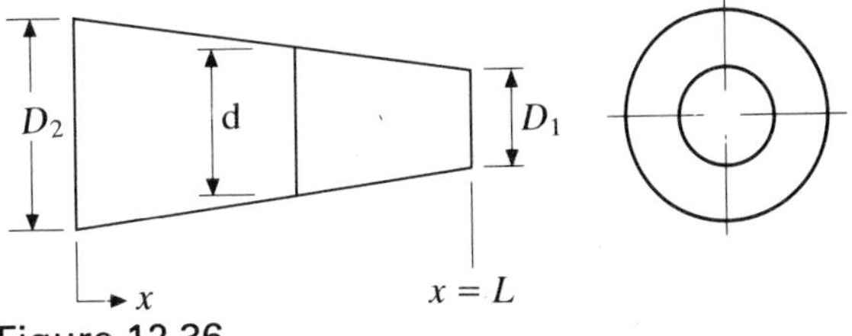

Figure 12.36

Solution From Fig. 12.36 we see that the diameter at a position x is

$$d = D_2 - \frac{(D_2 - D_1)x}{L}$$

and the cross-section area $A = \pi d^2/4$. The stress $\sigma = F/A$ and the strain

$$\varepsilon = \sigma/E = \frac{4F}{E\pi(D_2 - (D_2 - D_1)x/L)^2}$$

Now $\quad \varepsilon = \dfrac{\partial u}{\partial x} = \dfrac{du}{dx}$

so $\quad u = \displaystyle\int_0^L \frac{du}{dx}dx = -\frac{4F}{E\pi}\int_0^L \frac{dx}{(a - bx)^2}$

$$= \frac{F}{EA/\pi}\frac{1}{b(a - bx)}\Bigg|_0^L$$

where $a = D_2$ and $b = (D_2 - D_1)/L$

hence $\quad u = \dfrac{4F}{E\pi}\left[\dfrac{1}{b(a - bL)} - \dfrac{1}{ba}\right]$

$$= \frac{4FL}{\pi EA(D_2 - D_1)}\left[\frac{(D_2 - D_1)}{D_1 D_2}\right]$$

$$= \frac{4FL}{E\pi D_1 D_2}$$

Example 12.2

A load washer is a device which responds to a compressive load, producing an electrical output proportional to the applied force. In order to make a load cell capable of registering both compression and tension it is precompressed by a bolt as shown on Fig. 12.37. The stiffness of the load washer is k and the bolt is made out of a

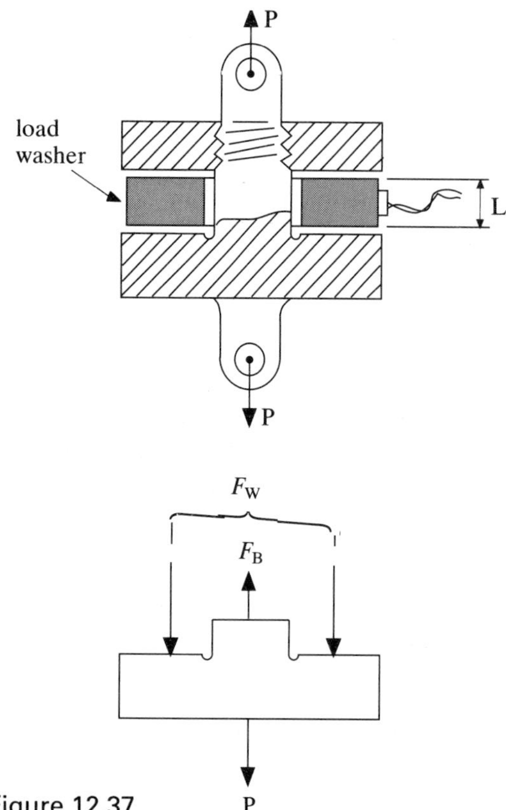

Figure 12.37

material whose Young's modulus is E.

a) If the lead of the thread on the bolt is λ determine the compressive load on the load washer if the nut is tightened by n turns.

b) Also find the change in load on the washer as a fraction of the change in load on the load cell for a pre-tightening of n turns.

Solution

a) We will assume that the head of the bolt and the region covered by the nut have negligible distortion. In this case there is an initial lack of fit of $n\lambda$, which means that in the assembled state the stretch in the bolt plus the compression of the washer must equal $n\lambda$. Also, by equilibrium, the tensile force in the bolt must equal the compressive force in the washer.

If the compressive force in the washer is F_W and the tensile force in the bolt is F_B then

$$n\lambda = F_W/k + F_B L/(AE) \qquad \text{(i)}$$

since $F_W = F_B$

$$F_W = \frac{n\lambda}{1/k + L/(AE)}$$

b) From the free-body diagram

$$P = F_B - F_W$$

substituting in (i)

$$n\lambda = F_W/k + (P + F_W)L/(AE)$$

gives $$F_W = \frac{n\lambda k - PkL/(AE)}{1 + kL/(AE)}$$

thus $$\frac{\Delta F_W}{\Delta P} = -\frac{k}{AE/L + k}.$$

The sensitivity of the load cell will be that of the load washer reduced by the ratio $k/(AE/L + k)$.

Note that the above equations are only valid whilst

$$P < n\lambda AE/L$$

Example 12.3

A flat steel plate with dimensions a, b, c in the x, y, z directions is under the action of a uniform stress in the x direction only, see Fig. 12.38. Show from first principles that if Poisson's ratio is 0.29 then the longitudinal strain is zero when $\theta = 61.7°$.

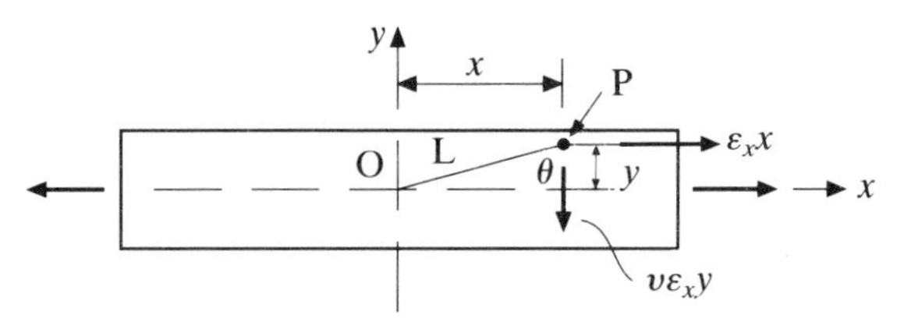

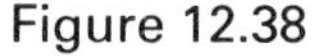
Figure 12.38

By drawing Mohr's circles for stress and for strain confirm the previous result. Show also that the maximum shear stress and the maximum shear strain occur when $\theta = 45°$. By further consideration of the diagrams at $\theta = 45°$, prove that $E = 2G(1 + \nu)$.

Solution In Fig. 12.38 the point P is situated a distance L from the origin of the axes, the line OP being at an angle θ to the x axis. The relative movement in the x direction is $\varepsilon_x x$ and in the y direction it is $-\nu\varepsilon_x y$. Resolving along OP

$$\varepsilon_x x \cos\theta - \nu\varepsilon_x y \sin\theta = \varepsilon_L L$$

where ε_L is, by definition, the strain in the direction of L. Now $x = L\cos\theta$ and $y = L\sin\theta$ so

$$\varepsilon_L = \varepsilon_x[\cos^2\theta - \nu\sin^2\theta].$$

For $\varepsilon_L = 0$

$$\tan^2\theta = 1/\nu = 1/0.29$$

therefore $\quad \theta = 61.7°$.

By measurement on the Mohr's circle for strain, Fig. 12.39, zero normal strain occurs at approximately 62°.

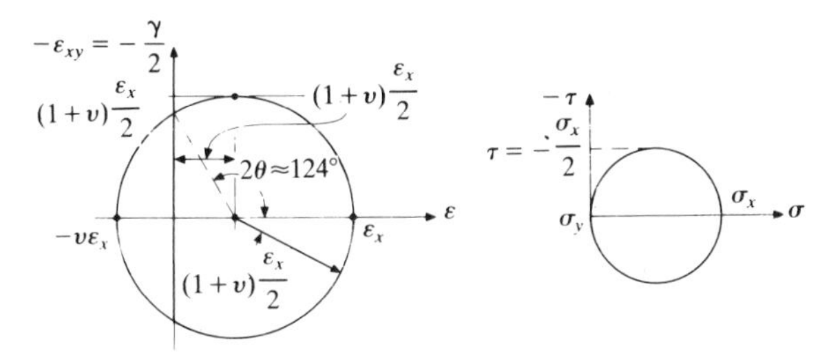

Figure 12.39

From the diagrams it is readily seen that the maximum shear stress occurs when $2\theta = 90°$ and has a value of $\sigma_x/2$. The maximum shear strain also occurs at $\theta = 45°$ and has a value of $(1+\nu)\,\varepsilon_x$.

From the values just quoted

$$G = \frac{\tau}{\gamma} = \frac{\sigma_x/2}{(1+\nu)\,\varepsilon_x} = \frac{\sigma_x}{\varepsilon_x}\frac{1}{2(1+\nu)} = E\frac{1}{2(1+\nu)}$$

or $\quad E = 2G(1+\nu)$

Example 12.4

Using the same data as in example 12.3 evaluate the values of the shear modulus and the bulk modulus, given that Young's modulus is 200 GN/m^2 and Poisson's ratio is 0.29.

Solution From example 12.3

$$G = \frac{E}{2(1+\nu)} = \frac{200\times10^9}{2(1+0.29)} = 77.5\ \text{GN/m}^2$$

The change in volume

$$\Delta V = a\varepsilon_x bc - b\nu\varepsilon_x ac - c\nu\varepsilon_x ab$$
$$= abc\varepsilon_x(1-2\nu)$$

thus the volumetric strain

$$\Delta V/V = \varepsilon_x(1-2\nu)$$

The mean pressure $-p = -$(average stress) $= -\sigma_x/3$ so by definition the bulk modulus

$$K = \frac{-p}{\Delta V/V} = \frac{\sigma_x/3}{\varepsilon_x(1-2\nu)} = \frac{E}{3(1-2\nu)}$$
$$= \frac{200\times10^9}{3(1-2\times0.29)} = 158.7\ \text{GN/m}^2$$

(Because all elastic moduli must be positive, it is clear from the general expression for K that $\nu<0.5$.)

Example 12.5

Obtain an expression for the strain energy per unit volume for an isotropic homogeneous material in terms of its principal stresses. Find also an expression for the strain energy associated with the change in volume, and hence find an expression for strain energy associated with distortion.

Solution The total strain energy per unit volume can be found from the work done by the normal forces acting on the surface of a unit cube.

The work done equals the total strain energy

$$U_t = \frac{\sigma_1}{2}\left[\frac{\sigma_1}{E} - \nu\frac{\sigma_2}{E} - \nu\frac{\sigma_3}{E}\right]$$
$$+\frac{\sigma_2}{2}\left[\frac{\sigma_2}{E} - \nu\frac{\sigma_3}{E} - \nu\frac{\sigma_1}{E}\right]$$
$$+\frac{\sigma_3}{2}\left[\frac{\sigma_3}{E} - \nu\frac{\sigma_2}{E} - \nu\frac{\sigma_1}{E}\right]$$
$$= \frac{1}{2E}[\sigma_1{}^2 + \sigma_2{}^2 + \sigma_3{}^2$$
$$-2\nu(\sigma_2\sigma_3 + \sigma_3\sigma_1 + \sigma_1\sigma_2)]$$

The work done in changing the volume of the unit cube is

$$U_v = \frac{1}{2}\left(-p\right)\frac{\Delta V}{V} = \frac{1}{2}\frac{p^2}{K}$$

In the general case $-p =$ the average stress $= (\sigma_1 + \sigma_2 + \sigma_3)/3$

$$U_v = \frac{1}{2}\frac{(\sigma_1+\sigma_2+\sigma_3)^2}{9K}$$
$$= \frac{1}{2\times9K}[\sigma_1{}^2 + \sigma_2{}^2 + \sigma_3{}^2$$
$$+2(\sigma_2\sigma_3 + \sigma_3\sigma_1 + \sigma_1\sigma_2)]$$

The total strain energy is the sum of the volumetric strain energy and the distortional or shear strain energy, i.e.

$$U_t = U_v + U_s$$

therefore

$$U_s = U_t - U_v$$

$$= \frac{1}{2}\left[\left(\frac{1}{E} - \frac{1}{9K}\right)(\sigma_1^2 + \sigma_2^2 + \sigma_3^2) - \left(\frac{2\nu}{E} + \frac{2}{9K}\right)(\sigma_1\sigma_2 + \sigma_2\sigma_3 + \sigma_3\sigma_1)\right]$$

Now $E = 2G(1+\nu)$ and $K = \dfrac{E}{3(1-2\nu)}$

so $$K = \frac{2G(1+\nu)}{3(1-2\nu)}.$$

Substituting these values into the previous equation

$$U_s = \frac{1}{4G(1+\nu)}\frac{2(1+\nu)}{3}[\sigma_1^2 + \sigma_2^2 + \sigma_3^2 - (\sigma_1\sigma_2 + \sigma_2\sigma_3 + \sigma_3\sigma_1)]$$

$$= \frac{1}{2G}\frac{1}{6}[(\sigma_1 - \sigma_2)^2 + (\sigma_2 - \sigma_3)^2 + (\sigma_3 - \sigma_1)^2]$$

Example 12.6

The von Mises–Hencky theory of failure for ductile materials suggests that, under the action of multi-axial stresses, failure will occur when the maximum shear strain energy is equal to that which occurs when failure occurs in a simple tensile test.

Show that this leads to the formula

$$(\sigma_1 - \sigma_2)^2 + (\sigma_2 - \sigma_3)^2 + (\sigma_3 - \sigma_1)^2 = 2\sigma_y^2$$

where σ_y = the yield stress.

Solution From example 12.4 the expression for shear strain energy is

$$U_s = \frac{1}{2G}\frac{1}{6}[(\sigma_1 - \sigma_2)^2 + (\sigma_2 - \sigma_3)^2 + (\sigma_3 - \sigma_1)^2]$$

In a simple tensile test $\sigma_2 = \sigma_3 = 0$, so

$$U_s = \frac{1}{12G}[2\sigma_1^2] = \frac{1}{6G}\sigma_1^2 \quad \text{where } \sigma_1 = \sigma_y$$

therefore

$$(\sigma_1 - \sigma_2)^2 + (\sigma_2 - \sigma_3)^2 + (\sigma_3 - \sigma_1)^2 = 2\sigma_y^2$$

It is interesting to note that the shear strain energy in the simple tensile test is greater than that calculated from the maximum shear stress and strain which occur at 45° to the bar axis. The strain energy based on this calculation is

$$U = \frac{1}{2G}[\sigma_1/2]^2 = \frac{1}{8G}\sigma_1^2$$

The difference is due to the fact that in this situation the element is still being distorted, because the transverse stress is still zero even though the other two normal stresses are equal to each other.

Example 12.7

A thin-walled cylindrical pressure vessel shown in Fig. 12.40 contains a gas at pressure p. The diameter of the shell is 0.6 m and the thickness of the steel is 6 mm. Given that the yield stress of steel is 300 MN/m^2 find the maximum allowable pressure based on the stresses on the curved surfaces remote from the ends.

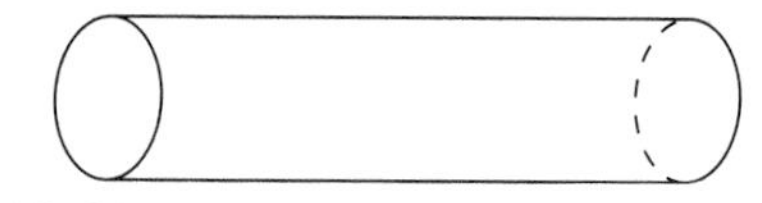

Figure 12.40

Solution The material is under the action of a longitudinal stress and a hoop stress as shown in Fig. 12.41. We shall consider the equilibrium of a unit length of the cylinder. Resolving in the y direction

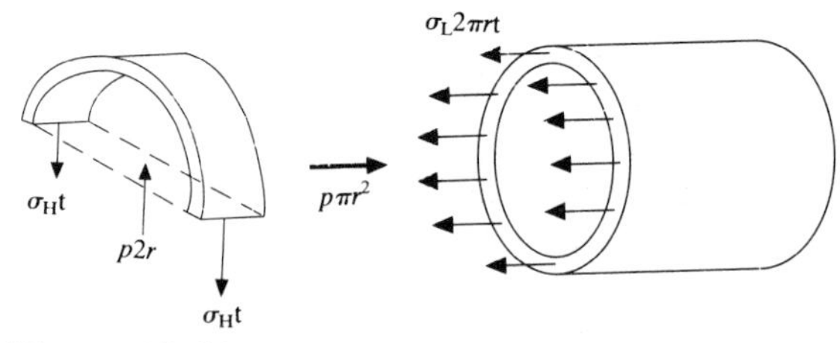

Figure 12.41

$$p2r - 2t\sigma_H = 0$$

where σ_H is known as the hoop stress therefore

$$\sigma_H = \frac{pr}{t}$$

Considering the equilibrium of one part of the cylinder

$$p\pi r^2 - \sigma_L 2\pi rt = 0$$

so the longitudinal stress

$$\sigma_L = \frac{pr}{2t}$$

As there is no shear on the element σ_H and σ_L are by definition the principal stresses σ_1 and σ_2. The third principal stress is taken to be zero, because with $r \gg t$ it follows that σ_1 and σ_2 are much greater than p. Substitution into the von Mises–Hencky equation,

$$\left(\frac{pr}{2t}\right)^2 [(2-1)^2 + (1-0)^2 + (0-2)^2] = 2\sigma_y{}^2$$

or $$\left(\frac{pr}{2t}\right)^2 (6) = 2\sigma_y{}^2$$

giving

$$p = \frac{2\sigma_y t}{\sqrt{3}r} = \frac{2 \times 300 \times 10^6 \times 0.006}{\sqrt{3} \times 0.300} = 6.9 \text{ MPa}$$

The corresponding hoop and longitudinal stresses are

$$\sigma_L = \frac{6.9 \times 10^6 \times 0.300}{2 \times 0.006} = 173 \text{ MPa}$$

$$\sigma_H = \frac{6.9 \times 10^6 \times 0.300}{0.006} = 346 \text{ MPa}$$

Note that $\sigma_H > \sigma_y$. This is because a high proportion of the strain energy is associated with a change in volume of the material while it is the shear strain energy which relates to failure.

Example 12.8

Figure 12.42 shows a simply supported beam carrying a uniformly distributed load of intensity w. The beam is made from a material with a Young's modulus E and the second moment of area of the beam cross-section is I.

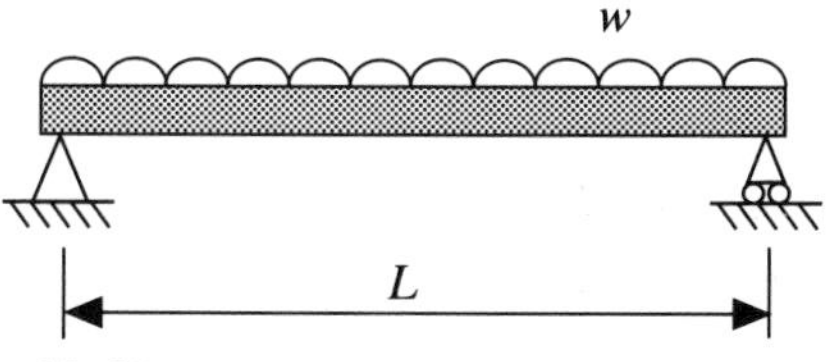

Figure 12.42

Obtain an expression for the shear force, bending moment and deflection of the beam. Use the area-moment method to check the expression for the deflection at the centre.

Solution The total load carried by the beam is wL and by symmetry the reactions at the supports will be $wL/2$. The shear force at $x = 0$ is therefore $+wL/2$. A sketch of the shear force diagram and of the bending moment diagram is shown in Fig. 12.43.

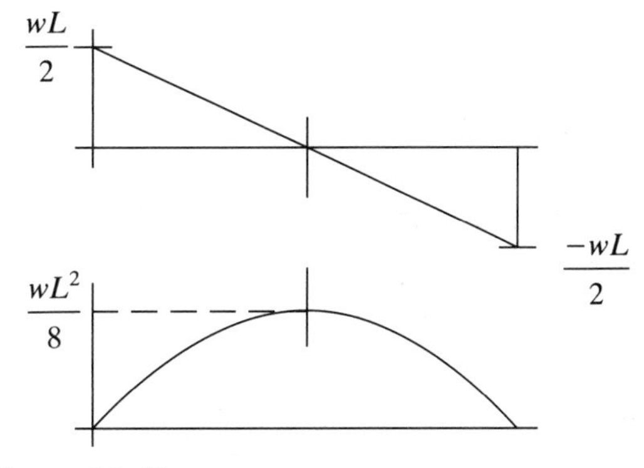

Figure 12.43

$$V = V_0 = \int_0^x (-w)\,dx = \frac{wL}{2} - wx$$

and the bending moment is

$$M = M_0 + \int_0^x V\,dx$$

$$= 0 + \frac{wLx}{2} - \frac{wx^2}{2}$$

The slope of the beam

$$\frac{dy}{dx} = \frac{dy}{dx_0} + \int_0^x \frac{M}{EI}\,dx$$

$$= \frac{dy}{dx_0} + \frac{1}{EI}\left[\frac{wLx^2}{4} - \frac{wx^3}{6}\right]$$

and the deflection

$$y = y_0 + \int_0^x \frac{dy}{dx}\,dx$$

$$= 0 + \frac{dy}{dx_0}x + \frac{1}{EI}\left[\frac{wLx^3}{12} - \frac{wx^4}{24}\right]$$

Now at $x = L$ the deflection is zero, so

$$0 = \frac{dy}{dx_0}L + \frac{1}{EI}\left[\frac{wL^4}{12} - \frac{wL^4}{24}\right]$$

therefore

$$\frac{dy}{dx_0} = -\frac{wL^3}{24EI}.$$

Finally

$$y = -\frac{wL^4}{24EI}\left[\left(\frac{x}{L}\right) - 2\left(\frac{x}{L}\right)^3 + \left(\frac{x}{L}\right)^4\right]$$

The area moment method states that the first moment of the area under the M/EI diagram about point 1 gives the change in deflection of the beam between points 1 and 2 relative to the tangent at point 2. In this case we may make use of the symmetry of the system from which it is seen that the slope at the centre is zero. Therefore the moment of the area between one end and the centre about the pin joint will equal the maximum deflection.

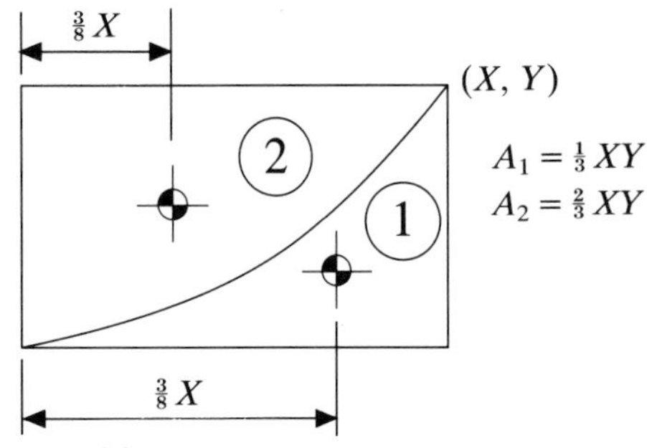

Figure 12.44

Figure 12.44 gives the standard properties of a parabola from which we can write the deflection

$$y_0 - y_{L/2} = -y_{L/2} = \frac{2}{3}\left(\frac{wL^2}{8} \times \frac{L}{2}\right)\left(\frac{5}{8} \times \frac{L}{2}\right)$$

$$= \frac{5wL^4}{384}$$

Example 12.9

Figure 12.45 shows a uniform beam resting on simple supports with a load of 12 kN applied 4 m from A. The second moment of area of the beam cross-section is 0.0004 m^4 and Young's modulus is 200 GN/m^2.

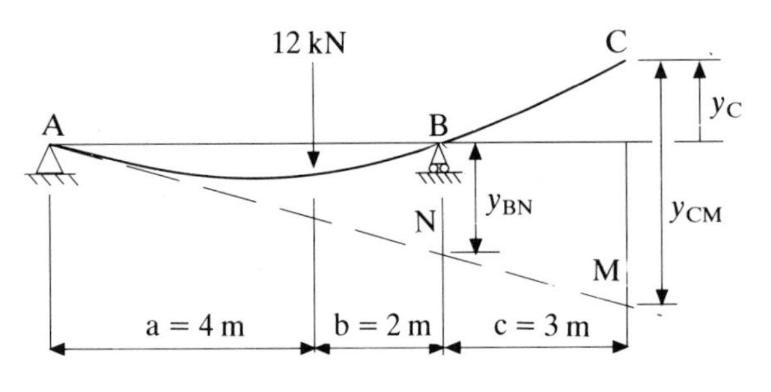

Figure 12.45

Using the area moment method determine the deflection at the point C.

Solution By taking moments about the point A the reaction at point B may be found

$$R_B L - Wa = 0$$

or $R_B = Wa/L$

and by resolving vertically upwards

$$R_A + R_B - W = 0$$

so $R_A = R_B - W = Wb/L$

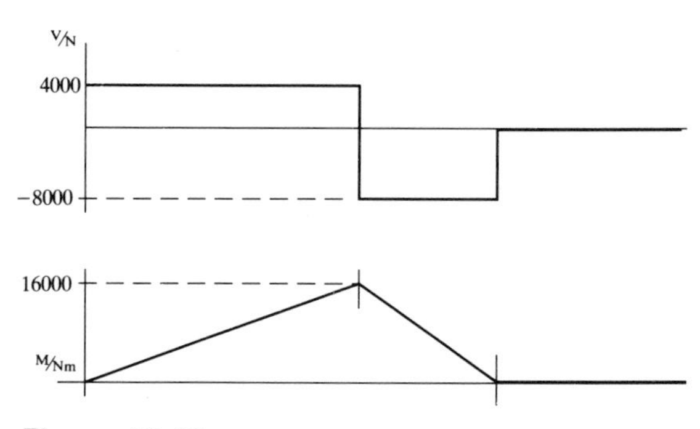

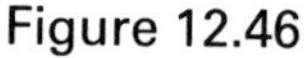

Figure 12.46

The shear force and bending moment diagrams may now be sketched as shown in Fig. 12.46. The peak bending moment is readily shown to be $Wab/L = 12000 \times 4 \times 2/6 = 16000$ Nm.

The application of the area moment method is a little more difficult because we do not know the position for zero slope. We know that the deflection at B is zero, so the moment about B of the area under the M/EI diagram between A and B will give the deflection at B relative to the tangent line at A. This deflection is also seen to be given by the product of the slope at A and the distance between A and B.

$$y_{BN} = \frac{Wab}{EIL}\frac{a}{2}\left[b + \frac{a}{3}\right] + \frac{Wab}{EIL}\frac{b}{2}\frac{2b}{3}$$

$$= \frac{16000}{2 \times 10^{11} \times 0.0004}$$

$$\times (4 \times 2/2 + 4 \times 4/6 + 2 \times 2 \times 2/6)$$

$$= 1.6 \times 10^{-3}\ \text{m}$$

Now $\dfrac{dy}{dx_0} = y_{BN}/L = 1.6 \times 10^{-3}/6$

$$= 0.267 \times 10^{-3}\ \text{rad.}$$

The deflection at C relative to the tangent at A is

$$y_{CM} = \frac{Wab}{EIL}\left[\frac{a}{2}\left(b + a/3 + c\right) + \frac{b}{2}\left(2b/3 + c\right)\right]$$

$$= \frac{16000}{2 \times 10^{11} \times 0.0004}$$

$$\times \left[\frac{4}{2}\left(2 + 4/3 + 3\right) + \frac{2}{2}\left(2 \times 2/3 + 3\right)\right]$$

$$= 3.4 \times 10^{-3}\ \text{m}$$

Thus $y_C = y_{CM} - \dfrac{dy}{dx_0}(a + b + c)$

$$= 3.4 \times 10^{-3} - 0.267 \times 10^{-3}(4 + 2 + 3)$$

$$y_C = 1.00\ \text{mm}$$

Example 12.10

A channel, the cross-section details of which are given in Fig. 12.47(a), is used for the beam shown in Fig. 12.47(b). This is loaded by a force parallel to the beam axis but offset by 0.3 m.

Determine the maximum tensile and compressive stresses in the beam due to bending.

Solution By taking moments about A, from the free-body diagram of Fig. 12.47(c)

$$10000 \times 0.3 - R_C 3 = 0$$

$$R_C = 1000\ \text{N}$$

and resolving vertically upwards

$$R_C + R_A = 0$$

so $$R_A = -1000\ \text{N}.$$

The alternative separate free-body diagrams of Fig. 12.47(d) show that a couple of 3 kNm is applied to the beam at B.

The shear force and bending moment diagrams are shown in Fig. 12.47(e). Note that the shear force is constant along the beam which usually implies a constant slope to the bending moment diagram. However, the couple applied at B causes a step of 3000 Nm into the bending moment diagram.

From the diagram we see that the maximum positive bending moment is 1000 Nm and the largest negative bending moment is −2000 Nm.

We now require to find the position of the neutral axis and the second moment of area about this axis. This will be done by breaking the cross-section shape into two rectangles for which we know the geometric properties. The two

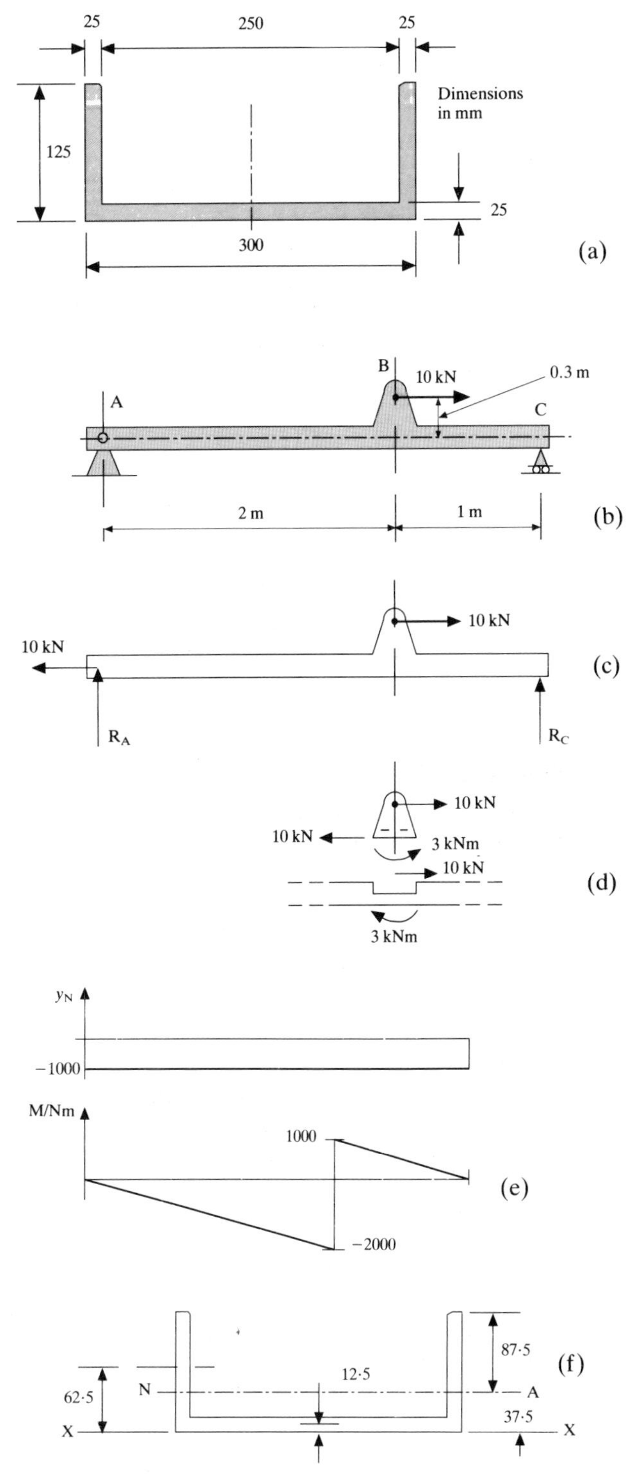

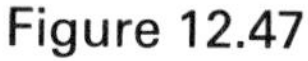

Figure 12.47

	Area A	$\bar{x}$	$A\bar{x}$	I_{xx}
1	$50 \times 125 = 6250$	62.5	390625	$50 \times 125^3/3 = 32.55 \times 10^6$
2	$250 \times 25 = 6250$	12.5	78125	$250 \times 25^3/3 = 1.30 \times 10^6$
Σ	12500		468750	33.85×10^6

flanges will be considered as a single one of twice the width, see Fig. 12.47(f).

In the table below $\bar{x}$ is the position of the centroid of the individual parts from the XX axis. All dimensions are in mm.

The intersection of the neutral layer with a given cross section is known as the neutral axis (NA).

The position of the centroid

$$x_{NA} = \frac{\Sigma A\bar{x}}{\Sigma A} = \frac{468750}{12500} = 37.5 \text{ mm}.$$

Using the parallel axes theorem

$$\begin{aligned} I_{NA} &= I_{xx} - (\Sigma A)x_{NA}^2 \\ &= 33.85 \times 10^6 - 12500 \times 37.5^2 \\ &= 16.25 \times 10^6 \text{ mm}^4. \end{aligned}$$

Just to the right of B the maximum compressive stress occurs in the top layers where $y = 87.5$ mm so

$$\sigma_c = -\frac{M}{I}y = -\frac{1000 \times 10^3}{16.25 \times 10^6} 87.5$$

$$= -5.83 \text{ N/mm}^2 = -5.38 \text{ MN/m}^2$$

and the maximum tensile stress occurs at the bottom fibres where $y = -37.5$ mm so

$$\sigma_t = -\frac{1000 \times 10^3}{16.25 \times 10^6}(-37.5) = 2.31 \text{ MN/m}^2.$$

Just to the left of B the bending moment is twice the magnitude of, but of opposite sign to, that just to the right therefore the bending stresses are

$$\sigma_c = 4.62 \text{ MN/m}^2 \quad \text{in the lower fibre}$$

and $\sigma_t = 10.76 \text{ MN/m}^2$ in the upper fibres.

In the section between A and B there is a tensile force of 10 kN so there is a uniform tensile stress of

$$10000/12500 = 0.8 \text{ N/mm}^2 = 0.8 \text{ MN/m}^2$$

and this has to be added to the bending stresses, so finally

$$\sigma_c = 4.62 - 0.80 = 3.82 \text{ MN/m}^2$$

and $\sigma_t = 10.76 + 0.80 = 11.56 \text{ MN/m}^2$.

Example 12.11

A shaft 0.5 m long is required to transmit 80 kW at 300 rev/min. It is specified that the twist shall not exceed 0.25° and the shear stress is not to be greater than 36 MN/m^2. The shear modulus of the material is 85 GN/m^2.

Determine (a) the minimum diameter of a solid shaft to satisfy the specification; and (b) the inside and outside diameters of a hollow shaft to meet the specification.

What is the weight ratio of the two designs?

Solution The torque to be transmitted is found from

$$\text{power} = \text{torque} \times \text{angular speed}$$

$$\text{torque } T = \frac{80 \times 1000}{300 \times 2\pi/60} = 2546 \text{ Nm}$$

and the allowable twist

$$\theta = 0.25 \times 2\pi/360 = 0.0044 \text{ rad}.$$

a) From

$$\frac{T}{J} = \frac{G\theta}{L} = \frac{\tau}{r}$$

$$\frac{J}{r} = \frac{J}{D/2} = \frac{\pi D^3}{16} > \frac{T}{\tau} = \frac{2546}{36 \times 10^6} = 70.7 \times 10^6$$

or $$D > \left[\frac{16 \times 70.7 \times 10^6}{\pi}\right]^{1/3} = 71.1 \text{ mm}$$

Also $$J > \frac{TL}{G\theta} = \frac{2546 \times 0.5}{85 \times 10^9 \times 0.0044} = 3.4 \times 10^{-6} \text{ m}^4$$

since $J = \pi D^4/32$

$$D > \left[\frac{32 \times 3.4 \times 10^{-6}}{\pi}\right]^{1/4} = 76.7 \text{ mm}$$

From the two calculations it is seen that the twist requirement is more demanding, hence a diameter of 76.7 mm will satisfy both criteria.

Because we are dealing with inequalities some care must be exercised on the choice of equations. We must not choose expressions where both twist and stress appear together, since both criteria cannot be satisfied simultaneously.

b) With two unknown diameters to be determined it is possible to satisfy both criteria simultaneously. Denoting the outer diameter by D and the inner by d we may write

$$r = \frac{D}{2} = \frac{\tau L}{G\theta} = \frac{36 \times 10^6 \times 0.5}{85 \times 10^9 \times 0.0044} = 48.1 \text{ mm}$$

so $D = 96.2$ mm

Also

$$J = \frac{TL}{G\theta} = \frac{2564 \times 0.5}{85 \times 10^9 \times 0.0044} = 3.43 \times 10^{-6} \text{ m}^4$$

$$J = \frac{\pi(D^4 - d^4)}{32} = 3.43 \times 10^{-6}$$

$$D^4 - d^4 = \frac{32 \times 3.43 \times 10^{-6}}{\pi} = 34.90 \times 10^{-6}$$

$$d = [0.0962^4 - 34.90 \times 10^{-6}]^{1/4}$$
$$= 84.4 \text{ mm}$$

The ratio of the weights of the two designs is the same as the ratio of the cross-section areas

$$\frac{76.7^2}{96.2^2 - 84.4^2} = 2.76$$

Example 12.12

A close-coiled helical spring, shown in Fig. 12.48, is loaded axially. The mean diameter of the coil is D and the diameter of the wire is d. The number of turns of the helix is N and G is the modulus of rigidity of the wire material.

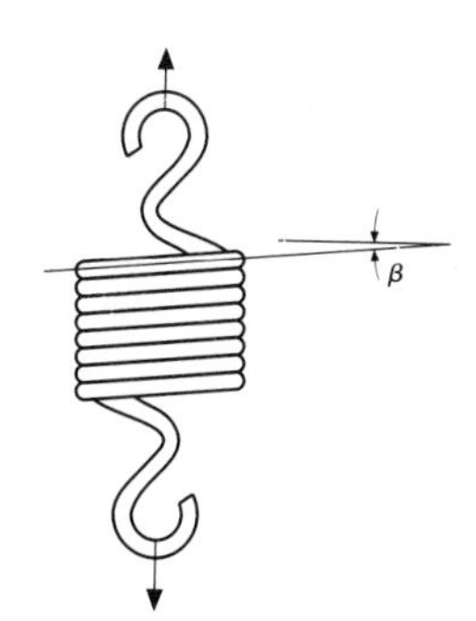

Figure 12.48

Derive (a) an expression for the stiffness of the spring; (b) an expression for the maximum axial tension; and (c) given the following data, determine the stiffness of the spring, the maximum tensile load and the extension at this load.

$G = 85$ GN/m^2, $\tau_{\text{allowable}} = 300$ MN/m^2, $D = 15$ mm, $d = 2$ mm and $N = 8$.

Solution The implication of the term close-coiled is that the helix angle β is small, which in turn means that the strain due to bending is small. It will also be assumed that the index of the spring (D/d) is large and this allows us to neglect distortion due to the shear force.

Consider the free-body diagram shown in Fig. 12.49(a). The downward force at the section must be equal in magnitude to P, and the couple of magnitude $P(D/2)$ about the axis perpendicular to the plane containing the two P forces is required for equilibrium. If the force and the couple at the section are then resolved into axial and transverse directions for the wire, as shown in Fig. 12.49(b), we obtain

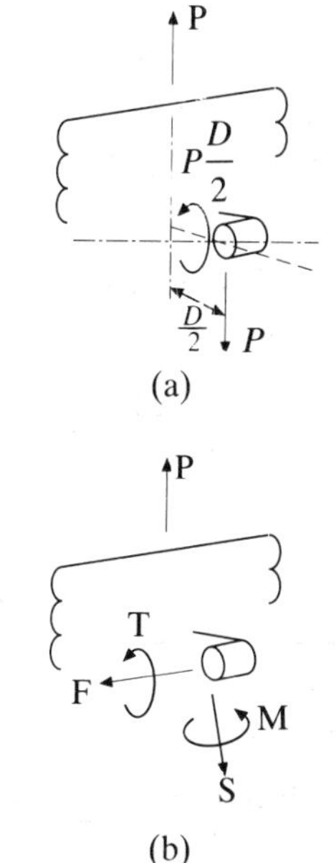

Figure 12.49

the torque $T = P(D/2)\cos\beta$
the bending moment $M = P(D/2)\sin\beta$
the axial force $F = P\sin\beta$
and the shear force $S = P\cos\beta$.

The full analysis is very complicated but very good agreement between theory and practice may be obtained for cases where β is small. We shall use a strain energy method and because of the assumptions stated only the strain energy due to torsion will be significant.

a) The strain energy due to torsion in a small length ds of wire is

$$dU = \frac{T\,d\theta}{2}$$

but $\quad \dfrac{T}{J} = G\dfrac{d\theta}{ds}$

therefore $\quad dU = \dfrac{T^2}{2GJ}ds.$

Integrating over the whole length S of the wire

$$U \doteq \frac{T^2 S}{2GJ}$$

Now the length of the wire, for small β, is πDN and the torque $T = PD/2$, so substitution gives

$$U = \frac{P^2 D^2 \pi DN}{8GJ} = \frac{P^2 D^3 N}{8Gd^4/32} = \frac{8P^2 D^3 N}{2Gd^4}$$

Since the strain energy stored must be equal to the external work done it follows that

$$U = \frac{P\delta}{2}$$

hence $\quad \delta = \dfrac{8PD^3 N}{Gd^4}$

so the stiffness

$$k = \frac{P}{\delta} = \frac{Gd^4}{8ND^3} = \frac{GD}{8N}\left(\frac{d}{D}\right)^4$$

b) The maximum allowable tensile load $\bar{P}$ is found from

$$\frac{\bar{P}D/2}{J} = \frac{\tau_{allowable}}{d/2}$$

thus $\quad \bar{P} = \dfrac{\tau_a J 4}{Dd} = \dfrac{4\tau_a \pi d^4/32}{Dd} = \dfrac{\pi d^3}{8D}\tau_{allowable}$

c) Inserting the values given

$k = 6.3$ N/mm

$\bar{P} = 62.8$ N

and the extension

$$\delta = 62.8/6.3 = 9.98 \text{ mm}$$

As a check on the helix angle let us assume that initially the coils of the spring are touching so that the distance between the coils is d. It follows that the helix angle is arctan $d/(\pi D) = 2.4°$. After the application of the maximum load the coil separation will be increased by $\delta/N = 9.88/8 = 1.24$, therefore $\beta = \arctan((2 + 1.24)/(\pi D)) = 3.93°$. This angle is still small.

Example 12.13

One common type of strain gauge is a small metal grid often formed from etched foil which is cemented to a surface for which the stress level is to be measured. To enable the principal strains and stresses to be determined three gauges may be arranged into a rosette as shown in Fig. 12.50. The change in resistance of the gauges when strained can be related to the strain along the gauge axis.

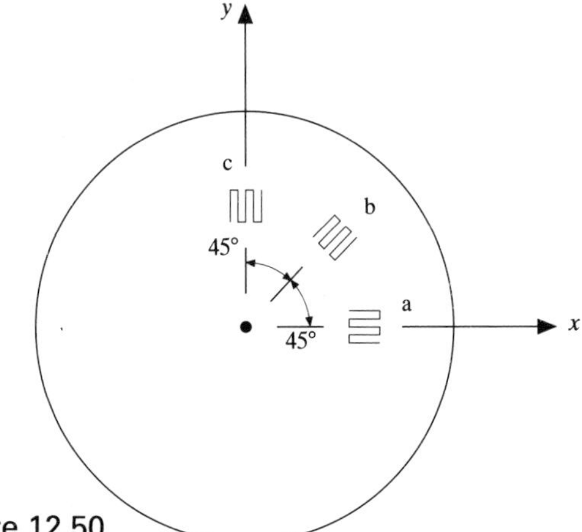

Figure 12.50

The rosette is cemented to the surface of a specimen and the specimen is then loaded. The strains indicated by the three strain gauges are

$$\varepsilon_a = 700\mu\varepsilon,\ \varepsilon_b = 200\mu\varepsilon \text{ and } \varepsilon_c = 100\mu\varepsilon.$$

Calculate the values of the principal strains and of the principal stresses, given that Young's modulus $E = 200$ GN/m^2 and Poisson's ratio $\nu = 0.29$.

Solution Figure 12.51 shows a sketch of a Mohr's circle for strain. The three strains, in this case all are positive, are laid out along the abscissa. Since gauges a and c are at 90° to each other the radii OA and OC will be at 180°, that is COA is a diameter of the circle the centre of which is at the mean of the two strains $(700 + 100)/2 = 400\mu\varepsilon$. The angle 2θ must be chosen such that the projection of the radius OB on the abscissa is the strain read from gauge b. From the geometry, $OA = 300/\cos(2\theta)$ and $OB = 200/\sin(2\theta)$ and since $OA = OB$ it follows that $\tan(2\theta) = 200/300$ giving $2\theta = 33.69°$ and $\theta = 16.85°$. The radius of the circle is 300/

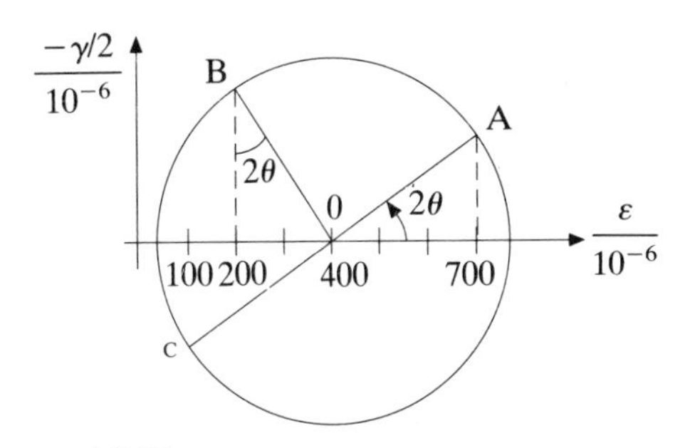

Figure 12.51

$(\cos 2\theta) = 360.56\mu\varepsilon$ hence the principal strains are

$$\varepsilon_1 = 400 + 360.56 = 760.56\mu\varepsilon$$

and $\quad \varepsilon_2 = 400 - 360.56 = 39.44\mu\varepsilon.$

To calculate the principal stresses we refer to Fig. 12.52 and note that, because the gauge is on the surface $\sigma_3 = 0$; this does not imply that ε_3 is necessarily zero.

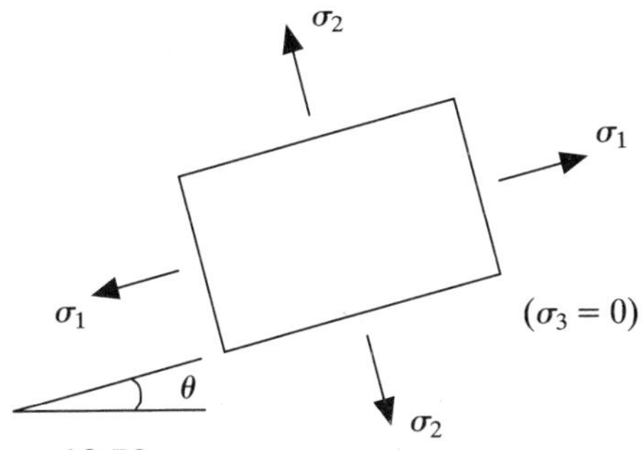

Figure 12.52

$$\varepsilon_1 = \frac{\sigma_1}{E} - \frac{\sigma_2 \nu}{E}$$

and $$\varepsilon_2 = \frac{\sigma_2}{E} - \frac{\sigma_1 \nu}{E}$$

therefore $$\varepsilon_1 + \nu\varepsilon_2 = \sigma_1 \frac{(1-\nu^2)}{E}$$

and $$\varepsilon_2 + \nu\varepsilon_1 = \sigma_2 \frac{(1-\nu^2)}{E}$$

leading to

$$\sigma_1 = \frac{E}{(1-\nu^2)}(\varepsilon_1 + \nu\varepsilon_2)$$

$$= \frac{200 \times 10^9}{(1-0.29^2)}(700 + 0.29 \times 100)10^{-6} = 159\ \text{MN/m}^2$$

$$\sigma_2 = \frac{E}{(1-\nu^2)}(\varepsilon_2 + \nu\varepsilon_1)$$

$$= \frac{200 \times 10^9}{(1-0.29^2)}(100 + 0.29 \times 700)10^{-6}$$

$$= 66.2\ \text{MN/m}^2$$

Example 12.14

Show that for a beam the strain energy stored due to bending is given by

$$U = \int_{x=0}^{x=L} \frac{M^2}{2EI} \mathrm{d}x$$

Where M is the bending moment, x is the distance measured along the beam axis, E is Young's modulus and I is the second moment of area.

Use this expression to show that the deflection of a simply supported beam with a central concentrated load W is

$$\delta = \frac{WL^3}{48EI}.$$

Solution Consider a length of the beam, ΔL, under the action of pure bending. This implies that the only loading is that due to couples M applied at each end. The work done is

$$U = \int M \mathrm{d}\theta$$

where θ is the difference in rotation between the ends of the beam.

From equation 12.67

$$\frac{M}{EI} = \frac{\mathrm{d}^2 y}{\mathrm{d}x^2} = \frac{d}{\mathrm{d}x}\left(\frac{\mathrm{d}y}{\mathrm{d}x}\right) = \frac{\mathrm{d}\theta}{\mathrm{d}x}$$

so if M is constant

$$\frac{M}{EI} = \frac{\mathrm{d}\theta}{\mathrm{d}x} = \frac{\theta}{\Delta L}$$

leading to

$$U = \int \frac{EI\theta}{\Delta L} \mathrm{d}\theta = \frac{EI\theta^2}{2\Delta L} = \frac{M^2 \Delta L}{2EI}.$$

This expression is based on the assumption that the bending moment is constant, however if the strain energy due to shear distortion in a beam is negligible then this expression may be used for cases where the bending moment is a function of

x. Thus by replacing ΔL by dx we have

$$U = \int_{x=0}^{x=L} \frac{M^2}{2EI} dx$$

For a simply supported beam with a central load the reaction at each end will be $W/2$, so between the end and the centre the bending moment is $(W/2)x$. The strain energy can be found for one half the beam and the result doubled thus

$$U = 2\int_{x=0}^{x=L/2} \frac{(W/2)^2 x^2}{2EI} = \frac{(W/2)^2 (L/2)^3}{EI3}$$

$$= \frac{W^2 L^3}{32 \times 3EI}$$

This must be equal to the work done by the applied load

$$U = W\delta/2.$$

Equating these expressions gives

$$\delta = \frac{WL^3}{48EI}$$

Problems

12.1 A hydraulic press exerts a force of 5 MN. This load is carried by two similar steel rods supporting the upper head of the press. Calculate the diameter of each rod and find the extension of each rod in a length of 2 m. The safe stress is 85 MN/m^2 and Young's modulus is 200 GN/m^2.

12.2 A steel and a brass wire of 3 m lengths and of diameter 2 mm and 2.5 mm respectively hang vertically from two points in the same horizontal plane and 125 mm apart. To the lower ends of the wires is attached a light rod which supports a weight of 0.45 kN hung midway between the wires. Find the angle at which the rod will set to the horizontal. Young's modulus for brass is 85 GN/m^2 and that for steel is 193 GN/m^2.

12.3 If the ultimate shearing stress of mild steel is 340 MN/m^2 calculate the force necessary to punch a 26 mm diameter hole in a plate 13 mm thick.

12.4 A square of material initially 160 mm × 160 mm is deformed to a rectangle 176 mm × 156 mm.

Determine from first principles (a) the longitudinal strains along directions 0°, 30°, 45°, 60° and 90° to the longer edge; and (b) the shear strains corresponding to the directions in part (a).

12.5 A flat plate of thickness t tapers from a width of b_1 to a greater width b_2 over a length L. The plate is subjected to an axial tensile load P. Assuming that the stress distribution is uniform across the width and that the limit of proportionality is not exceeded, show that the modulus of elasticity of the material is

$$E = \frac{PL}{\delta t(b_2 - b_1)} \ln(b_2/b_1)$$

where δ is the total extension.

12.6 A short horizontal rigid bar is supported by a vertical steel wire in the centre and a vertical brass wire at each end. The wires are attached to a rigid support and each has a cross-section area of 160 mm^2 and length 6 m. A load of 20 kN is applied to the centre of the bar.

Calculate (a) the stress in each wire; and (b) the extension of each wire.
Take $E_{steel} = 207$ GN/m^2 and $E_{brass} = 87$ GN/m^2.

12.7 Calculate the thickness of a spherical steel vessel 2 m internal diameter to sustain an internal pressure of 2 MN/m^2 with a tensile stress of 125 MN/m^2. Also find the change of volume due to the pressure.

Young's modulus = 210 GN/m^2 and Poisson's ratio = 0.28.

12.8 A cylinder is 2 m long and 0.75 m in diameter. Its wall thickness is 10 mm. It is closed at the ends by flat end plates and the end effects may be disregarded.

The internal pressure is raised by 1 MN/m^2, calculate for the cylinder (a) the increase in length; (b) the increase in diameter; and (c) the increase in volume. Take $E = 200$ GN/m^2 and $\nu = 0.30$.

12.9 Sketch in good proportion the shear force and bending moment diagrams for the beams shown in Figs 12.53(a) to 12.53(f). In each case state the maximum values of shear force and bending moment, also show the position of any points of contraflexure.

12.10 Find the bending moment which may be resisted by a cast iron pipe 200 mm external and 150 mm internal diameter when the greatest allowable stress due to bending is 10 MN/m^2.

12.11 A beam of rectangular cross-section, width b and depth d, is freely supported at its ends. It is just sufficiently strong to support its own weight of w/unit length. If the length, width and depth are all halved, what uniformly distributed load/unit length would the beam now support in addition to the new self-weight?

12.12 Evaluate the second moment of area about the XX axis through the centroid of the I section shown in Fig. 12.54.

12.13 An I section beam has flanges 75 mm × 10 mm and web 125 mm × 10 mm and rests on supports 4 m apart. The beam carries a concentrated load W at its centre and a load $W/2$ one quarter of the way along. Calculate the magnitude of W if the maximum stress induced by bending is 60 MN/m^2.

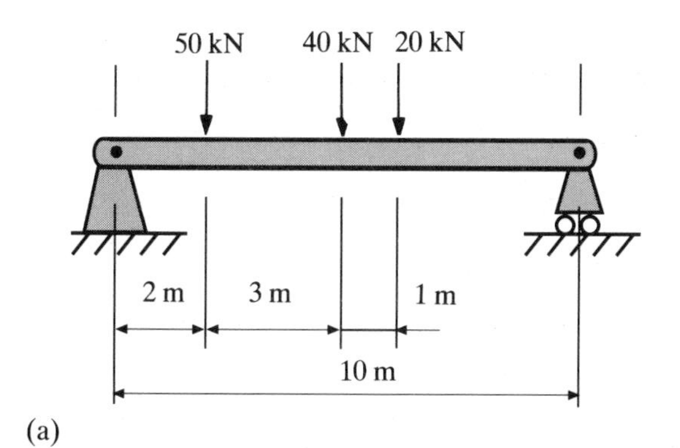

(a)

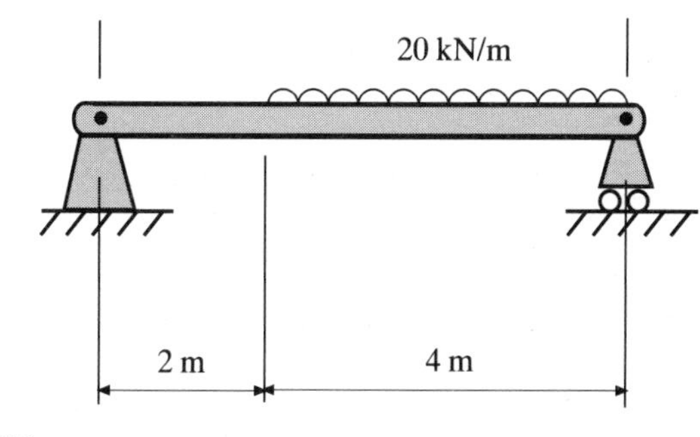

(b)

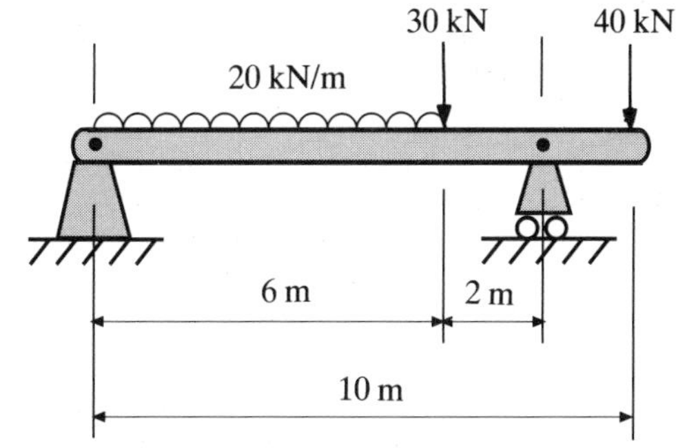

(c)

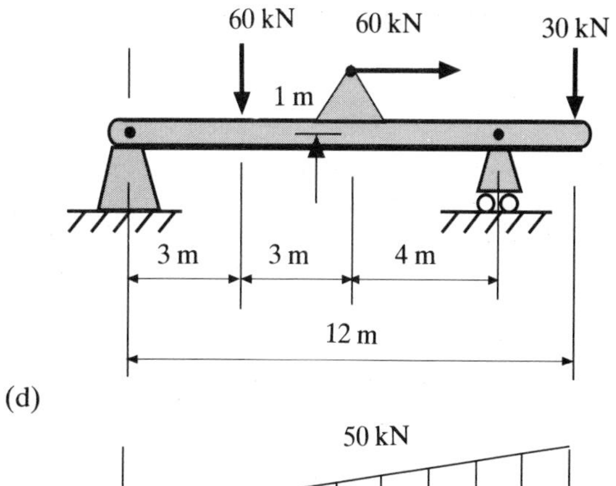

(d)

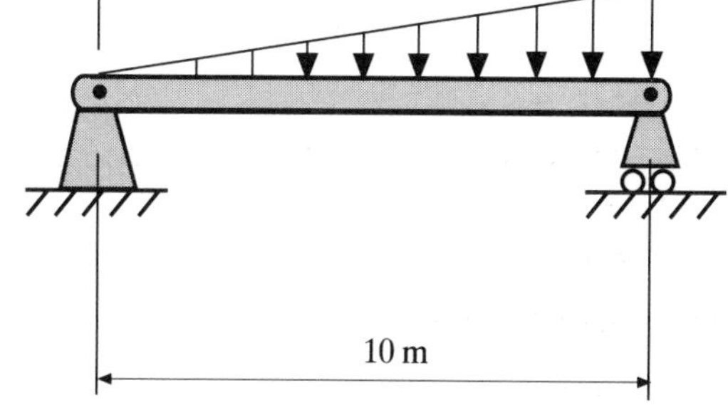

(e)

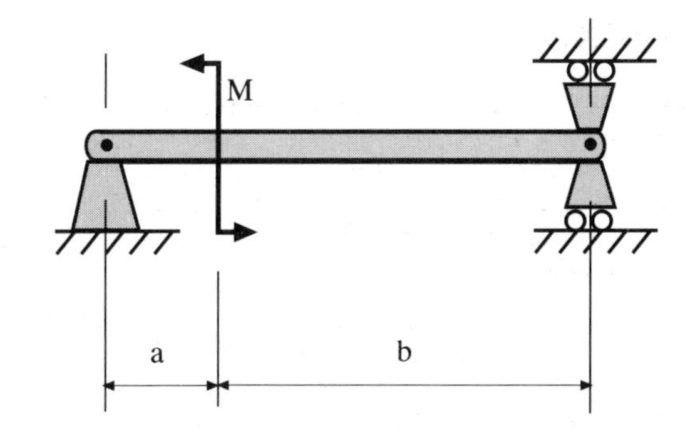

(f)

Figure 12.53

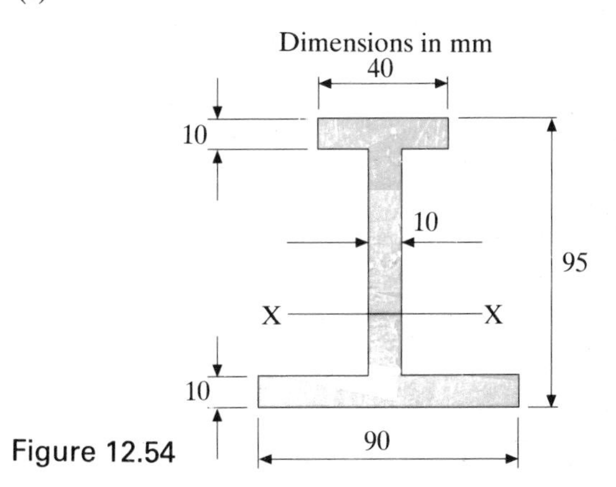

Figure 12.54

12.14 If the permissible stresses are 30 MN/m^2 in compression and 15 MN/m^2 in tension, evaluate the maximum uniformly distributed load which the beam of problem 12.13 can safely carry.

12.15 A compound girder is built up of two 330 mm deep rolled steel joists placed side by side and joined top and bottom by flange plates 350 mm wide and 12.5 mm thick. Determine the safe uniformly distributed load for a simply supported girder on a span of 6 m when the working stress is 120 MN/m^2. The relevant second moment of area for one joist is 0.0012 m^4.

12.16 A uniform cantilever of length L has a load W applied at a distance a from the fixed end. Show that the slope and deflection at the end of the cantilever are given by,

$$\theta = Wa^2/(2EI) \quad \text{and} \quad \delta = \frac{Wa^2}{2EI}[L - a/3].$$

If the load is uniformly distributed, derive new expressions for the slope and deflection at the end of the cantilever.

12.17 A cantilever of length L carries a concentrated load W at its free end and is propped at a distance a from the fixed end to the same level as the fixed end.

Find (a) the load in the prop; and (b) the point of contraflexure.
(*Hint*: Consider the force P in the prop to be an externally applied load. Calculate the deflection at the prop due to W alone and then apply P such that the deflection at the prop is again zero.)

12.18 A simply supported uniform beam, length L, carries a concentrated load W a distance b from the left hand end. Show that the maximum deflection occurs between 42.3% and 57.7% of the span of the beam as b varies from 0 to L.

12.19 A uniform beam AB, length $2L$, is simply and symmetrically supported at its ends on another uniform simply supported beam CD of length $4L$. Beam AB carries a concentrated load W at its centre. If the second moment of area of CD is three times that of AB, derive an expression for the total deflection at the mid-point of AB. Both beams are of the same material.

12.20 A hollow shaft is 125 mm external and 75 mm internal diameter. Compare the torsional strength of this shaft with that of a solid shaft of the same weight per unit run, the maximum shearing stresses being equal.

12.21 Find the least possible diameter of a solid shaft to transmit 7.46 kW at 30000 rev/min, if the shearing stress is not to exceed 90 MN/m^2.

12.22 A solid alloy shaft 60 mm diameter is to be coupled in series with a hollow steel shaft of the same external diameter. Find the internal diameter of the hollow shaft if its angle of twist is to be 80% of that of the alloy shaft for the same torque. Determine the power that can be safely transmitted at a speed of 800 rev/min if the limits of shearing stress are 50 and 80 MN/m^2 in the alloy and the steel respectively. The modulus of rigidity for steel is $2\frac{1}{2}$ times that for the alloy.

12.23 Find the maximum axial load which may be applied to a helical spring having a wire diameter 12.5 mm, a mean coil diameter of 75 mm and 18 turns if the maximum shear stress is not to exceed 420 MN/m^2.

Also find the corresponding deflection if the modulus of rigidity is 84 GN/m^2.

12.24 Two helical springs of the same height are made from the same 12.5 mm diameter circular cross-section wire and have the same number of turns. The mean diameters of the springs are 75 mm and 100 mm. They are nested together and are then compressed between two parallel planes. Determine the load in each spring for an applied load of 500 N.

12.25 A strain gauge rosette is made from three equally spaced strain gauges a, b and c. The rosette is cemented on to the surface of a specimen, and when the specimen is loaded the individual gauge readings are 364, −300 and 364 micro strain respectively.

Determine the magnitude and direction of the principal strains and the maximum shear strain. Find also the corresponding stresses.

Appendix 1
Vector algebra

A vector in the context of mechanics is defined as a quantity having magnitude and a direction and therefore may be represented by a line segment. A vector $\boldsymbol{V}$ may be written as $V\boldsymbol{e}$ where V is the scalar magnitude and $\boldsymbol{e}$ is a unit vector in the direction of $\boldsymbol{V}$. In this book a distinction is made between the always positive modulus $|\boldsymbol{V}|$ and the scalar magnitude V which may be positive or negative.

A.1 Addition of vectors

By definition two vectors are added by the parallelogram law as shown in Fig. A1.1.

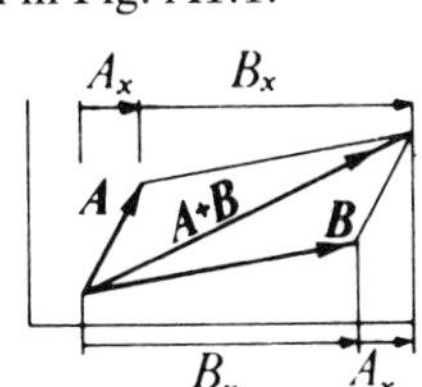

Figure A1.1

If $\boldsymbol{i}$, $\boldsymbol{j}$ and $\boldsymbol{k}$ are unit vectors in the x-, y- and z-directions respectively, then

$$\boldsymbol{V} = V_x\boldsymbol{i} + V_y\boldsymbol{j} + V_z\boldsymbol{k} \tag{A1.1}$$

where V_x, V_y and V_z are the scalar components of $\boldsymbol{V}$.

By Pythagoras's theorem,

$$|\boldsymbol{V}| = \sqrt{(V_x^2 + V_y^2 + V_z^2)} \tag{A1.2}$$

From Fig. A1.1 it is seen that

$$\boldsymbol{A} + \boldsymbol{B} = (A_x + B_x)\boldsymbol{i} + (A_y + B_y)\boldsymbol{j} + (A_z + B_z)\boldsymbol{k} \tag{A1.3}$$

$$= \boldsymbol{B} + \boldsymbol{A}$$

Since $\quad \boldsymbol{V} = V\boldsymbol{e}$,

$$\boldsymbol{e} = \frac{V_x\boldsymbol{i} + V_y\boldsymbol{j} + V_z\boldsymbol{k}}{\sqrt{(V_x^2 + V_y^2 + V_z^2)}} \tag{A1.4}$$

$$= l\boldsymbol{i} + m\boldsymbol{j} + n\boldsymbol{i}$$

where l, m and n are the direction cosines of the unit vector $\boldsymbol{e}$ relative to the x-, y- and z-axes respectively.

A1.2 Multiplication of vectors

Scalar product

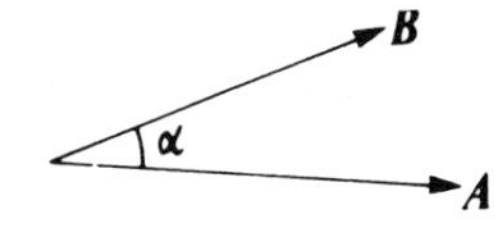

Figure A1.2

The scalar product of two vectors (Fig. A1.2) is defined as

$$\boldsymbol{A} \cdot \boldsymbol{B} = |\boldsymbol{A}||\boldsymbol{B}|\cos\alpha = \boldsymbol{B} \cdot \boldsymbol{A} \tag{A1.5}$$

Hence $\quad \boldsymbol{i}\cdot\boldsymbol{i} = \boldsymbol{j}\cdot\boldsymbol{j} = \boldsymbol{k}\cdot\boldsymbol{k} = 1$

and $\quad \boldsymbol{i}\cdot\boldsymbol{j} = \boldsymbol{j}\cdot\boldsymbol{k} = \boldsymbol{k}\cdot\boldsymbol{i} = 0$

therefore $\quad \boldsymbol{A}\cdot\boldsymbol{B} = A_xB_x + A_yB_y + A_zB_z \quad$ (A1.6)

If one vector is a unit vector $\boldsymbol{e}$, then

$$\boldsymbol{A} \cdot \boldsymbol{e} = |\boldsymbol{A}|\cos\alpha \tag{A1.7}$$

which is the component of $\boldsymbol{A}$ in the direction of $\boldsymbol{e}$.

The work done by a force $\boldsymbol{F}$ over a displacement $\mathrm{d}\boldsymbol{s}$ is

$$\mathrm{d}W = \boldsymbol{F} \cdot \mathrm{d}\boldsymbol{s} \tag{A1.8}$$

and the power is

$$\boldsymbol{F} \cdot \frac{\mathrm{d}\boldsymbol{s}}{\mathrm{d}t} = \boldsymbol{F} \cdot \boldsymbol{v} \tag{A1.9}$$

Vector product

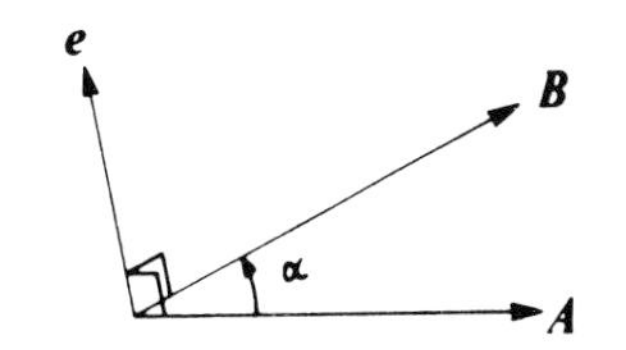

Figure A1.3

The vector product of two vectors (Fig. A1.3) is defined as

$$\boldsymbol{A} \times \boldsymbol{B} = |\boldsymbol{A}||\boldsymbol{B}|\sin\alpha\,\boldsymbol{e} = -\boldsymbol{B} \times \boldsymbol{A} \tag{A1.10}$$

In Cartesian co-ordinates,

$$A \times B = \begin{vmatrix} i & j & k \\ A_x & A_y & A_z \\ B_x & B_y & B_z \end{vmatrix} \quad \text{(A1.11)}$$

$$= i(A_y B_z - A_z B_y) + j(A_z B_x - A_x B_z) + k(A_x B_y - A_y B_x) \quad \text{(A1.12)}$$

If a body has an angular velocity of ω then the velocity of a point A relative to a point O is

$$v_{A/O} = \omega \times r_{A/O} \quad \text{(A1.13)}$$

The moment of a force F about a point O is

$$r \times F \quad \text{(A1.14)}$$

Triple scalar product

The triple scalar product is written $A \cdot B \times C$ and is the volume of the parallelepiped formed by the three vectors as shown in Fig. A1.4. The position of the dot and the cross may be interchanged provided that the same cyclic order is maintained. If the cyclic order is reversed then the sign of the result is changed. It is easily seen that if any two of the vectors are parallel then the product is zero.

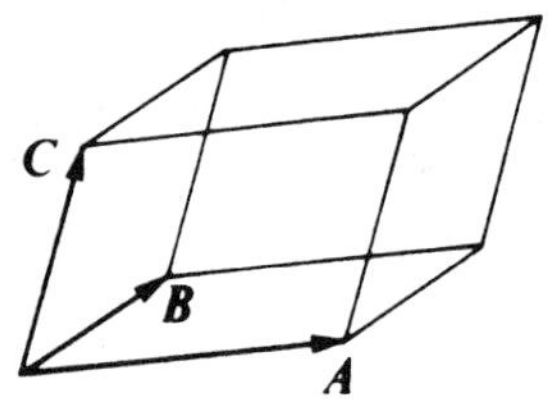

Figure A1.4

The component of the moment of a force in the direction of e is

$$e \cdot (r \times F) \quad \text{(A1.15)}$$

If e is parallel to r or F then the product is zero.

Triple vector product

The triple vector product $A \times (B \times C)$ can be shown to be equal to

$$B(A \cdot C) - C(A \cdot B) \quad \text{(A1.16)}$$

The moment of momentum of a particle due to rigid-body rotation is

$$r \times mv = r \times (m\omega \times r)$$
$$= \omega(mr \cdot r) - r(mr \cdot \omega) \quad \text{(A1.17)}$$

If r is normal to ω then the moment of momentum of the particle is

$$\omega m r^2 \quad \text{(A1.18)}$$

A1.3 Differentiation of vectors with respect to time

See Fig. A1.5. By definition

$$\frac{dV}{dt} = \lim_{\Delta t \to 0} \frac{V(t+\Delta t) - V(t)}{\Delta t} \quad \text{(A1.19)}$$

$$= \lim_{\Delta t \to 0} \frac{\Delta V}{\Delta t}$$

and is therefore a vector in the direction of ΔV.

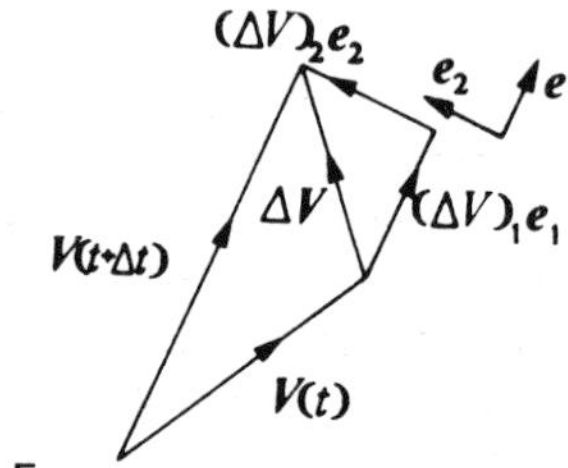

Figure A1.5

The vector ΔV may be written as the vector sum of two orthogonal components:

$$\Delta V = (\Delta V)_1 e_1 + (\Delta V)_2 e_2$$

so that

$$\frac{dV}{dt} = \lim_{\Delta t \to 0} \left[\frac{(\Delta V)_1}{\Delta t} e_1 + \frac{(\Delta V)_2}{\Delta t} e_2 \right]$$

Note that the magnitude of $V(t+\Delta t)$ is $(V + \Delta V)$, but $\Delta V \neq |\Delta V|$.

Differentiation of a product of two vectors

$$\frac{d}{dt}(A \bigcirc B)$$

$$= \lim_{\Delta t \to 0} \left[\frac{(A + \Delta A) \bigcirc (B + \Delta B) - A \bigcirc B}{\Delta t} \right]$$

$$= \lim_{\Delta t \to 0} \left[\frac{A \bigcirc \Delta B + \Delta A \bigcirc B + \Delta A \bigcirc \Delta B}{\Delta t} \right]$$

$$= A \bigcirc \frac{dB}{dt} + \frac{dA}{dt} \bigcirc B \quad \text{(A1.20)}$$

The symbol $\bigcirc$ signifies either scalar or vector multiplication; in the case of the vector product it is essential that the order of multiplication be preserved.

Appendix 2
Units

A physical quantity is expressed as the product of a pure number and a unit. Physical laws which exist between physical quantities are conveniently expressed in systems of consistent units such that the form of the equation is independent of the system of units chosen.

Four systems will be listed here, namely the Système International d'Unités (SI); the centimetre, gram, second system (c.g.s.); the British absolute system based on the foot, pound and second (f.p.s.); and the British Engineering system based on the foot, slug and second (f.s.s.).

In all these systems $F = ma$ and weight $W = mg$, where g is the gravitational field strength. A standard value of the field strength at the surface of the earth is given as 9.80665 N/kg (m/s^2) or approximately 31.174 ft/s^2 or pdl/lb or lbf/slug.

By definition,

$$1 \text{ slug} = 32.174 \text{ lb}$$

$$1 \text{ lbf} = 32.174 \text{ pdl}$$

also $1 \text{ kgf} = 9.80665 \text{ N}$

Note that these are exact relationships and do not vary with location as does g.

Conversion of British units to SI units is achieved using the following exact conversion factors:

$$1 \text{ ft} = 0.3048 \text{ m}$$
$$1 \text{ lb} = 0.45359237 \text{ kg}$$

Using these values,

$$1 \text{ pdl} = 1 \text{ lb}\frac{\text{ft}}{\text{s}^2} = 1 \text{ kg}\left(\frac{\text{lb}}{\text{kg}}\right)\text{m}\left(\frac{\text{ft}}{\text{m}}\right)\frac{1}{\text{s}^2}$$
$$= 1 \text{ kg } 0.4536 \text{ m } 0.3048 \text{ s}^{-2}$$
$$= 0.1383 \text{ kg m s}^{-2}$$
$$= 0.1383 \text{ N}$$

$$1 \text{ lbf} = 1\left(\frac{\text{lbf}}{\text{pdl}}\right)\text{pdl} = 1\left(\frac{\text{lbf}}{\text{pdl}}\right)\left(\frac{\text{pdl}}{\text{N}}\right)\text{N}$$
$$= 1 \times 32.174 \times 0.1383 \text{ N}$$
$$= 4.448 \text{ N}$$

Table A2.1

Quantity	*Unit and symbol* SI	c.g.s.	f.p.s.	f.s.s.
Mass	kilogram, kg	gram, g	pound, lb	slug
Length	metre, m	centimetre, cm	foot, ft	foot, ft
Time	second, s	second, s	second, s	second, s
Angle	radian, rad	radian, rad	radian, rad	radian, rad
Force	newton (kg m s^{-2}), N	dyne	poundal, pdl	pound force, lbf
Energy / Work	joule (m N), J	erg	foot poundal, ft pdl	foot pound force
Power	watt (J s^{-1}), W	erg s^{-1}	ft pdl s^{-1}	ft lbf s^{-1}
Pressure	pascal (N m^{-2}), Pa	dyne cm^{-2}	pdl ft^{-2}	16f ft^{-2}
Moment of force	N m	dyne cm	pdl ft	lbf ft
Moment of inertia	kg m^2	g cm^2	lb ft^2	slug ft^2
Velocity	m s^{-1}	cm s^{-1}	ft s^{-1}	ft s^{-1}
Acceleration	m s^{-2}	cm s^{-2}	ft s^{-2}	ft s^{-2}

[1 micron = 10^{-6} m, 1 litre = 10^{-3} m^3, 1 tonne = 10^3 kg, 1 bar = 10^5 Pa]

Similarly for work and energy:

1 ft pdl = 0.042 J
1 ft lbf = 1.356 J

also 1 h.p. = 550 ft lbf s^{-1} = 745.700 W

Other useful conversion factors are

Density 1 lb ft^{-3} = 16.0185 kg m^{-3}
Pressure 1 lbf in^{-2} = 6894.76 N m^{-2} (Pa)
1 atmosphere (atm) = 1.01325 $\times 10^5$ N m^{-2}

Note that in the above calculations the symbol for the unit is treated as if it were an ordinary algebraic quantity.

Table A2.2

Factor	*Prefix*	*Symbol*	
10^{12}	tera	T	
10^9	giga	G	
10^6	mega	M	
10^3	kilo	k	
10^2	hecto	h	These multiples are not encouraged
10^1	deca	da	
10^{-1}	deci	d	
10^{-2}	centi	c	
10^{-3}	milli	m	
10^{-6}	micro	μ	
10^{-9}	nano	n	
10^{-12}	pico	p	
10^{-15}	femto	f	
10^{-18}	atto	a	

The use of the prefixes is illustrated by the following:

0.000001 m = 1 μm (micron)
1000000 N = 1 MN (meganewton)
0.1 m = 100 mm
10^4 N = 10 kN
10^3 kg = 1 Mg (*not* 1 kkg)

Other systems of units are still seen in which the use of a mass unit is avoided by writing $m = (W/g)$ so that $F = (W/g)a$. Alternatively the acceleration may be expressed as multiples of g to give $F = W(a/g)$.

The use of variable units of force such as the pound weight (lb wt) and the gram weight (gm wt) is now moribund and must be discouraged. For practical purposes a force equal to the weight of the unit mass will often find favour in elementary applications, so the kilogram force (kgf), or its close equivalent 1 da N, may continue to be used in non-scientific applications.

Occasionally one sees the use of systems involving lbf, in, s; or kgf, cm, s. In these cases the corresponding unit masses are 386 lb and 981 kg respectively.

When labelling the axes of graphs or writing the headings for tables of values, the following scheme is unambiguous.

The approximate value for the density of steel (ρ) is 7850 kg/m^3, so

$$\rho = 7850 \text{ kg m}^{-3} = 7.850 \times 10^3 \text{ kg m}^{-3} = 7.850 \text{ Mg m}^{-3}$$

It follows that

$$\frac{\rho}{\text{kg m}^{-3}} = 7850$$

$$\frac{\rho}{10^3 \text{ kg m}^{-3}} = 7.850$$

$$\frac{\rho}{\text{Mg m}^{-3}} = 7.850$$

The practice of heading a list of numbers in the manner ρ 10^3 kg m^3 leaves a doubt as to whether the 10^3 refers to the physical quantity or to its unit of measurement. However

$$\frac{\rho}{10^3 \text{ kg m}^3},$$

for example, has no such ambiguity.

References

For further information, the following booklets should be consulted:

1. British Standards Institution, BS 5555:1993 (incorporating ISO 1000), *specification for SI units and recommendations for use of their multiples and of certain other units.*
2. The Symbol Committee of the Royal Society, *Quantities, units, and symbols*

Appendix 3
Approximate integration

The trapezoidal rule

The area under the curve shown shaded in Fig. A3.1 is divided into equal strips of width w and the ordinates are labelled a_1 to a_n. The curve is then approximated to straight lines between the ordinates.

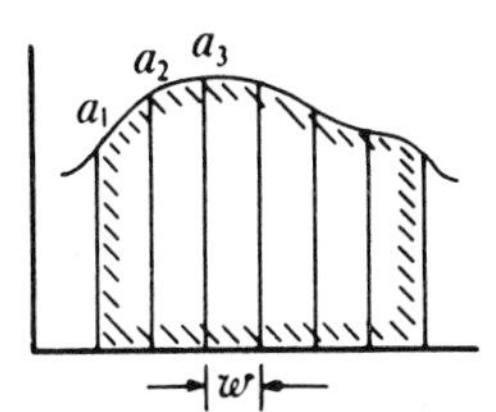

Figure A3.1

$$\text{The area is } \tfrac{1}{2}w\{(\text{first} + \text{last ordinate}) + 2(\text{remaining ordinates})\} \quad \text{(A3.1)}$$

Simpson's rule

The area is divided into an *even* number of strips but this time the curve is approximated to a parabola between three consecutive ordinates. In terms of the ordinates, the area is

$$\tfrac{1}{3}w[\text{first} + \text{last} + 4(\text{even}) + 2(\text{remaining odd})] \quad \text{(A3.2)}$$

Appendix 4
Conservative forces and potential energy

Notes on conservative forces

If the work done by a force in moving a particle from position 1 to postion 2 is independent of the path taken (see Fig. A4.1) then the force is said to be conservative.

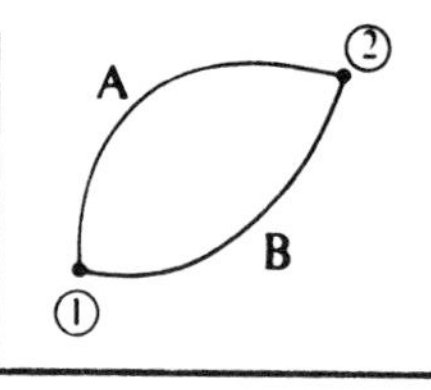

Figure A4.1

$$_A\int_1^2 \boldsymbol{F}\cdot \mathrm{d}\boldsymbol{s} = {}_B\int_1^2 \boldsymbol{F}\cdot \mathrm{d}\boldsymbol{s} = -{}_B\int_2^1 \boldsymbol{F}\cdot \mathrm{d}\boldsymbol{s}$$

thus $$_A\int_1^2 \boldsymbol{F}\cdot \mathrm{d}\boldsymbol{s} + {}_B\int_2^1 \boldsymbol{F}\cdot \mathrm{d}\boldsymbol{s} = 0$$

or the integral around a closed path is zero:

$$\oint \boldsymbol{F}\cdot \mathrm{d}\boldsymbol{s} = 0 \qquad \text{(A4.1)}$$

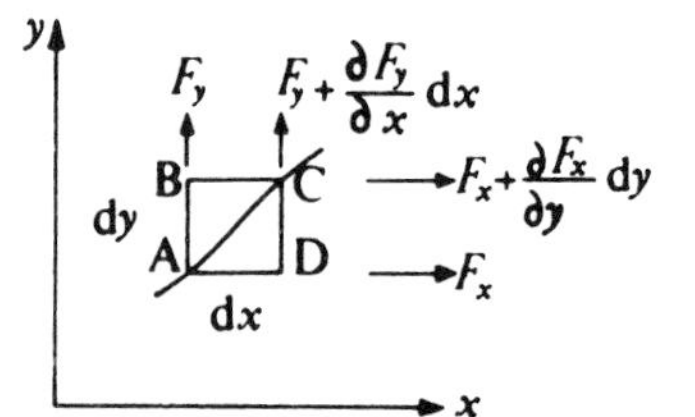

Figure A4.2

For an element of path (Fig. A4.2), the work done along path ABC must equal that along path ADC, thus

$$F_y \mathrm{d}y + \left(F_x + \frac{\partial F_x}{\partial y}\mathrm{d}y\right)\mathrm{d}x$$

$$= \mathrm{F_x} \mathrm{d}x + \left(F_y + \frac{\partial F_y}{\partial x}\mathrm{d}x\right)\mathrm{d}y$$

therefore $$\frac{\partial F_y}{\partial x} = \frac{\partial F_x}{\partial y} \qquad \text{(A4.2)}$$

In polar co-ordinates (Fig. A4.3),

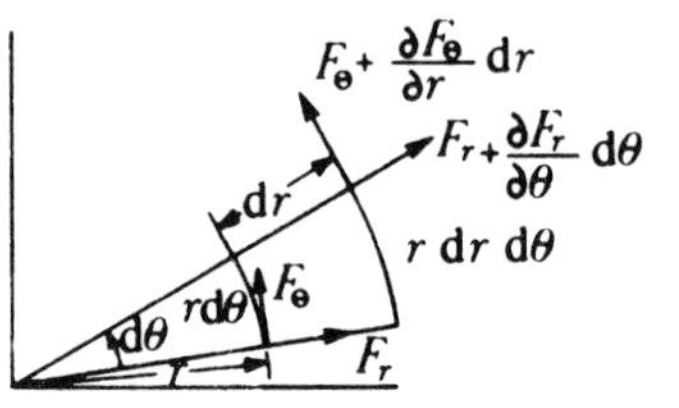

Figure A4.3

$$F_r \mathrm{d}r + \left(F_\theta + \frac{\partial F_\theta}{\partial r}\mathrm{d}r\right)(r + \mathrm{d}r)\,\mathrm{d}\theta$$

$$= F_\theta r\,\mathrm{d}\theta + \left(F_r + \frac{\partial F_r}{\partial \theta}\mathrm{d}\theta\right)\mathrm{d}r$$

therefore $$F_\theta + \left(\frac{\partial F_\theta}{\partial r}\right)r = \frac{\partial F_r}{\partial \theta} \qquad \text{(A4.3)}$$

Potential energy

Potential energy is defined as

$$V = -\int \boldsymbol{F}\cdot \mathrm{d}\boldsymbol{s} + \text{constant}$$

$$= -\int F_x \mathrm{d}x - \int F_y \mathrm{d}y + \text{constant}$$

$$\mathrm{d}V = -\boldsymbol{F}\cdot \mathrm{d}\boldsymbol{s} = -F_x \mathrm{d}x - F_y \mathrm{d}y$$

But, from the theory of differentials,

$$\mathrm{d}V = \frac{\partial V}{\partial x}\mathrm{d}x + \frac{\partial V}{\partial y}\mathrm{d}y$$

hence

$$F_x = -\frac{\partial V}{\partial x}$$

and

$$F_y = -\frac{\partial V}{\mathrm{d}y} \qquad \text{(A4.4)}$$

In polar co-ordinates,

$$dV = -F_r dr - F_\theta r d\theta$$

but $$dV = \frac{\partial V}{\partial r} dr + \frac{\partial V}{\partial \theta} d\theta$$

hence
$$\left.\begin{aligned} F_r &= -\frac{\partial V}{\partial r} \\ \text{and} \quad & \\ F_\theta &= -\frac{1}{r}\frac{\partial V}{\partial \theta} \end{aligned}\right\} \quad \text{(A4.5)}$$

For a uniform gravitation field $-g\boldsymbol{j}$,

$$V = mgy + \text{constant} \quad \text{(A4.6)}$$

For an inverse-square-law field, $-(\mu/r^2)\boldsymbol{e}_r$ where μ = a constant,

$$V = -\frac{\mu}{r} + \text{constant} \quad \text{(A4.7)}$$

For a linear spring with a stiffness k,

$$V = \tfrac{1}{2}k\,\delta^2 \quad \text{(A4.8)}$$

where δ is the extension of the spring.

Appendix 5
Properties of plane areas and rigid bodies

Centroid

The position of the centroid of a plane area is given by

$$x_G = \frac{\int x \, dA}{\int dA} \quad \text{and} \quad y_G = \frac{\int y \, dA}{\int dA} \tag{A5.1}$$

where dA is an elemental area and $\int dA$ is the area, A.

Second moment of area

The second moment of area about the x axis is

$$I_{Ox} = \int x^2 dA = A k_{Ox}^2 \tag{A5.2}$$

Parallel-axes theorem

$$I_{Ox} = I_{OG} + A x_G^2 \tag{A5.3}$$

Perpendicular-axes theorem

$$I_{Oz} = I_{Ox} + I_{Oy} \tag{A5.4}$$

Centre of mass

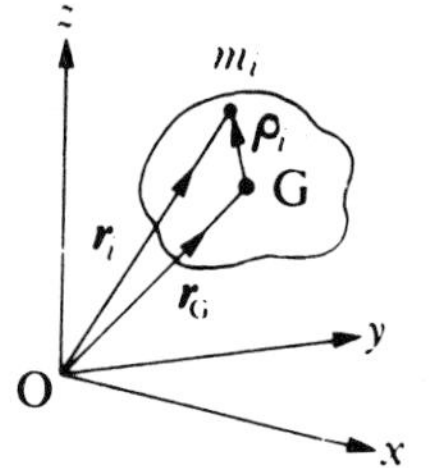

Figure A5.1

The position of the centre of mass G (Fig. A5.1) is defined by

$$\boldsymbol{r}_G = \frac{\sum m_i \boldsymbol{r}_i}{M} \tag{A5.5}$$

or by the three scalar equations

$$x_G = \frac{\sum m_i x_i}{M} \tag{A5.6a}$$

$$y_G = \frac{\sum m_i y_i}{M} \tag{A5.6b}$$

$$z_G = \frac{\sum m_i z_i}{M} \tag{A5.6c}$$

Also $\sum m_i \boldsymbol{\rho}_i = \boldsymbol{0}$ or, in scalar form, $\sum m_i \rho_{xi} = 0$, etc.

Moment of inertia

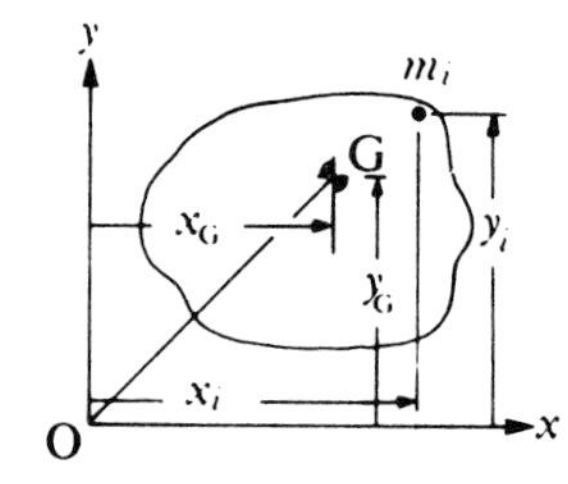

Figure A5.2

In Fig. A5.2, the moment of inertia about the z-axis is

$$I_{Oz} = M k_{Oz}^2 = \sum m_i (x_i^2 + y_i^2) \tag{A5.7}$$

$$I_{Oxy} = \sum m_i x_i y_i \tag{A5.8}$$

Parallel-axes theorem

$$I_{Oz} = I_{Gz} + M(x_G^2 + y_G^2) \tag{A5.9}$$

$$I_{Oxy} = I_{Gxy} + M x_G y_G \tag{A5.10}$$

Perpendicular-axes theorem

For a thin lamina in the xy-plane,

$$I_{Oz} = I_{Ox} + I_{Oy} \tag{A5.11}$$

Properties of plane areas and rigid bodies

Table A5.1

Body	*Centre of mass*	*Moment of inertia*	*Second moment of area*
Circular lamina	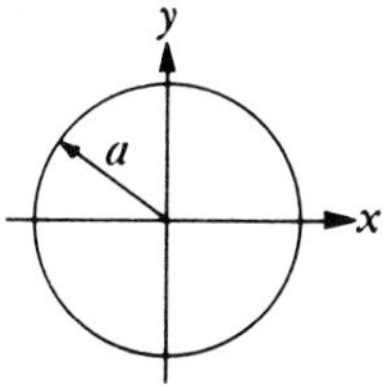	$I_{Oz} = M\frac{a^2}{2}$ $I_{Ox} = M\frac{a^2}{4}$	$I_{Oz} = A\frac{a^2}{2}$ $I_{Ox} = A\frac{a^2}{4}$
Semicircular lamina	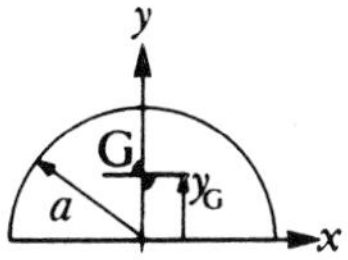 $y_G = \frac{4a}{3\pi}$	$I_{Ox} = M\frac{a^2}{4}$ $I_{Oy} = M\frac{a^2}{4}$	$I_{Ox} = A\frac{a^2}{4}$ $I_{Oy} = A\frac{a^2}{4}$
Triangular lamina	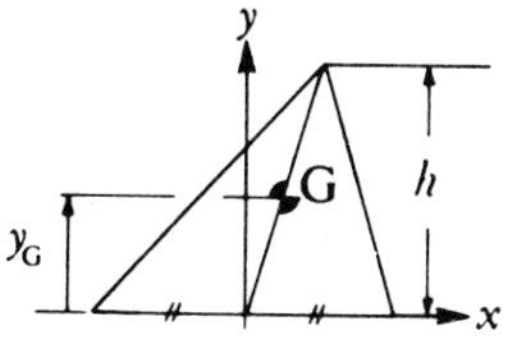 $y_G = \frac{1}{3}h$	$I_{Ox} = M\frac{h^2}{2}$	$I_{Ox} = A\frac{h^2}{2}$
Rectangular lamina	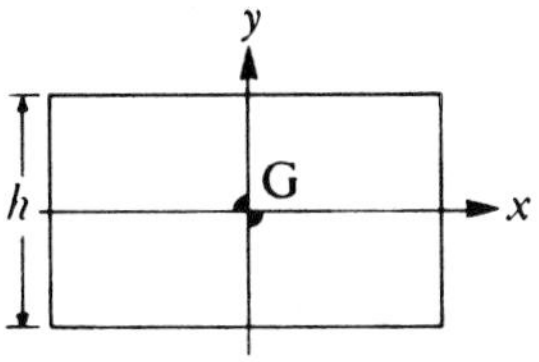	$I_{Gx} = M\frac{h^2}{12}$	$I_{Gx} = A\frac{h^2}{12}$
Solid cylinder	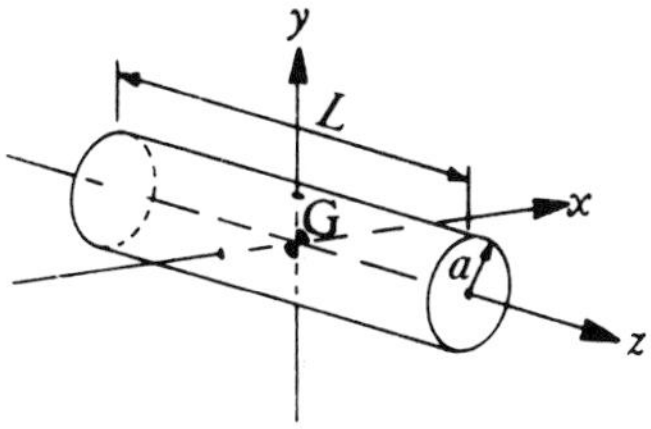	$I_{Gz} = M\frac{a^2}{2}$ $I_{Gx} = \frac{M}{12}(3a^2 + L^2)$	
Solid sphere	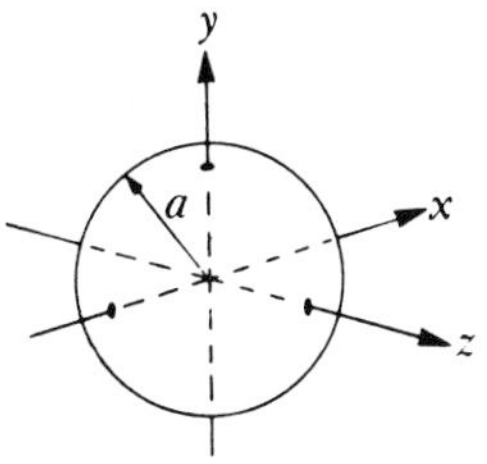	$I_{Gz} = \frac{2}{5}Ma^2$	

Table A5.1 – continued

Body	*Centre of mass*	*Moment of inertia*	*Second moment of area*
Solid hemisphere	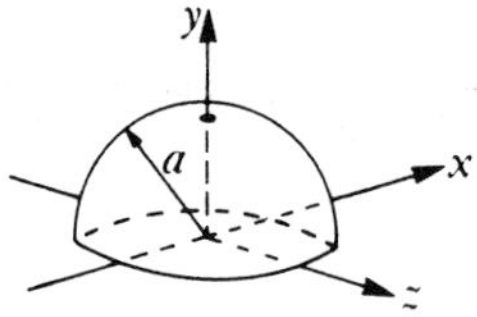 $y_G = \frac{3}{8}a$	$I_{Oz} = I_{Ox} = \frac{2}{5}Ma^2$	
Rectangular bar	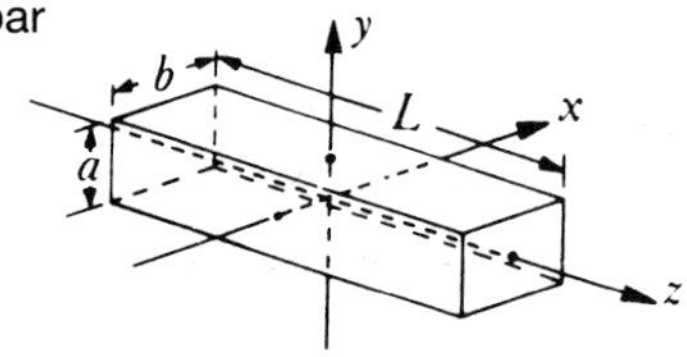	$I_{Gz} = \frac{M}{12}(a^2 + b^2)$ $I_{Gx} = \frac{M}{12}(a^2 + L^2)$	

Appendix 6
Summary of important relationships

Kinematics

a) Cartesian co-ordinates:

$$\boldsymbol{v} = \dot{x}\boldsymbol{i} + \dot{y}\boldsymbol{j} + \dot{z}\boldsymbol{k} \quad \text{(A6.1)}$$

$$\boldsymbol{a} = \ddot{x}\boldsymbol{i} + \ddot{y}\boldsymbol{j} + \ddot{z}\boldsymbol{k} \quad \text{(A6.2)}$$

b) Cylindrical co-ordinates:

$$\boldsymbol{v} = \dot{R}\boldsymbol{e}_r + R\dot{\theta}\boldsymbol{e}_\theta + \dot{z}\boldsymbol{k} \quad \text{(A6.3)}$$

$$\boldsymbol{a} = (\ddot{R} - \dot{\theta}^2 R)\boldsymbol{e}_r + (R\ddot{\theta} + 2\dot{R}\dot{\theta})\boldsymbol{e}_\theta + \ddot{z}\boldsymbol{k} \quad \text{(A6.4)}$$

c) Path co-ordinates:

$$\boldsymbol{v} = \dot{s}\boldsymbol{e}_t \quad \text{(A6.5)}$$

$$\boldsymbol{a} = \ddot{s}\boldsymbol{e}_t + \frac{\dot{s}^2}{\rho}\boldsymbol{e}_n \quad \text{(A6.6)}$$

d) Spherical co-ordinates

$$\boldsymbol{v} = \dot{r}\boldsymbol{e}_r + r\dot{\theta}\cos\phi\,\boldsymbol{e}_\theta + r\dot{\phi}\boldsymbol{e}_\phi \quad \text{(A6.7)}$$

$$\begin{aligned}\boldsymbol{a} = &(\ddot{r} - r\dot{\phi}^2 - r\dot{\theta}^2\cos^2\phi)\boldsymbol{e}_r \\ &+ (r\ddot{\theta}\cos\phi - 2r\dot{\theta}\dot{\phi}\sin\phi + 2\dot{r}\dot{\theta}\cos\phi)\boldsymbol{e}_\theta \\ &+ (r\ddot{\phi} + 2\dot{r}\dot{\phi} + r\dot{\theta}^2\sin\phi\cos\phi)\boldsymbol{e}_\phi\end{aligned} \quad \text{(A6.8)}$$

Kinetics (Planar motion)

$$\left.\begin{aligned}\sum F_x &= M\ddot{x}_G \\ \sum F_y &= M\ddot{y}_G\end{aligned}\right\} \quad \boldsymbol{F} = \frac{\mathrm{d}}{\mathrm{d}t}\left(\sum m_i \boldsymbol{v}_i\right) \quad \text{(A6.9)}$$

$$\sum M_G = I_G\ddot{\theta} \quad \text{(A6.10)}$$

Work–energy

Kinetic energy: $\frac{1}{2}I_G\omega^2 + \frac{1}{2}Mv_G{}^2$ (A6.11)

Potential energy:

i) gravitational, mgy (A6.12)

ii) strain, for simple spring, $\frac{1}{2}k\delta^2$ (A6.13)

Work done by non-conservative forces
= (k.e. + p.e.)$_2$ − (k.e. + p.e.)$_1$ + 'losses' (A6.14)

Free vibration of a linear damped system

If the equation of motion is of the form

$$m\ddot{x} + c\dot{x} + kx = 0 \quad \text{(A6.15)}$$

undamped natural frequency
$= \omega_n = (k/m)^{1/2}$ (A6.16)

critical damping $= c_{crit.} = 2(km)^{1/2}$ (A6.17)

damping ratio $= \zeta = c/c_{crit.}$ (A6.18)

Equation A6.15 may be rewritten

$$\ddot{x} + 2\zeta\omega_n\dot{x} + \omega_n{}^2 x = 0 \quad \text{(A6.19)}$$

For $\zeta < 1$,

$$x = \mathrm{e}^{-\zeta\omega_n t}(A\cos\omega_d t + B\sin\omega_d t) \quad \text{(A6.20)}$$

where $\omega_d = \omega_n(1 - \zeta^2)^{1/2}$.

For $\zeta = 1$,

$$x = \mathrm{e}^{-\omega_n t}(A + Bt) \quad \text{(A6.21)}$$

For $\zeta > 1$,

$$\begin{aligned}x = &A\exp[-\zeta - \sqrt{(\zeta^2 - 1)}]\,\omega_n t \\ &+ B\exp[-\zeta + \sqrt{(\zeta^2 - 1)}]\,\omega_n t\end{aligned} \quad \text{(A6.22)}$$

or

$$\begin{aligned}x = \mathrm{e}^{-\zeta\omega_n t}\{&A\cosh[\omega_n\sqrt{(\zeta^2 - 1)}]t \\ &+ B\sinh[\omega_n\sqrt{(\zeta^2 - 1)}]t\}\end{aligned} \quad \text{(A6.23)}$$

Logarithmic decrement

$$\delta = 2\pi\zeta/(1 - \zeta^2)^{1/2} \quad \text{(A6.24)}$$

Steady-state forced vibration

If the equation of motion is of the form

$$m\ddot{x} + c\dot{x} + kx = F_0\cos\omega t \quad \text{(A6.25)}$$

or

$$\ddot{x} + 2\zeta\omega_n\dot{x} + \omega_n{}^2 x = \mathrm{Re}\,F_0\exp(\mathrm{j}\omega t) \quad \text{(A6.26)}$$

then the steady-state solution is

$$x = X\cos(\omega t - \phi) = X\mathrm{Re}\{\exp[\mathrm{j}(\omega t - \phi)]\}$$

where

$$X = \frac{(F_0/k)}{\{[1 - (\omega/\omega_n)^2]^2 + 4\zeta^2(\omega/\omega_n)^2]\}^{1/2}} \quad \text{(A6.27)}$$

and

$$\tan\phi = 2\zeta(\omega/\omega_n)/[1 - (\omega/\omega_n)^2]^{1/2} \quad \text{(A6.28)}$$

Vibration of many degrees-of-freedom systems

The general matrix equation is

$$[\mathbf{m}](\ddot{\boldsymbol{x}})+[\mathbf{k}](\boldsymbol{x})=(0) \quad \text{(A6.29)}$$

which has solutions of the form

$$(\boldsymbol{x})=(\boldsymbol{A})\mathrm{e}^{\lambda t} \quad \text{(A6.30)}$$

The characteristic equation is

$$\mathrm{Det}[\lambda^2[\mathbf{m}]+[\mathbf{k}]]=0 \quad \text{(A6.31)}$$

Principle of orthogonality

$$(\boldsymbol{A}_1)[\mathbf{m}](\boldsymbol{A}_2)=0 \quad \text{(A6.32)}$$

and $(\boldsymbol{A}_1)[\mathbf{k}](\boldsymbol{A}_2)=0$ (A6.33)

Stability of linear system

Systems up to the fifth order, described by an equation of the form

$$(a_5\mathrm{D}^5+a_4\mathrm{D}^4+a_3\mathrm{D}^3+a_2\mathrm{D}^2+a_1\mathrm{D}+a_0)x=f(t)$$

where $\mathrm{D}=\mathrm{d}/\mathrm{d}t$, are stable provided that

$$a_5>0,\ a_1>0,\ a_2>0\ a_3>0,\ a_4>0,\ a_5>0$$

$$a_2a_1>a_3a_0 \quad \text{and}$$

$$(a_5a_0+a_3a_2)a_1>a_1{}^2a_4+a_3{}^2a_0 \quad \text{(A6.34)}$$

Differentiation of a vector

$$\mathrm{d}V/\mathrm{d}t=\partial V/\partial t+\boldsymbol{\omega}\times V \quad \text{(A6.35)}$$

where $\boldsymbol{\omega}$ is the angular velocity of the moving frame of reference.

Kinetics of a rigid body

For a body rotating about a fixed point,

$$\boldsymbol{M}_\mathrm{O}=\mathrm{d}\boldsymbol{L}_\mathrm{O}/\mathrm{d}t=\partial\boldsymbol{L}_\mathrm{O}/\partial t+\boldsymbol{\omega}\times\boldsymbol{L}_\mathrm{O} \quad \text{(A6.36)}$$

also $\boldsymbol{M}_\mathrm{G}=\mathrm{d}\boldsymbol{L}_\mathrm{G}/\mathrm{d}t=\partial\boldsymbol{L}_\mathrm{G}/\partial t+\boldsymbol{\omega}\times\boldsymbol{L}_\mathrm{G}$ (A6.37)

Referred to principal axes, the moment of momentum is

$$\boldsymbol{L}=I_{xx}\omega_x\boldsymbol{i}+I_{yy}\omega_y\boldsymbol{j}+I_{zz}\omega_z\boldsymbol{k} \quad \text{(A6.38)}$$

Euler's equations are

$$\left.\begin{aligned}M_x&=I_{xx}\dot{\omega}_x-(I_{yy}-I_{zz})\omega_y\omega_z\\M_y&=I_{yy}\dot{\omega}_y-(I_{zz}-I_{xx})\omega_z\omega_x\\M_z&=I_{zz}\dot{\omega}_z-(I_{xx}-I_{yy})\omega_x\omega_y\end{aligned}\right\} \quad \text{(A6.39)}$$

Kinetic energy

For a body rotating about a fixed point,

$$\begin{aligned}\text{k.e.}&=\tfrac{1}{2}\boldsymbol{\omega}\cdot\boldsymbol{L}_\mathrm{O}\\&=\tfrac{1}{2}\{\boldsymbol{\omega}\}^\mathrm{T}[I]\{\boldsymbol{\omega}\}\end{aligned} \quad \text{(A6.40)}$$

Referred to principal axes,

$$\text{k.e.}=\tfrac{1}{2}I_{xx}\omega_x{}^2+\tfrac{1}{2}I_{yy}\omega_y{}^2+I_{zz}\omega_z^2 \quad \text{(A6.41)}$$

In general,

$$\text{k.e.}=\tfrac{1}{2}\boldsymbol{\omega}\cdot\boldsymbol{L}_\mathrm{G}+\tfrac{1}{2}m\boldsymbol{v}_\mathrm{G}\cdot\boldsymbol{v}_\mathrm{G} \quad \text{(A6.42)}$$

Continuum mechanics

Wave equation

$$E\frac{\partial^2u}{\partial x^2}=\rho\frac{\partial^2u}{\partial t^2} \quad \text{(A6.43)}$$

Wave speed

$$c=\surd(E/\rho) \quad \text{(A6.44)}$$

Continuity equation

$$\frac{\Delta m}{\Delta t}=\int_S\rho\boldsymbol{v}\cdot\mathrm{d}\boldsymbol{S}+\int_V\frac{\partial\rho}{\partial t}\mathrm{d}V=0. \quad \text{(A6.45)}$$

Equation of motion for a fluid

$$F=\lim_{\Delta t\to0}\frac{\Delta p}{\Delta t}$$

$$=\int_S\rho\boldsymbol{v}(\boldsymbol{v}\cdot\mathrm{d}\boldsymbol{S})+\int_V\frac{\partial(\rho\boldsymbol{v})}{\partial t}\mathrm{d}V \quad \text{(A6.46)}$$

Euler's equation

$$-g\cos\alpha-\frac{1}{\rho}\frac{\partial p}{\partial s}=v\frac{\partial v}{\partial s}+\frac{\partial v}{\partial t} \quad \text{(A6.47)}$$

Bernoulli's equation

$$\frac{p}{\rho}+\frac{v^2}{2}+gz=\text{constant} \quad \text{(A6.48)}$$

Plane stress and strain

$$\varepsilon'_{xx}=\varepsilon_{xx}\cos^2\theta+\varepsilon_{yy}\sin^2\theta+\varepsilon_{xy}2\cos\theta\sin\theta \quad \text{(A6.49)}$$

$$\varepsilon'_{yy}=\varepsilon_{yy}\cos^2\theta+\varepsilon_{xx}\sin^2\theta-\varepsilon_{xy}2\cos\theta\sin\theta \quad \text{(A6.50)}$$

$$\begin{aligned}\varepsilon'_{xy}&=(\varepsilon_{yy}-\varepsilon_{xx})\sin\theta\cos\theta+\varepsilon_{xy}(\cos^2\theta-\sin^2\theta)\\&=\frac{(\varepsilon_{yy}-\varepsilon_{xx})}{2}\sin2\theta+\varepsilon_{xy}\cos2\theta\end{aligned} \quad \text{(A6.51)}$$

$$\sigma'_{xx}=\sigma_{xx}\cos^2\theta+\sigma_{yy}\sin^2\theta+\sigma_{xy}2\cos\theta\sin\theta \quad \text{(A6.52)}$$

$$\sigma'_{yy}=\sigma_{yy}\cos^2\theta+\sigma_{xx}\sin^2\theta-\sigma_{xy}2\cos\theta\sin\theta \quad \text{(A6.53)}$$

$$\sigma'_{xy} = (\sigma_{yy} - \sigma_{xx}) \sin\theta \cos\theta + \sigma_{xy}(\cos^2\theta - \sin^2\theta)$$

$$= \frac{(\sigma_{yy} - \sigma_{xx})}{2} \sin 2\theta + \varepsilon_{xy} \cos 2\theta \quad \text{(A6.54)}$$

$$\sigma_1 = \lambda\Delta + 2\mu\varepsilon_1 \quad \text{(A6.55)}$$

$$\sigma_2 = \lambda\Delta + 2\mu\varepsilon_2 \quad \text{(A6.56)}$$

$$\sigma_3 = \lambda\Delta + 2\mu\varepsilon_3 \quad \text{(A6.57)}$$

Elastic constants

$$E = 2\mu(1+\nu) = 2G(1+\nu) \quad \text{(A6.58)}$$

$$K = \lambda + 2\mu/3 = E/3(1-2\nu) \quad \text{(A6.59)}$$

Strain energy

$$U = \frac{\sigma_{xx}\varepsilon_{xx}}{2} + \frac{\sigma_{yy}\varepsilon_{yy}}{2} + \frac{\sigma_{zz}\varepsilon_{zz}}{2} + \frac{\tau_{xy}\gamma_{xy}}{2} + \frac{\tau_{yz}\gamma_{yz}}{2} + \frac{\tau_{zx}\gamma_{zx}}{2} \quad \text{(A6.60)}$$

Torsion of circular cross-section shafts

$$\frac{T}{J} = \frac{G\theta}{L} = \frac{\tau}{r} \quad \text{(A6.61)}$$

Shear force and bending moment

$$V = \int w \, dx \quad \text{(A6.62)}$$

and $M = \int\int w \, dx \, dx = \int V \, dx$ (A6.63)

Bending of beams

$$-\frac{\sigma}{y} = \frac{M}{I} = \frac{E}{R} \quad \text{(A6.64)}$$

Deflection of beams

$$\frac{dy}{dx} = \int \frac{M}{EI} dx \quad \text{(A6.65)}$$

and $y = \iint \frac{M}{EI} dx \, dx$ (A6.66)

Area moment method

$$\frac{dy}{dx_2} - \frac{dy}{dx_1} = \int_{x_1}^{x_2} \frac{M}{EI} dx \quad \text{(A6.67)}$$

$$y_2 - y_1 = \theta_2(x_2 - x_1) - \int_0^{(x_2 - x_1)} x \frac{M}{EI} dx \quad \text{(A6.68)}$$

Appendix 7
Matrix methods

A7.1 Matrices

A matrix is a rectangular array of numbers. A matrix with m rows and n columns is said to be of order $m \times n$ and is written

$$\begin{bmatrix} a_{11} & a_{12} \dots & & \dots a_{1n} \\ a_{21} & a_{22} & & \cdot \\ \cdot & \cdot & & \cdot \\ \cdot & \cdot & a_{ij} & \cdot \\ \cdot & \cdot & & \cdot \\ a_{m1} & a_{m2} \dots & & \dots a_{mn} \end{bmatrix} = [A]$$

Special matrices

a) Row matrix

$$[a_1 \quad a_2 \dots a_n] = \lfloor A \rfloor$$

b) Column matrix

$$\begin{bmatrix} a_1 \\ \cdot \\ \cdot \\ \cdot \\ a_m \end{bmatrix} = \{A\}$$

c) Square matrix, one for which $m = n$

d) Diagonal matrix, a square matrix such that non-zero elements occur only on the leading diagonal:

$$\begin{bmatrix} a_{11} & 0 & 0 & \cdot & 0 \\ 0 & a_{22} & 0 & \cdot & \cdot \\ 0 & \cdot & \cdot & \cdot & \cdot \\ \cdot & \cdot & \cdot & \cdot & a_{nn} \end{bmatrix}$$

e) Unit matrix or identity matrix, where

$$a_{11} = a_{22} = \dots = a_{nn} = 1,$$

all other elements being zero.

e.g. $$\begin{bmatrix} 1 & 0 & 0 \\ 0 & 1 & 0 \\ 0 & 0 & 1 \end{bmatrix} = [I]_3$$

N.B. $[I][A] = [A][I] = [A]$

f) Symmetric matrix, where $a_{ij} = a_{ji}$

g) Null matrix, $[\mathbf{0}]$, all elements are zero

A7.2 Addition of matrices

The addition of matrices of the same order is defined as the addition of corresponding elements, thus

$$[A] + [B] = [B] + [A]$$

$$= \begin{bmatrix} (a_{11} + b_{11}) & (a_{12} + b_{12}) & \dots \\ \cdot & \cdot & \cdot \\ \cdot & \cdot & \cdot \\ \cdot & \cdot & \cdot \\ \cdot & \cdot & (a_{mn} + b_{mn}) \end{bmatrix} \quad \text{(A7.1)}$$

A7.3 Multiplication of matrices

If $[C] = [A][B]$ then the elements of $[C]$ are defined by

$$c_{ij} = \sum a_{ik} b_{kj}$$

where k equals the number of columns in $[A]$, which must also equal the number of rows in $[B]$. This is illustrated by the following scheme which can be used when evaluating a product.

$$\begin{matrix} \begin{bmatrix} & & \\ & \text{square} & \\ & & \end{bmatrix} & \begin{bmatrix} b_{11} & b_{12} & b_{13} & b_{14} \\ b_{21} & b_{22} & b_{23} & b_{24} \\ b_{31} & b_{32} & b_{33} & b_{34} \end{bmatrix} \Leftarrow [B] \\ \begin{bmatrix} a_{11} & a_{12} & a_{13} \\ a_{21} & a_{22} & a_{23} \end{bmatrix} & \begin{bmatrix} c_{11} & c_{12} & c_{13} & c_{14} \\ c_{21} & c_{22} & c_{23} & c_{24} \end{bmatrix} \\ \Uparrow & \Uparrow \\ [A] & [C] = [A][B] \end{matrix} \quad \text{(A7.2)}$$

e.g. $c_{13} = a_{11}b_{13} + a_{12}b_{23} + a_{13}b_{33}$

In general, $[A][B] \neq [B][A]$

A7.4 Transpose of a matrix

The transpose of a matrix $[A]$, written $[A]^T$, is a matrix such that its ith row is the ith column of the original matrix e.g.

$$\begin{bmatrix} a_{11} & a_{12} & a_{13} \\ a_{21} & a_{22} & a_{23} \end{bmatrix}^{\mathrm{T}} = \begin{bmatrix} a_{11} & a_{21} \\ a_{12} & a_{22} \\ a_{13} & a_{23} \end{bmatrix} \quad \text{(A7.3)}$$

A7.5 Inverse of a matrix

The inverse of a matrix $[A]$, written $[A]^{-1}$, is defined by

$$[A][A]^{-1} = [I]$$
$$[A]^{-1}[A] = [I] \quad \text{(A7.4)}$$

The inverse can be defined only for a square matrix and even for these matrices there are cases where the inverse does not exist. In this book we need not be concerned with the various methods for inverting a matrix.

A7.6 Matrix representation of a vector

By the definition of matrix multiplication, the vector

$$\boldsymbol{V} = v_x\boldsymbol{i} + v_y\boldsymbol{j} + v_z\boldsymbol{k}$$

may be written as either

$$[v_x \quad v_y \quad v_z]\begin{bmatrix} \boldsymbol{i} \\ \boldsymbol{j} \\ \boldsymbol{k} \end{bmatrix} = \lfloor V \rfloor \{e\}$$

or

$$[\boldsymbol{i} \quad \boldsymbol{j} \quad \boldsymbol{k}]\begin{bmatrix} v_x \\ v_y \\ v_z \end{bmatrix} = \lfloor e \rfloor \{V\}$$

Thus, noting that $\lfloor V \rfloor = \{V\}^{\mathrm{T}}$,

$$\boldsymbol{V} = \{V\}^{\mathrm{T}}\{e\} = \{e\}^{\mathrm{T}}\{V\} \quad \text{(A7.5)}$$

A7.7 Change of co-ordinate system

A vector may be represented in terms of a set of orthogonal unit vectors $\boldsymbol{i}'$, $\boldsymbol{j}'$, and $\boldsymbol{k}'$ which are orientated relative to a set $\boldsymbol{i}, \boldsymbol{j}$, and $\boldsymbol{k}$; thus

$$\boldsymbol{V} = [v_x \quad v_y \quad v_z]\begin{bmatrix} \boldsymbol{i} \\ \boldsymbol{j} \\ \boldsymbol{k} \end{bmatrix} = \{V\}^{\mathrm{T}}\{e\}$$

$$= [v_x' \quad v_y' \quad v_z']\begin{bmatrix} \boldsymbol{i}' \\ \boldsymbol{j}' \\ \boldsymbol{k}' \end{bmatrix} = \{V'\}^{\mathrm{T}}\{e'\}$$

The unit vectors of one set of co-ordinates is expressible in terms of the unit vectors of another set of co-ordinates; thus

$$\boldsymbol{i}' = a_{11}\boldsymbol{i} + a_{12}\boldsymbol{j} + a_{13}\boldsymbol{k}$$
$$\boldsymbol{j}' = a_{21}\boldsymbol{i} + a_{22}\boldsymbol{j} + a_{23}\boldsymbol{k}$$
$$\boldsymbol{k}' = a_{31}\boldsymbol{i} + a_{32}\boldsymbol{j} + a_{33}\boldsymbol{k}$$

where for example a_{11}, a_{12}, and a_{13} are the components of the unit vector $\boldsymbol{i}'$ and are therefore the direction cosines between $\boldsymbol{i}'$ and the x-, y-, z-axes respectively.

In matrix notation,

$$\begin{bmatrix} \boldsymbol{i}' \\ \boldsymbol{j}' \\ \boldsymbol{k}' \end{bmatrix} = \begin{bmatrix} a_{11} & a_{12} & a_{13} \\ a_{21} & a_{22} & a_{23} \\ a_{31} & a_{32} & a_{33} \end{bmatrix}\begin{bmatrix} \boldsymbol{i} \\ \boldsymbol{j} \\ \boldsymbol{k} \end{bmatrix} \quad \text{(A7.6)}$$

or $\{e'\} = [A]\{e\}$

If we assume $\{V'\} = [Q]\{V\}$, where $[Q]$ is some transformation matrix, then

since $\quad \boldsymbol{V} = \{V'\}^{\mathrm{T}}\{e\}' = \{V\}^{\mathrm{T}}\{e\}$

$$\{V\}^{\mathrm{T}}[Q]^{\mathrm{T}}[A]\{e\} = \{V\}^{\mathrm{T}}\{e\}$$

and because this is true for any arbitrary $\{V\}$ it follows that

$$[Q]^{\mathrm{T}}[A] = [I]$$

or $\quad [Q]^{\mathrm{T}} = [A]^{-1} \quad$ (A7.6)

The magnitude of a vector is a scalar independent of the co-ordinate system, so

$$\boldsymbol{V}\cdot\boldsymbol{V} = V^2 = \{V\}^{\mathrm{T}}\{V\} = \{V'\}^{\mathrm{T}}\{V'\}$$
$$= \{V\}^{\mathrm{T}}[Q]^{\mathrm{T}}[Q]\{V\}$$

thus $\quad [Q]^{\mathrm{T}}[Q] = [I] \quad$ or $\quad [Q]^{\mathrm{T}} = [Q]^{-1} \quad$ (A7.7)

showing that the inverse of $[Q]$ is its transpose. Such transformations are called orthogonal. From equations A7.6 and A7.7 we see that

$$[A]^{-1} = [Q]^{-1} \quad \text{or} \quad [A] = [Q] \quad \text{(A7.8)}$$

Summarising, we have

$$\{e'\} = [A]\{e\} \qquad \{V'\} = [A]\{V\}$$
$$\{e\} = [A]^{\mathrm{T}}\{e'\} \qquad \{V\} = [A]^{\mathrm{T}}\{V'\}$$

From equation A7.6

$$\boldsymbol{i}'\cdot\boldsymbol{i}' = a_{11}{}^2 + a_{12}{}^2 + a_{13}{}^2 = 1 \quad \text{(A7.9)}$$

with similar expressions for $\boldsymbol{j}'\cdot\boldsymbol{j}'$ and $\boldsymbol{k}'\cdot\boldsymbol{k}$.

Also, from equations A7.6 and A7.9,

$$\boldsymbol{i}'\cdot\boldsymbol{j}' = a_{11}a_{21} + a_{12}a_{22} + a_{13}a_{23} = 0 \quad \text{(A7.10)}$$

with similar expressions for $\boldsymbol{j}'\cdot\boldsymbol{k}'$ and $\boldsymbol{k}'\cdot\boldsymbol{i}'$.

Rotation about the z-axis
From Fig. A7.1, it is seen that

$$x' = x\cos\theta + y\sin\theta$$
$$y' = -x\sin\theta + y\cos\theta$$
$$z' = z$$

$$\text{or}\quad \{V'\} = \begin{bmatrix} x' \\ y' \\ z' \end{bmatrix} = \begin{bmatrix} \cos\theta & \sin\theta & 0 \\ -\sin\theta & \cos\theta & 0 \\ 0 & 0 & 1 \end{bmatrix} \begin{bmatrix} x \\ y \\ z \end{bmatrix}$$

$$\{V'\} = [A]\{V\} \qquad \text{(A7.11)}$$

A7.8 Change of axes for moment of inertia

In this section $[J]$ will be used for moment of inertia, to avoid confusion with the identity matrix $[I]$.

The kinetic energy of a rigid body rotating about a fixed point (or relative to its centre of mass) is given by equation 11.83 which can be written as

$$\tfrac{1}{2}\{\boldsymbol{\omega}\}^{\mathrm{T}}[\boldsymbol{J}]\{\boldsymbol{\omega}\} = \tfrac{1}{2}\{\boldsymbol{\omega}'\}^{\mathrm{T}}[\boldsymbol{J}']\{\boldsymbol{\omega}'\}$$

This is a scalar quantity and therefore independent of the choice of axes so

$$\text{if}\quad \{\boldsymbol{\omega}'\} = [A]\{\boldsymbol{\omega}\}$$

$$\text{then}\quad \{\boldsymbol{\omega}'\}^{\mathrm{T}}[\boldsymbol{J}']\{\boldsymbol{\omega}'\} = \{\boldsymbol{\omega}\}^{\mathrm{T}}[A]^{\mathrm{T}}[\boldsymbol{J}'][A\{\boldsymbol{\omega}\} = \{\boldsymbol{\omega}\}^{\mathrm{T}}[\boldsymbol{J}]\{\boldsymbol{\omega}\}$$

$$\text{thus}\quad [\boldsymbol{J}] = [A]^{\mathrm{T}}[\boldsymbol{J}'][A]$$

$$\text{or}\quad [\boldsymbol{J}'] = [A][\boldsymbol{J}][A]^{\mathrm{T}} \qquad \text{(A7.12)}$$

If the x'- and the y'-axes have direction cosines of l, m, n and l', m', n' respectively,

by multiplication

$$b_{11} = lJ_{xx} - mJ_{yx} - nJ_{zx}$$
$$b_{12} = -lJ_{xy} + mJ_{yy} - nJ_{zy}$$
$$b_{13} = -lJ_{xz} + mJ_{yz} + nJ_{zz}$$

$$\text{and}\quad J_{xx}' = l^2J_{xx} + m^2J_{yy} + n^2J_{zz} - 2(J_{xy}lm + J_{xz}ln + J_{yz}mn) \qquad \text{(A7.14)}$$

$$J_{xy}' = -(ll'J_{xx} + mm'J_{yy} + nn'J_{zz}) + (lm' + ml')J_{xy} + (ln' + nl')J_{xz} + (mn' + nm')J_{yz} \qquad \text{(A7.15)}$$

A7.9 Transformation of the components of a vector

a) Cylindrical to Cartesian co-ordinates:

$$\begin{matrix} V_x \\ V_y \\ V_z \end{matrix} = \begin{matrix} \cos\theta & -\sin\theta & 0 \\ \sin\theta & \cos\theta & 0 \\ 0 & 0 & 1 \end{matrix} \quad \begin{matrix} V_R \\ V_\theta \\ V_z \end{matrix} \qquad \text{(A7.16)}$$

$$\{V\}_{\mathrm{C}} = [A]_\theta\{V\}_{\mathrm{cyl.}}$$

b) Spherical to cylindrical co-ordinates (see Figs 1.5 and 1.6(a)):

$$\begin{matrix} V_R \\ V_\theta \\ V_z \end{matrix} = \begin{matrix} \cos\phi & 0 & -\sin\phi \\ 0 & 1 & 0 \\ \sin\phi & 0 & \cos\phi \end{matrix} \quad \begin{matrix} V_r \\ V_\theta \\ V_\phi \end{matrix}$$

$$\{V\}_{\mathrm{cyl.}} = [A]_\phi\{V\}_{\mathrm{sph.}} \qquad \text{(A7.17)}$$

$$\text{then}\quad [A] = \begin{bmatrix} l & m & n \\ l' & m' & n' \\ l'' & m'' & n'' \end{bmatrix}$$

Using the following multiplication scheme:

$$[A]^{\mathrm{T}} \Downarrow$$

$$[J] \Rightarrow \begin{bmatrix} J_{xx} & -J_{xy} & -J_{xz} \\ -J_{yx} & J_{yy} & -J_{yz} \\ -J_{zx} & -J_{zy} & J_{zz} \end{bmatrix} \begin{bmatrix} l & l' & l'' \\ m & m' & m'' \\ n & n' & n'' \end{bmatrix}$$

$$[A] \Downarrow$$

$$\begin{bmatrix} l & m & n \\ l' & m' & n' \\ l'' & m'' & n'' \end{bmatrix} \begin{bmatrix} b_{11} & b_{12} & b_{13} \\ & \text{etc.} & \\ & & \end{bmatrix} \begin{bmatrix} J_{xx}' & -J_{xy}' & -J_{xz}' \\ & \text{etc.} & \\ & & \end{bmatrix}$$

$$\Uparrow \qquad\qquad \Uparrow$$
$$[\boldsymbol{A}][\boldsymbol{J}] = [\boldsymbol{B}] \qquad [\boldsymbol{J}'] \qquad \text{(A7.13)}$$

c) Spherical to Cartesian co-ordinates:

$$\underbrace{\begin{bmatrix} V_x \\ V_y \\ V_z \end{bmatrix}}_{\{V\}_C} = \underbrace{\begin{bmatrix} \cos\theta & -\sin\theta & 0 \\ \sin\theta & \cos\theta & 0 \\ 0 & 0 & 1 \end{bmatrix}}_{[A]_\theta} \times \underbrace{\begin{bmatrix} \cos\phi & 0 & -\sin\phi \\ 0 & 1 & 0 \\ \sin\phi & 0 & \cos\phi \end{bmatrix}}_{[A]_\phi} \underbrace{\begin{bmatrix} V_r \\ V_\theta \\ V_\phi \end{bmatrix}}_{\{V\}_{sph.}}$$

$$= \underbrace{\begin{bmatrix} \cos\theta\cos\phi & -\sin\theta & -\cos\theta\sin\phi \\ \sin\theta\cos\phi & \cos\theta & -\sin\theta\sin\phi \\ \sin\phi & 0 & \cos\phi \end{bmatrix}}_{[A]_{\theta\phi} = [A]_\theta [A]_\phi} \begin{bmatrix} V_r \\ V_\theta \\ V_\phi \end{bmatrix}$$

$$\{V\}_C = [A]_{\theta\phi} \{V\}_{sph.} \qquad \text{(A7.18)}$$

Appendix 8
Properties of structural materials

Our attention here is centred mainly on ferrous and non-ferrous metals. However the principles apply to other solid materials.

A8.1 Simple tensile test

In principle the tensile test applies an axial strain to a standard specimen and measurements are taken of the change in length between two specified marks, defined as the gauge length, and also of the resulting tensile load. Alternatively, the test could be carried out by applying a dead load and recording the subsequent strain.

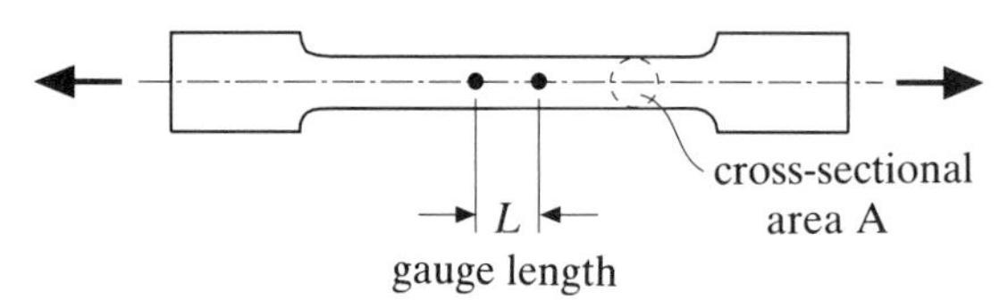

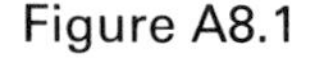
Figure A8.1

Figure A8.1 shows a typical specimen where A is the original cross-section area. Figure A8.2 shows the load-extension plot for a mild steel specimen. Note that load/original-cross-section area is the nominal stress and extension/gauge length is the strain so the shape of the stress-strain curve is the same. The extension axis is shown broken since the extensions at e and f are very much greater than that at points a to d.

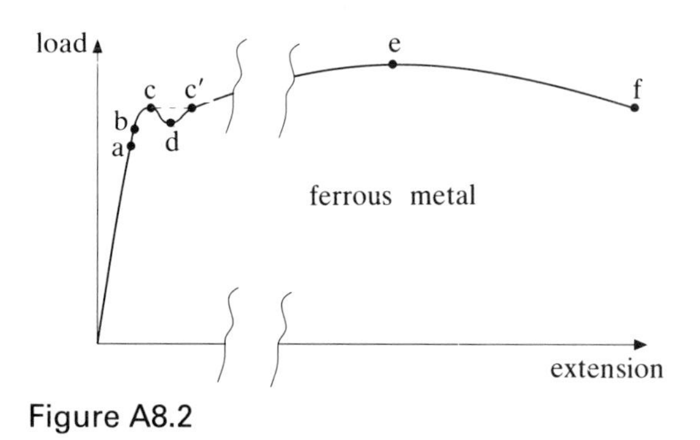

Figure A8.2

The point a is the *limit of proportionality*, i.e. up to this point the material obeys Hooke's law. Point b is the *elastic limit*, this means that any loading up to this point is reversible and the unloading curve retraces the loading curve. In practice the elastic limit occurs just after the limit of proportionality. After this point any unloading curve is usually a straight line parallel to the elastic line. Point c is known as the *yield point*, sometimes called the upper yield point. Point d is called the *lower yield point.* If the test is carried out by applying a load rather than an extension then the extension will increase from point c without any increase in load to the point c'. Further straining will cause plastic deformation to take place until the maximum load is reached at point e. This is known as the *ultimate tensile load*. After this a 'neck' will form in the specimen resulting in a large reduction in the cross-section area until failure occurs at point f.

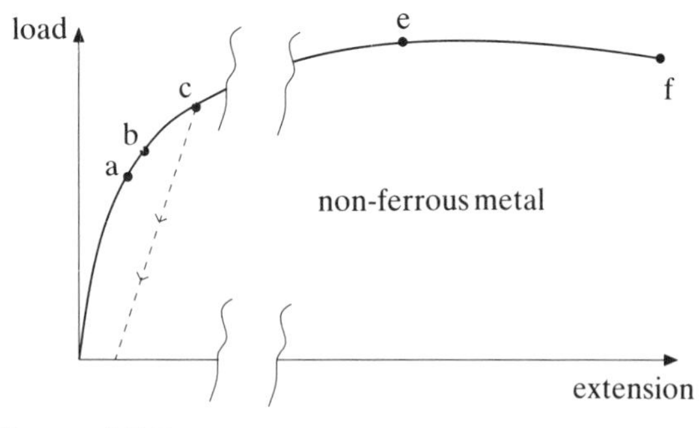

Figure A8.3

Figure A8.3 shows a similar plot for a non-ferrous metal where it is noticed that no well-defined yield point appears. At the point c the stress is known as a *proof stress*. For example a 0.2% proof stress is one which when removed leaves a permanent strain of 0.002.

A strain of 0.002 can also be referred to as 2 milli-strain (mε) or as 2000 micro-strain ($\mu\varepsilon$).

Both the above cases are for ductile materials and the degree of ductility is measured either by quoting the final strain in the form of a percentage elongation, or in the form of the percentage reduction of area at the neck.

For brittle materials failure occurs just after the elastic limit there being little or no plastic deformation.

Answers to problems

1.1 $0.80\boldsymbol{i}+0.53\boldsymbol{j}+0.27\boldsymbol{k}$
1.2 $(4\boldsymbol{i}+4\boldsymbol{j}+2\boldsymbol{k})$ m
1.3 $(7, 2, 6)$ m
1.4 $(0.87\boldsymbol{i}+0.35\boldsymbol{j}+0.35\boldsymbol{k})$
1.5 $(-3\boldsymbol{i}-4\boldsymbol{j}-\boldsymbol{k})$ m, $(3\boldsymbol{i}+4\boldsymbol{j}+\boldsymbol{k})$ m
1.6 $(3\boldsymbol{i}-\boldsymbol{j}-\boldsymbol{k})$ m, $(0.90\boldsymbol{i}-0.30\boldsymbol{j}-0.30\boldsymbol{k})$
1.7 $(0, 2, 8)$ m
1.8 $(3, 2.8, 2.8)$ m, (4.104 m, 43.03°, 2.8 m)
1.10 (16.25, 10.84, 4.33) km, (19.53 km, 33.7°, 4.33 km)
1.11 75.6°, 128.3°, 41.9°
1.12 79.62°
1.13 3 m, 2.92 m
1.14 9.2 m, 8.6 m, 7.8 m
1.15 8.17 m, 97.7°

2.1 $(-27\boldsymbol{i}+223\boldsymbol{j})$ m, $(-24\boldsymbol{i}+216\boldsymbol{j})$ m/s, $(-10\boldsymbol{i}+144\boldsymbol{j})$ m/s^2
2.2 $(6.25\boldsymbol{i}+11.17\boldsymbol{j})$ m, $(3\boldsymbol{i}+10\boldsymbol{j})$ m/s
2.3 5 m/s^2
2.4 $(8.66\boldsymbol{i}+5.0\boldsymbol{j})$ m/s, $(-10\boldsymbol{i}+17.32\boldsymbol{j})$ m/s^2
2.5 7.368 knots, W 16° 19′ N
2.6 $(-0.384\boldsymbol{i}+2.66\boldsymbol{j})$ m/s, $(-15.83\boldsymbol{i}+2.41\boldsymbol{j})$ m/s^2
2.7 $(6.62\boldsymbol{i}+4.66\boldsymbol{j})$ m/s, $(-11.83\boldsymbol{i}+8.41\boldsymbol{j})$ m/s^2, 8.09 m/s, 14.51 m/s^2
2.8 $(2.0\boldsymbol{i}+3.45\boldsymbol{j})$ m/s, $(-0.12\boldsymbol{i}+5.58\boldsymbol{j})$ m/s
2.9 17.89 m/s
2.10 a) 0.6 m/s^2, 17.8 m, b) 0.8 m/s^2, 12.0 m
2.11 a) 1.0 m/s, b) 1.6 s

3.1 $(-8.33\boldsymbol{i}+3.33\boldsymbol{j})$ m/s
3.2 5.08 m/s, 2.18 m/s
3.3 a) $\sqrt{[2R_0x_1/m]}\boldsymbol{i}$, b) $\sqrt{[R_0x_1/m]}\boldsymbol{i}$, c) $\sqrt{[3R_0x_1/(2m)]}\boldsymbol{i}$, d) $\sqrt{[2R_0x_1/(3m)]}\boldsymbol{i}$
3.4 $(61\boldsymbol{i}+19\boldsymbol{j})$ m/s
3.5 $(320\boldsymbol{i}-160\boldsymbol{j})$N
3.6 64 m/s, 320 m
3.7 a) No motion, b) 0.657 m/s^2, c) 2.55 m/s^2
3.8 3.29 m/s^2, 15.52 kN
3.11 14.82 N
3.12 0.163 m/s^2, 6.5 m/s, 260 m
3.13 1.24 m/s^2
3.14 24.0 s
3.15 6.4 m/s
3.16 0.65
3.17 17.5 s, 1 in 7, 0.91 m/s^2
3.18 41.5 m/s^2, 39.5 m/s
3.19 544 m
3.20 1.385 m/s^2, 0.436 m/s^2

4.1 87.0 N m anticlockwise
4.2 39 N, 22 N, tension, 0.92 N m
4.3 21.0 kN, 3.49 kN, 14.4 kN
4.4 a) $(-210\boldsymbol{i}+5050\boldsymbol{j})$ N, b) 5830 N, compression
4.5 a) 190 N ∠ 52°, b) 285 N m clockwise
4.7 $-30\,\boldsymbol{k}$ N, $(-10\boldsymbol{i}+30\boldsymbol{j})$ N m
4.8 228.8 N, 102 N m, 192 N m
4.9 363.3 N m, 9323 N m, $\alpha=62.63°$, $\beta=88.81°$, $\gamma=-27.39°$
4.10 b) $(46\boldsymbol{i}+20\boldsymbol{j}+30\boldsymbol{k})$ N, c) e.g. (1.433, 1.667, 0)
4.11 500 N, 1500 N, 120 N m, 1700 N m
4.12 a) $(29.32\boldsymbol{j}-10\boldsymbol{k})$ N, $0.4\boldsymbol{i}$ N m, b) 0.59 N m, 28.5 N m, 394 N, 29.3 N
4.13 $F_A=(-2\boldsymbol{i}+247.5\boldsymbol{j})$ N, $F_B=(-252.5\boldsymbol{j}-11.9\boldsymbol{k})$ N, $F_C=(5\boldsymbol{j}+1.9\boldsymbol{k})$ N
4.14 4204 N
4.15 7000 kg, 69.4 kN

5.1 $-6.64\boldsymbol{k}$ rad/s, $-0.998\boldsymbol{i}$ m/s, $(-0.898\boldsymbol{i}+0.399\boldsymbol{j})$ m/s
5.2 3.71 m/s, 4.47 rad/s anticlockwise
5.3 $v=-e\sin\theta\omega$, $a=-e\cos\theta\omega^2$
5.4 a) 3.95 anticlockwise, b) 0.934 m/s →
5.5 900 rad/s^2 anticlockwise
5.7 a) $(-7.80\boldsymbol{i})$ m/s, $(-6.75\boldsymbol{i}+7.5\boldsymbol{j})$ m/s, $(-75.8\boldsymbol{k})$rad/s, b) $(-3980\boldsymbol{i})$m/s^2, $(-3590\boldsymbol{i}-1360\boldsymbol{j})$ m/s^2, $(12900\boldsymbol{k})$ rad/s^2
5.8 $v_C=30.8\boldsymbol{i}$ m/s, $v_E=-24.2\boldsymbol{i}$ m/s, $\boldsymbol{a}_C=3080\boldsymbol{i}$ m/s^2, $\boldsymbol{a}_E=-4630\boldsymbol{i}$ m/s^2
5.9 a) 0.5 rad/s anticlockwise, b) 1.02 m/s ∠ 73°, c) 15 rad/s
5.10 7.3 m/s ↗ 8°, 910 m/s^2 ↖ 20°
5.11 $39\boldsymbol{k}$ rad/s, $3330\boldsymbol{k}$ rad/s^2, 2.15 m/s, $150\boldsymbol{k}$ rad/s, -1450 rad/s^2
5.12 a) 0.72 rad/s anticlockwise, b) 2.39 rad/s^2, anticlockwise
5.14 $25\boldsymbol{k}$ rev/s
5.15 $\omega_A/\omega_F=-9.68$

6.4 (6.67, 14.17) mm
6.10 220 N, 1133 N
6.11 a) 297 kN, b) 204 kN
6.12 a) 17.68 N, 1.25 N m clockwise

6.13 308.2 rad/s
6.14 2.077 m/s^2, 5.194 kN, 4.616 kN
6.15 18.0 m
6.16 15.46 m/s (tipping)
6.19 3.94 m/s
6.20 a) 42.86 m/s^2, b) 7637 N
6.21 20.44 kN
6.22 a) 2.81 kN, b) 98.0 N$\rightarrow$
6.23 b) 1804 N m
6.24 $(2348\boldsymbol{i} - 540\boldsymbol{j} + 3924\boldsymbol{k})$ N
6.25 a) $0.518\surd(g/l)$, zero, b) $T_{AO1} = 1.268\,cmg/(b+c)$, $T_{BO2} = 1.268\,bmg/(b+c)$
6.26 a) 11.4 kN, b) 209 N m anticlockwise
6.27 a) $(-11.31\boldsymbol{i} + 4.69\boldsymbol{j})$ m/s^2, $(23.4\boldsymbol{k})$ rad/s^2, b) $(-54.4\boldsymbol{i} + 0.25\boldsymbol{j})$ N, $(9.37\boldsymbol{i} + 12.5\boldsymbol{j})$ N
6.28 263.2 kN/m
6.29 a) 99 kg m^2, c) $(-12.60\boldsymbol{i} + 12.12\boldsymbol{j})$ m/s^2, $3.464\boldsymbol{k}$ rad/s, $3.465\boldsymbol{k}$ rad/s^2
6.30 $1.776\,mg$
6.31 $(17.32\boldsymbol{i} - 10\boldsymbol{j})$ N, $(4.73\boldsymbol{i} - 6.09\boldsymbol{j})$ N

7.1 0.45 m
7.2 1.01 m/s
7.3 a) 3.52 m/s, b) $(-0.44\boldsymbol{i} + 3.96\boldsymbol{k})$ N
7.4 5668 m/s
7.5 a) 593.1 m/s, b) 166.7 m/s
7.6 93.7 rad/s
7.10 a) $15\boldsymbol{k}$ N m, b) 2.356 kW, c) $-10\boldsymbol{k}$ N m
7.11 107.8 rad/s^2
7.12 $\dfrac{1}{2\pi}\times\sqrt{\left[\dfrac{(n-1)\{mg/r + k_1(n-1)\}}{(mr^2/2)\{n^2 - 2n + 2\}}\right]}$ where $n = R/r$, $\dfrac{1}{2\pi}\sqrt{\left[\dfrac{ka^2 - mg\cos 30°\, l/2}{ml^2/3}\right]}$
7.13 a) 0.0105 m/s^2, b) 26.8 m/s
7.14 10.9 N m clockwise
7.16 $0.268\,g/l$, zero
7.17 54.2 N m
7.19 70.0°
7.21 0.056 m (stable)
7.22 $\pm 108.6°$ from vertical
7.23 $k_1 > \tfrac{1}{2}m_1 g l_1$, $k + k_1 > (\tfrac{1}{2}m + m_1)gl$, $[k + k_1 - (\tfrac{1}{2}m + m_1)gl] \times [k_1 - \tfrac{1}{2}m_1 g l_1] > k_1{}^2$

8.1 2.25 m
8.4 $(I_A + mR^2)\omega_O/I_A$
8.5 $(mv_O a/I)\boldsymbol{k}$
8.7 2.154 m/s, 28.3 J
8.8 84.24 N, 231.5 N, 18.0 m
8.9 31.76 N
8.10 433 N, 750 N, 100 N m
8.11 a) $2\rho R^2\omega\rightarrow$, b) $2\rho R^2\alpha$
8.16 9880 kg, 194.2 m/s
8.17 176.7 m/s, 1392 m, 2933 m

9.1 a) $(1/2\pi)\surd(k/m)$, b) $(1/2\pi)\surd(k/m)$
9.2 $mga/(2\pi v)^2$
9.3 $(a/2\pi b)\surd(k/m)$
9.4 a) $(1/2\pi)\surd(4k/m)$, b) $(1/2\pi)\surd(4ka^2/I_O)$
9.5 0.37 Hz, 1.47 m/s
9.6 $(1/2\pi)\surd[5ga^2/7(R-a)]$
9.8 $25/2\pi$ Hz, 22.2 mm
9.9 2.08 mm
9.10 80 Hz, 49°
9.11 $\pi c\omega/[1 + (c\omega/m)^2]$, $c = m\omega$
9.12 25 mm
9.13 2.9 mm, 2.79 m/s^2
9.15 1.6°
9.16 20%
9.17 0.37 mm
9.18 1.5°, 5%
9.19 0.17 V
9.20 $2[\pi + 2\beta_0/(2\alpha_0 - \beta_0)]\surd(I/k)$
9.21 a) 0.94 s, b) 0.02 N m
9.23 13.47 Hz, 66.89 Hz
9.24 1.79 Hz, 0.60 Hz
9.25 59.22 kN/m, 30.8 Hz, 80.8 Hz
9.26 $0.136\surd(g/L)$, $0.365\surd(g/L)$ Hz
9.27 92 μm, 41.5 μm, 6.0 Hz, 11.3 Hz

10.1 G/(1 + GH)
10.2 $(\mathrm{ABC}+1)\theta_e = \theta_i + \mathrm{C}y$, $(\mathrm{ABC}+1)x = \mathrm{A}\theta_i + \mathrm{AC}_y$, $(\mathrm{ABC}+1)w = \mathrm{AB}\theta_i + \mathrm{ABC}y$, $(\mathrm{ABC}+1)z = \mathrm{AB}\theta_i - y$, $(\mathrm{ABC}+1)\theta_o = \mathrm{ABC}\theta_i - \mathrm{C}y$
10.4 a) $[(I_m + I_L)\mathrm{D}^2 + C\mathrm{D} + K_1K_2]\theta_o = K_1K_2\theta_i$, $(C/2)[(I_m + I_L)K_1K_2]^{1/2}$, b) $[(I_m + I_L)\mathrm{D}^2 + C\mathrm{D} + K_1K_2]\theta_o = K_1K_2\theta_i - Q_L$, $(C/2)[(I_m + I_L)K_1K_2]^{1/2}$, c) $[(n^2I_m + I_L)\mathrm{D}^2 + C\mathrm{D} + nK_1K_2] = nK_1K_2\theta_i$, $(C/2)[(n^2I_m + I_L)nK_1K_2)^{1/2}$
10.6 l_2/l_1, $(l_1 + l_2)/(l_1 k)$
10.7 $A = A_1/A_2$ where $A_1 = l_2C$, $A_2 = l_1C + S(l_1 + l_2)/k$, $\tau = C(l_1 + l_2)/(kA_2)$, $\tau_i = C/k$
10.10 4 N m/rad, 0.8 rad
10.11 $t = 1$ s, 0.147 rad/s^2
10.14 a) 0.94 s, b) 0.1 rad
10.15 zero, zero, $2A(I_D + I_L)/K$
10.16 a) $[I_AI_B\mathrm{D}^4 + C(I_A + I_B)\mathrm{D}^3 + SI_A\mathrm{D}^2 + SC\mathrm{D}]\theta_A = [I_B\mathrm{D}^2 + C\mathrm{D} + S]Q$, b) $[I_AI_B\mathrm{D}^4 + C(I_A + I_B)\mathrm{D}^3 + (SI_A + KI_B)\mathrm{D}^2 + C(K+S)\mathrm{D} + SK]\theta_A = K(I_B\mathrm{D}^2 + C\mathrm{D} + S]\theta_i$
10.17 a) $T_d = 6 + 0.01\omega_e$, 10 s, b) 156 rad/s
10.18 360 N m, $(0.6\mathrm{D}^2 + 4.9\mathrm{D} + 36)\omega_o = 8640$
10.19 $a_1a_2 = a_0a_3$, $(1/2\pi)\surd(a_0a_2)$
10.20 $gB\theta_1/(A - g)$
10.22 a) 10.46, b) 25
10.23 50
10.24 a) $\dfrac{10(1 + 0.2\mathrm{j}\omega)}{\mathrm{j}\omega}$, b) $\dfrac{5}{\mathrm{j}\omega(1 + 0.1\mathrm{j}\omega)}$
10.25 a) 3543 Nm/rad, b) infinity, c) 0.42

11.1 a) Parallel to $(\boldsymbol{i} - \boldsymbol{j} + \boldsymbol{k})$, 120°, (b) No
11.2 5.457 km, 21.50°, 32.13°, -98.24 m/s, 7.092×10^{-3} rad/s, 3.890×10^{-2} rad/s,

7.422 m/s^2,
-2.604×10^{-4} rad/s^2,
1.618×10^{-3} rad/s^2

11.3 $(6.928\boldsymbol{i} + 3\boldsymbol{j} + 4\boldsymbol{k})$ m/s, $(15.359\boldsymbol{i} + 153.56\boldsymbol{j} - 158.56\boldsymbol{k})$ m/s^2

11.4 a) $(r\omega_w - a\omega_A)\boldsymbol{i} - b\omega_A\boldsymbol{j}$,
b) $(r\dot{\omega}_w - a\dot{\omega}_A + b\omega_A{}^2)\boldsymbol{i} + (2r\omega_A\omega_w - b\dot{\omega}_A - a\omega_A{}^2)\boldsymbol{j} - r\omega_w\boldsymbol{k}$,
c) $\omega_w\boldsymbol{j} + \omega_A\boldsymbol{k}$, $-\omega_A\omega_w\boldsymbol{i} + \dot{\omega}_w\boldsymbol{j} + \dot{\omega}_A\boldsymbol{k}$

11.5 $(3.584\boldsymbol{j} + 1.369\boldsymbol{k})$ m/s, $(51.37\boldsymbol{i} - 1.745\boldsymbol{j} - 0.667\boldsymbol{k})$ m/s^2

11.6 a) $m\{(\dot{\theta}^2 + \dot{\phi}^2)r\cos\theta\boldsymbol{i} + r\dot{\theta}^2\sin\theta\boldsymbol{j} + (a\dot{\phi}^2 - 2\dot{\theta}\dot{\phi}r\sin\theta)\boldsymbol{k}\}$,
b) $m\{[(\dot{\theta}^2 + \dot{\phi}^2)r\cos\theta\cos\phi + (a\dot{\phi}^2 - 2\dot{\theta}\dot{\phi}r\sin\theta)\sin\phi]\boldsymbol{I} + r\dot{\theta}^2\sin\theta\boldsymbol{J} - [(\dot{\theta}^2 + \dot{\phi}^2)r\cos\theta\sin\phi - (a\dot{\phi}^2 - 2\dot{\theta}\dot{\phi}r\sin\theta)\cos\phi]\boldsymbol{k}\}$

11.7 $(4\boldsymbol{i} - 2\boldsymbol{j})$ m/s, 2.4 m/s, $-(3.2\boldsymbol{i} + 6.4\boldsymbol{j} + 2.4\boldsymbol{k})$ rad/s

11.9 $\omega_x = \omega r[br\sin\theta\cos\theta + (b^2 + z^2)\cos\theta]/(l^2z)$
$\omega_y = \omega r[-br\cos^2\theta + z^2\sin\theta]/(l^2z)$
$\omega_z = \omega r[r + b\sin\theta]/l^2$
$\dot{\omega}_x = -[a_B(r\sin\theta + b) + \omega^2rz\sin\theta]/l^2$
$\dot{\omega}_y = [a_Br\cos\theta + \omega^2rz\cos\theta]/l^2$
$\dot{\omega}_z = \omega^2rb\cos\theta/l^2$

11.10 $(0.52\boldsymbol{i} - 1.04\boldsymbol{j} - 0.78\boldsymbol{k})$ m/s, $(-0.31\boldsymbol{i} - 0.16\boldsymbol{j} + 1.2\boldsymbol{k})$ rad/s, $(2.2\boldsymbol{i} - 4.3\boldsymbol{j} - 3.2\boldsymbol{k})$ m/s^2

11.11 $M(b^2 + c^2)/3$, $M(a^2 + c^2)/3$, $M(a^2 + b^2)/3$, $Mab/4$, $Mac/4$, $Mbc/4$

11.12 $11\rho a^3/3$, $\rho a^3/2$, $\rho a^3/2$

11.14 $11\rho a^3\dot{\Omega}_z/3$, $\sqrt{2}\rho a^3(\dot{\Omega}_z{}^2 + \Omega_z{}^4)^{1/2}/2$

11.15 0.0112 kg m^2, -0.0167 kg m^2, 1.222 N m, 1.853 N m

11.16 $1.41K$ N m, 3.50 N m

11.18 1.425 kN

11.19 46.1 s

11.20 $C_x = \frac{1}{2}MR^2\dot{\phi}\dot{\psi} + mr(l\dot{\phi}^2 - 2r\dot{\phi}\dot{\psi})$,
$C_y = MR^2\ddot{\phi}/4$,
$C_z = \frac{1}{2}MR^2\ddot{\psi} + mr(l\ddot{\phi} - r\ddot{\psi})$ where $M = \pi R^2\rho$ and $m = \pi d^2\rho/4$

11.22 a) $6\boldsymbol{i}$ m/s, zero, $3(-\boldsymbol{j} + \boldsymbol{k})$ rad/s,
b) $-36\boldsymbol{k}$ m/s^2,
c) 148.2 N (tension)

12.1 193 mm, 0.85 mm

12.2 0.231°

12.3 0.36 MN

12.4 a) 0.100, 0.069, 0.038, 0.006, -0.025,
b) 0.000, -0.11, -0.12, -0.11, 0.00

12.6 a) $\sigma_B = 28.54$ MN/m^2, $\sigma_S = 67.93$ MN/m^2,
b) 1.97 mm

12.7 8 mm, 5.39×10^{-3} m^3

12.8 a) 0.075 mm,
b) 0.12 mm,
c) 314.8×10^{-6} m^3

12.9

	Max S.F.	Max B.M.	Point of contraflexure
a	68	190	—
b	-53.3	71.1	—
c	-117.5	132	6.87
d	$+/-30$	90	7.0
e	-33.3	64.17	—
f	M/L	$-bM/L\,(a<b)$	a

12.10 5.39 kNm

12.11 $w' = w/4$

12.12 2.49×10^6 mm^4

12.13 5.61 kN

12.14 10.086 kN/m

12.15 75 kN/m

12.16 $\theta = WL^2/(6EI)$, $\delta = WL^3/(8EI)$

12.17 $P = W(3L/a - 1)/2$, $a/3$

12.19 $\delta = -7WL^3/(6I_{CD})$

12.20 Ratio = 1.7

12.21 5.1 mm

12.22 50.5 mm, 142 kW

12.23 4.3 kN, 127.6 mm

12.24 352 N, 148 N

12.25 $\varepsilon_1 = 500\mu$ at 30° to a,
$\varepsilon_2 = -300\mu$, $\gamma = 800\mu$,
$\sigma_1 = 9.02$ MN/m^2, $\sigma_2 = -2.86$ MN/m^2,
$\tau = 5.94$ MN/m^2

Index